RENEWALS 458-4574

WITHDRAWN
UTSA LIBRARIES

Systems & Control: Foundations & Applications

Founding Editor

Christopher I. Byrnes, Washington University

Richard Vinter

Optimal Control

With 12 Figures

Birkhäuser
Boston • Basel • Berlin

Library
University of Texas
of San Antonio

Richard Vinter
Department of Electrical Engineering
Imperial College of Science, Technology,
    and Medicine
London SW7 2BT
UK

Library of Congress Cataloging-in-Publication Data
Vinter, R.B. (Richard B.)
    Optimal control / Richard Vinter.
        p. cm. — (Systems and control)
    Includes bibliographical references and index.
    ISBN 0-8176-4075-4 (alk. paper)
        1. Automatic control—Mathematical models.   2. Control theory.   I. Title.   II. Systems &
control.
    TJ213 .V5465   2000
    629.8—dc21
                                                                99-057788
                                                                CIP

Printed on acid-free paper.
© 2000 Birkhäuser Boston   *Birkhäuser*  ®

All rights reserved. This work may not be translated or copied in whole or in part without
the written permission of the publisher (Birkhäuser Boston, c/o Springer-Verlag New York,
Inc., 175 Fifth Avenue, New York, NY 10010, USA), except for brief excerpts in connection
with reviews or scholarly analysis. Use in connection with any form of information storage
and retrieval, electronic adaptation, computer software, or by similar or dissimilar methodology now known or hereafter developed is forbidden.
The use of general descriptive names, trade names, trademarks, etc., in this publication, even
if the former are not especially identified, is not to be taken as a sign that such names, as
understood by the Trade Marks and Merchandise Marks Act, may accordingly be used freely
by anyone.

ISBN 0-8176-4075-4
ISBN 3-7643-4075-4   SPIN 10748579

Typeset by the author in LaTeX.
Printed and bound by Edwards Brothers, Inc., Ann Arbor, MI.
Printed in the United States of America.

9 8 7 6 5 4 3 2 1

Library
University of Texas
at San Antonio

To Donna

Contents

Preface

"Where shall I begin, please your Majesty?" he asked. "Begin at the beginning," the King said, gravely, "and go on till you come to the end: then stop."

– Lewis Carroll, *Alice in Wonderland*

What control strategy will transfer a space vehicle from one circular orbit to another in minimum time or in such a manner as to minimize fuel consumption? How should a batch distillation column be operated to maximize the yield, subject to specified constraints on purity of the product? Practical questions such as these underlie the field of Optimal Control. In the language of mathematics, Optimal Control concerns the properties of control functions that, when inserted into a differential equation, give solutions which minimize a "cost" or measure of performance.[1] In engineering applications the control function is a control strategy. The differential equation describes the dynamic response of the mechanism to be controlled, which, of course, depends on the control strategy employed.

Systematic study of optimal control problems dates from the late 1950s, when two important advances were made. One was the the Maximum Principle, a set of necessary conditions for a control function to be optimal. The other was Dynamic Programming, a procedure that reduces the search for an optimal control function to finding the solution to a partial differential equation (the Hamilton–Jacobi Equation).

In the following decade, it became apparent that progress was being impeded by a lack of suitable analytic tools for investigating local properties of functions which are nonsmooth; i.e., not differentiable in the traditional sense.

Nonsmooth functions were encountered at first attempts to put Dynamic Programming on a rigorous footing, specifically attempts to relate value functions and solutions to the Hamilton–Jacobi Equation. It was found that, for many optimal control problems of interest, the only "solutions" to

---

[1] This is a simplification: the field also concerns optimization problems with dynamic constraints which might be functional differential equations, difference equations, partial differential equations, or take other forms.

the Hamilton–Jacobi Equation have discontinuous derivatives. How should we interpret these solutions? New ideas were required to answer this question since the Hamilton–Jacobi Equation of Optimal Control is a nonlinear partial differential equation for which traditional interpretations of generalized solutions, based on the distributions they define, are inadequate.

Nonsmooth functions surfaced once again when efforts were made to extend the applicability of necessary conditions such as the Maximum Principle. A notable feature of the Maximum Principle (and one that distinguishes it from necessary conditions derivable using classical techniques), is that it can take account of pathwise constraints on values of the control functions. For some practical problems, the constraints on values of the control depend on the vector state variable. In flight mechanics, for example, the maximum and minimum thrust of a jet engine (a control variable) will depend on the altitude (a component of the state vector). The Maximum Principle in its original form is not in general valid for problems involving state dependent control constraints. One way to derive necessary conditions for these problems, and others not covered by the Maximum Principle, is to reformulate them as generalized problems in the Calculus of Variations, the cost integrands for which include penalty terms to take account of the constraints. The reformulation comes at a price, however. To ensure equivalence with the original problems it is necessary to employ penalty terms with discontinuous derivatives. So the route to necessary conditions via generalized problems in the Calculus of Variations can be followed only if we know how to adapt traditional necessary conditions to allow for nonsmooth cost integrands.

Two important breakthroughs occurred in the 1970s. One was the end product of a long quest for effective, local descriptions of "non-smooth" functions, based on generalizations of the concept of the "subdifferential" of a convex function, to larger function classes. F. H. Clarke's theory of generalized gradients, by achieving this goal, launched the field of Nonsmooth Analysis and provided a bridge to necessary conditions of optimality for nonsmooth variational problems (and in particular optimal control problems reformulated as generalized problems in the Calculus of Variations). The other breakthrough, a somewhat later development, was the concept of Viscosity Solutions, due to M. G. Crandall and P.-L. Lions, that provides a framework for proving existence and uniqueness of generalized solutions to Hamilton–Jacobi equations arising in Optimal Control.

Nonsmooth Analysis and Viscosity Methods, introduced to clear a bottleneck in Optimal Control, have had a significant impact on Nonlinear Analysis as a whole. Nonsmooth Analysis provides an important new perspective: useful properties of functions, even differentiable functions, can be proved by examining related nondifferentiable functions, in the same way that trigonometric identities relating to real numbers can sometimes simply be derived by a temporary excursion into the field of complex numbers. Viscosity Methods, on the other hand, provide a fruitful approach to studying generalized solutions to broad classes of nonlinear partial differential equations which extend beyond Hamilton–Jacobi equations of Optimal Control and their approximation for computational purposes.

The Calculus of Variations (in its modern guise as Optimal Control) continues to uphold a long tradition, as a stimulus to research in other fields of mathematics.

Clarke's influential book, *Optimization and Nonsmooth Analysis*, [38] of 1982 covered many important advances of the preceding decade in Nonsmooth Analysis and its applications to the derivation of necessary conditions in Optimal Control. Since then, Optimal Control has remained an active area of research. In fact, it has been given fresh impetus by a number of developments. One is widespread interest in nonlinear controller design methods, based on the solution of an optimal control problem at each controller update time and referred to as *Model Predictive Control* [117],[103]. Another is the role of Optimal Control and Differential Games in generalizations of H–infinity controller design methods to nonlinear systems [90]. There has been a proliferation of new nonsmooth necessary conditions, distinguished by the hypotheses under which they are valid and by their ability to eliminate from consideration certain putative minimizers that are not excluded by rival sets of necessary conditions. However recent work has helped to clarify the relationships between them. This has centered on the Extended Euler–Lagrange Condition, a generalization of Euler's Equation of the classical Calculus of Variations. The Extended Euler–Lagrange Condition subsumes the Hamilton Inclusion, a key early nonsmooth necessary condition, and is valid under greatly reduced hypotheses. A number of other necessary conditions, such as the Nonsmooth Maximum Principle, follow as simple corollaries. Necessary conditions for problems involving pathwise state constraints have been significantly improved, and degenerate features of earlier necessary conditions for such problems have been eliminated. Sharper versions of former nonsmooth transversality conditions have been introduced into the theory. Techniques for deriving improved necessary conditions for free time problems are now also available. Extensions of Tonelli's work on the structure of minimizers have been carried out and used to derive new necessary conditions for variational problems in cases when traditional necessary conditions do not even make sense. In fact, there have been significant advances in virtually all areas of nonsmooth Optimal Control since the early 1980s.

Over the last two decades, viscosity techniques have had a growing following. Applications of Viscosity Methods are now routine in Stochastic Control, Mathematical Finance, Differential Games and other fields besides Optimal Control.

Dynamic Programming is well served by a number of up-to-date expository texts, including Fleming and Soner's book *Controlled Markov Processes and Viscosity Solutions* [66], Barles's lucid introductory text, *Solutions de Viscosité des Equations de Hamilton–Jacobi*, and Bardi and Capuzzo Dolcetta's comprehensive monograph, *Optimal Control and Viscosity Solutions of Hamilton-Jacobo-Bellman Equations* [14] . This cannot be said of applications of Nonsmooth Analysis to the derivation of necessary conditions in Optimal Control. Expository texts such as Loewen's *Optimal Control Via Nonsmooth Analysis* [94] and Clarke's *Methods of Dynamic and Nonsmooth Optimization* [38] give the flavor of contemporary thinking in these areas of Optimal Control, as do the relevant sections of Clarke et al.'s recent monograph, *Nonsmooth Analysis and Control Theory* [54]. But details of recent advances, dispersed as they are over a wide literature and written in a wide variety of styles, are by no means easy to follow.

The main purpose of this book is to bring together as a single publication many major developments in Optimal Control based on Nonsmooth Analysis of recent years, and thereby render them accessible to a broader audience. Necessary conditions receive special attention. But other topics are covered as well. Material on the important, and unjustifiably neglected, topic of minimizer regularity provides a showcase for the application of nonsmooth necessary conditions to derive qualitative information about solutions to variational problems. The chapter on Dynamic Programming stands a little apart from other sections of the book, as it is complementary to recent mainstream research in the area based on Viscosity Methods (and which in any case is already the subject matter of substantial expository texts). Instead we concentrate on aspects of Dynamic Programming well matched to the analytic techniques of this book, notably the characterization (in terms of the Hamilton–Jacobi Equation) of extended-valued value functions associated with problems having endpoint and state constraints, inverse verification theorems, sensitivity relationships and links with the Maximum Principle.

A subsidiary purpose is to meet the needs of readers with little prior exposure to modern Optimal Control who seek quick answers to the questions: what are the main results, what were the deficiencies of the "classical" theory and to what extent have they been overcome? Chapter 1 provides, for their benefit, a lengthy overview, in which analytical details are suppressed and the emphasis is placed instead on communicating the underlying ideas.

To render this book self-contained, preparatory chapters are included on Nonsmooth Analysis, measurable multifunctions, and differential inclusions. Much of this material is implicit in the recent books of R. T.

Rockafellar and J. B. Wets [125] and Clarke et al. [53], and the somewhat older book by J.-P. Aubin and H. Frankowska [12]. It is expected, however, that readers, whose main interest is in Optimization rather than in broader application areas of Nonsmooth Analysis, which require additional techniques, will find these chapters helpful, because of the strong focus on topics relevant to Optimization.

Optimal Control is now a large field, and our choice of material for inclusion in this book is necessarily selective. The techniques used here to derive necessary conditions of optimality are within a tradition of research pioneered and developed by Clarke, Ioffe, Loewen, Mordukhovich, Rockafellar, Vinter, and others, based on perturbation, elimination of constraints and passage to the limit. The necessary conditions are "state of the art," as far as this tradition is concerned. Alternative approaches are not addressed, such as that of H. Sussmann [132], a synthesis of traditional ideas for approximating reachable sets and of extensions to the Warga's theory of derivate containers, which permit a relaxation of hypotheses under which the Maximum Principle is valid in some respects and leads to different kinds of necessary conditions for problems in which the dynamic constraint takes the form of a differential inclusion. The topic of higher order necessary conditions, addressed for example in [155] and [102], is not entered into, nor are computational aspects, examined for example in [113], discussed.

I wish to thank Dorothée Bessis, Fernando Fontes, Thomas Städler, and Harry Zheng for their assistance in preparing the manuscript, and my colleagues at Imperial College, including John Allwright, Alessandro Astolfi, Martin Clark, Mark Davis, Imad Jaimoukha, David Limebeer, Sandro Machietto, David Mayne, John Perkins, Roger Sargent, and George Weiss for their part in creating a working environment in which writing this book has been possible. Many people have influenced my thinking on contents and presentation. I offer special thanks to Francis Clarke, for generously sharing his insights and ideas with me and for making our research collaboration, the fruits of which feature prominently in this publication, so rewarding and enjoyable.

Above all, I express my profound gratitude to my wife, Donna, for her encouragement and support throughout this seemingly endless writing project and for helping me, finally, bring it to a close.

The webpage http.//www.ps.ic.ac.uk/~rbv/oc.html will record errors, ambiguities, etc., in the book, as they come to light. Readers' contributions, via the e-mail address r.vinter@ic.ac.uk, are welcome.

London, United Kingdom   *Richard Vinter*

Notation

B	Closed unit ball in Euclidean space
$|x|$	Eulidean norm of x
$d_C(x)$	Euclidean distance of x from C
$\operatorname{int} C$	Interior of C
$\operatorname{bdy} C$	Boundary of C
$\bar{C}$	Closure of C
$N_C^P(x)$	Proximal normal cone to C at x
$\hat{N}_C(x)$	Strict normal cone to C at x
$N_C(x)$	Limiting normal cone to C at x
$T_C(x)$	Bouligand tangent cone to C at x
$\bar{T}_C(x)$	Clarke tangent cone to C at x
$\operatorname{epi} f$	Epigraph of f
$\partial^P f(x)$	Proximal subdifferential of f at x
$\hat{\partial} f(x)$	Strict subdifferential of f at x
$\partial f(x)$	Limiting subdifferential of f at x
$\partial_P^\infty f(x)$	Asymptotic proximal subdifferential of f at x
$\hat{\partial}_P^\infty f(x)$	Asymptotic strict subdifferential of f at x
$\partial^\infty f(x)$	Asymptotic Limiting Subdifferential of f at x
$\operatorname{dom} f$	(Effective) domain of f
$\operatorname{Gr} F$	Graph of F
$\operatorname{epi} f$	Epigraph of f
$f^0(x; v)$	Generalized directional derivative of f at x in the direction v
$\Psi_C(x)$	Indicator function of the set C at the point x
$\nabla f(x)$	Gradient vector of f at x
$x_i \xrightarrow{C} x$	$x_i \to x$ and $x_i \in C \quad \forall i$
$x_i \xrightarrow{f} x$	$x_i \to x$ and $f(x_i) \to f(x) \quad \forall i$
$\operatorname{supp} \mu$	Support of the measure μ
$W^{1,1}(I; R^n)$	Absolutely continuous functions $f : I \to R^n$
$H, \mathcal{H}$	Hamiltonian, Unmaximized Hamiltonian

Chapter 1

Overview

Everything should be made as simple as possible, but not any simpler.

– Albert Einstein

1.1 Optimal Control

Optimal Control emerged as a distinct field of research in the 1950s, to address in a unified fashion optimization problems arising in scheduling and the control of engineering devices, beyond the reach of traditional analytical and computational techniques. Aerospace engineering is an important source of such problems, and the relevance of Optimal Control to the American and Russian space programs gave powerful initial impetus to the field. A simple example is:

The Maximum Orbit Transfer Problem. A rocket vehicle is in a circular orbit. What is the radius of the largest possible coplanar orbit to which it can be transferred over a fixed period of time? See Figure 1.1.

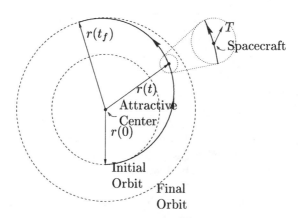

FIGURE 1.1. The Maximal Orbit Transfer Problem.

The motion of the vehicle during the maneuver is governed by the rocket thrust and by the rocket thrust orientation, both of which can vary with time. The variables involved are

$$
\begin{aligned}
r &= \text{radial distance of vehicle from attracting center,} \\
u &= \text{radial component of velocity,} \\
v &= \text{tangential component of velocity,} \\
m &= \text{mass of vehicle,} \\
T_r &= \text{radial component of thrust, and} \\
T_t &= \text{tangential component of thrust.}
\end{aligned}
$$

The constants are

$$
\begin{aligned}
r_0 &= \text{initial radial distance,} \\
m_0 &= \text{initial mass of vehicle,} \\
\gamma_{\max} &= \text{maximum fuel consumption rate,} \\
T_{\max} &= \text{maximum thrust,} \\
\mu &= \text{gravitational constant of attracting center, and} \\
t_f &= \text{duration of maneuver.}
\end{aligned}
$$

A precise formulation of the problem, based on an idealized point mass model of the space vehicle is as follows.

$$
\left\{
\begin{aligned}
&\text{Minimize } -r(t_f) \\
&\text{over radial and tangential components of the thrust history,} \\
&\qquad\qquad (T_r(t), T_t(t)),\ 0 \le t \le t_f,\ \text{satisfying} \\
&\dot{r}(t) = u, \\
&\dot{u}(t) = v^2(t)/r(t) - \mu/r^2(t) + T_r(t)/m(t), \\
&\dot{v}(t) = -u(t)v(t)/r(t) + T_t(t)/m(t), \\
&\dot{m}(t) = -(\gamma_{\max}/T_{max})(T_r^2(t) + T_t^2(t))^{1/2}, \\
&(T_r^2(t) + T_t^2(t))^{1/2} \le T_{\max}, \\
&m(0) = m_0,\ r(0) = r_0,\ u(0) = 0,\ v(0) = \sqrt{\mu/r_0}, \\
&u(t_f) = 0,\ v(t_f) = \sqrt{\mu/r(t_f)}.
\end{aligned}
\right.
$$

Here $\dot{r}(t)$ denotes $dr(t)/dt$, etc. It is standard practice in Optimal Control to formulate optimization problems as minimization problems. Accordingly, the problem of maximizing the radius of the terminal orbit $r(t_f)$ is replaced by the equivalent problem of minimizing the "cost" $-r(t_f)$. Notice that knowledge of the *control function* or *strategy* $(T_r(t), T_t(t))$, $0 \le t \le t_f$ permits us to calculate the cost $-r(t_f)$: we solve the differential equations, for the specified boundary conditions at time $t = 0$, to obtain the corresponding *state trajectory* $(r(t), u(t), v(t), m(t))$, $0 \le t \le t_f$, and thence determine $-r(t_f)$. The control strategy therefore has the role of choice variable in the optimization problem. We seek a control strategy that minimizes the cost, from among the control strategies whose associated state trajectories satisfy the specified boundary conditions at time $t = t_f$.

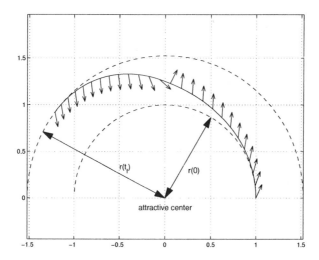

FIGURE 1.2. An Orbit Transfer Strategy.

Figure 1.2 shows a control strategy for the following values of relevant dimensionless parameters:

$$\frac{T_{max}/m_0}{\mu/r_0^2} = 0.1405, \quad \frac{\gamma_{max}}{T_{max}/\sqrt{\mu/r_0}} = 0.07487, \quad \frac{t_f}{\sqrt{r_0^3/\mu}} = 3.32 .$$

For this strategy, the radius of the terminal circular orbit is

$$r(t_f) = 1.5 r_0.$$

The arrows indicate the magnitude and orientation of the thrust at times $t = 0, 0.1 t_f, 0.2 t_f, \ldots, t_f$. As indicated, full thrust is maintained. The thrust is outward for (approximately) the first half of the maneuver and inward for the second.

Suppose, for example, that the attracting center is the Sun, the space vehicle weighs $10,000$ lb, the initial radius is 1.50 million miles (the radius of a circle approximating the Earth's orbit), the maximum thrust is 0.85 lb (i.e., a force equivalent to the gravitational force on a 0.85 lb mass on the surface of the earth), the maximum rate of fuel consumption is 1.81 lb/day, and the transit time is 193 days. Corresponding values of the constants are

$$\begin{aligned}
&T_{max} = 3.778\,\text{N}, &&m_0 = 4.536 \times 10^3\,\text{kg}, \\
&r_0 = 1.496 \times 10^{11}\,\text{m}, &&\gamma_{max} = 0.9496 \times 10^{-5}\,\text{kg s}^{-1}, \\
&t_f = 1.6675 \times 10^7\,\text{s}, &&\mu = 1.32733 \times 10^{20}\,\text{m}^3\text{s}^{-2}.
\end{aligned}$$

Then the terminal radius of the orbit is 2.44 million miles. (This is the radius of a circle approximating the orbit of the planet Mars.)

Numerical methods, inspired by necessary conditions of optimality akin to the Maximum Principle of Chapter 6, were used to generate the above control strategy.

Optimal Control has its origins in practical flight mechanics problems. But now, 40 years on, justification for research in this field rests not only on aerospace applications, but on applications in new areas such as process control, resource economics, and robotics. Equally significant is the stimulus Optimal Control has given to research in related branches of mathematics (convex analysis, nonlinear analysis, functional analysis, and dynamical systems).

From a modern perspective, Optimal Control is an outgrowth of the Calculus of Variations which takes account of new kinds of constraints (differential equation constraints, pathwise constraints on control functions "parameterizing" the differential equations, etc.) encountered in advanced engineering design. A number of key recent developments in Optimal Control have resulted from marrying old ideas from the Calculus of Variations and modern analytical techniques. For purposes both of setting Optimal Control in its historical context and of illuminating current developments, we pause to review relevant material from the classical Calculus of Variations.

1.2   The Calculus of Variations

The Basic Problem in the Calculus of Variations is that of finding an arc $\bar{x}$ which minimizes the value of an integral functional

$$J(x) = \int_S^T L(t, x(t), \dot{x}(t))dt$$

over some class of arcs satisfying the boundary condition

$$x(S) = x_0 \quad \text{and} \quad x(T) = x_1.$$

Here $[S, T]$ is a given interval, $L : [S, T] \times R^n \times R^n \to R$ is a given function, and x_0 and x_1 are given points in R^n.

The Brachistochrone Problem. An early example of such a problem was the *Brachistochrone Problem* circulated by Johann Bernoulli in the late 17th century. Positive numbers s_f and x_f are given. A frictionless bead, initially located at the point $(0, 0)$, slides along a wire under the force of gravity. The wire, which is located in a fixed vertical plane, joins the points $(0, 0)$ and (s_f, x_f). What should the shape of the wire be, in order that the bead arrive at its destination, the point (s_f, x_f), in minimum time?

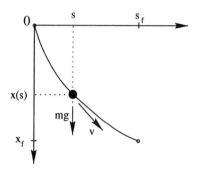

FIGURE 1.3. The Brachistochrone Problem.

There are a number of possible formulations of this problem. We now describe one of them. (See Figure 1.3.) Denote by s and x the horizontal and vertical distances of a point on the path of the bead (vertical distances are measured downward). We restrict attention to wires describable as the graph of a suitably regular function $x(s)$, $0 \leq s \leq s_f$. For any such function x, the velocity $v(s)$ is related to the downward displacement $x(s)$, when the horizontal displacement is s, according to

$$mgx(s) = \frac{1}{2}mv^2(s)$$

("loss of potential energy equals gain of kinetic energy"). But, denoting the time variable as t, we have

$$v(s) = \frac{\sqrt{1 + |dx(s)/ds|^2}}{dt(s)/ds}.$$

It follows that the transit time is

$$\int_0^{s_f} dt = \int_0^{s_f} \frac{\sqrt{1 + |dx(s)/ds|^2}}{v(s)} ds.$$

Eliminating $v(s)$ from the preceding expressions, we arrive at a formula for the transit time:

$$J(x) = \int_0^{s_f} L(s, x(s), \dot{x}(s)) ds,$$

in which

$$L(s, x, w) := \frac{\sqrt{1 + |w|^2}}{\sqrt{2gx}}.$$

The problem is to minimize $J(x)$ over some class of arcs x satisfying

$$x(0) = 0 \quad \text{and} \quad x(s_f) = x_f.$$

This is an example of the Basic Problem of the Calculus of Variations, in which $(S, x_0) = (0, 0)$ and $(T, x_1) = (s_f, x_f)$.

Suppose that we seek a minimizer in the class of absolutely continuous arcs. It can be shown that the minimum time t^* and the minimizing arc $(x(t), s(t))$, $0 \leq t \leq t^*$ (expressed in parametric form with independent variable time t) are given by the formulae

$$x(t) = a\left(1 - \cos\sqrt{\frac{g}{a}}\,t\right) \quad \text{and} \quad s(t) = a\left(\sqrt{\frac{g}{a}}\,t - \sin\sqrt{\frac{g}{a}}\,t\right).$$

Here, a and t^* are constants that uniquely satisfy the conditions

$$x(t^*) = x_f,$$
$$s(t^*) = t_f,$$
$$0 \leq \sqrt{\frac{g}{a}}\,t^* \leq 2\pi.$$

The minimizing curve is a cycloid, with infinite slope at the point of departure: it coincides with the locus of a point on the circumference of a disc of radius a, which rolls without slipping along a line of length t_f.

Problems of this kind, the minimization of integral functionals, may perhaps have initially attracted attention as individual curiosities. But throughout the 18th and 19th centuries their significance became increasingly evident, as the list of the laws of physics that identified states of nature with minimizing curves and surfaces lengthened. Examples of "Rules of the Minimum" are as follows.

Fermat's Principle in Optics. The path of a light ray achieves a local minimum of the transit times over paths between specified endpoints that visit the relevant reflecting and refracting boundaries. The principle predicts Snell's Laws of Reflection and Refraction, and the curved paths of light rays in inhomogeneous media. See Figure 1.4.

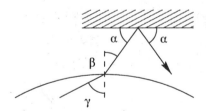

FIGURE 1.4. Fermat's Principle Predicts Snell's Laws.

Dirichlet's Principle. Take a bounded, open set $\Omega \subset R^2$ with boundary $\partial\Omega$, in which a static two-dimensional electric field is distributed. Denote by $V(x)$ the voltage at point $x \in \Omega$. Then $V(x)$ satisfies Poisson's Equation

$$\Delta V(x) = 0 \text{ for } x \in \Omega$$
$$V(x) = \bar{V}(x) \text{ for } x \in \partial\Omega.$$

Here, $\bar{V} : \partial\Omega \to R$ is a given function, which supplies the boundary data.

Dirichlet's Principle characterizes the solution to this partial differential equation as the solution of a minimization problem

$$\begin{cases} \text{Minimize } \int_\Omega \nabla V(x) \cdot \nabla V(x) dx \\ \text{over surfaces } V \text{ satisfying } V(x) = \bar{V}(x) \text{ on } \partial\Omega. \end{cases}$$

This optimization problem involves finding a *surface* that minimizes a given integral functional. See Figure 1.5.

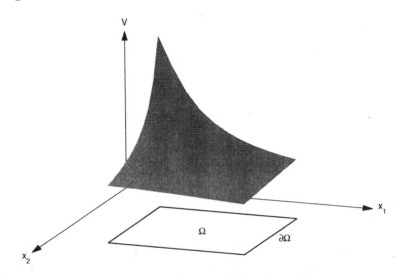

FIGURE 1.5. A Mimimizer for the Dirichlet Integral.

Dirichlet's Principle and its generalizations are important in many respects. They are powerful tools for the study of existence and regularity of solutions to boundary value problems. Furthermore, they point the way to Galerkin methods for computing solutions to partial differential equations, such as Poisson's equation: the solution is approximated by the minimizer of the above Dirichlet integral over some finite-dimensional subspace $\mathcal{S}_N$ of the domain of the original optimization problem, spanned by a finite collection of "basis" functions $\{\phi_i\}_{i=1}^N$,

$$\mathcal{S}_N = \{\sum_{i=1}^N \alpha_i \phi_i(x) : \alpha \in R^N\}.$$

The widely used finite element methods are modern implementations of Galerkin's method.

The Action Principle. Let $x(t)$ be the vector of generalized coordinates of a conservative mechanical system. The Action Principle asserts that $x(t)$

evolves in a manner to minimize (strictly speaking, to render stationary) the "action," namely,

$$\int [T(x(t), \dot{x}(t)) - V(x(t))]dt.$$

Here $T(x, \dot{x})$ is the kinetic energy and $V(x)$ is the potential energy. Suppose, for example, $x = (r, \theta)$, the polar coordinates of an object of mass m moving in a plane under the influence of a radial field (the origin is the center of gravity of a body, massive in relation to the object). See Figure 1.6. Then

$$T(x, \dot{x}) = \frac{1}{2}m(\dot{r}^2 + r^2\dot{\theta}^2)$$

and

$$V(r) = -K/r,$$

for some constant K. The action in this case is

$$\int \left(\frac{1}{2}m[\dot{r}^2(t) + r^2(t)\dot{\theta}^2(t)] - K/r(t) \right) dt.$$

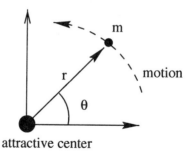

attractive center

FIGURE 1.6. Motion in the Plane for a Radial Field.

The Action Principle has proved a fruitful starting point for deriving the dynamical equations of complex interacting systems, and for studying their qualitative properties (existence of periodic orbits with prescribed energy, etc.).

Necessary Conditions

Consider the optimization problem

$$(CV) \quad \begin{cases} \text{Minimize } J(x) := \int_S^T L(t, x(t), \dot{x}(t))dt \\ \text{over arcs } x \text{ satisfying} \\ x(S) = x_0 \text{ and } x(T) = x_1, \end{cases}$$

in which $[S, T]$ is a fixed interval, $L : R \times R^n \times R^n \to R$ is a given C^2 function, and x_0 and x_1 are given points in R^n. Precise formulation of problem (CV) requires us to specify the domain of the optimization problem. We take this to be $W^{1,1}([S, T]; R^n)$, the class of absolutely continuous R^n-valued arcs on $[S, T]$, for reasons that are discussed presently.

The systematic study of minimizers $\bar{x}$ for this problem was initiated by Euler, whose seminal paper of 1744 provided the link with the equation:

$$\frac{d}{dt} L_v(t, \bar{x}(t), \dot{\bar{x}}(t)) = L_x(t, \bar{x}(t), \dot{\bar{x}}(t)). \tag{1.1}$$

(In this equation, L_x and L_v are the gradients of $L(t, x, v)$ with respect to the second and third arguments, respectively.)

The Euler Equation (1.1) is, under appropriate hypotheses, a necessary condition for an arc $\bar{x}$ to be a minimizer. Notice that, if the minimizer $\bar{x}$ is a C^2 function, then the Euler Equation is a second-order, n-vector differential equation:

$$L_{vt}(t, \bar{x}(t), \dot{\bar{x}}(t)) + L_{vx}(t, \bar{x}(t), \dot{\bar{x}}(t)) \dot{\bar{x}}(t)$$
$$+ L_{vv}(t, \bar{x}(t), \dot{\bar{x}}(t)) \ddot{\bar{x}}(t) = L_x(t, \bar{x}(t), \dot{\bar{x}}(t)).$$

A standard technique for deriving the Euler Equation is to reduce the problem to a scalar optimization problem, by considering a one-parameter family of *variations*. The Calculus of Variations, incidentally, owes its name to these ideas. (*Variations* of the minimizing arc cannot reduce the cost; conditions on minimizers are then derived by processing this information, with the help of a suitable *calculus* to derive necessary conditions of optimality.) Because of its historical importance and its continuing influence on the derivation of necessary conditions, we now describe the technique in detail.

Fix attention on a minimizer $\bar{x}$. Further hypotheses are required to derive the Euler Equation. We assume that there exists some number K such that

$$|L(t, x, v) - L(t, y, w)| \leq K(|x - y| + |v - w|) \tag{1.2}$$

for all $x, y \in R^n$ and all $v, w \in R^n$.

Take an arbitrary C^1 arc y, which satisfies the homogeneous boundary conditions

$$y(S) = y(T) = 0.$$

Then, for any $\epsilon > 0$, the "variation" $x + \epsilon y$, which satisfies the endpoint constraints, must have cost not less than that of $\bar{x}$. It follows that

$$\epsilon^{-1}[J(\bar{x} + \epsilon y) - J(\bar{x})] \geq 0.$$

Otherwise expressed

$$\int_S^T \epsilon^{-1}[L(t, \bar{x}(t) + \epsilon y(t), \dot{\bar{x}}(t) + \epsilon \dot{y}(t)) - L(t, \bar{x}(t), \dot{\bar{x}}(t))]dt \geq 0.$$

Under Hypothesis (1.2), the Dominated Convergence Theorem permits us to pass to the limit under the integral sign. We thereby obtain the equation

$$\int_S^T [L_x(t, \bar{x}(t), \dot{\bar{x}}(t)) \cdot y(t) + L_v(t, \bar{x}(t), \dot{\bar{x}}(t)) \cdot \dot{y}(t)] dt \geq 0.$$

This relationship holds, we note, for all continuously differentiable functions y satisfying the boundary conditions $y(S) = 0$ and $y(T) = 0$. By homogeneity, the inequality can be replaced by equality.

Now apply integration by parts to the first term on the left. This gives

$$\int_S^T [- \int_S^t L_x(s, \bar{x}(s), \dot{\bar{x}}(s)) ds + L_v(t, \bar{x}(t), \dot{\bar{x}}(t))] \cdot \dot{y}(t) dt = 0.$$

Take any continuous function $w : [S, T] \to R^n$ that satisfies

$$\int_S^T w(t) dt = 0. \tag{1.3}$$

Then the continuously differentiable arc $y(t) \equiv \int_S^t w(s) ds$ vanishes at the endtimes. Consequently,

$$\int_S^T [- \int_S^t L_x(s, \bar{x}(s), \dot{\bar{x}}(s)) ds + L_v(t, \bar{x}(t), \dot{\bar{x}}(t))] \cdot w(t) dt = 0, \tag{1.4}$$

a relationship that holds for all continuous arcs w satisfying (1.3). To advance the analysis, we require

Lemma (Raymond Dubois) *Take a function $a \in L^2([S, T]; R^n)$. Suppose that*

$$\int_S^T a(t) \cdot w(t) \, dt = 0 \tag{1.5}$$

for every continuous function w that satisfies

$$\int_S^T w(t) dt = 0. \tag{1.6}$$

Then there exists some vector $d \in R^n$ such that

$$a(t) = d \quad \text{for a.e. } t \in [S, T].$$

Proof. We give a contemporary proof (based on an application of the Separation Principle in Hilbert space) of this classical lemma, a precursor of 20th century theorems on the representation of distributions.

By hypothesis, for any continuous function w, (1.6) implies (1.5). By using the fact that the continuous functions are dense in $L^2([S, T]; R^n)$

with respect to the strong L^2 topology, we readily deduce that (1.6) implies (1.5) also when w is a L^2 function. This fact is used shortly.

Define the constant elements $e^j \in L^2([S,T]; R^n)$, $j = 1, \ldots, n$ to have components:

$$e_i^j(t) = \begin{cases} 1 & \text{if } i = j \\ 0 & \text{if } i \neq j \end{cases}, \quad i = 1, \ldots, n.$$

Since a function is a constant function if and only if it can be expressed as a linear combination of the $e_i^j(t)$s, the properties asserted in the lemma can be rephrased:

$$a \in V,$$

where V is the (closed) n-dimensional subspace of $L^2([S,T]; R^n)$ spanned by the e^js. Suppose that the assertions are false; i.e., $a \notin V$. Since V is a closed subspace, we deduce from the Separation Theorem that there exists a nonzero element $w \in L^2$ and $\epsilon > 0$ such that

$$< a, w >_{L^2} \; \leq \; < v, w >_{L^2} - \epsilon.$$

for all $v \in V$. Because V is a subspace, it follows that

$$< v, w >_{L^2} = 0$$

for all $v \in V$. But then

$$< a, w >_{L^2} \leq -\epsilon. \tag{1.7}$$

Observe that, for each j,

$$< e^j, w >_{L^2} = 0,$$

since $e^j \in V$. This last condition can be expressed as

$$\int_S^T w(t) dt = 0.$$

In view of our earlier comments,

$$< a, w >_{L^2} = 0.$$

This contradicts (1.7). The lemma is proved. $\square$

Return now to the derivation of the Euler Equation. We identify the function $a(.)$ of the lemma with

$$t \to -\int_S^t L_x(s, \bar{x}(s), \dot{\bar{x}}(s)) ds + L_v(t, \bar{x}(t), \dot{\bar{x}}(t)).$$

In view of (1.4), the lemma informs us that there exists a vector d such that

$$-\int_S^t L_x(s, \bar{x}(s), \dot{\bar{x}}(s))ds + L_v(t, \bar{x}(t), \dot{\bar{x}}(t)) = d \quad \text{a.e.} \qquad (1.8)$$

Since $L_x(t, \bar{x}(t), \dot{\bar{x}}(t))$ is integrable, it follows that $t \to L_v(t, \bar{x}(t), \dot{\bar{x}}(t))$ is almost everywhere equal to an absolutely continuous function and

$$\frac{d}{dt}L_v(t, \bar{x}(t), \dot{\bar{x}}(t)) = L_x(t, \bar{x}(t), \dot{\bar{x}}(t)) \quad \text{a.e.}$$

We have verified the Euler Equation and given it a precise interpretation, when the domain of the optimization problem is the class of absolutely continuous arcs.

The above analysis conflates arguments assembled over several centuries. The first step was to show that smooth minimizers $\bar{x}$ satisfy the pointwise Euler Equation. Euler's original derivation made use of discrete approximation techniques. Lagrange's alternative derivation introduced variational methods similar to those outlined above (though differing in the precise nature of the "integration by parts" step). Erdmann subsequently discovered that, if the domain of the optimization problem is taken to be the class of piecewise C^1 functions (i.e., absolutely continuous functions with piecewise continuous derivatives) then

"the function $t \to L_v(t, \bar{x}(t), \dot{\bar{x}}(t))$ has removable discontinuities."

This condition is referred to as the *First Erdmann Condition*.

For piecewise C^1 minimizers, the integral version of the Euler Equation (1.8) was first regarded as a convenient way of combining the pointwise Euler Equation and the First Erdmann Condition. We refer to it as the *Euler–Lagrange Condition*. An analysis in which absolutely continuous minimizers substitute for piecewise C^1 minimizers is an early 20th century development, due to Tonelli.

Another important property of minimizers can be derived in situations when L is independent of the t (write $L(x, v)$ in place of $L(t, x, v)$). In this "autonomous" case the second-order n-vector differential equation of Euler can be integrated. There results a first-order differential equation involving a "constant of integration" c:

$$L_v(\bar{x}(t), \dot{\bar{x}}(t)) \cdot \dot{\bar{x}}(t) - L(\bar{x}(t), \dot{\bar{x}}(t)) = c. \qquad (1.9)$$

This condition is referred to as the *Second Erdmann Condition* or *Constancy of the Hamiltonian Condition*. It is easily deduced from the Euler–Lagrange Condition when $\bar{x}$ is a C^2 function. Fix t. We calculate in this case

$$\frac{d}{dt}(L_v \cdot \dot{\bar{x}}(t) - L) = \frac{d}{dt}L_v \cdot \dot{\bar{x}}(t) + L_v \cdot \ddot{\bar{x}}(t) - L_x \cdot \dot{\bar{x}}(t) - L_v \cdot \ddot{\bar{x}}(t)$$

$$= (\frac{d}{dt}L_v - L_x) \cdot \dot{\bar{x}}(t)$$
$$= 0.$$

(In the above relationships, L, L_v, etc., are evaluated at $(\bar{x}(t), \dot{\bar{x}}(t))$.) We deduce (1.9).

A more sophisticated analysis leads to an "almost everywhere"' version of this condition for autonomous problems, when the minimizer $\bar{x}$ in question is assumed to be merely absolutely continuous.

Variations of the type $\bar{x}(t) + \epsilon y(t)$ lead to the Euler–Lagrange Condition. Necessary conditions supplying additional information about minimizers have been derived by considering other kinds of variations. We note in particular the *Weierstrass Condition* or the *Constancy of the Hamiltonian Condition*:

$$L_v(t, \bar{x}(t), \dot{\bar{x}}(t)) \cdot \dot{\bar{x}}(t) - L(t, \bar{x}(t), \dot{\bar{x}}(t))$$
$$= \max_{v \in R^n} \{L_v(t, \bar{x}(t), \dot{\bar{x}}(t)) \cdot v - L(t, \bar{x}(t), v)\}.$$

Suppressing the $(t, \bar{x}(t))$ argument in the notation and expressing the Weierstrass Condition as

$$L(v) - L(\dot{\bar{x}}(t)) \geq L_v(\dot{\bar{x}}) \cdot (v - \dot{\bar{x}}(t)) \text{ for all } v \in R^n,$$

we see that it conveys no useful information when $L(t, x, v)$ is convex with respect to the v variable: in this case it simply interprets L_v as a subgradient of L in the sense of convex analysis. In general however, it tells us that L coincides with its "convexification" (with respect to the velocity variable) along the optimal trajectory; i.e.,

$$L(t, \bar{x}(t), \dot{\bar{x}}(t)) = \tilde{L}(t, \bar{x}(t), \dot{\bar{x}}(t)).$$

Here $\tilde{L}(t, x, .)$ is the function with epigraph set co $\{\text{epi } L(t, x, .)\}$. See Figure 1.7.

The above necessary conditions (the Euler–Lagrange Condition, the Weierstrass Condition, and the Second Erdmann Condition) have convenient formulations in terms of the "adjoint arc"

$$p(t) = L_v(t, \bar{x}(t), \dot{\bar{x}}(t)).$$

They are

$$(\dot{p}(t), p(t)) = \nabla_{x,v} L(t, \bar{x}(t), \dot{\bar{x}}(t)), \tag{1.10}$$

$$p(t) \cdot \dot{\bar{x}}(t) - L(t, \bar{x}(t), \dot{\bar{x}}(t)) = \max_{v \in R^n}[p(t) \cdot v - L(t, \bar{x}(t), v)],$$

and, in the case when L does not depend on t,

$$p(t) \cdot \dot{\bar{x}}(t) - L(t, \bar{x}(t), \dot{\bar{x}}(t)) = c$$

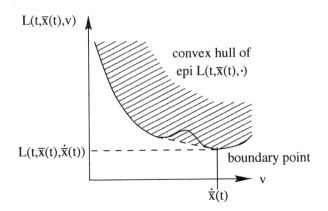

FIGURE 1.7. The Weierstrass Condition.

for some constant c.

To explore the qualitative properties of minimizers it is often helpful to reduce the Euler–Lagrange Condition to a system of specially structured first-order differential equations. This was first achieved by Hamilton for Lagrangians L arising in mechanics. In this analysis the *Hamiltonian*,

$$H(t,x,p) := \max_{v \in R^n}\{p \cdot v - L(t,x,v)\},$$

has an important role.

Suppose that the right side has a unique maximizer v_{max}. This will depend on (t,x,p) and so we can use it to define a function $\chi : [S,T] \times R^n \times R^n \to R^n$:

$$\chi(t,x,p) := v_{\max}.$$

Then

$$H(t,x,p) = p \cdot \chi(t,x,p) - L(t,x,\chi(t,x,p))$$

and

$$\nabla_v(p \cdot v - L(t,x,v))|_{v=\chi(t,x,p)} = 0.$$

The last relationship implies

$$p = L_v(t,x,\chi(t,x,p)). \tag{1.11}$$

(The mapping from x vectors to p vectors implicit in this relationship, for fixed t, is referred to as the *Legendre Transformation*.)

Now consider an arc $\bar{x}$ and associated adjoint arc p that satisfy the Euler–Lagrange and Weierstrass Conditions, namely

$$(\dot{p}(t), p(t)) = \nabla_{x,v} L(t, \bar{x}(t), \dot{\bar{x}}(t)) \tag{1.12}$$

and

$$p(t) \cdot \dot{\bar{x}}(t) - L(t, \bar{x}(t), \dot{\bar{x}}(t)) = \max_{v \in R^n} \{p(t) \cdot v - L(t, \bar{x}(t), v)\}.$$

Since it is assumed that the "maximizer" in the definition of the Hamiltonian is unique, it follows from the Weierstrass condition that

$$\dot{\bar{x}}(t) = \chi(t, \bar{x}(t), p(t)).$$

Fix t. Let us assume that $\chi(t, \cdot, \cdot)$ is differentiable. Then we can calculate the gradients of $H(t, \cdot, \cdot)$:

$$
\begin{aligned}
\nabla_x H(t, x, p)|_{x=\bar{x}(t), p=p(t)} \\
&= \nabla_x (p \cdot \chi(t, x, p) - L(t, x, \chi(t, x, p)))|_{x=\bar{x}(t), p=p(t)} \\
&= p \cdot \chi_x(t, x, p) - L_x(t, x, \chi(t, x, p)) \\
&\quad - L_v(t, x, \chi(t, x, p)) \cdot \chi_x(t, x, p)|_{x=\bar{x}(t), p=p(t)} \\
&= (p - L_v(t, x, \chi(t, x, p))) \cdot \chi_x(t, x, p) - L_x(t, x, \chi(t, x, p))|_{x=\bar{x}(t), p=p(t)} \\
&= 0 - \dot{p}(t).
\end{aligned}
$$

(The last step in the derivation of these relationships makes use of (1.11) and (1.12).) We have evaluated the x-derivative of H:

$$\nabla_x H(t, \bar{x}(t), p(t)) = -\dot{p}(t).$$

As for the p-derivative, we have

$$
\begin{aligned}
\nabla_p H(t, x, p)|_{x=\bar{x}(t), p=p(t)} \\
&= \nabla_p (p \cdot \chi(t, x, p) - L(t, x, \chi(t, x, p)))|_{x=\bar{x}(t), p=p(t)} \\
&= \chi(t, x, p) + p \cdot \chi_p(t, x, p) - L_v(t, x, \chi(t, x, p)) \cdot \chi_p(t, x, p)|_{x=\bar{x}(t), p=p(t)} \\
&= \dot{\bar{x}}(t) + (p(t) - L_v(t, \bar{x}(t), \dot{\bar{x}}(t))) \cdot \chi_p(t, \bar{x}(t), p(t)) \\
&= \dot{\bar{x}}(t) + 0.
\end{aligned}
$$

Combining these relationships, we arrive at the system of first-order differential equations of interest, namely, the *Hamilton Condition*

$$(-\dot{p}(t), \dot{\bar{x}}(t)) = \nabla H(t, \bar{x}(t), p(t)), \tag{1.13}$$

in which ∇H denotes the gradient of $H(t, \cdot, \cdot)$.

So far, we have limited attention to the problem of minimizing an integral functional of arcs with fixed endpoints. A more general problem is that in which the arcs are constrained to satisfy the boundary condition

$$(x(S), x(T)) \in C, \tag{1.14}$$

for some specified subset $C \subset R^n \times R^n$. (The fixed endpoint problem is a special case, in which C is chosen to be $\{(x_0, x_1)\}$. The above necessary conditions remain valid when we pass to this more general endpoint constraint, since a minimizer $\bar{x}$ over arcs satisfying (1.14) is also a minimizer for the fixed endpoint problem in which $C = \{\bar{x}(S), \bar{x}(T)\}$. But something

more is required: replacing the fixed endpoint constraint by other types
of endpoint constraint that allow endpoint variation introduces extra degrees of freedom into the optimization problem, which should be reflected
in supplementary necessary conditions. These conditions are conveniently
expressed in terms of the adjoint arc p:

$$(p(S), -p(T)) \text{ is an outward normal to } C \text{ at } (\bar{x}(S), \bar{x}(T)).$$

They are collectively referred to as the *Transversality Condition*, because
they assert that a vector composed of endpoints of the adjoint arc is orthogonal or "transverse" to the tangent hyperplane to C at $(\bar{x}(S), \bar{x}(T))$.
Figure 1.8 illustrates the transversality condition, in relation to an arc x
and adjoint arc p.

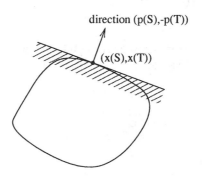

direction (p(S),-p(T))

(x(S),x(T))

FIGURE 1.8. The Transversality Condition.

The Transversality Condition too can be derived by considering suitable variations (variations which, on this occasion, allow perturbations of
the endpoint values of the minimizer $\bar{x}$). In favorable circumstances the
Transversality Condition combines with the boundary condition (1.14) to
supply the appropriate number of $2n$ boundary conditions to accompany
the system of $2n$ equations (1.13).

The following theorem brings together these classical conditions and gives
precise hypotheses under which they are satisfied.

Theorem 1.2.1 *Let $\bar{x}$ be a minimizer for*

$$\begin{cases} \text{Minimize } J(x) := \int_S^T L(t, x(t), \dot{x}(t))dt \\ \text{over absolutely continuous arcs } x \text{ satisfying} \\ (x(S), x(T)) \in C, \end{cases}$$

*for some given interval $[S, T]$, function $L : [S, T] \times R^n \times R^n \to R$, and
closed set $C \subset R^n \times R^n$. We list the hypotheses:*

(i) $L(.,x,v)$ is measurable for each (x,v) and $L(t,.,.)$ is continuously differentiable for each $t \in [S,T]$;

(ii) there exist $k(.) \in L^1$, $\beta > 0$ and $\epsilon > 0$ such that

$$|L(t,x,v) - L(t,x',v)| \le (k(t) + \beta|v|)|x - x'|$$
$$\text{for all } x, x' \in \bar{x}(t) + \epsilon B, \ v \in R^n;$$

(iii) $L(t,x,.)$ is convex for all $(t,x) \in [S,T] \times R^n$;

(iv) $H(t,.,.)$ is continuously differentiable on a neighborhood of $(\bar{x}(t), \dot{\bar{x}}(t))$ for each $t \in [S,T]$ where

$$H(t,x,p) := \sup_{v \in R^n} \{p \cdot v - L(t,x,v)\}.$$

Suppose that Hypotheses (i) and (ii) are satisfied. Then there exists an absolutely continuous arc $p : [S,T] \to R^n$ satisfying the following conditions.

The Euler–Lagrange Condition

$$(\dot{p}(t), p(t)) = \nabla_{x,v} L(t, \bar{x}(t), \dot{\bar{x}}(t)) \quad a.e.$$

The Weierstrass Condition

$$p(t) \cdot \dot{\bar{x}}(t) - L(t, \bar{x}(t), \dot{\bar{x}}(t)) = \max_{v \in R^n} \{p(t) \cdot v - L(t, \bar{x}(t), v)\} \quad a.e.$$

The Transversality Condition

$$(p(S), -p(T)) \in N_C(\bar{x}(T), \bar{x}(t)).$$

If additionally Hypotheses (iii) and (iv) are satisfied, then p can be chosen also to satisfy

The Hamilton Condition

$$(-\dot{p}(t), \dot{\bar{x}}(t)) = \nabla_{x,p} H(t, \bar{x}(t), p(t)) \quad a.e.$$

Finally, suppose that the hypothesis

(v) $L(t,x,v)$ is independent of t

is satisfied. Then all the above assertions can be strengthened to require that p also satisfies

Constancy of the Hamiltonian Condition

$$p(t) \cdot \dot{\bar{x}}(t) - L(t, \bar{x}(t), \dot{\bar{x}}(t)) = c \quad a.e.$$

for some constant c.

The theorem is a special case of general necessary conditions which are proved in Chapter 7.

In this theorem, the Transversality Condition is expressed in a way that makes sense when C is assumed to be a closed set. It makes reference to the "limiting normal cone" $N_C(\bar{x}(S), \bar{x}(T))$. This is defined in the overview of Nonsmooth Analysis provided in Section 1.8.

1.3   Existence of Minimizers and Tonelli's Direct Method

Deriving the Euler–Lagrange Condition and related conditions would appear to reduce the problem of finding a minimizer to one merely of solving a differential equation. But this overlooks the fact that satisfaction of the Euler–Lagrange Condition is merely a *necessary* condition for optimality; we cannot be certain, without further analysis, that an arc satisfying these conditions is truly a minimizer.

Earlier this century, Tonelli recognized the significance of establishing the *existence* of minimizers before using necessary conditions of optimality to try to identify them. Tonelli's *Direct Method* for obtaining a minimizer consists of the following steps.

1: Show that a minimizer exists.

2: Search among arcs satisfying the necessary conditions for an arc with lowest cost.

These steps, when successfuly carried out, are guaranteed to yield a minimizer. Notice however that, if we neglect Step 1, then Step 2 alone can be positively misleading. The point is illustrated by the following example.

$$\left\{ \begin{array}{l} \text{Minimize } \int_0^1 x(t)\dot{x}^2(t)dt \\ \text{over absolutely continuous arcs } x \text{ satisfying} \\ x(0) = 0 \text{ and } x(1) = 0. \end{array} \right.$$

This example is not precisely matched to the necessary conditions of Theorem 1.2.1 because the integrand L fails to satisfy Hypothesis (ii). It can nevertheless be shown that any *Lipschitz continuous* minimizer $\bar{x}$ satisfies the Euler–Lagrange, Weierstrass, and Transversality Conditions. In this case the Euler–Lagrange Condition takes the form:

$$2\frac{d}{dt}\left\{\bar{x}(t)\dot{\bar{x}}(t)\right\} = \dot{\bar{x}}^2(t) \quad \text{a.e.}$$

Obviously the arc

$$\bar{x} \equiv 0$$

satisfies the Euler Equation. It is in fact the *unique* Lipschitz continuous function so doing. To see this note that, for any Lipschitz continuous arc y which vanishes at the endtimes and satisfies the Euler–Lagrange Condition, we have

$$\frac{d^2}{dt^2}\, y^2(t) \;=\; \dot y^2(t) \quad \text{a.e.}$$

It follows that, for some constant c,

$$\frac{d}{dt} y^2(t) \;=\; c + \int_0^t \dot y^2(s)ds \ \text{ for all } t.$$

Since y vanishes at the left endpoint and is assumed to have bounded slope, we deduce from this last condition that $c = 0$. If $y \not\equiv 0$ then $\dot y$ must be nonzero on a set of positive measure, in order to satisfy the endpoint constraints. But then

$$y(1)^2 = 0 + \int_0^1 \int_0^s \dot y^2(t)\,dsdt \;>\; 0\,,$$

in violation of the endpoint constraints. We have confirmed that $\bar x \equiv 0$ is the unique Lipschitz continuous arc satisfying the Euler–Lagrange Condition.

Notice, however, that $\bar x$ is not a minimizer, not even in a local sense. Indeed by choosing the parameter $\epsilon > 0$ sufficently small we can arrange that the arc $t \to -\epsilon t(1-t)$, which has cost strictly less than 0, is arbitrarily close to $\bar x$ with respect to the supremum norm.

To summarize, in this example there is a unique Lipschitz continuous arc satisfying the constraints of the problem and also the Euler–Lagrange, Weierstrass, and Transversality Conditions. Yet it is not a minimizer. This is a case in which a naive belief in the power of necessary conditions to identify minimizers leads us astray. The pathological feature of this example is, of course, that there are no minimizers in the class of Lipschitz continuous functions. The collection of arcs satisfying necessary conditions for a Lipschitz continuous arc to be a minimizer will not therefore contain a minimizer, even though it is nonempty.

Clearly, it makes sense to speak of a minimizer only if we specify the class of functions from which minimizers are chosen. Unfortunately, hypotheses under which we can guarantee existence of minimizers in the "elementary" function spaces (C^1 functions, piecewise C^1 functions, etc.) are, for many purposes, unacceptably restrictive. One of Tonelli's most significant contributions in the Calculus of Variations was to show that, by enlarging the domain of (CV) to include $W^{1,1}([S,T]; R^n)$ arcs, the existence of minimizers could be guaranteed under broad, directly verifiable hypotheses on the Lagrangian L. An existence theorem, in the spirit of Tonelli's pioneering work in this field, is:

Theorem 1.3.1 *Consider the minimization problem*

$$\begin{cases} \textit{Minimize } \int_S^T L(t, x(t), \dot{x}(t))dt \textit{ over } x \in W^{1,1}([S,T]; R^n) \\ \textit{satisfying} \\ x(S) = x_0 \textit{ and } x(T) = x_1, \end{cases}$$

for given $[S,T] \subset R$, $L : [S,T] \times R^n \times R^n \to R$, $x_0 \in R^n$, and $x_1 \in R^n$. Assume that

(i) L is continuous,

(ii) $L(t, x, .)$ is convex for all $t \in [S,T]$ and $x \in R^n$, and

(iii) there exist constants $c > 0$, $d > 0$, and $\alpha > 0$ such that

$$L(t, x, v) \geq c|v|^{1+\alpha} - d \textit{ for all } t \in [S,T], \; x \in R^n, \textit{ and } v \in R^n.$$

Then there exists a minimizer (in the class $W^{1,1}([S,T]; R^n)$).

A more general version of this theorem is proved in Chapter 2.

Nowadays, existence of minimizers is recognized as an important topic in its own right, which gives insights into whether variational problems have been properly formulated and also into the regularity properties of minimizers. But, as we have already observed, existence of minimizers was first studied as an ingredient in the Direct Method, i.e., to justify seeking a minimizer from the class of arcs satisfying the necessary conditions.

Unfortunately, existence theorems such as Theorem 1.3.1 are not entirely adequate for purposes of applying the Direct Method. The difficulty is this: implicit in the Direct Method is the assumption that a single set of hypotheses on the data for the optimization problem at hand simultaneously guarantees existence of minimizers and also validity of the necessary conditions. Yet for certain variational problems of interest, even though existence of a minimizer $\bar{x}$ is guaranteed by Theorem 1.3.1, it is not clear a priori that the hypotheses are satisfied under which necessary conditions, such as those listed in Theorem 1.2.1, are valid at $\bar{x}$. For these problems we cannot be sure that a search over arcs satisfying the Euler–Lagrange Condition and related conditions will yield a minimizer.

There are two ways out of this dilemma. One is to refine the existence theory by showing that $W^{1,1}$ minimizers are confined to some subset of $W^{1,1}$ for which the standard necessary conditions are valid. The other is to replace the standard necessary conditions by conditions that are valid under hypotheses similar to those of existence theory. Tonelli's achievements along these lines, together with recent significant developments, provide much of the subject matter for Chapter 11.

1.4   Sufficient Conditions and the Hamilton–Jacobi Equation

We have discussed pitfalls in seeking minimizers among arcs satisfying the necessary conditions. The steps initiated by Tonelli for dealing with them, namely, investigating hypotheses under which minimizers exist, is a relatively recent development. They were preceded by various procedures, some of a rather ad hoc nature, for testing whether an arc satisfying known necessary conditions, which we have somehow managed to find, is truly a minimizer. For certain classes of variational problems, an arc $\bar{x}$ satisfying the necessary conditions of Theorem 1.2.1 is a minimizer (at least in some "local" sense) if it can be shown to satisfy also certain higher-order conditions (such as the strengthened Legendre or Jacobi Conditions). In other cases we might hope to carry out calculations, "completing the square" or construction of a "verification function" for example, which make it obvious that an arc we have obtained and have reason to believe is a minimizer, is truly a minimizer. This ragbag of "indirect" procedures, none of which may apply, contrasts with Tonelli's Direct Method, which is in some sense a more systematic approach to finding minimizers.

However one "indirect" method for confirming that a given arc is a minimizer figures prominently in current research and we discuss it here.

It is based on the relationship between the variational problem

$$(Q) \begin{cases} \text{Minimize } \int_S^T L(t, x(t), \dot{x}(t))dt + g(x(T)) \\ \text{over Lipschitz continuous arcs } x \text{ satisfying} \\ x(S) = x_0 \text{ and } x(T) \in R^n \end{cases}$$

(in which $L : [S,T] \times R^n \times R^n \to R$ and $g : R^n \to R$ are given continuous functions and x_0 is a given n-vector) and Hamilton–Jacobi partial differential equation:

$$(\text{HJE}) \begin{cases} \phi_t(t,x) + \min_{v \in R^n} \{\phi_x(t,x) \cdot v + L(t,x,v)\} = 0 \\ \qquad\qquad\qquad\qquad\qquad \text{for all } t \in (S,T), \, x \in R^n \\ \phi(T,x) = g(x) \text{ for all } x \in R^n. \end{cases}$$

We sometimes find it convenient to express the first of these equations in terms of the Hamiltonian H:

$$H(t,x,p) := \sup_{v \in R^n} \{p \cdot v - L(t,x,v)\};$$

thus

$$\phi_t(t,x) - H(t,x,-\phi_x(t,x)) = 0 \quad \text{for all } t \in (S,T), x \in R^n.$$

(It is helpful in the present context to consider a variational problem in which the right endpoint is unconstrained and a function of the terminal value of the arc is added to the cost.)

The following proposition summarizes the approach, which is referred to as Carathéodory's Verification Technique.

Proposition 1.4.1 *Let $\bar{x}$ be a Lipschitz continuous arc satisfying $\bar{x}(S) = x_0$. Suppose that a continuously differentiable function $\phi : R \times R^n \to R$ can be found satisfying (HJE) and also*

$$\phi_x(t, \bar{x}(t)) \cdot \dot{\bar{x}}(t) + L(t, \bar{x}(t), \dot{\bar{x}}(t))$$
$$= \text{Min}_{v \in R^n} \{\phi_x(t, \bar{x}(t)) \cdot v + L(t, \bar{x}(t), v)\} \quad a.e. \ t \in [S, T].$$

Then $\bar{x}$ is a minimizer for (Q) and

$$\phi(S, x_0) = \inf (Q).$$

Proof. Take any Lipschitz continuous arc $x : [S, T] \to R^n$ for which $x(S) = x_0$. Then $t \to \phi(t, x(t))$ is a Lipschitz continuous function and

$$\frac{d\phi}{dt}(t, x(t)) = \phi_t(t, x(t)) + \phi_x(t, x(t)) \cdot \dot{x}(t) \quad a.e.$$

Now express $\phi(t, x(t))$ as the integral of its derivative:

$$\phi(S, x_0) = -\int_S^T \frac{d}{dt}\phi(t, x(t))dt + g(x(T))$$

(we have used the facts that $x(S) = x_0$ and $\phi(T, x) = g(x)$)

$$= -\int [\phi_t(t, x(t)) + \phi_x(t, x(t)) \cdot \dot{x}(t) + L(t, x(t), \dot{x}(t))]dt$$
$$+ \int_S^T L(t, x(t), \dot{x}(t))dt + g(x(T)).$$

But the first term on the right satisfies

$$-\int_S^T [\phi_t(t, x(t)) + \phi_x(t, x(t)) \cdot \dot{x}(t) + L(t, x(t), \dot{x}(t))]dt$$
$$\leq -\int_S^T [\phi_t(t, x(t)) + \inf_v\{\phi_x(t, x(t)) \cdot v + L(t, x(t), v)\}]dt$$
$$= 0. \tag{1.15}$$

It follows that

$$\phi(S, x_0) \leq \int_S^T L(t, x(t), \dot{x}(t))dt + g(x(T)). \tag{1.16}$$

Now repeat the above arguments with $\bar{x}$ in place of x. Since, by hypothesis,

$$\phi_x(t, \bar{x}(t)) \cdot \dot{\bar{x}}(t) + L(t, \bar{x}(t), \dot{\bar{x}}(t)) = \min_{v \in R^n} \{\phi_x(t, \bar{x}(t)) \cdot v + L(t, \bar{x}(t), v)\},$$

we can replace inequality (1.15) by equality:

$$\int_S^T [-\phi_t(t, \bar{x}(t)) + \phi_x(t, \bar{x}(t)) \cdot \dot{\bar{x}}(t) + L(t, \bar{x}(t), \dot{\bar{x}}(t))]dt = 0.$$

Consequently

$$\phi(S, x_0) = \int_S^T L(t, \bar{x}(t), \dot{\bar{x}}(t))dt + g(\bar{x}(T)). \tag{1.17}$$

We see from (1.16) and (1.17) that $\bar{x}$ has cost $\phi(S, x_0)$ and, on the other hand, any other Lipschitz continuous arc satisfying the constraints of problem (Q) has cost not less than $\phi(S, x_0)$. It follows that $\phi(S, x_0)$ is the minimum cost and $\bar{x}$ is a minimizer. $\square$

In ideal circumstances, it is possible to find a continuously differentiable solution ϕ to the Hamilton–Jacobi Equation such that the function

$$v \rightarrow \phi_x(t, x) \cdot v + L(t, x, v)$$

has a unique minimizer (write it $d(t, x)$ to emphasize that it will depend on (t, x)) and the differential equation

$$\begin{aligned} \dot{x}(t) &= d(t, x(t)) \quad \text{a.e.} \\ x(0) &= x_0 \end{aligned}$$

has a Lipschitz continuous solution $\bar{x}$. Then, Proposition 1.4.1 informs us, $\bar{x}$ must be a minimizer.

Success of this method for finding minimizers hinges of course on finding a solution ϕ to the Hamilton–Jacobi Equation. Recall that, if a solution ϕ confirms the optimality of some putative minimizer, then $\phi(S, x_0)$ coincides with the minimum cost for (Q). (This is the final assertion of Proposition 1.4.1.) It follows that the natural candidate for solution to the Hamilton–Jacobi Equation is the *value function* $V : [S, T] \times R^n \rightarrow R$ for (Q):

$$V(t, x) = \inf(Q_{t,x}) .$$

Here, the right side denotes the infimum cost of a variant of problem (Q), in which the initial data (S, x_0) is replaced by (t, x):

$$(Q_{t,x}) \begin{cases} \text{Minimize } \int_t^T L(s, y(s), \dot{y}(s))ds + g(y(T)) \\ \text{over Lipschitz continuous functions } y \text{ satisfying} \\ y(t) = x. \end{cases}$$

The following proposition gives precise conditions under which the value function is a solution to (HJE).

Proposition 1.4.2 *Let V be the value function for (Q). Suppose that*

(i) *V is a continuously differentiable function;*

(ii) *for each $(t, x) \in [S, T] \times R^n$, the optimization problem $(Q_{t,x})$ has a minimizer that is continuously differentiable.*

Then V is a solution to (HJE).

Proof. Take $(t, x) \in [S, T] \times R^n$, $\tau \in [t, T]$, a continuously differentiable function $y : [t, T] \to R^n$, and a continuously differentiable minimizer $\bar{y}$ for $(Q_{t,x})$. Simple contradiction arguments lead to the following relationships.

$$V(t, x) \leq \int_t^\tau L(s, y(s), \dot{y}(s)) ds + V(\tau, y(\tau)),$$

$$V(t, x) = \int_t^\tau L(s, \bar{y}(s), \dot{\bar{y}}(s)) ds + V(\tau, \bar{y}(\tau)).$$

Take arbitrary $v \in R^n$, $\epsilon \in (0, T - t)$, and choose $y(s) = x + (s - t)v$. Then

$$\epsilon^{-1}[V(t + \epsilon, x + \epsilon v) - V(t, x)] + \epsilon^{-1} \int_t^{t+\epsilon} L(s, x + sv, v) ds \geq 0$$

and

$$\epsilon^{-1}\left[V\left(t + \epsilon, x + \int_t^{t+\epsilon} \dot{\bar{y}}(s) ds\right) - V(t, x)\right] + \epsilon^{-1} \int_t^{t+\epsilon} L(s, \bar{y}(s), \dot{\bar{y}}(s)) ds = 0.$$

Passage to the limit as $\epsilon \downarrow 0$ gives

$$V_t(t, x) + V_x(t, x) \cdot v + L(t, x, v) \geq 0$$

and

$$V_t(t, x) + V_x(t, x) \cdot \dot{\bar{y}}(t) + L(t, x, \dot{\bar{y}}(t)) = 0.$$

Bearing in mind that v is arbitrary, we conclude that

$$V_t(t, x) + \min_{v \in R^n} \{V_x(t, x) \cdot v + L(t, x, v)\} = 0.$$

We have confirmed that V satisfied (HJE). The fact that V also satisfies the boundary condition $V(T, x) = g(x)$ follows directly from the definition of the value function. $\square$

How might we find solutions to the Hamilton–Jacobi Equation? One approach, "construction of a field of extremals," is inspired by the above interpretation of the value function. The idea is to generate a family of continuously differentiable arcs $\{y_{t,x}\}$, parameterized by points $(t,x) \in [S,T] \times R^n$, satisfying the necessary conditions. We then choose

$$\phi(t,x) = \int_t^T L(t, y_{t,x}(s), \dot{y}_{t,x}(s))ds + g(y_{t,x}(T)).$$

If it turns out that ϕ is a continuously differentiable function and the extremal $y_{t,x}$ really is a minimizer for $(Q_{t,x})$ (a property of which we have no a priori knowledge), then $V \equiv \phi$ and ϕ is a solution to the Hamilton–Jacobi Equation.

The Hamilton–Jacobi Equation is a nonlinear partial differential equation of hyperbolic type. From a classical perspective, constructing a field of extremals (a procedure for building up a solution to a partial differential equation from the solutions to a family of ordinary differential equations) amounts to solving the Hamilton–Jacobi Equation by the method of characteristics.

1.5   The Maximum Principle

A convenient framework for studying minimization problems, encountered in the optimal selection of flight trajectories and other areas of advanced engineering design and operation, is to regard them as special cases of the problem:

$$(P) \begin{cases} \text{Minimize } g(x(S), x(T)) \\ \text{over measurable functions } u : [S,T] \to R^n \text{ and arcs} \\ \qquad\qquad\qquad\qquad x \in W^{1,1}([S,T]; R^n) \text{ satisfying} \\ \dot{x}(t) = f(t, x(t), u(t)) \quad \text{a.e.,} \\ u(t) \in U \quad \text{a.e.,} \\ (x(S), x(T)) \in C, \end{cases}$$

the data for which comprise an interval $[S,T]$, functions $f : [S,T] \times R^n \times R^m \to R^n$ and $g : R^n \times R^n \to R$, and sets $C \subset R^n$ and $U \subset R^m$.

This formulation of the Optimal Control Problem (or variants on it in which, for example, the cost is augmented by an integral term, or the end-points of the time interval $[S,T]$ are included among the choice variables, or pathwise constraints are imposed on values of x) is referred to as the Pontryagin formulation. The importance of this formulation is that it embraces a wide range of significant optimization problems which are beyond the reach of traditional variational techniques and, at the same time, it is very well suited to the derivation of general necessary conditions of optimality.

In (P), the n-vector dependent variable x is called the *state*. The function describing its time evolution, $x(t)$, $S \leq t \leq T$, is called the *state trajectory*.

The state trajectory depends on our choice of control function $u(t)$, $S \leq t \leq T$, and the initial state $x(S)$. The object is to choose a control function $u(.)$ and initial state to minimize the value of the cost $g(x(S), x(T))$ resulting from our choice of $u(.)$ and $x(S)$.

Frequently, the initial state is fixed; i.e., C takes the form

$$C = \{x_0\} \times C_1 \text{ for some } x_0 \text{ and some } C_1 \subset R^n.$$

In this case, (P) is a minimization problem over control functions. However, allowing freedom in the choice of initial state introduces a flexibility into the formulation that is useful in some applications. A case in point is application of Optimal Control to some problems of exploitation of renewable resources, where an endpoint constraint set of interest is

$$C = \{(x_0, x_1) : x_0 = x_1\}.$$

This "periodic" endpoint constraint corresponds to the requirement that the population which is being harvested is the same at the beginning and the end of the harvesting cycle ("sustainable harvesting"). The size of the initial population is determined by optimization.

There follows a statement of the Maximum Principle, whose discovery by L. S. Pontryagin et al. in the 1950s was an important milestone in the emergence of Optimal Control as a distinct field of research.

Theorem 1.5.1 (The Maximum Principle) *Let $(\bar{u}, \bar{x})$ be a minimizer for (P). Assume that*

(i) g is continuously differentiable,

(ii) C is a closed set,

(iii) f is continuous, $f(t, ., u)$ is continuously differentiable for each (t, u), and there exist $\epsilon > 0$ and $k(.) \in L^1$ such that

$$|f(t, x, u) - f(t, x', u)| \leq k(t)|x - x'|$$

for all $x, x' \in \bar{x}(t) + \epsilon B$ and $u \in U$, a.e., and

(iv) Gr U is a Borel set.

Then there exist an arc $p \in W^{1,1}([S, T]; R^n)$ and $\lambda \geq 0$, not both zero, such that the following conditions are satisfied:

The Adjoint Equation:

$$-\dot{p}(t) = p(t) \cdot f_x(t, \bar{x}(t), \bar{u}(t)), \quad a.e.;$$

The Generalized Weierstrass Condition:

$$p(t) \cdot f(t, \bar{x}(t), \bar{u}(t)) = \max_{u \in U} p(t) \cdot f(t, \bar{x}(t), u) \quad a.e.;$$

The Transversality Condition:

$$(p(S), -p(T)) = \lambda g_x(\bar{x}(S), \bar{x}(T)) + \eta$$

for some $\eta \in N_C(\bar{x}(S), \bar{x}(T))$.

The limiting normal cone N_C of the endpoints constraint set C featured in the above Transversality Condition is defined presently; in the case that C is a smooth manifold it reduces to the set of outward-pointing normals.

An alternative statement of the Maximum Principle is in terms of the *Unmaximized Hamiltonian*

$$\mathcal{H}(t, x, u) := p \cdot f(t, x, u).$$

The adjoint equation (augmented by the state equation $\dot{x} = f$) and the Generalized Weierstrass Condition can be written

$$(\dot{\bar{x}}(t), -\dot{p}(t)) = \nabla_{x,p} \mathcal{H}(t, \bar{x}(t), p(t), \bar{u}(t)) \quad \text{a.e.}$$

and

$$\mathcal{H}(t, \bar{x}(t), p(t), \bar{u}(t)) = \max_{u \in U} \mathcal{H}(t, \bar{x}(t), p(t), u) \quad \text{a.e.},$$

a form of the conditions that emphasizes their affinity with Hamilton's system of equations in the Calculus of Variations.

In favorable circumstances, we are justified in setting the cost multiplier $\lambda = 1$, and the Generalized Weierstrass Condition permits us to express u as a function of x and p

$$u = u^*(x, p).$$

The Maximum Principle then asserts that a minimizing arc $\bar{x}$ is the first component of a pair of absolutely continuous functions $(\bar{x}, p)$ satisfying the differential equation

$$(-\dot{p}(t), \dot{\bar{x}}(t)) = \nabla_{x,p} \mathcal{H}(t, \bar{x}(t), p(t), u^*(\bar{x}(t), p(t))) \quad \text{a.e.} \qquad (1.18)$$

and the endpoint conditions

$$(\bar{x}(S), \bar{x}(T)) \in C \quad \text{and} \quad (p(S), -p(T)) \in N_C(\bar{x}(S), \bar{x}(T)).$$

The minimizing control is given by the formula

$$\bar{u}(t) = u^*(\bar{x}(t), p(t)).$$

Notice that the (vector) differential equation (1.18) is a system of $2n$ scalar, first-order differential equations. If C is a k-dimensional manifold, specified by k scalar functional constraints on the endpoints of $\bar{x}$, then the Transversality Condition imposes $2n - k$ scalar functional constraints on the endpoints of p. Thus the "two-point boundary value problem" which we must solve to obtain $\bar{x}$ has the "right" number of endpoint conditions.

To explore the relationship between the Maximum Principle and classical conditions in the Calculus of Variations, let us consider the following refinement of the Basic Problem of the Calculus of Variations, in which a pathwise constraint on the velocity variable $\dot{x}(t) \in U$ is imposed. (U is a given Borel subset of R^n.)

$$\begin{cases} \text{Minimize } \int_S^T L(t, x(t), \dot{x}(t))dt \\ \text{over arcs } x \text{ satisfying} \\ \dot{x}(t) \in U, \\ (x(S), x(T)) \in C. \end{cases} \qquad (1.19)$$

The problem can be regarded as the following special case of (P):

$$\begin{cases} \text{Minimize } z(T) \\ \text{over measurable functions } u \\ \qquad \text{and arcs } (x, z) \in W^{1,1}([S, T]; R^{n+1}) \text{ satisfying} \\ (\dot{x}(t), \dot{z}(t)) = (u, L(t, x, u)) \text{ a.e.,} \\ u(t) \in U \text{ a.e.,} \\ z(S) = 0, (x(S), x(T)) \in C. \end{cases} \qquad (1.20)$$

To be precise, $\bar{x}$ is a minimizer for (1.19) if and only if

$$\left(u \equiv \dot{\bar{x}}, \bar{x}, \bar{z}(t) = \int_S^t L(\tau, \bar{x}(\tau), \dot{\bar{x}}(\tau))d\tau \right)$$

is a minimizer for (1.20).

Let $\bar{x}$ be a minimizer for (1.19). Under appropriate hypotheses, the Maximum Principle, applied to Problem (1.20), supplies $\lambda \geq 0$ and an adjoint arc (which, for convenience, we partition as $(q, -\alpha)$) satisfying

$$(q, \alpha, \lambda) \neq 0 \qquad (1.21)$$
$$-\dot{q}(t) = -\alpha(t)L_x(t, \bar{x}(t), \dot{\bar{x}}(t)), \quad \dot{\alpha}(t) = 0, \qquad (1.22)$$
$$(q(S), -q(T)) = \xi', \quad \alpha(T) = \lambda,$$

for some $\xi' \in N_C(\bar{x}(S), \bar{x}(T))$, and

$$q(t) \cdot \dot{\bar{x}}(t) - \alpha(t)L(t, \bar{x}, \dot{\bar{x}}(t)) = \max_{v \in U}\{q(t) \cdot v - \alpha(t)L(t, \bar{x}, v)\}, \qquad (1.23)$$

We now impose the "constraint qualification":

(CQ): λ can be chosen strictly positive.

Two special cases, in either of which (CQ) is automatically satisfied, are

(i) $\dot{\bar{x}}(t) \in \text{int } U$ on a subset of $[S, T]$ having positive measure;

(ii) $C = C_0 \times R^n$ for some subset $C_0 \subset R^n$.

(It is left to the reader to check that, if the condition $\lambda > 0$ is violated, then, in either case, (1.22) and (1.23) imply $(q, \alpha, \lambda) = 0$, in contradiction of (1.21).)

It is known that $\dot{\alpha} \equiv 0$. It follows that $\alpha(.) \equiv \lambda$. In terms of $p(t) := \lambda^{-1} q(t)$ and $\xi := \lambda^{-1} \xi'$, these conditions can be expressed

$$\dot{p}(t) = L_x(t, \bar{x}(t), \dot{\bar{x}}(t)) \quad \text{a.e.} \tag{1.24}$$

$$(p(S), -p(T)) = \xi, \tag{1.25}$$

for some $\xi \in N_C(\bar{x}(S), \bar{x}(T))$ and

$$p(t) \cdot \dot{\bar{x}}(t) - L(t, \bar{x}, \dot{\bar{x}}(t)) = \max_{v \in U} \{p(t) \cdot v - L(t, \bar{x}, v)\}. \tag{1.26}$$

Notice that, when $U = R^n$, the Generalized Weierstrass Condition (1.26) implies

$$p(t) \in L_v(t, \bar{x}(t), \dot{\bar{x}}(t)) \quad \text{a.e.}$$

(L_v denotes the derivative of $L(t, x, .)$.) This combines with (1.24) to give

$$(\dot{p}(t), p(t)) = \nabla_{x,v} L(t, \bar{x}(t), \dot{\bar{x}}(t)) \quad \text{a.e.}$$

We have shown that the Maximum Principle subsumes the Euler–Lagrange Condition and the Weierstrass Condition. But is has much wider implications, because it covers problems with pathwise velocity constraints.

The innovative aspects of the Maximum Principle, in relation to classical optimality conditions, are most clearly revealed when it is compared with the Hamilton Condition.

For Problem (1.19), the Maximum Principle associates with a minimizer $\bar{x}$ an adjoint arc p satisfying relationships (1.24) to (1.26) above. These can be expressed in terms of the Unmaximized Hamiltonian

$$\mathcal{H}(t, x, p, v) = p \cdot v - L(t, x, v)$$

as

$$(-\dot{p}(t), \dot{\bar{x}}(t)) = \nabla_{x,v} \mathcal{H}(t, \bar{x}, p(t), \dot{\bar{x}}(t))$$

and

$$\mathcal{H}(t, \bar{x}(t), p(t), \dot{\bar{x}}(t)) = \max_{v \in U} \{\mathcal{H}(t, \bar{x}(t), p(t), v)\}.$$

On the other hand, classical conditions in the form of Hamilton's equations, which are applicable in the case $U = R^n$ are also a set of differential equations satisfied by $\bar{x}$ and an adjoint arc p,

$$(-\dot{p}(t), \dot{\bar{x}}(t)) = \nabla_{x,v} H(t, \bar{x}(t), p(t)),$$

where H is the Hamiltonian

$$H(t, x, p) = \max_{v \in R^n} \{p \cdot v - L(t, x, v)\}.$$

Both sets of necessary conditions are intimately connected with the function $\mathcal{H}$ and its supremum H with respect to the velocity variable. However there is a crucial difference. In Hamilton's classical necessary conditions, the supremum-taking operation is applied *before* the function $\mathcal{H}$ is differentiated and inserted into the vector differential equation for p and $\bar{x}$. In the Maximum Principle applied to Problem (1.19), by contrast, supremum-taking is carried out only *after* the differential equations for p and $\bar{x}$ have been assembled.

One advantage of postponing supremum-taking is that we can dispense with differentiability hypotheses on the Hamiltonian $H(t, ., .)$. (If the data $L(t, ., .)$ are differentiable, it follows immediately that $\mathcal{H}(t, ., ., .)$ is differentiable, as required for derivation of the Maximum Principle, but it does *not* follow in general that the derived function $H(t, ., .)$ is differentiable, as required for derivation of the Hamilton Condition.)

Other significant advantages are that conditions (1.24) to (1.26) are valid in circumstances when the velocity $\bar{x}(t)$ is constrained to lie in some set U, and that they may be generalized to cover problems involving differential equation constraints. These are the features that make the Maximum Principle so well suited to problems of advanced engineering design.

1.6   Dynamic Programming

We have seen how necessary conditions from the Calculus of Variations evolved into the Maximum Principle, to take account of pathwise constraints encountered in advanced engineering design. What about optimality conditions related to the Hamilton–Jacobi Equation? Here, too, long-established techniques in the Calculus of Variations can be adapted to cover present day applications.

It is convenient to discuss these developments, referred to as Dynamic Programming, in relation to a variant of the problem studied earlier, in which an integral cost term is added to the cost, and the right endpoint of the state trajectory is unconstrained.

$$(I) \begin{cases} \text{Minimize } J(x) := \int_S^T L(t, x(t), u(t))dt + g(x(T)) \\ \text{over measurable functions } u : [S, T] \to R^m \\ \qquad\qquad \text{and } x \in W^{1,1}([S, T]; R^n) \text{ satisfying} \\ \dot{x}(t) = f(t, x(t), u(t)) \text{ a.e.,} \\ u(t) \in U \text{ a.e.,} \\ x(S) = x_0, \end{cases}$$

the data for which comprise an interval $[S, T]$, a set $U \subset R^m$, a point $x_0 \in R^n$, and functions $f : [S, T] \times R^n \times R^m \to R^n$, $L : [S, T] \times R^n \times R^m \to R$, and $g : R^n \to R$.

It is assumed that f, L, and g are continuously differentiable functions and that f satisfies additional assumptions ensuring that, in particular, the differential equation $\dot{y}(s) = f(s, y(s), u(s))$, $y(t) = \xi$, has a unique solution $y(.)$ on $[t, T]$ for an arbitrary control function $u(.)$, initial time $t \in [S, T]$, and initial state $\xi \in R^n$.

The Hamilton–Jacobi Equation for this problem is:

$$\text{(HJE)}' \quad \begin{cases} \phi_t(t, x) + \min_{u \in U} \{\phi_x(t, x) \cdot f(t, \xi, u) + L(t, \xi, u)\} = 0 \\ \qquad\qquad\qquad\qquad\qquad\qquad\qquad \text{for all } (t, x) \in (S, T) \times R^n, \\ \phi(T, x) = g(x) \text{ for all } x \in R^n. \end{cases}$$

Finding a suitable solution to this equation is one possible approach to verifying optimality of a putative minimizer. This extension of Carathéodory's Verification Technique for problem (I) is summarized as the following optimality condition: the derivation is along very similar lines to the proof of Proposition 1.4.1.

Proposition 1.6.1 *Let* $(\bar{x}, \bar{u})$ *satisfy the constraints of optimal control problem* (I). *Suppose that there exists* $\phi \in C^1$ *satisfying (HJE)$'$ and also*

$$\phi_x(t, \bar{x}(t)) \cdot f(t, \bar{x}(t), \bar{u}(t)) + L(t, \bar{x}(t), \bar{u}(t))$$
$$= \min_{u \in U} \{\phi_x(t, \bar{x}) \cdot f(t, \bar{x}(t), u) + L(t, \bar{x}(t), u)\}. \qquad (1.27)$$

Then

(a) $(\bar{x}, \bar{u})$ *is a minimizer for* (I);

(b) $\phi(S, x_0)$ *is the minimum cost for* (I).

The natural candidate for "verification function" ϕ is the value function $V : [S, T] \times R^n \to R$ for (I). For $(t, x) \in [S, T] \times R^n$ we define

$$V(t, x) := \inf(I_{t,x})$$

where the right side indicates the infimum cost for a modified version of (I), in which the initial time and state (S, x_0) are replaced by (t, x):

$$(I_{t,x}) \quad \begin{cases} \text{Minimize } \int_t^T L(s, y(s), u(s))ds + g(x(T)) \\ \text{over measurable functions } u : [t, T] \to R^m \\ \qquad\qquad\qquad \text{and } y \in W^{1,1}([t, T]; R^n) \text{ satisfying} \\ \dot{y}(s) = f(s, y(s), u(s)) \text{ a.e. } s \in [t, T], \\ u(s) \in U \text{ a.e. } s \in [t, T], \\ x(t) = x. \end{cases}$$

Indeed, we can mimic the proof of Proposition 1.4.2 to show

Proposition 1.6.2 *Let* V *be the value function for* (I). *Suppose that*

(i) V is a continuously differentiable function.

(ii) For each $(t,x) \in [S,T] \times R^n$, the optimization problem $(I_{t,x})$ has a minimizer with continuous control function.

Then V is a solution to (HJE).

One reason for the special significance of the value function in Optimal Control is its role in the solution of the "Optimal Synthesis Problem." A feature of many engineering applications of Optimal Control is that knowledge of an optimal control strategy is required for a variety of initial states and times. This is because the initial state typically describes the deviation of plant variables from their nominal values, due to disturbances; the requirements of designing a control system to correct this deviation, regardless of the magnitudes and time of occurrence of the disturbances, make necessary consideration of more than one initial state and time. The *Optimal Synthesis Problem* is that of obtaining a solution to $(I)_{t,x}$ in feedback form such that the functional dependence involved is independent of the datapoint $(t,x) \in [S,T] \times R^n$. To be precise, we seek a function $G : [S,T] \times R^n \to R^m$ such that the feedback equation

$$u(s) = G(s, y(s)),$$

together with the dynamic constraints and initial condition,

$$\begin{cases} \dot{y}(s) = f(s, y(s), u(s)) & \text{for a.e. } s \in [\tau, T], \\ y(\tau) = \xi \end{cases}$$

can be solved to yield a minimizer (y, u) for $(I)_{\tau, \xi}$. Furthermore, it is required that G be independent of (τ, ξ). We mention that feedback implementation, which is a dominant theme in Control Engineering, has many significant benefits besides the scope it offers for treatment of arbitrary initial data.

Consider again the Dynamic Programming sufficient condition of optimality. Let ϕ be a continuously differentiable function that satisfies (HJE)'. For each (t,x) define the set $Q(s,x) \subset R^m$:

$$Q(t,x) := \{v' : \phi_x(t,x) \cdot f(t,x,v') + L(t,x,v') = $$
$$= \min_{v \in U} \{\phi_x(t,x) \cdot f(t,x,v) + L(t,x,v)\}.$$

Let us assume that Q is single-valued and that, for each $(t,x) \in [S,T] \times R^n$, the equations

$$\begin{aligned} \dot{y}(s) &= f(s, y(s), u(s)) \quad \text{a.e. } s \in [t,T], \\ u(s) &= Q(s, y(s)) \quad \text{a.e. } s \in [t,T], \\ y(t) &= x, \end{aligned}$$

have a solution y on $[t, T]$. Then, by the sufficient condition, $(y, u(s) = Q(s, x(s)))$ is a minimizer for $(I_{t,x})$, whatever the initial datapoint (t, x) happens to be. Evidently, Q solves the synthesis problem.

We next discuss the relationship between Dynamic Programming and the Maximum Principle. The question arises, if ϕ is a solution to (HJE)' and if $(\bar{x}, \bar{u})$ is a minimizer for (I), can we interpret adjoint arcs for $(\bar{x}, \bar{u})$ in terms of ϕ? The nature of the relationship can be surmised from the following calculations.

Assume that ϕ is twice continuously differentiable. We deduce from (HJE)' that, for each $t \in [S, T]$,

$$\phi_x(t, \bar{x}(t)) \cdot f(t, \bar{x}(t), \bar{u}(t)) + L(t, \bar{x}(t), \bar{u}(t))$$
$$= \min_{v \in U}\{\phi_x(t, \bar{x}(t)) \cdot f(t, \bar{x}(t), v) + L(t, \bar{x}(t), v)\}, \qquad (1.28)$$

and

$$\phi_t(t, \bar{x}(t)) + \phi_x(t, \bar{x}(t)) \cdot f(t, \bar{x}(t), \bar{u}(t)) + L(t, \bar{x}(t), \bar{u}(t))$$
$$= \min_{x \in R^n}\{\phi_t(t, x) + \phi_x(t, x) \cdot f(t, x, \bar{u}(t)) + L(t, x, \bar{u}(t))\}. \quad (1.29)$$

Since ϕ is assumed to be twice continuously differentiable, we deduce from (1.29) that

$$\phi_{tx} + \phi_{xx} \cdot f + \phi_x \cdot f_x + L_x = 0. \qquad (1.30)$$

(In this equation, all terms are evaluated at $(t, \bar{x}(t), \bar{u}(t))$.)

Now define the continuously differentiable functions p and h as follows:

$$p(t) := -\phi_x(t, \bar{x}(t)), \qquad h(t) := \phi_t(t, \bar{x}(t)). \qquad (1.31)$$

We have

$$-\dot{p}(t) = \phi_{tx}(t, \bar{x}(t)) + \phi_{xx}(t, \bar{x}(t)) \cdot f(t, \bar{x}(t), \bar{u}(t))$$
$$= -\phi_x(t, \bar{x}(t)) \cdot f_x(t, \bar{x}(t), \bar{u}(t)) - L_x(t, \bar{x}(t), \bar{u}(t)),$$

by (1.30). It follows from (1.31) that

$$-\dot{p}(t) = p(t) \cdot f_x(t, \bar{x}(t), \bar{u}(t)) - L_x(t, \bar{x}(t), \bar{u}(t)) \quad \text{for all } t \in [S, T]. \ (1.32)$$

We deduce from the boundary condition for (HJE)' and the definition of p that

$$-p(T) \ (= \phi_x(T, \bar{x}(T))) \ = \ g_x(\bar{x}(T)). \qquad (1.33)$$

From (1.28) and (1.31),

$$p(t) \cdot f(t, \bar{x}(t), \bar{u}(t)) - L(t, \bar{x}(t), \bar{u}(t))$$
$$= \max_{v \in U}\{p(t) \cdot f(t, \bar{x}(t), v) - L(t, \bar{x}(t), v)\}. \qquad (1.34)$$

We deduce from (1.28) and (1.31) that

$$h(t) = H(t, \bar{x}(t), p(t)), \tag{1.35}$$

where, as usual, H is the Hamiltonian

$$H(t, x, p) := \max_{v \in U} \{ p(t) \cdot f(t, \bar{x}(t), v) - L(t, \bar{x}(t), v) \}.$$

Conditions (1.32) to (1.35) tell us that, when a smooth verification function ϕ exists, then a form of the Maximum Principle is valid, in which the adjoint arc p is related to the gradient $\nabla \phi$ of ϕ according to

$$(H(t, \bar{x}(t), p(t)), -p(t)) = \nabla \phi(t, \bar{x}(t)) \quad \text{for all } t \in [S, T]. \tag{1.36}$$

As a proof of the Maximum Principle, the above analysis is unsatisfactory because the underlying hypothesis that there exists a smooth verification function is difficult to verify outside simple special cases.

However the relationship (1.36) is of considerable interest, for the following reasons. It is sometimes useful to know how the minimum cost of an optimal control problem depends on parameter values. In engineering applications, for example, such knowledge helps us to assess the extent to which optimal performance is degraded by parameter drift (variation of parameter values due to ageing of components, temperature changes, etc.). If the parameters comprise the initial time and state components (t, x), the relevant information is supplied by the value function which, under favorable circumstances, can be obtained by solving the Hamilton–Jacobi Equation. However for high-dimensional, nonlinear problems it is usually not feasible to solve the Hamilton–Jacobi Equation. On the other hand, computational schemes are available for calculating pairs of functions $(\bar{x}, \bar{u})$ satisfying the conditions of the Maximum Principle and accompanying adjoint arcs p, even for high-dimensional problems. Of course the adjoint arc is calculated for the specified initial data $(t, x) = (S, x_0)$ and does not give a full picture of how the minimum cost depends on the initial data in general. The adjoint arc does however at least supply gradient (or "sensitivity") information about this dependence near the nominal initial data $(t, x) = (S, x_0)$: in situations where there is a unique adjoint arc associated with a minimizer and the value function V is a smooth solution to the Hamilton–Jacobi Equation, we have from (1.36) that

$$\nabla V(S, x_0) = (H(S, x_0, p(S)), -p(S)).$$

A major weakness of classical Dynamic Programming, as summarized above, is that it is based on the hypothesis that the value function is continuously differentiable. (For some aspects of the classical theory, even stronger regularity properties must be invoked.) This hypothesis is violated in many cases of interest.

The historical development of necessary conditions of optimality and of Dynamic Programming in Optimal Control could not be more different. Necessary conditions made a rapid start, with the early conceptual breakthroughs involved in the formulation and rigorous derivation of the Maximum Principle derivation. The Maximum Principle remains a key optimality condition to this day. Dynamic Programming, by contrast, was a slow beginner. Early developments (in part reviewed above) were fairly obvious extensions of well-known techniques in the Calculus of Variations. In the 1960s Dynamic Programming was widely judged to lack a rigorous foundation: it aimed to provide a general approach to the solution of Optimal Control Problems yet Dynamic Programming, as originally formulated, depended on regularity properties of the value function that frequently it did not possess. The conceptual breakthroughs, required to elevate the status of Dynamic Programming from that of a useful heuristic tool to a cornerstone of modern Optimal Control, did not occur until the 1970s. They involved a complete reappraisal of what we mean by "solution" to the Hamilton–Jacobi Equation.

1.7   Nonsmoothness

By the early 1970s, it was widely recognized that attempts to broaden the applicability of available "first-order" necessary conditions and also to put Dynamic Programming on a rigorous footing were being impeded by a common obstacle: a lack of suitable techniques for analyzing local properties of nondifferentiable functions and of sets with nondifferentiable boundaries.

Consider first Dynamic Programming, where the need for new analytic techniques is most evident. This focuses attention on the relationship between optimal control problems such as

$$(I) \begin{cases} \text{Minimize } J(x) := \int_S^T L(t, x(t), \dot{x}(t))dt + g(x(T)) \\ \text{over measurable functions } u : [S, T] \to R^m \\ \qquad\qquad \text{and absolutely continuous arcs } x \text{ satisfying} \\ \dot{x}(t) = f(t, x(t), u(t)) \text{ a.e.,} \\ u(t) \in U \text{ a.e.,} \\ x(S) = x_0 \end{cases}$$

and the Hamilton–Jacobi Equation

$$(\text{HJE})' \begin{cases} \phi_t(t, x) + \min_{u \in U} \{\phi_x(t, x) \cdot f(t, x, u) + L(t, x, u)\} = 0 \\ \qquad\qquad\qquad\qquad \text{for all } (t, x) \in (S, T) \times R^n, \\ \phi(T, x) = g(x) \text{ for all } x \in R^n. \end{cases}$$

The elementary theory tells us that, if the value function V is continuously differentiable (and various other hypotheses are satisfied) then V is a solution to (HJE)'. One shortcoming is that it does not exclude the existence of other solutions to the (HJE)' and, therefore, does not supply a full characterization of the value function in terms of solutions to (HJE)'. Another, more serious, shortcoming is the severity of the hypothesis "V is continuously differentiable," which is extensively invoked in the elementary analysis. In illustration of problems in which this hypothesis is violated, consider the following example.

$$\begin{cases} \text{Minimize } x(1) \\ \text{over measurable functions } u : [0,1] \to R \\ \qquad\qquad \text{and } x \in W^{1,1}([0,1]; R) \text{ satisfying} \\ \dot{x}(t) = xu \text{ a.e.,} \\ u(t) \in [-1,+1] \text{ a.e.,} \\ x(0) = 0. \end{cases}$$

The Hamilton–Jacobi Equation in this case takes the form

$$\begin{cases} \phi_t(t,x) - |\phi_x(t,x)x| = 0 & \text{for all } (t,x) \in (0,1) \times R, \\ \phi(1,x) = x \text{ for all } x \in R. \end{cases}$$

By inspection, the value function is

$$V(t,x) = \begin{cases} x \exp^{+(1-t)} & \text{if } x \geq 0 \\ x \exp^{-(1-t)} & \text{if } x < 0. \end{cases}$$

We see that V satisfies the Hamilton–Jacobi Equation on $\{(t,x) \in (0,1) \times R : x \neq 0\}$. However V cannot be said to be a classical solution because V is nondifferentiable on the subset $\{(t,x) \in (0,1) \times R : x = 0\}$.

What is significant about this example is that the nondifferentiability of the value function here encountered is by no means exceptional. It is a simple instance of a kind of optimal control problem frequently encountered in engineering design, in which the differential equation constraint

$$\dot{x} = f(t,x,u)$$

has right side affine in the u variable, the control variable is subject to simple magnitude constraints, and the cost depends only on the terminal value of the state variable; for these problems, we *expect* the value function to be nondifferentiable.

How can the difficulties associated with nondifferentiable value functions be overcome? Of course we can choose to define nondifferentiable solutions of a given partial differential equation in a number of different ways. However the particular challenge in Dynamic Programming is to come up with a definition of solution, according to which the value function is a unique solution.

Unfortunately we must reject traditional interpretations of nondifferentiable solution to partial differential equations based on distributional derivatives, since the theory of "distribution sense solutions" is essentially a *linear* theory that is ill-matched to the *nonlinear* Hamilton–Jacobi Equation. It might be thought, on the other hand, that a definition based on almost everywhere satisfaction of the Hamilton–Jacobi Equation would meet our requirements. But such a definition is simply too coarse to provide a characterization of the value function under hypotheses of any generality.

Clearing the "nondifferentiable value functions" bottleneck in Dynamic Programming, encountered in the late 1970s, was a key advance in Optimal Control. Success in relating the value function and the Hamilton–Jacobi Equation was ultimately achieved under hypotheses of considerable generality by introducing new solution concepts, "viscosity" solutions, and related notions of a generalized solution, based on a fresh look at local approximation of nondifferentiable functions.

The need to "differentiate the undifferentiable" arises also when we attempt to derive necessary conditions for optimal control problems not covered by the Maximum Principle. Prominent among these are problems

$$\begin{cases}
\text{Minimize } g(x(S), x(T)) \\
\text{over measurable functions } u : [S, T] \to R^n \\
\qquad \text{and } x \in W^{1,1}([S, T]; R^n) \text{ satisfying} \\
\dot{x}(t) = f(t, x(t), u(t)) \quad \text{a.e.,} \\
u(t) \in U(t, x(t)) \quad \text{a.e.,} \\
(x(S), x(T)) \in C,
\end{cases} \tag{1.37}$$

involving a *state-dependent* control constraint set $U(t, x) \subset R^m$. Problems with state-dependent control constraint sets are encountered in flight mechanics applications where, for example, the control variable $u(t)$ is the engine thrust. For large excursions of the state variables, it is necessary to take account of the fact that the upper and lower bounds on the thrust depend on atmospheric pressure and therefore on altitude. Since altitude is a state component, the control function is required to satisfy a constraint of the form

$$u(t) \in U(t, x)$$

in which

$$U(t, x) := \{u : a^-(t, x) \le u \le a^+(t, x)\}$$

and $a^-(t, x)$ and $a^+(t, x)$ are, respectively, the state-dependent lower and upper bounds on the thrust at time t. In this simple case, it is possible to reduce to the state-independent control constraint case by redefining the problem. (The state dependence of the control constraint set can be absorbed into the dynamic constraint.) But in other cases, involving several interacting control variables, the constraints on which depend on the value of the state variable, this may no longer be possible or convenient.

A natural framework for studying problems with state-dependent control constraints is provided by:

$$\begin{cases} \text{Minimize } g(x(S), x(T)) \\ \text{over arcs } x \in W^{1,1} \text{ satisfying} \\ \dot{x}(t) \in F(t, x(t)) \quad \text{a.e.,} \\ (x(S), x(T)) \in C. \end{cases} \tag{1.38}$$

Now, the dynamic constraint takes the form of a *differential inclusion*:

$$\dot{x}(t) \in F(t, x(t)) \quad \text{a.e.}$$

in which, for each (t, x), $F(t, x)$ is a given subset of R^n. The perspective here is that, in dynamic optimization, the fundamental choice variables are state trajectories, not control function/state trajectory pairs, and the essential nature of the dynamic constraint is made explicit by identifying the set $F(t, x(t))$ of allowable values of $\dot{x}(t)$. The earlier problem, involving a state-dependent control constraint set can be fitted to this framework by choosing F to be the multifunction

$$F(t, x) := \{f(t, x, u) : u \in U(t, x)\}. \tag{1.39}$$

Indeed it can be shown, under mild hypotheses on the data for Problem (1.37), that the two problems (1.37) and (1.38) are equivalent for this choice of F in the sense that $\bar{x}$ is a minimizer for (1.38) if and only if $(\bar{x}, \bar{u})$ is a minimizer for (1.37) (for some $\bar{u}$).

By what means can we derive necessary conditions for Problem (1.38)? One approach is to seek necessary conditions for a related problem, from which the dynamic constraint has been eliminated by means of a penalty function:

$$\begin{cases} \text{Minimize } \int_S^T L(t, x(t), \dot{x}(t)) dt + g(x(S), x(T)) \text{ over } x \in W^{1,1}([S, T]; R^n) \\ \text{satisfying} \\ (x(S), x(T)) \in C. \end{cases}$$
$$\tag{1.40}$$

Here

$$L(t, x, v) := \begin{cases} 0 & \text{if } v \in F(t, x) \\ +\infty & \text{if } v \notin F(t, x). \end{cases}$$

(Notice that Problems (1.38) and (1.40) have the same cost at an arc x satisfying the constraints of (1.38). On the other hand, if x violates the constraints of (1.38), it is excluded from consideration as a minimizer for (1.40), since the penalty term in the cost ensures that it has infinite cost.)

We have arrived at a problem in the Calculus of Variations, albeit one with discontinuous data. It is natural then to seek necessary conditions for a minimizer $\bar{x}$ in the spirit of classical conditions, including the Euler–Lagrange and Hamilton Conditions:

$$(\dot{p}(t), p(t)) \in \text{``}\partial_{x,v}\text{''} L(t, \bar{x}(t), \dot{\bar{x}}(t))$$

and

$$(-\dot{p}(t), \dot{\bar{x}}(t)) \in \text{``}\partial_{x,p}\text{''}\, H(t, \bar{x}(t), p(t))$$

for some p. Here

$$H(t, x, p) := \max_{v \in R^n} \{p \cdot v - L(t, x, v)\}.$$

These conditions involve derivatives of the functions L and H. Yet, in the present context, L is discontinuous and H is, in general, nondifferentiable. How then should these relationships be interpreted?

These difficulties were decisively overcome in the 1970s and a new chapter of Optimal Control was begun, with the introduction of new concepts of local approximation of nondifferentiable functions and of a supporting calculus to justify their use in Variational Analysis.

1.8   Nonsmooth Analysis

Nonsmooth Analysis is a branch of nonlinear analysis concerned with the local approximation of nondifferentiable functions and of sets with nondifferentiable boundaries. Since its inception in the early 1970s, there has been a sustained and fruitful interplay between Nonsmooth Analysis and Optimal Control. A familiarity with Nonsmooth Analysis is, therefore, essential for an in-depth understanding of present day research in Optimal Control. This section provides a brief, informal review of material on Nonsmooth Analysis, which is covered in far greater detail in Chapters 3 and 4.

The guiding question is:

> *How should we adapt classical concepts of outward normals to subsets of vector spaces with smooth boundaries and of gradients of differentiable functions, to cover situations in which the boundaries are nonsmooth and the functions are nondifferentiable?*

There is no single answer to this question. The need to study local approximations of nonsmooth functions arises in many branches of analysis and a wide repertoire of techniques has been developed to meet the requirements of different applications.

In this book emphasis is given to proximal normal vectors, proximal subgradients, and limits of such vectors. This reflects their prominence in applications to Optimal Control.

Consider first generalizations of "outward normal vector" to general closed sets (in a finite dimensional space).

The Proximal Normal Cone. *Take a closed set $C \subset R^k$ and a point $\bar{x} \in C$. A vector $\eta \in R^k$ is said to be a* proximal normal vector *to C at $\bar{x}$ if there exists $M \geq 0$ such that*

$$\eta \cdot (x - \bar{x}) \leq M|x - \bar{x}|^2 \quad \text{for all } x \in C. \tag{1.41}$$

The cone of all proximal normal vectors to C at $\bar{x}$ is called the proximal normal cone *to C at $\bar{x}$ and is denoted by $N_C^P(\bar{x})$:*

$$N_C^P(\bar{x}) := \{\eta \in R^k : \exists M \geq 0 \text{ such that } (1.41) \text{ is satisfied}\}.$$

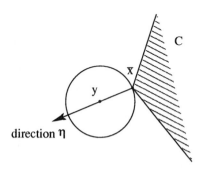

FIGURE 1.9. A Proximal Normal Vector.

Figure 1.9 provides a geometrical interpretation of proximal normal vectors: η is a proximal normal vector to C at $\bar{x}$ if there exists a point $y \in R^k$ such that $\bar{x}$ is the closest point to y in C and η is a scaled version of $y - \bar{x}$; i.e., there exists $\alpha \geq 0$ such that

$$\eta = \alpha(y - \bar{x}).$$

The Limiting Normal Cone. *Take a closed set $C \subset R^k$ and a point $\bar{x} \in C$. A vector η is said to be a* limiting normal vector *to C at $\bar{x} \in C$ if there exist sequences $x_i \xrightarrow{C} \bar{x}$ and $\eta_i \to \eta$ such that*

$$\eta_i \in N_C^P(x_i) \text{ for all } i.$$

The cone of limiting normal vectors to C at $\bar{x}$ is denoted $N_C(\bar{x})$:

$$N_C(\bar{x}) := \{\eta \in R^k : \exists x_i \xrightarrow{C} x \text{ and } \eta_i \to \eta \text{ such that } \eta_i \in N_C^P(x_i) \text{ for all } i\}.$$

(The notation $x_i \xrightarrow{C} x$ indicates that $x_i \to x$ and $x_i \in C$ for all i.)

Figure 1.10 illustrates limiting normal cones at various points in a set with nonsmooth boundary.

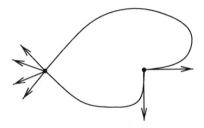

FIGURE 1.10. Limiting Normal Cones.

It can be shown that, if $\bar{x}$ is a boundary point of C, then the limiting normal cone contains nonzero elements. Limiting normal cones are closed cones, but they are not necessarily convex.

We consider next generalizations of the concept of "gradient" to functions that are not differentiable.

The Proximal Subdifferential. *Take an extended-valued, lower semi-continuous function $f : R^k \to R \cup \{+\infty\}$ and a point $\bar{x} \in$ dom $\{f\}$. A vector $\eta \in R^k$ is said to be a* proximal subgradient *of f at $\bar{x}$ if there exist $\epsilon > 0$ and $M \geq 0$ such that*

$$\eta \cdot (x - \bar{x}) \leq f(x) - f(\bar{x}) + M|x - \bar{x}|^2 \qquad (1.42)$$
for all points x that satisfy $|x - \bar{x}| \leq \epsilon$.

The set of all proximal subgradients of f at $\bar{x}$ is called the proximal subdifferential *of f at $\bar{x}$ and is denoted by $\partial^P f(\bar{x})$:*

$$\partial^P f(\bar{x}) := \{\exists \, \epsilon > 0 \ \text{and} \ M \geq 0 \ \text{such that} \ (1.42) \ \text{is satisfied}\,\}.$$

(The notation dom $\{f\}$ denotes the set $\{y : f(y) < +\infty\,\}$.)

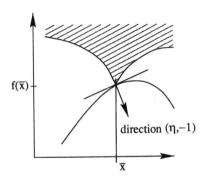

FIGURE 1.11. A Geometric Interpretation of Proximal Subgradients.

Figure 1.11 provides a geometric interpretation of proximal subgradients: a proximal subgradient to f at $\bar{x}$ is the slope at $x = \bar{x}$ of a paraboloid,

$$y = \eta \cdot (x - \bar{x}) + f(\bar{x}) - M|x - \bar{x}|^2,$$

that coincides with f at $x = \bar{x}$ and lies on or below the graph of f on a neighborhood of $\bar{x}$.

The Limiting Subdifferential. *Take an extended-valued, lower semicontinuous function $f : R^k \to R \cup \{+\infty\}$ and a point $\bar{x} \in \operatorname{dom}\{f\}$. A vector $\eta \in R^k$ is said to be a* limiting subgradient *of f at $\bar{x}$ if there exist sequences $x_i \xrightarrow{f} \bar{x}$ and $\eta_i \to \eta$ such that*

$$\eta \in \partial^P f(x_i) \quad \text{for all } i.$$

The set of all limiting subgradients of f at $\bar{x}$ is called the limiting subdifferential *and is denoted by $\partial f(\bar{x})$:*

$$\partial f(\bar{x}) := \{\eta : \exists\, x_i \xrightarrow{f} x \text{ and } \eta_i \to \eta \text{ such that } \eta_i \in \partial^P f(x_i) \text{ for all } i\}.$$

(The notation $x_i \xrightarrow{f} x$ indicates that $x_i \to x$ and $f(x_i) \to f(x)$ as $i \to \infty$.)

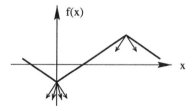

Graph of limiting subdifferential

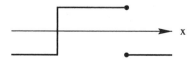

FIGURE 1.12. Limiting Subdifferentials.

Figure 1.12 illustrates how the limiting subdifferential depends on the base point for a nonsmooth function.

The limiting subdifferential is a closed set, but it need not be convex. It can happen that

$$\partial f(\bar{x}) \neq -\partial f(-\bar{x}).$$

This is because the inequality (1.42) in the definition of proximal subgradients does not treat positive and negative values of the function f in a symmetrical fashion.

We mention that there are, in fact, a number of equivalent ways of defining limiting subgradients. As so often in mathematics, it is a a matter of expository convenience what we regard as definitions and what we regard as consequences of these definitions. The "difference quotient" definition above has been chosen for this review chapter because it is often the simplest to use in applications. Another, equivalent, definition based on limiting normals to the epigraph set is chosen in Chapter 4 below, as a more convenient starting point for an in-depth analysis.

We now list a number of important properties of limiting normal cones and limiting subdifferentials.

(A): (Upper Semicontinuity)

(i) *Take a closed set $C \subset R^k$ and a point $x \in C$. Then for any convergent sequences $x_i \overset{C}{\to} x$ and $\eta_i \to \eta$ such that*

$$\eta_i \in N_C(x_i) \quad \textit{for all } i,$$

we have $\eta \in N_C(x)$.

(ii) *Take a lower semicontinuous function $f : R^k \to R \cup \{+\infty\}$ and a point $x \in \mathrm{dom}\,\{f\}$. Then for any convergent sequences $x_i \overset{f}{\to} x$ and $\xi_i \to \xi$ such that*

$$\xi_i \in \partial f(x_i) \quad \textit{for all } i,$$

we have $\xi \in \partial f(x)$.

(B): (Links between Limiting Normals and Subdifferentials) *Take a closed set $C \subset R^k$ and a point $\bar{x} \in C$. Then*

$$N_C(\bar{x}) = \partial \Psi_C(\bar{x}).$$

and

$$N_C(\bar{x}) \cap B = \partial d_C(\bar{x}).$$

Here, $\Psi_C : R^k \to R \cup \{+\infty\}$ is the indicator function of the set C

$$\Psi_C(x) := \begin{cases} 0 & \textit{if } x \in C \\ +\infty & \textit{if } x \notin C \end{cases}$$

and $d_C : R^k \to R$ is the distance function of the set C

$$d_C(x) := \inf_{y \in C} |x - y|.$$

It is well known that a Lipschitz continuous function f on R^k is differentiable almost everywhere with respect to Lebesgue measure. (This is Rademacher's Theorem.) It is natural then to ask, what is the relationship between the limiting subdifferential and limits of gradients at neighboring points? An answer is provided by the next property of the limiting subdifferential or, more precisely, its convex hull.

(C): (Limits of Derivatives) *Consider a lower semicontinuous function* $f : R^k \to R \cup \{+\infty\}$ *and a point* $x \in \text{dom } f$. *Suppose that* f *is Lipschitz continuous on a neighborhood of* $\bar{x}$. *Then, for any subset* $S \subset R^k$ *of zero* k-*dimensional Lebesgue measure, we have*

$$\text{co} \, \partial f(\bar{x})$$
$$= \text{co} \, \{\eta : \exists \, x_i \to x \text{ such that } \nabla f(x_i) \text{ exists and}$$
$$x_i \notin S \text{ for all } i \text{ and } \nabla f(x_i) \to \eta\}.$$

Calculations of the limiting subdifferential for a specific function, based on the defining relationships, is often a challenge, because of their indirect nature. If the function in question is Lipschitz continuous and we are content to estimate the limiting subdifferential $\partial f(\bar{x})$ by its convex hull $\text{co} \, \partial f(\bar{x})$, then Property (C) can simplify the task. A convenient feature of this characterization is that, in examining limits of neighboring gradients, we are allowed to exclude gradients on an arbitrary nullset S. When f is piecewise smooth, for example, we can exploit this flexibility by hiding in S problematical points (on ridges, at singularities, etc.) and calculating $\text{co} \, \partial f(\bar{x})$ merely in terms of gradients of f on subdomains where f is smooth.

Property (C) is significant also in other respects. It tells us, for example, that if a lower semicontinuous function $f : R^n \to R \cup \{+\infty\}$ is continuously differentiable near a point $\bar{x}$, then $\partial f(\bar{x}) = \{\nabla f(\bar{x})\}$. Since f is differentiable at x if and only if $-f$ is differentiable at x and $\nabla f(x) = -\nabla(-f)(x)$, we also deduce:

(D): (Homogeneity of the Convexified Limiting Subdifferential) *Consider a lower semicontinuous function* $f : R^k \to R \cup \{+\infty\}$ *and a point* $\bar{x} \in \text{dom } f$. *Suppose that* f *is Lipschitz continuous on a neighborhood of* $\bar{x}$. *Then*

$$\text{co} \, \partial f(\bar{x}) = -\text{co} \, \partial(-f)(\bar{x}).$$

Another useful property of Lipschitz continuous functions is

(E): (Bounds on Limiting Subdifferentials) *Consider a lower semicontinuous function* $f : R^k \to R \cup \{+\infty\}$ *and a point* $x \in \text{dom } f$. *Suppose that* f *is Lipschitz continuous on a neighborhood of* $\bar{x}$ *with Lipschitz constant* K. *Then*

$$\partial f(\bar{x}) \subset KB.$$

A fundamental property of the gradient of a differentiable function is that it vanishes at a local minimizer of the function. The proximal subdifferential has a similar role in identifying possible minimizers.

(F): (Limiting Subdifferentials at Minima) *Take a lower semicontinuous function* $f : R^k \to R \cup \{+\infty\}$ *and a point* $\bar{x} \in \mathrm{dom}\, f$. *Assume that, for some* $\epsilon > 0$

$$f(\bar{x}) \leq f(x) \quad \text{for all } x \in \bar{x} + \epsilon B.$$

Then

$$0 \in \partial^P f(\bar{x}).$$

In particular then, $0 \in \partial f(\bar{x})$.

In applications, it is often necessary to estimate limiting subdifferentials of composite functions in terms of limiting subgradients of their constituent functions. Fortunately, an extensive calculus is available for this purpose. Some important calculus rules are as follows.

(G): (Positive Homogeneity) *Take a lower semicontinuous function* $f : R^k \to R \cup \{+\infty\}$, *a point* $\bar{x} \in \mathrm{dom}\, f$, *and* $\alpha \geq 0$. *Then*

$$\partial(\alpha f)(\bar{x}) = \alpha \partial f(\bar{x}).$$

(H): (Sum Rule) *Consider a collection* $f_i : R^k \to R \cup \{+\infty\}$, $i = 1, \ldots, n$ *of lower semicontinuous extended-valued functions, and a point* $\bar{x} \in \cap_{i=1,\ldots,n} \mathrm{dom}\, f_i$. *Assume that all the* f_is *except possibly one are Lipschitz continuous on a neighborhood of* $\bar{x}$. *Then the limiting subdifferential of* $(f_1 + \ldots + f_n)(x) = f_1(x) + \ldots + f_n(x)$ *satisfies*

$$\partial(f_1 + \ldots + f_n)(\bar{x}) \subset \partial f_1(\bar{x}) + \ldots + \partial f_n(\bar{x}).$$

(I): (Max Rule) *Consider a collection* $f_i : R^k \to R \cup \{+\infty\}$, $i = 1, \ldots, n$ *of lower semicontinuous extended-valued functions and a point* $\bar{x} \in \cap_{i=1,\ldots,n} \mathrm{dom}\, f_i$. *Assume that all the* f_is, *except possibly one, are Lipschitz continuous on a neighborhood of* $\bar{x}$. *Then the limiting subdifferential of* $(\max\{f_1, \ldots, f_n\})(x) = \max\{f_1(x), \ldots, f_n(x)\}$ *satisfies*

$$\partial(\max_{i=1,\ldots,n} f_i)(\bar{x}) \subset$$

$$\{ \sum_{i=1,\ldots,n} \alpha_i \partial f_i(\bar{x}) : \alpha_i \geq 0 \text{ for all } i \in J \text{ and } \sum_{i \in J} \alpha_i = 1\}$$

in which

$$J := \{i \in \{1, \ldots, n\} : f_i(x) = \max_j f_j(\bar{x})\}.$$

(J): (Chain Rule) *Take Lipschitz continuous functions* $G : R^k \to R^m$ *and* $g : R^m \to R$ *and a point* $\bar{x} \in R^k$. *Then the limiting subgradient* $\partial(g \circ G)(\bar{x})$ *of the composite function* $x \to g(G(x))$ *at* $\bar{x}$ *satisfies:*

$$\partial(g \circ G)(\bar{x}) \subset \{\eta : \eta \in \partial(\gamma \cdot G)(\bar{x}) \text{ for some } \gamma \in \partial g(G(\bar{x}))\}.$$

Many of these rules are extensions of familiar principles of differential calculus. (Notice however that, in most cases, equality has been replaced by set inclusion.) An exception is the Max Rule above. The Max Rule is a truly nonsmooth principle; it has no parallel in traditional analysis because the operation of taking the pointwise maximum of a collection of smooth functions usually generates a nonsmooth function. ("Corners" typically occur when there is a change in the function on which the maximum is achieved.)

We illustrate the use of the calculus rules by proving a multiplier rule for a mathematical programming problem with functional inequality constraints:

$$\begin{cases} \text{Minimize } f(x) \\ \text{over } x \in R^k \text{ satisfying} \\ g_i(x) \le 0, \text{ for } i = 1, \ldots, n. \end{cases} \tag{1.43}$$

Here $f : R^k \to R$ and $g_i : R^k \to R$, $i = 1, \ldots, n$, are given functions.

Multiplier Rule I (Inequality Constraints) *Let* $\bar{x}$ *be a minimizer for Problem (1.43). Assume that* f, g_i, $i = 1, \ldots, n$, *are Lipschitz continuous. Then there exist nonnegative numbers* $\lambda_0, \lambda_1, \ldots, \lambda_n$ *such that*

$$\sum_{i=0}^{n} \lambda_i = 1,$$

$$\lambda_i = 0 \quad \text{if } g_i(\bar{x}) < 0 \text{ for } i = 1, \ldots, n$$

and

$$0 \in \lambda_0 \partial f(\bar{x}) + \sum_{i=1}^{n} \lambda_i \partial g_i(\bar{x}). \tag{1.44}$$

(A related, but sharper, multiplier rule is proved in Chapter 5.)

Proof. Consider the function

$$\phi(x) = \max\{f(x) - f(\bar{x}), g_1(x), \ldots, g_n(x)\}.$$

We claim that $\bar{x}$ minimizes ϕ over $x \in R^k$. If this were not the case, it would be possible to find x' such that

$$\phi(x') < \phi(\bar{x}) = 0.$$

It would then follow that

$$f(x') - f(\bar{x}) < 0, \quad g_1(x') < 0, \dots, g_n(x') < 0,$$

in contradiction of the minimality of $\bar{x}$. The claim has been confirmed.

According to Rule (F)

$$0 \in \partial\phi(\bar{x}).$$

We deduce from Rule (I) that there exist nonnegative numbers $\lambda_0, \dots, \lambda_n$ such that

$$\sum_{i=0}^{n} \lambda_i = 1,$$

and

$$0 \in \lambda_0 \partial f(\bar{x}) + \sum_{i=1}^{n} \lambda_i \partial g_i(\bar{x}).$$

Notice that, if $g_i(\bar{x}) < 0$, then

$$g_i(\bar{x}) < \phi(\bar{x}).$$

So Rule (I) supplies the supplementary information that

$$\lambda_i = 0 \quad \text{if } g_i(\bar{x}) < 0.$$

The multiplier rule is proved. $\square$

In the case when $f, g_1, \dots, g_n$ are continuously differentiable, inclusion (1.44) implies

$$0 \in \lambda_0 \nabla f(\bar{x}) + \sum_{i=1}^{n} \lambda_i \nabla g_i(\bar{x}),$$

the traditional Fritz–John Condition for problems with functional inequality constraints. Nonsmooth Analysis supplies a new simple proof of a classical theorem in the case when the functions involved are continuously differentiable and extends it, by supplying information even in the situations when the functions are not differentiable. Application of nonsmooth calculus rules to some derived nonsmooth function (in this case ϕ) is a staple of "nonsmooth" methodology. Notice that, in the above proof, we can expect ϕ to be nonsmooth, even if the functions from which it is constructed are smooth.

Calculus rules are some of the tools of the trade in Nonsmooth Analysis. In many applications however, these are allied to variational principles. A variational principle is an assertion that some quantity is minimized. Two which are widely used in Nonsmooth Analysis are the Exact Penalization

Theorem and Ekeland's Theorem. Suppose that the point $\bar{x} \in R^k$ is a minimizer for the constrained minimization problem

$$\begin{cases} \text{Minimize } f(x) \text{ over } x \in R^k \\ \text{satisfying } x \in C, \end{cases}$$

for which the data comprise a function $f : R^k \to R$ and a set $C \subset R^k$. For both analytical and computational reasons, it is often convenient to replace this by an unconstrained problem, in which the new cost

$$f(x) + p(x)$$

is the sum of the former cost and a term $p(x)$ that penalizes violation of the constraint $x \in C$. In the case that f is Lipschitz continuous, the Exact Penalization Theorem says that $\bar{x}$ remains a minimizer when the penalty function p is chosen to be the nonsmooth function

$$p(x) = \hat{K} d_C(x),$$

for some constant $\hat{K}$ sufficiently large. Notice that the penalty term, while nonsmooth, is Lipschitz continuous.

The Exact Penalization Theorem *Take a function $f : R^k \to R$ and a closed set $C \subset R^k$. Assume that f is Lipschitz continuous, with Lipschitz constant K. Let $\bar{x}$ be a minimizer for*

$$\begin{cases} \text{Minimize } f(x) \text{ over } x \in R^k \\ \text{satisfying } x \in C. \end{cases}$$

Then, for any $\hat{K} \geq K$, $\bar{x}$ is a minimizer also for the unconstrained problem

$$\begin{cases} \text{Minimize } f(x) + \hat{K} d_C(x) \\ \text{over points } x \in R^k. \end{cases}$$

A more general version of this theorem is proved in Chapter 3.

One important consequence of the Exact Penalization Theorem is that, if $\bar{x}$ is a minimizer for the optimization problem of the theorem statement then, under the hypotheses of the Exact Penalization Theorem,

$$0 \in \partial f(\bar{x}) + N_C(\bar{x}).$$

(It is a straightforward matter to deduce this necessary condition of optimality for an optimization problem with an "implicit" constraint from the fact that $\bar{x}$ also minimizes $f(x) + K d_C(x)$ over R^n, and Rules (F), (H), and (B) above.)

We turn next to Ekeland's Theorem. In plain terms, this states that, if a point x_0 approximately minimizes a function f, then there is a point $\bar{x}$ close to x_0 which is a minimizer for some new function, obtained by adding a small perturbation term to the original function f. See Figure 3.1 on page 112.

Ekeland's Theorem Take a complete metric space $(X, d(.,.))$, a lower semicontinuous function $f : X \to R \cup \{+\infty\}$, a point $x_0 \in \operatorname{dom} f$, and a number $\epsilon > 0$ such that

$$f(x_0) \leq \inf_{x \in X} f(x) + \epsilon.$$

Then there exists $\bar{x} \in X$ such that

$$d(\bar{x}, x_0) \leq \epsilon^{1/2}$$

and

$$f(x) + \epsilon^{1/2} d(x, \bar{x})|_{x=\bar{x}} = \inf_{x \in X} \{f(x) + \epsilon^{1/2} d(x, \bar{x})\}.$$

Ekeland's Theorem and the preceding nonsmooth calculus rules can be used, for example, to derive a multiplier rule for a mathematical programming problem involving a functional equality constraint:

$$\begin{cases} \text{Minimize } f(x) \\ \text{over } x \in R^k \text{ satisfying} \\ G(x) = 0. \end{cases} \qquad (1.45)$$

The data for this problem comprise functions $f : R^k \to R$ and $G : R^k \to R^m$.

A Multiplier Rule (Equality Constraints) *Let x_0 be a minimizer for problem (1.45). Assume that f and G are Lipschitz continuous.*
Then there exists a number $\lambda_0 \geq 0$ and a vector $\lambda_1 \in R^m$ such that

$$\lambda_0 + |\lambda_1| = 1$$

and

$$0 \in \lambda_0 \partial f(x_0) + \partial(\lambda_1 \cdot G)(x_0). \qquad (1.46)$$

Proof. The simple proof above of the Multiplier Rule for inequality constraints suggests that equality constraints can be treated in the same way. We might hope to use the fact that, for the problem with equality constraints, x_0 is an unconstrained minimizer of the function

$$\gamma(x) := \max\{f(x) - f(x_0), |G(x)|\}.$$

By Rule (F), $0 \in \partial\gamma(x_0)$. We deduce from the Max Rule (I) that there exist multipliers $\lambda_0 \geq 0$ and $\nu_1 \geq 0$ such that $\lambda_0 + \nu_1 = 1$ and

$$0 \in \lambda_0 \partial f(x_0) + \nu_1 \partial |G(x)| \,|_{x=x_0}. \tag{1.47}$$

We aim then to estimate vectors in $\partial |G(x)| \,|_{x=x_0}$ using nonsmooth calculus rules, in such a manner as to deduce the asserted multiplier rule. The flaw in this approach is that $0 \in \partial |G(x)| \,|_{x=x_0}$ with the result that (1.47) is trivially satisfied by the multipliers $(\lambda_0, \nu_1) = (0, 1)$. In these circumstances, (1.47) cannot provide useful information about x_0.

Ekeland's Theorem can be used to patch up the argument, however. The difficulty we have encountered is that $0 \in \partial |G(x)| \,|_{x=x_0}$. The idea is to *perturb* the function $\gamma(.)$ such that, along some sequence $x_i \to x_0$, (1.47) is approximately satisfied at x_i but, significantly, $0 \notin \partial |G(x)| \,|_{x=x_i}$. The perturbed version of (1.47) is no longer degenerate and we can expect to recover the multiplier rule in the limit as $i \to \infty$.

Take a sequence $\epsilon_i \downarrow 0$. For each i define the function

$$\phi_i(x) := \max\{f(x) - f(x_0) + \epsilon_i, |G(x)|\}.$$

Note that $\phi_i(x_0) = \epsilon_i$ and $\phi_i(x) \geq 0$ for all x. It follows that

$$\phi_i(x_0) \leq \inf_x \phi(x) + \epsilon_i.$$

In view of the fact that ϕ_i is lower semicontinuous (indeed Lipschitz continuous) with respect to the metric induced by the Eulidean norm, we deduce from Ekeland's theorem that there exists $x_i \in R^k$ such that

$$|x_i - x_0| \leq \epsilon_i^{1/2}$$

and

$$\tilde{\phi}_i(x_i) \leq \inf_{x \in R^k} \tilde{\phi}_i(x),$$

where

$$\tilde{\phi}_i(x) := \max\{f(x) - f(x_0) + \epsilon_i, |G(x)|\} + \epsilon_i^{1/2}|x - x_i|.$$

Since the function $x \to |x - x_i|$ is Lipschitz continuous with Lipschitz constant 1, we know that its limiting subgradients are contained in the closed unit ball B. By Rules (E), (F), and (H) then,

$$0 \in \partial\tilde{\phi}_i(x_i) + \epsilon_i^{1/2}B.$$

Otherwise expressed,

$$0 \in z_i + \epsilon_i^{1/2}B$$

for some point

$$z_i \in \partial \max\{f(x) - f(x_0) + \epsilon_i, |G(x)|\}|_{x=x_i}.$$

We now make an important observation: for each i,

$$\max\{f(x) - f(x_0) + \epsilon_i, |G(x)|\}|_{x=x_i} > 0.$$

Indeed, if this were not the case, we would have $f(x_i) \leq f(x_0) - \epsilon_i$ and $G(x_i) = 0$, in violation of the optimality of x_0.

It follows that

$$|G(x_i)| = 0 \quad \text{implies} \quad f(x) - f(x_0) + \epsilon_i > |G(x_i)|. \tag{1.48}$$

Now apply the Max Rule (I) to $\tilde{\phi}_i$. We deduce the existence of nonnegative numbers λ_0^i and ν_1^i such that

$$\lambda_0^i + \nu_1^i = 1,$$

$$z_i \in \lambda_0^i \partial f(x_i) + \nu_1^i \partial |G(x_i)|.$$

Furthermore, by (1.48),

$$|G(x_i)| = 0 \quad \text{implies} \quad \nu_1^i = 0. \tag{1.49}$$

We note next that, if $y' \neq 0$, then all limiting subgradients of $y \to |y - y'|$ at y' have unit Euclidean norm. (This follows from the fact that $\nabla |y| = y/|y|$ for $y \neq 0$ and Rule (C).) We deduce from the Chain Rule and (1.49) that there exists a vector $a_i \in R^m$ such that $|a_i| = 1$ and

$$z_i \in \lambda_0^i \partial f(x_i) + \nu_1^i \partial(a_i \cdot G)(x_i).$$

Writing $\lambda_1^i := \nu_1^i a_i$, we have, by positive homogeneity,

$$\lambda_0^i + |\lambda_1^i| = 1 \tag{1.50}$$

and

$$z_i = \lambda_0^i \partial f(x_i) + \partial(\lambda_1^i \cdot G)(x_i).$$

It follows that, for each i, there exist points

$$\gamma_0^i \in \partial f(x_i), \quad \gamma_1^i \in \partial(\lambda_1^i \cdot G)(x_i) \quad \text{and} \quad e_i \in \epsilon_i^{1/2} B$$

such that

$$0 = \lambda_0^i \gamma_0^i + \gamma_1^i + e_i. \tag{1.51}$$

By rule (E), and in view of (1.50) and of the Lipschitz continuity of f and G, the sequences

$$\{\gamma_0^i\}, \{\gamma_1^i\}, \{\lambda_0^i\} \text{ and } \{\lambda_1^i\}$$

are bounded. We can therefore arrange, by extracting subsequences, that they have limits $\gamma_0, \gamma_1, \lambda_0, \lambda_1$, respectively. Since the λ_0^is are nonnegative, it follows from (1.50) that

$$\lambda_0 \geq 0 \quad \text{and} \quad \lambda_0 + |\lambda_1| = 1.$$

But $e_i \to 0$. We deduce from (1.51), in the limit as $i \to \infty$, that

$$0 = \lambda_0 \gamma_0 + \gamma_1.$$

It is a straightforward matter to deduce from the upper semicontinuity properties of the limiting subdifferential, the Sum Rule (H), and Rule (E) that

$$\gamma_0 \in \partial f(x_0) \quad \text{and} \quad \gamma_1 \in \partial(\lambda_1 \cdot G)(x_0).$$

If follows that

$$0 \in \lambda_0 \partial f(x_0) + \partial(\lambda_1 \cdot G)(x_0).$$

This is what we set out to show. $\square$

In the case when f and g are continuously differentiable, Condition (1.46) implies

$$0 = \lambda_0 \nabla f(x_0) + \lambda_1 \nabla G(x_0).$$

Once again, the tools of Nonsmooth Analysis provide both an alternative derivation of a well-known Multiplier Rule in Nonlinear Programming for differentiable data and also extensions to nonsmooth cases. $\square$

The above proof is somewhat complicated, but it is no more so (if we have the above outlined techniques of Nonsmooth Analysis at our disposal) than traditional proofs of Multiplier Rules for problems with equality constraints (treating the smooth case), based on the classical Inverse Function Theorem, say. The Multiplier Rule for problems with equality constraints, even for differentiable data, is not an elementary result and we must expect to work for it.

When we seek to generalize classical results involving differentiable data to nonsmooth cases, variational principles often play a key role. The proof of the Multiplier Rule above is included in this review chapter as an illuminating prototype. The underlying ideas are used repeatedly in future chapters to derive necessary conditions of optimality in Optimal Control. In many applications, variational principles such as Ekeland's Theorem (which concern properties of nondifferentiable functions) carry the burden traditionally borne by the Inverse Function Theorem and its relatives in classical analysis (which concern properties of differentiable functions).

1.9 Nonsmooth Optimal Control

The analytical techniques of the preceding section were developed primarily for the purpose of applying them to Optimal Control. We now briefly describe how they have been used to overcome fundamental difficulties encountered earlier in this field.

Consider first necessary conditions. As discussed, the following optimization problem, in which the dynamic constraint takes the form of a differential inclusion, serves as a convenient paradigm for a broad class of optimal control problems, including some problems to which the Maximum Principle is not directly applicable.

$$
\begin{cases}
\text{Minimize } g(x(S), x(T)) \\
\text{over arcs } x \in W^{1,1} \text{ satisfying} \\
\dot{x}(t) \in F(t, x(t)) \text{ a.e.,} \\
(x(S), x(T)) \in C.
\end{cases}
\tag{1.52}
$$

Here, $g : R^n \times R^n \to R$ is a given function, $F : [S, T] \times R^n \rightsquigarrow R^n$ is a given multifunction, and $C \subset R^n \times R^n$ is a given subset. We proposed reformulating it as a generalized problem in the Calculus of Variations:

$$
\begin{cases}
\text{Minimize } \int_S^T L(t, x(t), \dot{x}(t))dt + g(x(S), x(T)) \\
\text{over absolutely continuous arcs } x \text{ satisfying} \\
(x(S), x(T)) \in C,
\end{cases}
$$

in which L is the extended valued function

$$
L(t, x, v) := \begin{cases}
0 & \text{if } v \in F(t, x) \\
+\infty & \text{if } v \notin F(t, x),
\end{cases}
$$

and deriving generalizations of classical necessary conditions to take account of nonsmoothness of the data. This can be done, with recourse to Nonsmooth Analysis. The details are given in future chapters.

We state merely one set of necessary conditions of this nature. The following generalization of the Hamilton Condition in the Calculus of Variations due to Clarke, was an early landmark in Nonsmooth Optimal Control. As usual we write

$$
H(t, x, p) = \max_{v \in F(t,x)} p \cdot v .
$$

Theorem (The Hamilton Inclusion) *Let $\bar{x}$ be a minimizer for Problem (1.52). Assume that, for some $\epsilon > 0$, the data for the problem satisfy:*

(a) the multifunction $F : [S, T] \times R^n \rightsquigarrow R^n$ has as values closed, nonempty convex sets, and F is $\mathcal{L} \times \mathcal{B}^n$ measurable;

(b) there exists $c(.) \in L^1$ such that

$$
F(t, x) \subset c(t)B \quad \forall \, x \in \bar{x}(t) + \epsilon B \quad a.e.;
$$

(c) there exist $k(.) \in L$ such that

$$
F(t, x) \subset F(t, x') + k(t)|x - x'|B \quad \forall \, x, x' \in \bar{x}(t) + \epsilon B \quad a.e.;
$$

(d) C is closed and g is Lipschitz continuous on $(\bar{x}(S), \bar{x}(T)) + \epsilon B$.

Then there exist $p \in W^{1,1}([S,T];R^n)$ and $\lambda \geq 0$ such that

(i) $(p, \lambda) \neq 0$.

(ii) $(-\dot{p}(t), \dot{\bar{x}}(t)) \in \operatorname{co} \partial H(t, \bar{x}(t), p(t))$ a.e.

(iii) $(p(S), -p(T)) \in \lambda \partial g(\bar{x}(S), \bar{x}(T)) + N_C(\bar{x}(S), \bar{x}(T))$.

(The sigma algebra $\mathcal{L} \times \mathcal{B}^n$, referred to in Hypothesis (a), is defined in Chapter 2. ∂H denotes the limiting subdifferential of $H(t, ., ., .)$.)

The essential feature of this extension is that the adjoint "equation," Condition (ii), is now interpreted in terms of a generalized subdifferential of the Hamiltonian H, namely, the Clarke generalized gradient of $H(t, ., .)$, which coincides with the convex hull of the limiting subdifferential. The limiting subdifferential and the limiting normal cone are called upon to give meaning to the transversality condition (iii). This theorem is a special case of Theorem 7.7.1 below.

We stress that the Hamilton Inclusion is an intrinsically nonsmooth condition. Even when the multifunction F is the velocity set of some smooth control system

$$\dot{x}(t) = f(t, x(t), \dot{x}(t)), \quad u(t) \in U \quad \text{a.e.,}$$

("smooth" in the sense that f is a smooth function and U has a smooth boundary), i.e.,

$$F(t, x) = f(t, x, U),$$

we can still expect that $H(t, ., .)$ will be nonsmooth and that the apparatus of Nonsmooth Analysis will be required to make sense of the Hamilton Inclusion.

Other nonsmooth necessary conditions are derived in future chapters. These include a generalization of the Euler–Lagrange Condition for Problem (1.52) and a generalization of the Maximum Principle for a nonsmooth version of the optimal control problem of Section 1.5.

Consider next Dynamic Programming. This, we recall, concerns the relationship between the value function V, which describes how the minimum cost of an optimal control problem depends on the initial data, and the Hamilton– Jacobi Equation.

For now, we focus on the optimal control problem

$$\begin{cases} \text{Minimize } g(x(S), x(T)) \\ \text{over measurable functions } u, \text{ and arcs } x \in W^{1,1} \text{ satisfying} \\ \dot{x}(t) \in f(t, x(t), u(t)) \text{ a.e.,} \\ u(t) \in U \text{ a.e.,} \\ x(S) = x_0, \ x(T) \in C, \end{cases}$$

to which corresponds the Hamilton–Jacobi Equation:

$$\begin{cases} \phi_t(t,x) + \min_{u \in U} \phi_x(t,x) \cdot f(t,\xi,u) = 0 \\ \qquad \qquad \qquad \text{for all } (t,x) \in (S,T) \times R^n \\ \phi(T,x) = g(x) + \Psi_C(x) \quad \text{for all } x \in R^n. \end{cases}$$

(Ψ_C denotes the indicator function of the set C.)

It has been noted that the value function is often nonsmooth. It cannot therefore be identified with a classical solution to the Hamilton–Jacobi Equation above, under hypotheses of any generality. However the machinery of Nonsmooth Analysis provides a characterization of the value function in terms of the Hamilton–Jacobi Equation, even in cases when V is discontinuous. There are a number of ways to do this. One involves the concept of Viscosity Solution. Another is based on properties of proximal subgradients. In this book we follow this latter approach because it is better integrated with our treatment of other topics and because it is particularly well suited to problems with endpoint constraints.

Specifically, we show that, under mild hypotheses, V is the unique lower semicontinuous function which is a generalized solution of the Hamilton–Jacobi Equation. Here, "generalized solution" means a function ϕ taking values in the extended real line $R \cup \{+\infty\}$, such that the following conditions are satisfied.

(i): For each $(t,x) \in ((S,T) \times R^n) \cap \text{dom}\,\phi$

$$\eta^0 + \inf_{v \in F(t,x)} \eta^1 \cdot v = 0 \quad \forall\, (\eta^0, \eta^1) \in \partial^P \phi(t,x);$$

(ii): $\phi(T,x) = g(x) + \Psi_C(x) \quad \forall\, x \in R^n;$

(iii): $\phi(S,x) = \liminf_{t' \downarrow S, x' \to x} \phi(t',x') \quad \forall\, x \in R^n,$
$\phi(T,x) = \liminf_{t' \uparrow T, x' \to x} \phi(t',x') \quad \forall\, x \in R^n.$

Of principal interest here is Condition (i); (ii) merely reproduces the earlier boundary condition; (iii) is a regularity condition on ϕ, which is automatically satisfied when ϕ is continuous.

If ϕ is a C^2 function on $(S,T) \times R^n$, Condition (i) implies that ϕ is a classical solution to the Hamilton–Jacobi Equation. This follows from the fact that, in this case, $\partial^P V(t,x) = \{\nabla V(t,x)\}$.

What is remarkable about this characterization is the unrestrictive nature of the conditions it imposes on V for it to be a value function. Condition (i) merely requires us to test certain relationships on the subset of $(S,T) \times R^n$:

$$\mathcal{D} = \{(t,x) \in (S,T) \times R^n \cap \text{dom}\, V \, : \, \partial^P V(t,x) \neq \emptyset\},$$

a subset that can be very small indeed (in relation to $(S,T) \times R^n$); it need not even be dense.

Another deficiency of classical Dynamic Programming is its failure to provide precise sensitivity information about the value function V, although our earlier formal calculations suggest that, if p is the adjoint arc associated with a minimizer $(\bar{x}, \bar{u})$, then

$$(\max_{u \in U} p(t) \cdot f(t, \bar{x}, u), -p(t)) \; = \; \nabla V(t, \bar{x}(t)) \quad \text{a.e. } t \in [S, T]. \tag{1.53}$$

Indeed for many cases of interest V fails to be differentiable and (1.53) does not even make sense. Once again, Nonsmooth Anaysis fills the gap. The sensitivity relation (1.53) can be given precise meaning as an inclusion in which the gradient ∇V is replaced by a suitable "nonsmooth" subdifferential, namely, the convexified limiting subdifferential $\text{co}\,\partial V$.

1.10   Epilogue

Our overview of Optimal Control is concluded. To summarize, Optimal Control concerns the properties of minimizing arcs. Its roots are in the investigation of minimizers of integral functionals, which has been an active area of research for over 200 years, in the guise of the Calculus of Variations. The distinguishing feature of Optimal Control is that it can take account of dynamic and pathwise constraints, of a nature encountered in the calculation of optimal flight trajectories and other areas of advanced engineering design and operations.

Key early advances were the Maximum Principle and an intuitive understanding of the relationship between the value function and the Hamilton–Jacobi Equation of Dynamic Programming. The antecedents of Dynamic Programming in the branch of the Calculus of Variations known as Field Theory were apparent. On the other hand the nature of the Maximum Principle, and the techniques first employed to prove it, based on approximation of reachable sets, suggested that essentially new kinds of necessary conditions were required to deal with the constraints of optimal control problems and new techniques to derive them.

Recent developments in Optimal Control, aimed at extending the range of application of available necessary conditions of optimality, emphasize its similarities rather than its differences with the Calculus of Variations. Here, the constraints of Optimal Control are replaced by extended-valued penalty terms in the integral functional to be minimized. Problems in Optimal Control are thereby reformulated as extended problems in the Calculus of Variations with nonsmooth data. It is then possible to derive, in the context of Optimal Control Theory, optimality conditions of remarkable generality, analogous to classical necessary conditions in the Calculus of Variations, in which classical derivatives are replaced by "generalized" derivatives of nonsmooth functions. Nonsmooth Analysis, which gives meaning to generalized derivatives has, of course, had a key role in these developments. Nonsmooth

Analysis, by supplying the appropriate machinery for interpreting generalized solutions to the Hamilton–Jacobi Equation, can be used also to clarify the relationship between the value function and the Hamilton–Jacobi Equation, hinted at in the 1950s.

This overview stresses developments in Optimal Control that provide the background to material in this book. It ignores a number of important topics, such as higher-order conditions of optimality, optimality conditions for a wide variety of nonstandard optimal control problems (impulsive control problems, problems involving functional differential equations, etc.), computational optimal control, and an analysis of the dependence of minimizers on parameters, such as that provided in [100]. It also follows only one possible route to optimality conditions of recent interest, which highlights their affinity with classical conditions. A complete picture would include others, such as those of Neustadt [110], Halkin [79], Warga [147], Zhu [156], and Sussmann [132], which are centered on refinements of the Maximum Principle.

Optimal Control has been an active research field for over 40 years, much longer if you regard it as an outgrowth of the Calculus of Variations. Newcomers will, with reason, expect to find an armory of techniques for computing optimal controls and an extensive library of solved problems. They will be disappointed. The truth is, solving optimal control problems of scientific or engineering interest is often extremely difficult. Analytical tools of Optimal Control, such as the Maximum Principle, provide equations satisfied by minimizers. They do not tell us how to solve them or, if we do succeed in solving them, that the "solutions" are the right ones. Over two centuries ago Euler wrote of fluid mechanics (he could have commented thus equally about the Calculus of Variations),

If it is not permitted to us to penetrate to a complete knowledge concerning the motion of fluids, it is not to mechanics or to the insufficiency of the known principles of motion, that we must attribute the cause. It is analysis itself which abandons us here.

As regards Optimal Control, the picture remains the same today. It is quite clear that general methodologies giving "complete knowledge" of solutions to large classes of optimal contol problems will never be forthcoming, if by "solution" is meant formulae describing the solution in terms of standard functions.

Even in cases when we can completely solve a variational problem by analysis, translating general theory into a minimizer for a specific problem is often a major undertaking. Often we find that for seemingly simple classical problems of longstanding interest, the hypotheses are violated, under which the general theory guarantees existence of minimizers and supplies information about them. In cases when optimality conditions are valid, the information they supply can be indirect and incomplete.

Some of these points are illustrated by the Brachistochrone Problem of Section 1.1. It is easy to come up with a candidate $\bar{x}(.)$ for a minimizing path by solving the Euler–Lagrange Condition, namely, a cycloid satisfying the boundary conditions. But is it truly a minimizer? If we adopt the formulation of Section 1.1, in which we seek a minimizer among curves describable as the graph of an absolutely continuous function in the vertical plane, the answer is yes. But this is by no means easy to establish. The cost integrand violates the Tonelli criteria for existence of minimizers (although other criteria can be invoked in this case to establish existence of a minimum time path). Then it is not even clear at the outset that minimizers will satisfy the Euler–Lagrange Condition because the standard Lipschitz continuity hypothesis commonly invoked to derive the condition, namely, Hypothesis (ii) of Theorem 1.2.1, is violated. Various arguments can be used to overcome these difficulties. (We can study initially a simpler problem in which, to eliminate the troublesome infinite slope at the left endpoint, the ball is assumed to have a small initial velocity, or a pathwise constraint is imposed on the slope of the arcs considered, for example, and subsequently limits are examined.) But none of them are straightforward. Solution of the Brachistochrone Problem is discussed at length in, for example, [136] and [133].

The Brachistrochrone Problem can be solved by analysis, albeit with some hard work. The Maximal Orbit Transfer Problem of Section 1.1 is more typical of advanced engineering applications. We made no claims that this problem has been solved. But numerical schemes, inspired by the Maximum Principle, generate the control strategy of Figure 1.2, which is *probably* optimal for this problem.

It is only when we adopt a more realistic view of the scope of Optimal Control that we can truly appreciate the benefits of research in this field:

Simple Problems.   While we cannot expect Optimal Control Theory alone to provide complete solutions to complex optimal control problems of advanced engineering design, it does often provide optimal strategies for simplified versions of these problems. Optimal strategies for a simpler problem can serve as adequate close-to-optimal strategies (synonymously "suboptimal" strategies) for the original problem. Alternatively, they can be used as initial guesses at an optimal strategy for use in numerical schemes to solve the original problem. This philosophy underlies some applications of Optimal Control to Resource Economics, where simplifying assumptions about the underlying dynamics lead to problems that can be solved explicitly by analytical means [38]. In a more general context, many optimal control problems of practical importance, in which there are only small excursions of the state variables, can be approximated by the linear quadratic control problem

$$
\begin{cases}
\text{Minimize } \int_0^T (x(t) \cdot Qx(t) + u(t) \cdot Ru(t))dt + x(T) \cdot Px(T) \\
\text{over } u \in L^2 \text{ and } x \in W^{1,1} \text{ satisfying} \\
\dot{x}(t) = Ax(t) + Bu(t) \quad \text{a.e.,} \\
x(0) = x_0,
\end{cases}
$$

with data the number $T > 0$, the $n \times n$ symmetric matrices $Q \geq 0$, $R > 0$, and $P \geq 0$, the $n \times n$ matrix A, the $n \times m$ matrix B, and the n-vector x_0. The linear state equation here corresponds to linearization of nonlinear dynamics about an operating point. Weighting matrices in the cost are chosen to reflect economic criteria and to penalize constraint violations. In engineering applications, solving (LQG) (or related problems involving an approximate, linear, model) is a common ingredient in controller design.

Numerical Methods. For many optimal control problems that cannot be solved by the application of analysis alone, we can at least seek a minimizer with the help of numerical methods. Among the successes of Optimal Control Theory is the impact it has had on Computational Optimal Control, in the areas of both algorithm design and algorithm convergence analysis. Ideas such as exact penalization, for example, and techniques for local approximation of nonsmooth functions, so central to the theory, have taken root also in general computational procedures for solving optimal control problems. We mention in particular work of Polak and Mayne, who early exploited nonsmooth constructs and analytic methods from Optimal Control in a computational context [101]. (See [116] and [113] for numerous references). Global convergence analysis, which amounts to establishing that limit points satisfy first-order conditions of optimality is, of course, intimately connected with Optimal Control Theory. The significance of Global Convergence Analysis is that it makes explicit classes of problems for which the algorithm is effective and, on the other hand, it points to possible modifications to existing algorithms to increase the likelihood of convergence.

Qualitative Properties. Optimal Control can also provide useful qualitative information about minimizers. Often there are various ways of formulating an engineering design problem as an optimization problem. Knowledge of criteria for existence of solutions gives insights into whether a sensible formulation has been adopted. Optimal control supplies a priori information about regularity of minimizers. This is of benefit for computation of optimal controls, since a "best choice" of optimization algorithm can depend on minimizer regularity, as discussed, for example, in [78]. Minimizer regularity is also relevant to the modeling of physical phenomena. In applications to Nonlinear Elasticity, for example, existence of minimizers with infinite gradients has been linked to material failure and so an analysis identifying classes of problems for which minimizers have bounded derivatives gives insights into mechanisms of fracture [13]. Finally, and perhaps most significantly from the point of view of solving engineering problems, necessary conditions can supply structural information that greatly simplifies the computation of optimal controls. In the computation of optimal flight trajectories, for instance, information obtained from known necessary conditions of optimality concerning bang-bang subarcs, relationships governing dependent variables on interior arcs, bounds on the number of switches, etc., is routinely used to reduce the computational burden [26].

1.11 Notes for Chapter 1

This chapter highlights major themes of the book and links them to earlier developments in mathematics. It is not intended as a comprehensive historical review of Optimal Control.

Troutman [136] covers much of the material on the Calculus of Variations in Section 1.2, in lucid fashion. L. C. Young's inspirational, if individualistic, book [152] on the subject and its relationship with early developments in Optimal Control is also recommended. See, too, [28].

From numerous older books on classical Optimal Control, we single out for clarity and accessibility [81], [19], [65], and [88]. A recent expository text is [153]. Warga [147] provides more detailed coverage, centered on his own contributions to the theory of existence of minimizers and necessary conditions. Bryson and Ho's widely read book [25], which follows a traditional applied mathematics approach to Optimal Control, includes a wealth of engineering-inspired examples. See also [93].

Rockafellar and Wet's book [125] provides extensive, up-to-date coverage of Nonsmooth Analysis in finite dimensions.

As regards Nonsmooth Optimal Control, Clarke's book [38], which had a major role in winning an audience for the field, remains a standard reference. Expository accounts of this material (and extensions) are to be found in [37] and [94]. The recent book by Clarke et al. [54] covers important developments in infinite-dimensional Nonsmooth Analysis and their applications not only in Optimal Control, but also in stabilizing control system design.

The Maximum Orbital Transfer Problem discussed in Section 1.1 is a refinement of a problem studied by Kopp and McGill, and reported in [25], to allow for thrusts of variable magnitude. A second-order, feasible directions algorithm due to R. Pytlak, documented in [115], was used in the computations. The Brachistochrone Problem of Section 1.2 is analyzed in many publications, including [136] and [152].

Chapter 2

Measurable Multifunctions and Differential Inclusions

You cannot argue with someone who denies the first principles.

– Anonymous Medieval Proposition

2.1 Introduction

Differential inclusions

$$\dot{x}(t) \in F(t, x(t)) \quad \text{a.e. } t \in I \tag{2.1}$$

feature prominently in modern treatments of Optimal Control. This has come about for several reasons. One is that Condition (2.1), summarizing constraints on allowable velocities, provides a convenient framework for stating hypotheses under which optimal control problems have solutions and optimality conditions may be derived. Another is that, even when we choose not to formulate an optimal control problem in terms of a differential inclusion, in cases when the data are nonsmooth, often the very statement of optimality conditions makes reference to differential inclusions. It is convenient then at this stage to highlight important properties of multifunctions and differential inclusions of particular relevance in Optimal Control.

Section 2.3 deals primarily with measurability issues. The material is based on a by now standard definition of a measurable set-valued function, which is a natural analogue of the familiar concept of a measurable point-valued function. Criteria are given for a specified set-valued function to be measurable. These are used to establish that measurability is preserved under various operations on measurable multifunctions (composition, taking limits, etc.) frequently encountered in applications.

A key consideration is whether a set-valued function $\Gamma(t)$ has a *measurable selection* $\gamma(t)$. This means that γ is a Lebesgue measurable function satisfying

$$\gamma(t) \in \Gamma(t) \quad \text{a.e. } t \in I .$$

The concept of a measurable selection has many uses, not least to define solutions to differential inclusions: in the case when $I = [a, b]$, a solution x to (2.1) is an absolutely continuous function satisfying

$$x(t) = x(a) + \int_a^t \gamma(s)ds,$$

where $\gamma(t)$ is some measurable selection of $F(t, x(t))$.

Sections 2.4 and 2.5 concern solutions to the differential inclusion (2.1). The first important result is an existence theorem (the Generalized Filippov Existence Theorem), which gives conditions under which a solution (called an F-trajectory) may be found "near" an arc that approximately satisfies the differential inclusion. Of comparable significance is the Compactness of Trajectories Theorem which describes the closure properties of the set of F-trajectories. These results are used extensively in future chapters.

In Sections 2.6 through 2.8, we examine questions of when, and in what sense, Optimal Control problems have minimizers. The Compactness of Trajectories Theorem leads directly to simple criteria for existence of minimizers, for Optimal Control problems formulated in terms of the differential inclusion (2.1), when F takes as values convex sets. If F fails to be convex-valued, there may be no minimizers in a traditional sense. But existence of minimizers can still be guaranteed, if the domain of the optimal control problem is enlarged to include additional elements. This is the theme of relaxation, which is taken up in Section 2.7. Finally, in Section 2.8, broad criteria are established for existence of minimizers to the Generalized Bolza Problem (GBP). (GBP) subsumes a large class of Optimal Control problems, and is a natural framework for studying existence of minimizers under hypotheses of great generality.

2.2 Convergence of Sets

Take a sequence of sets $\{A_i\}$ in R^n. There are number of ways of defining limit sets. For our purposes, "Kuratowski sense" limit operations are the most useful. The set

$$\liminf_{i \to \infty} A_i$$

(the Kuratowski lim inf) comprises all points $x \in R^n$ satisfying the condition: there exists a sequence $x_i \to x$ such that $x_i \in A_i$ for all i.

The set

$$\limsup_{i \to \infty} A_i$$

(the Kuratowski lim sup) comprises all points $x \in R^n$ satisfying the condition: there exist a subsequence $\{A_{i_j}\}$ of $\{A_i\}$ and a sequence $x_j \to x$ such that $x_j \in A_{i_j}$ for all j.

$\liminf_{i \to \infty} A_i$ and $\limsup_{i \to \infty} A_i$ are (possibly empty) closed sets, related according to

$$\liminf_{i \to \infty} A_i \subset \limsup_{i \to \infty} A_i.$$

In the event $\liminf_{i \to \infty} A_i$ and $\limsup_{i \to \infty} A_i$ coincide, we say that $\{A_i\}$ has a limit (in the Kuratowski sense) and write

$$\lim_{i \to \infty} A_i := \liminf_{i \to \infty} A_i \, (= \limsup_{i \to \infty} A_i).$$

The sets $\liminf_{i\to\infty} A_i$ and $\limsup_{i\to\infty} A_i$ are succinctly expressed in terms of the distance function

$$d_A(x) = \inf_{y \in A} |x - y|,$$

thus

$$\liminf_{i\to\infty} A_i = \{x \;:\; \limsup_{i\to\infty} d_{A_i}(x) = 0\},$$

and

$$\limsup_{i\to\infty} A_i = \{x \;:\; \liminf_{i\to\infty} d_{A_i}(x) = 0\}.$$

Moving to a more general context, we take a set $D \subset R^k$ and a family of sets $\{S(y) \subset R^n : y \in D\}$, parameterized by points in $y \in D$. Fix a point $x \in R^k$. The set

$$\liminf_{y \overset{D}{\to} x} S(y)$$

(the Kuratowski lim inf) comprises all points ξ satisfying the condition: corresponding to any sequence $y_i \overset{D}{\to} x$, there exists a sequence $\xi_i \to \xi$ such that $\xi_i \in S(y_i)$ for all i. The set

$$\limsup_{y \overset{D}{\to} x} S(y)$$

(the Kuratowski lim sup) comprises all points ξ satisfying the condition: there exist sequences $y_i \overset{D}{\to} x$ and $\xi_i \to \xi$ such that $\xi_i \in S(y_i)$ for all i. (In the above, $y_i \overset{D}{\to} x$ means $y_i \to x$ and $y_i \in D$ for all i.)

If D is a neighborhood of x, we write $\liminf_{y\to x} S(y)$ in place of

$$\liminf_{y \overset{D}{\to} x} S(y),$$

etc.

Here, too, we have convenient characterizations of the limit sets in terms of the distance function on R^n:

$$\liminf_{y \overset{D}{\to} x} S(y) = \{\xi \in R^n : \limsup_{y \overset{D}{\to} x} d_{S(y)}(\xi) = 0\}$$

and

$$\limsup_{y \overset{D}{\to} x} S(y) = \{\xi \in R^n : \liminf_{y \overset{D}{\to} x} d_{S(y)}(\xi) = 0\}.$$

We observe that $\liminf_{y \overset{D}{\to} x} S(y)$ and $\limsup_{y \overset{D}{\to} x} S(y)$ are closed (possibly empty) sets, related according to

$$\liminf_{y \overset{D}{\to} x} S(y) \subset \limsup_{y \overset{D}{\to} x} S(y).$$

To reconcile these definitions, we note that, given a sequence of sets $\{A_i\}$ in R^n,

$$\limsup A_i = \limsup_{y \xrightarrow{D} x} S(y) \text{ etc.},$$

when we identify D with the subset $\{1, 1/2, 1/3, \ldots\}$ of the real line, choose $x = 0$ and define $S(y) = A_i$ when $y = i^{-1}$, $i = 1, 2, \ldots$.

2.3 Measurable Multifunctions

Take a set Ω. A multifunction $\Gamma : \Omega \rightsquigarrow R^n$ is a mapping from Ω into the space of subsets of R^n. For each $\omega \in \Omega$ then, $\Gamma(\omega)$ is a subset of R^n. We refer to a multifunction as convex, closed, or nonempty depending on whether $\Gamma(\omega)$ has the referred-to property for all $\omega \in \Omega$.

Recall that a measurable space $(\Omega, \mathcal{F})$ comprises a set Ω and a family $\mathcal{F}$ of subsets of Ω which is a σ-field, i.e.,

(i) $\emptyset \in \mathcal{F}$,

(ii) $F \in \mathcal{F}$ implies that $\Omega \backslash F \in \mathcal{F}$, and

(iii) $F_1, F_2, \ldots \in \mathcal{F}$ implies $\cup_{i=1}^{\infty} F_i \in \mathcal{F}$.

Definition 2.3.1 *Let $(\Omega, \mathcal{F})$ be a measurable space. Take a multifunction $\Gamma : \Omega \rightsquigarrow R^n$. Γ is measurable when the set*

$$\{x \in \Omega : \Gamma(x) \cap C \neq \emptyset\}$$

is $\mathcal{F}$ measurable for every open set $C \subset R^n$.

Fix a Lebesgue subset $I \subset R$. Let $\mathcal{L}$ denote the Lebesgue subsets of I. If Ω is the set I then "$\Gamma : I \rightsquigarrow R^n$ is measurable" is taken to mean that the multifunction is $\mathcal{L}$ measurable.

Denote by $\mathcal{B}^k$ the Borel subsets of R^k. The product σ-algebra $\mathcal{L} \times \mathcal{B}^k$ (that is, the smallest σ-algebra of subsets of $I \times R^k$ that contains all product sets $A \times B$ with $A \in \mathcal{L}$ and $B \in \mathcal{B}^k$) is often encountered in hypotheses invoked to guarantee measurability of multifunctions, validity of certain representations for multifunctions, etc.

A first taste of such results is provided by:

Proposition 2.3.2 *Take an $\mathcal{L} \times \mathcal{B}^m$ measurable multifunction $F : I \times R^m \rightsquigarrow R^k$ and a Lebesgue measurable function $u : I \to R^m$. Then $G : I \rightsquigarrow R^k$ defined by*

$$G(t) := F(t, u(t))$$

is an $\mathcal{L}$ measurable multifunction.

Proof. For an arbitrary choice of set $A \in \mathcal{L}$ and $B \in \mathcal{B}^m$, the set

$$\{t \in I : (t, u(t)) \in A \times B\}$$

is a Lebesgue subset because it is expressible as $A \cap u^{-1}(B)$ and u is Lebesgue measurable. Denote by $\mathcal{D}$ the family of subsets $E \subset I \times R^m$ for which the set

$$\{t \in I : (t, u(t)) \in E\}$$

is Lebesgue measurable. $\mathcal{D}$ is a σ-field, as is easily checked. We have shown that it contains all product sets $A \times B$ with $A \in \mathcal{L}$, $B \in \mathcal{B}^m$. So $\mathcal{D}$ contains the σ-field $\mathcal{L} \times \mathcal{B}^m$.

Take any open set $W \subset R^k$. Then, since F is $\mathcal{L} \times \mathcal{B}^m$ measurable, $E := \{(t, u) : F(t, u) \cap W \neq \emptyset\}$ is $\mathcal{L} \times \mathcal{B}^m$ measurable. But then

$$\{t \in I : F(t, u(t)) \cap W \neq \emptyset\} = \{t \in I : (t, u(t)) \in E\}$$

is a Lebesgue measurable set since $E \in \mathcal{D}$. Bearing in mind that W is an arbitrary open set, we conclude that $t \rightsquigarrow G(t) := F(t, u(t))$ is a Lebesgue measurable multifunction. $\square$

Specializing to the point-valued case we obtain:

Corollary 2.3.3 *Consider a function $g : I \times R^m \to R^k$ and a Lebesgue measurable function $u : I \to R^m$. Suppose that g is $\mathcal{L} \times \mathcal{B}^m$ measurable. Then the mapping $t \to g(t, u(t))$ is Lebesgue measurable.*

Functions $g(t, u)$ arising in Optimal Control to which Corollary 2.3.3 are often applied are composite functions of a nature covered by the following proposition.

Proposition 2.3.4 *Consider a function $\phi : I \times R^n \times R^m \to R^k$ satisfying the following hypotheses.*

(a) $\phi(t, . , u)$ is continuous for each $(t, u) \in I \times R^m$;

(b) $\phi(. , x, \cdot)$ is $\mathcal{L} \times \mathcal{B}^m$ measurable for each $x \in R^n$.

Then for any Lebesgue measurable function $x : I \to R^n$, the mapping $(t, u) \to \phi(t, x(t), u)$ is $\mathcal{L} \times \mathcal{B}^m$ measurable.

Proof. Let $\{r_j\}$ be an ordering of the set of n-vectors with rational coefficients. For each integer k define

$$\phi_k(t, u) := \phi(t, r_j, u),$$

where j is chosen (j will depend on k and t) such that

$$|x(t) - r_j| \leq 1/k \quad \text{and} \quad |x(t) - r_i| > 1/k \quad \text{for all } i \in \{1, 2, \ldots, j - 1\}.$$

(These conditions uniquely define j.)

Since $\phi(t, ., u)$ is continuous,

$$\phi_k(t, u) \to \phi(t, x(t), u) \text{ as } k \to \infty$$

for every $(t, u) \in I \times R^m$. It suffices then to show that ϕ_k is $\mathcal{L} \times \mathcal{B}^m$ measurable for an arbitrary choice of k.

For any open set $V \subset R^k$,

$$
\begin{aligned}
\phi_k^{-1}(V) &= \{(t, u) \in I \times R^m : \phi_k(t, u) \in V\} \\
&= \cup_{j=1}^{\infty} \left(\{(t, u) \in I \times R^m : \phi(t, r_j, u) \in V\} \right. \\
&\qquad \cap \ \{(t, u) \in I \times R^m : |x(t) - r_j| \le 1/k \\
&\qquad\qquad \text{and } |x(t) - r_i| > 1/k \text{ for } i = 1, \dots, j-1\}.)
\end{aligned}
$$

Since the set on the right side is a countable union of $\mathcal{L} \times \mathcal{B}^m$ measurable

sets, we have established that ϕ_k is an $\mathcal{L} \times \mathcal{B}^m$ measurable function. $\square$

The $\mathcal{L} \times \mathcal{B}^m$ measurability hypothesis of Proposition 2.3.4 is unrestrictive. It is satisfied, for example, by the Carathéodory functions.

Definition 2.3.5 *A function $g : I \times R^m \to R^k$ is said to be a Carathéodory function if*

(a) $g(., u)$ is Lebesgue measurable for each $u \in R^m$,

(b) $g(t, .)$ is continuous for each $t \in I$.

Proposition 2.3.6 *Consider a function $g : I \times R^m \to R^k$. Assume that g is a Carathéodory function. Then g is $\mathcal{L} \times \mathcal{B}^m$ measurable.*

Proof. Let $\{r_1, r_2, \dots\}$ be an ordering of the set of m-vectors with rational components. For every positive integer k, $t \in I$, and $u \in R^m$, define

$$g_k(t, u) := g(t, r_j),$$

in which the integer j is uniquely defined by the relationships:

$$|r_j - u| \le 1/k \quad \text{and} \quad |r_i - u| > 1/k \quad \text{for } i = 1, \dots, j-1.$$

Since $g(t, .)$ is assumed to be continuous, we have

$$g_k(t, u) \to g(t, u)$$

as $k \to \infty$, for each fixed $(t, u) \in I \times R^m$. It suffices then to show that g_k is $\mathcal{L} \times \mathcal{B}^m$ measurable for each k. However this follows from the fact that, for any open set $V \subset R^k$, we have

$$
\begin{aligned}
g_k^{-1}(V) &= \{(t, u) \in I \times R^m : g_k(t, u) \in V\} \\
&= \cup_{j=1}^{\infty} \{(t, u) \in I \times R^m : g(t, r_j) \in V, |u - r_j| \le 1/k \\
&\qquad \text{and } |u - r_i| > 1/k, \ i = 1, \dots, j-1\} \\
&= \cup_{j=1}^{\infty} (\{t \in I : g(t, r_j) \in V\} \times \{u \in R^m : |u - r_j| \le 1/k, \\
&\qquad \text{and } |u - r_i| > 1/k \text{ for } i = 1, 2, \dots, j-1\})
\end{aligned}
$$

and this last set is $\mathcal{L} \times \mathcal{B}^m$ measurable, since it is expressible as a countable union of sets of the form $A \times B$ with $A \in \mathcal{L}$ and $B \in \mathcal{B}^m$. $\square$

The preceding propositions combine incidentally to provide an answer to the following question concerning an appropriate framework for the formulation of variational problems: under what hypotheses on the function $L : I \times R^n \times R^n \to R$ is the integrand of the Lagrange functional

$$
\int L(t, x(t), \dot{x}(t)) dt
$$

Lebesgue measurable for an arbitrary absolutely continuous arc $x \in W^{1,1}$? The two preceding propositions guarantee that the integrand is Lebesgue measurable if

(i) $L(., x, .)$ is $\mathcal{L} \times \mathcal{B}^n$ measurable for each $x \in R^n$ and

(ii) $L(t, ., u)$ is continuous for each $(t, u) \in I \times R^n$.

This is because Proposition 2.3.4 tells us that $(t, u) \to L(t, x(t), u)$ is $\mathcal{L} \times \mathcal{B}^n$ measurable, in view of assumptions (i) and (ii), and since $x : I \to R^n$ is Lebesgue measurable. Corollary 2.3.3 permits us to conclude, since $t \to \dot{x}(t)$ is Lebesgue measurable, that $t \to L(t, x(t), \dot{x}(t))$ is indeed Lebesgue measurable.

The following theorem, a proof of which is to be found in [27], lists important characterizations of closed multifunctions that are measurable. (Throughout, I is a Lebesgue subset of R.)

Theorem 2.3.7 *Take a multifunction* $\Gamma : I \rightsquigarrow R^n$ *and define* $D := \{t \in I : \Gamma(t) \ne \emptyset\}$. *Assume that* Γ *is closed. Then the following statements are equivalent.*

(a) Γ is an $\mathcal{L}$ measurable multifunction.

(b) Gr Γ is an $\mathcal{L} \times \mathcal{B}^n$ measurable set.

(c) D is a Lebesgue subset of I and there exists a sequence $\{\gamma_k : D \to R^n\}$ of Lebesgue measurable functions such that

$$\Gamma(t) = \overline{\cup_{k=1}^{\infty}\{\gamma_k(t)\}} \quad \text{for all } t \in D. \tag{2.2}$$

The representation of a multifunction in terms of a countable family of Lebesgue measurable functions according to (2.2) is called the *Castaing Representation* of Γ.

Our aim now is to establish the measurability of a number of frequently encountered multifunctions derived from other multifunctions.

Proposition 2.3.8 *Take a measurable space $(\Omega, \mathcal{F})$ and a measurable multifunction $\Gamma : \Omega \rightsquigarrow R^n$. Then the multifunction $\tilde{\Gamma} : \Omega \rightsquigarrow R^n$ is also measurable in each of the following cases:*

(i) $\tilde{\Gamma}(y) := \overline{\Gamma(y)}$ for all $y \in \Omega$.

(ii) $\tilde{\Gamma}(y) := \operatorname{co}\Gamma(y)$ for all $y \in \Omega$.

Proof.

(i): $\tilde{\Gamma}$ is measurable in this case since, for any open set $W \in R^n$,

$$\{y \in \Omega : \overline{\Gamma(y)} \cap W \neq \emptyset\} = \{y \in \Omega : \Gamma(y) \cap W \neq \emptyset\}.$$

(ii): Define the multifunction $\Gamma^{(n+1)} : \Omega \rightsquigarrow R^{n \times (n+1)}$ to be

$$\Gamma^{(n+1)}(y) := \Gamma(y) \times \ldots \times \Gamma(y) \quad \text{for all } y \in \Omega.$$

Then $\Gamma^{(n+1)}$ is measurable, since $\{y \in \Omega : \Gamma^{(n+1)}(y) \cap W \neq \emptyset\}$ is obviously measurable for any set W in $R^{n \times (n+1)}$ that is a product of open sets of R^n, and therefore for any open set W, since an arbitrary open set in the product space can be expressed as a countable union of such sets.

Define also

$$\Lambda := \left\{ (\lambda_0, \ldots, \lambda_n) : \lambda_i \geq 0 \text{ for all } i, \sum_{i=0}^{n} \lambda_i = 1 \right\}.$$

Take any open set W in R^n and define

$$W^{(n+1)} := \left\{ (w_0, \ldots, w_n) : \sum_i \lambda_i w_i \in W, (\lambda_0, \ldots, \lambda_n) \in \Lambda \right\}.$$

Obviously, $W^{(n+1)}$ is an open set.

We must show that the set

$$\{y \in \Omega : \operatorname{co}\Gamma(y) \cap W \neq \emptyset\}$$

is measurable. But this follows immediately from the facts that $\tilde{\Gamma}^{(n+1)}$ is measurable and $W^{(n+1)}$ is open, since

$$\{y : \operatorname{co}\Gamma(y) \cap W \neq \emptyset\} = \{y : \Gamma^{(n+1)}(y) \cap W^{(n+1)} \neq \emptyset\}.$$

$\square$

The next proposition concerns the measurability properties of limits of sequences of multifunctions. We make reference to Kuratowski sense limit operations, defined in Section 2.2.

Theorem 2.3.9 *Consider closed multifunctions $\Gamma_j : I \rightsquigarrow R^n$, $j = 1, 2, \ldots$ Assume that Γ_j is $\mathcal{L}$ measurable for each j. Then the closed multifunction $\Gamma : I \rightsquigarrow R^n$ is also $\mathcal{L}$ measurable when Γ is defined in each of the following ways.*

(a) $\Gamma(t) := \overline{\cup_{j \geq 1} \Gamma_j(t)}$;

(b) $\Gamma(t) := \cap_{j \geq 1} \Gamma_j(t)$;

(c) $\Gamma(t) := \limsup_{j \to \infty} \Gamma_j(t)$;

(d) $\Gamma(t) := \liminf_{j \to \infty} \Gamma_j(t)$.

(c) and (d) imply in particular that if $\{\Gamma_j\}$ has a limit as $j \to \infty$, then $\lim_{j \to \infty} \Gamma_j$ is measurable.

Proof. (a): ($\Gamma(t) = \overline{\cup_{j \geq 1} \Gamma_j(t)}$)

Take any open set $W \subset R^n$. Then, since W is an open set, we have

$$
\begin{aligned}
\{t \in I : \Gamma(t) \cap W \neq \emptyset\} &= \{t \in I : \overline{\cup_{j \geq 1} \Gamma_j(t)} \cap W \neq \emptyset\} \\
&= \{t \in I : \cup_{j \geq 1} \Gamma_j(t) \cap W \neq \emptyset\} \\
&= \cup_{j \geq 1} \{t \in I : \Gamma_j(t) \cap W \neq \emptyset\}.
\end{aligned}
$$

This establishes the measurability of $t \rightsquigarrow \Gamma(t)$ since the set on the right side, a countable union of Lebesgue measurable sets, is Lebesgue measurable.

(b): ($\Gamma(t) = \cap_{j \geq 1} \Gamma_j(t)$)

In this case,

$$\operatorname{Gr}\Gamma = \operatorname{Gr}\{t \rightsquigarrow \cap_{j \geq 1} \Gamma_j(t)\} = \cap_{j \geq 1} \operatorname{Gr}\Gamma_j.$$

$\operatorname{Gr}\Gamma$ then is $\mathcal{L} \times \mathcal{B}^n$ measurable since each $\operatorname{Gr}\Gamma_j$ is $\mathcal{L} \times \mathcal{B}^n$ measurable by Theorem 2.3.7. Now apply again Theorem 2.3.7.

(c): $(\Gamma(t) = \limsup_{j\to\infty} \Gamma_j(t))$

The measurablility of $t \rightsquigarrow \Gamma(t)$ in this case follows from (a) and (b) and the following characterization of $\limsup_{j\to\infty} \Gamma_j(t)$.

$$\limsup_{j\to\infty} \Gamma_j(t) = \cap_{J\geq 1}\overline{\cup_{j\geq J}\Gamma_j(t)}.$$

(d): $(\Gamma(t) = \liminf_{j\to\infty} \Gamma_j(t))$

Define

$$\Gamma_j^k(t) := \Gamma_j(t) + (1/k)B.$$

Notice that Γ_j^k is measurable since, for any closed set $W \subset R^n$,

$$\{t : \Gamma_j^k(t) \cap W \neq \emptyset\} = \{t : \Gamma_j(t) \cap (W + (1/k)B) \neq \emptyset\}$$

and the set $W + (1/k)B$ is closed.

The measurability of Γ in this case too follows from (a) and (b) in view of the identity

$$\liminf_{j\to\infty} \Gamma_j(t) = \cap_{k\geq 1}\overline{\cup_{J\geq 1} \cap_{j\geq J} \Gamma_j^k(t)}.$$

$\square$

Proposition 2.3.10 *Take a multifunction* $F : I \times R^n \rightsquigarrow R^k$ *and a Lebesgue measurable function* $\bar{x} : I \to R^n$. *Assume that*

(a) *for each* x, $F(.,x) : I \rightsquigarrow R^k$ *is an* $\mathcal{L}$ *measurable, nonempty, closed multifunction;*

(b) *for each* t, $F(t,.)$ *is continuous at* $x = \bar{x}(t)$, *in the sense that*

$$y_i \to \bar{x}(t) \quad \text{implies} \quad F(t, \bar{x}(t)) = \lim_{i\to\infty} F(t, y_i).$$

Then $G : I \rightsquigarrow R^k$ *defined by*

$$G(t) := F(t, \bar{x}(t))$$

is a closed $\mathcal{L}$ *measurable multifunction.*

Proof. Let $\{r_i\}$ be an ordering of n-vectors with rational entries. For each integer l and for each $t \in I$, define

$$F_l(t) := F(t, r_j),$$

in which j is chosen according to the rule

$$|\bar{x}(t) - r_j| \leq 1/l \quad \text{and} \quad |\bar{x}(t) - r_i| > 1/l \quad \text{for } i = 1, \ldots, j - 1.$$

In view of the continuity properties of $F(t,.)$,

$$F(t, \bar{x}(t)) = \lim_{l \to \infty} F_l(t).$$

By Theorem 2.3.9 then it suffices to show that F_l is measurable for arbitrary l.

We observe however that

$$\text{Gr } F_l = \cup_j (\text{Gr } F(., r_j) \cap (A_j \times R^k)),$$

where

$$A_j := \{t : |\bar{x}(t) - r_j| \leq 1/l, |\bar{x}(t) - r_i| > 1/l \quad \text{for } i = 1, \ldots, j-1\}.$$

Since $F(., r_j)$ has an $\mathcal{L} \times \mathcal{B}^k$ measurable graph (see Theorem 2.3.7) and $\bar{x}$ is Lebesgue measurable, we see that Gr F_l is $\mathcal{L} \times \mathcal{B}^k$ measurable. Applying Theorem 2.3.7 again, we see that the closed multifunction F_l is measurable. $\square$

Take a multifunction $\Gamma : I \rightsquigarrow R^k$. We say that a function $x : I \to R^k$ is a *measurable selection* for Γ if

(i) x is Lebesgue measurable, and

(ii) $x(t) \in \Gamma(t)$ a.e.

We obtain directly from Theorem 2.3.7 the following conditions for Γ to have a measurable selection.

Theorem 2.3.11 *Let $\Gamma : I \rightsquigarrow R^k$ be a nonempty multifunction. Assume that Γ is closed and measurable. Then Γ has a measurable selection.*

(In fact Theorem 2.3.7 tells us rather more than this: not only does there exist a measurable selection under the stated hypotheses, but the measurable selections are sufficiently numerous to "fill out" the values of the multifunction.)

The above measurable selection theorem is inadequate for certain applications in which the multifunction is not closed. An important extension (see [27], [146]) is

Theorem 2.3.12 (Aumann's Measurable Selection Theorem) *Let $\Gamma : I \rightsquigarrow R^k$ be a nonempty multifunction. Assume that*

$$Gr \, \Gamma \text{ is } \mathcal{L} \times \mathcal{B}^k \text{ measurable.}$$

Then Γ has a measurable selection.

This can be regarded as a generalization of Theorem 2.3.11 since if Γ is closed and measurable then, by Theorem 2.3.7, $\mathrm{Gr}\,\Gamma$ is automatically $\mathcal{L} \times \mathcal{B}^k$ measurable.

Of particular significance in applications to Optimal Control is the following measurable selection theorem involving the composition of a function and a multifunction.

Theorem 2.3.13 (The Generalized Filippov Selection Theorem)
Consider a nonempty multifunction $U : I \rightsquigarrow R^m$ and a function $g : I \times R^m \to R^n$ satisfying

(a) the set $\mathrm{Gr}\,U$ is $\mathcal{L} \times \mathcal{B}^m$ measurable;

(b) the function g is $\mathcal{L} \times \mathcal{B}^m$ measurable.

Then for any measurable function $v : I \to R^n$, the multifunction $U' : I \rightsquigarrow R^m$ defined by

$$U'(t) := \{u \in U(t) : g(t, u) = v(t)\}$$

has an $\mathcal{L} \times \mathcal{B}^m$ measurable graph. Furthermore, if

$$v(t) \in \{g(t, u) : u \in U(t)\} \qquad a.e. \tag{2.3}$$

then there exists a measurable function $u : I \to R^m$ satisfying

$$u(t) \in U(t) \qquad a.e. \tag{2.4}$$

$$g(t, u(t)) = v(t) \qquad a.e. \tag{2.5}$$

Notice that Condition (2.3) is just a rephrasing of the hypothesis

$$U'(t) \text{ is nonempty for a.e. } t \in I$$

and the final assertion can be expressed in measurable selection terms as

the multifunction U' has a measurable selection,

a fact that follows directly from Aumann's Selection Theorem.

We mention that the name Filippov's Selection Theorem usually attaches to the final assertion of the theorem concerning existence of a measurable function $u : I \to R^m$ satisfying (2.4) and (2.5) under (2.3) and strengthened forms of Hypotheses (a) and (b), namely,

(a)$'$ U is a closed measurable multifunction, and

(b)$'$ g is a Carathéodory function.

Proof. By redefining $U'(t)$ on a nullset if required, we can arrange that $U'(t)$ is nonempty for all $t \in I$. In view of the preceding discussion it is required to show merely that Gr U' is $\mathcal{L} \times \mathcal{B}^m$ measurable. But this follows directly from the relationship

$$\text{Gr } U' = \phi^{-1}(\{0\}) \cap \text{Gr } U$$

in which $\phi(t, u) := g(t, u) - v(t)$, since under the hypotheses both $\phi^{-1}(\{0\})$ and Gr U are $\mathcal{L} \times \mathcal{B}^m$ measurable sets. $\square$

The relevance of Filippov's Theorem in a Control Systems context is illustrated by the following application. Take a function $f : [S, T] \times R^n \times R^m \to R^n$ and a multifunction $U : [S, T] \rightsquigarrow R^m$. The class of state trajectories for the control system

$$\begin{aligned} \dot{x}(t) &= f(t, x(t), u(t)) \quad \text{a.e. } t \in [S, T] \\ u(t) &\in U(t) \quad \text{a.e. } t \in [S, T] \end{aligned}$$

comprises every absolutely continuous function $x : [S, T] \to R^n$ for some measurable $u : [S, T] \to R^m$. It is often desirable to interpret the state trajectories as solutions of the differential inclusion $\dot{x}(t) \in F(t, x(t))$ with

$$F(t, x) := \{f(t, x, u) : u \in U(t)\}.$$

The question then arises whether the state trajectories for the control system are *precisely* the absolutely continuous functions x satisfying

$$\dot{x}(t) \in F(t, x(t)) \qquad \text{a.e.} \tag{2.6}$$

Clearly a necessary condition for an absolutely continuous function to be a state trajectory for the control system is that (2.6) is satisfied. Filippov's Theorem tells us that (2.6) is also a sufficient condition (i.e., the differential inclusion provides an equivalent description of state trajectories) under the hypotheses:

(i) $f(., x, .)$ is $\mathcal{L} \times \mathcal{B}^m$ measurable and $f(t, ., u)$ is continuous;

(ii) Gr U is $\mathcal{L} \times \mathcal{B}^m$ measurable.

To see this, apply the Generalized Filippov Selection Theorem with $g(t, u) = f(t, x(t), u)$ and $v = \dot{x}$. (The relevant hypotheses, (b) and (2.3), are satisfied in view of Proposition 2.3.4 and since $\dot{x}(t) \in F(t, x(t))$ a.e.) This yields a measurable function $u : [S, T] \to R^m$ satisfying $\dot{x}(t) = f(t, x(t), u(t))$ and $u(t) \in U(t)$ a.e. thereby confirming that x is a state trajectory for the control system.

Having once again an eye for future Optimal Control applications, we now establish measurability of various derived functions and the existence of measurable selections for related multifunctions. The source of a number of useful results is the following theorem concerning the measurability of a "marginal" function.

Theorem 2.3.14 *Consider a function* $g : I \times R^k \to R$ *and a closed nonempty multifunction* $\Gamma : I \rightsquigarrow R^k$. *Assume that*

(a) g *is a Carathéodory function;*

(b) Γ *is a measurable multifunction.*

Define the extended-valued function $\eta : I \to R \cup \{-\infty\}$

$$\eta(t) = \inf_{\gamma \in \Gamma(t)} g(t, \gamma) \qquad for \ t \in I.$$

Then η *is a Lebesgue measurable function. Furthermore, if we define*

$$I' := \{t \in I : \inf_{\gamma' \in \Gamma(t)} g(t, \gamma') = g(t, \gamma) \ for \ some \ \gamma \in \Gamma(t)\}$$

(i.e., I' *is the set of points* t *for which the infimum of* $g(t, .)$ *is achieved over* $\Gamma(t)$*) then* I' *is a Lebesgue measurable set and there exists a measurable function* $\gamma : I' \to R^k$ *such that*

$$\eta(t) = g(t, \gamma(t)) \qquad a.e. \ t \in I'. \tag{2.7}$$

Proof. Since Γ is closed, nonempty, and measurable, it has a Castaing representation in terms of some countable family of measurable functions $\{\gamma_i : I \to R^k\}$. Since $g(t, .)$ is continuous and $\{\gamma_i(t)\}$ is dense in $\Gamma(t)$,

$$\eta(t) = \inf\{g(t, \gamma_i(t)) : i \ \text{an integer}\}$$

for all $t \in I$.

Now according to Proposition 2.3.6 and Corollary 2.3.3, $t \to g(t, \gamma_i(t))$ is a measurable function. It follows from a well-known property of measurable functions that η, which we have expressed as the pointwise infimum of a countable family of measurable functions, is Lebesgue measurable and $\mathrm{dom}\{\eta\} := \{t \in I : \eta(t) > -\infty\}$ is a Lebesgue measurable set.

Now apply the Generalized Filippov Selection Theorem (identifying η with v and Γ with U, and replacing I by $\mathrm{dom}\{\eta\}$). If $I' = \emptyset$, there is nothing to prove. Otherwise, since

$$I' = \{t \in \mathrm{dom}\{\eta\} : \{\gamma \in \Gamma(t) : g(t, \gamma) = \eta(t)\} \neq \emptyset\},$$

I' is a nonempty, Lebesgue measurable set and there exists a measurable function $\gamma : I' \to R^k$ such that

$$\eta(t) = g(t, \gamma(t)) \qquad a.e. \ t \in I'.$$

$\square$

2.4   Existence and Estimation of F-Trajectories

Fix an interval $[S, T]$ and a relatively open set $\Omega \subset [S, T] \times R^n$. For $t \in [S, T]$, define

$$\Omega_t := \{x : (t, x) \in \Omega\}.$$

Take a continuous function $y : [S, T] \to R^n$ and $\epsilon > 0$. Then the ϵ tube about y is the set

$$T(y, \epsilon) := \{(t, x) \in [S, T] \times R^n : t \in [S, T], |x - y(t)| \leq \epsilon\}.$$

Consider a multifunction $F : \Omega \leadsto R^n$. An arc $x \in W^{1,1}([S, T]; R^n)$ is an F-trajectory if $\mathrm{Gr}\, x \in \Omega$ and

$$\dot{x}(t) \in F(t, x(t)) \quad \text{a.e. } t \in [S, T].$$

Naturally we would like to know when F-trajectories exist. We make extensive use of a local existence theorem that gives conditions under which an F-trajectory exists near a nominal arc $y \in W^{1,1}([S, T]; R^n)$. This theorem provides important supplementary information about how "close" to y the F-trajectory x may be chosen. Just how close will depend on the extent to which the nominal arc y fails to satisfy the differential inclusion, as measured by the function Λ_F,

$$\Lambda_F(y) := \int_S^T \rho_F(t, y(t), \dot{y}(t)) dt.$$

Here

$$\rho_F(t, x, v) := \inf\{|\eta - v| : \eta \in F(t, x)\}.$$

(We make use of function $\Lambda_F(y)$ only when the integrand above is Lebesgue measurable and F is a closed multifunction; then $\Lambda_F(y)$ is a nonnegative number that is zero if and only if y is an F-trajectory.)

We pause for a moment however to list some relevant properties of ρ_F that, among other things, give conditions under which the integral $\Lambda_F(y)$ is well-defined.

Proposition 2.4.1 *Take a multifunction $F : [S, T] \times R^n \leadsto R^n$.*

(a) Fix $(t, x) \in \mathrm{dom}\, F$. Then

$$|\rho_F(t, x, v) - \rho_F(t, x, v')| \leq |v - v'| \quad \text{for all } v, v' \in R^n.$$

(b) Fix $t \in [S, T]$. Suppose that, for some $\epsilon > 0$ and $k > 0$, $\bar{x} + \epsilon B \subset \mathrm{dom}\, F$ and

$$F(t, x) \subset F(t, x') + k|x - x'|B \quad \text{for all } x, x' \in \bar{x} + \epsilon B.$$

Then

$$|\rho_F(t, x, v) - \rho_F(t, x', v')| \leq |v - v'| + k|x - x'|$$

for all $v, v' \in R^n$ and $x, x' \in \bar{x} + \epsilon B$.

(c) *Assume that F is $\mathcal{L} \times \mathcal{B}^n$ measurable. Then for any Lebesgue measurable functions $y : [S,T] \to R^n$ and $v : [S,T] \to R^n$ such that $\mathrm{Gr}\, y \subset \mathrm{dom}\, F$, we have that $t \to \rho_F(t, y(t), v(t))$ is a Lebesgue measurable function on $[S,T]$.*

Proof.

(a) Choose any $\epsilon > 0$ and v, $v' \in R^n$. Since $F(t,x) \neq \emptyset$, there exists $\eta \in F(t,x)$ such that

$$
\begin{aligned}
\rho_F(t,x,v) \quad &\geq \quad |v - \eta| - \epsilon \\
&\geq \quad |v' - \eta| - |v - v'| - \epsilon \\
&\geq \quad \rho_F(t,x,v') - |v - v'| - \epsilon \,.
\end{aligned}
$$

(The second line follows from the triangle inequality.) Since $\epsilon > 0$ is arbitrary, and v and v' are interchangeable, it follows that

$$
|\rho_F(t,x,v') - \rho_F(t,x,v')| \leq |v - v'|.
$$

(b) Choose any x, $x' \in \bar{x}B$ and v, $v' \in R^n$. Take any δ. Since $F(t,x') \neq \emptyset$, there exists $\eta' \in F(t,x')$ such that

$$
\rho_F(t,x',v') > |v' - \eta'| - \delta.
$$

Under the hypotheses, there exists $\eta \in F(t,x)$ such that $|\eta - \eta'| \leq k|x - x'|$. Of course, $\rho_F(t,x,v) \leq |v - \eta|$. It follows from these relationships and the triangle inequality that

$$
\begin{aligned}
\rho_F(t,x,v) - &\rho_F(t,x',v') \\
&\leq \quad |v - \eta| - |v' - \eta'| + \delta \leq |v - v'| + |v' - \eta| - |v' - \eta'| + \delta \\
&\leq \quad |v - v'| + |\eta - \eta'| + \delta \leq |v - v'| + k|x - x'| + \delta.
\end{aligned}
$$

Since the roles of (x,v) and (x',v') are interchangeable and $\delta > 0$ is arbitrary, we conclude that

$$
|\rho_F(t,x,v) - \rho_F(t,x',v')| \leq |v - v'| + k|x - x'|.
$$

(c) For each v, the function $t \to \rho_F(t, y(t), v)$ is measurable. This follows from the identity

$$
\{t \in [S,T] : \rho_F(t, y(t), v) < \alpha\} = \{t : F(t, y(t)) \cap (v + \alpha \,\mathrm{int}\, B) \neq \emptyset\} \,,
$$

valid for any $\alpha \in R$, and the fact that $t \rightsquigarrow F(t, y(t))$ is measurable (see Proposition 2.3.2).

By Part (a) of the lemma, the function $\rho_F(t, y(t), .)$ is continuous, for each $t \in [S,T]$. It follows that $(t,v) \to \rho_F(t, y(t), v)$ is a Carathéodory function. But then $t \to \rho_F(t, y(t), v(t))$ is (Lebesgue) measurable on $[S, T]$, by Proposition 2.3.6 and Corollary 2.3.3. $\square$

The following proposition, concerning regularity properties of projections onto continuous convex multifunctions, is also required.

Proposition 2.4.2 *Take a continuous multifunction* $\Gamma : [S, T] \rightsquigarrow R^n$ *and a function* $u : [S, T] \to R^n$. *Assume that*

(i) u *is continuous,*

(ii) $\Gamma(t)$ *is nonempty, compact, and convex for each* $t \in [S, T]$, *and* Γ *is continuous; i.e., there exists* $o : R^+ \to R^+$ *with* $\lim_{\alpha \downarrow 0} o(\alpha) = 0$ *such that*
$$\Gamma(s) \subset \Gamma(t) + o(|t - s|)B \quad \text{for all } t, s \in [S, T].$$

Let $\hat{u} : [S, T] \to R^n$ *be the function defined according to*
$$|u(t) - \hat{u}(t)| = \min_{u' \in \Gamma(t)} |u(t) - u'| \quad \text{for all } t.$$

(There is a unique minimizer for each t since $\Gamma(t)$ is nonempty, closed, and convex.)
Then $\hat{u}$ is a continuous function.

Proof. Suppose that $\hat{u}$ is not continuous. Then there exist $\epsilon > 0$ and sequences $\{s_i\}$ and $\{t_i\}$ in $[S, T]$ such that $|t_i - s_i| \to 0$ and
$$|\hat{u}(t_i) - \hat{u}(s_i)| > \epsilon, \quad \text{for all } i. \tag{2.8}$$

Since the multifunction Γ is compact-valued and continuous and has bounded domain, $\mathrm{Gr}\, \Gamma$ is bounded. It follows that, for some $K > 0$,
$$u(s_i),\, u(t_i),\, \hat{u}(s_i),\, \hat{u}(t_i) \in KB.$$

Since Γ is continuous, there exist $\{y_i\}$ and $\{z_i\}$ such that
$$y_i \in \Gamma(t_i),\, z_i \in \Gamma(s_i) \quad \text{for all } i \tag{2.9}$$

and
$$|y_i - \hat{u}(s_i)| \to 0,\, |z_i - \hat{u}(t_i)| \to 0 \quad \text{as } i \to \infty. \tag{2.10}$$

Since Γ is a convex multifunction, $u(s_i) - \hat{u}(s_i)$ and $u(t_i) - \hat{u}(t_i)$ are normal vectors to $\Gamma(s_i)$ and $\Gamma(t_i)$, respectively. Noting (2.9), we deduce from the normal inequality for convex sets that
$$(u(s_i) - \hat{u}(s_i)) \cdot (z_i - \hat{u}(s_i)) \leq 0$$
$$(u(t_i) - \hat{u}(t_i)) \cdot (y_i - \hat{u}(t_i)) \leq 0.$$

It follows that

$$(u(s_i) - \hat{u}(s_i)) \cdot (\hat{u}(t_i) - \hat{u}(s_i)) \leq (u(s_i) - \hat{u}(s_i)) \cdot (\hat{u}(t_i) - z_i)$$
$$(u(t_i) - \hat{u}(t_i)) \cdot (\hat{u}(s_i) - \hat{u}(t_i)) \leq (u(t_i) - \hat{u}(t_i)) \cdot (\hat{u}(s_i) - y_i).$$

Adding these inequalities gives

$$(u(s_i) - u(t_i)) \cdot (\hat{u}(t_i) - \hat{u}(s_i)) + |\hat{u}(t_i) - \hat{u}(s_i)|^2$$
$$\leq (u(s_i) - \hat{u}(s_i)) \cdot (\hat{u}(t_i) - z_i) + (u(t_i) - \hat{u}(t_i)) \cdot (\hat{u}(s_i) - y_i).$$

We deduce that

$$|\hat{u}(t_i) - \hat{u}(s_i)|^2 \leq |u(s_i) - u(t_i)| \cdot |\hat{u}(t_i) - \hat{u}(s_i)|$$
$$+ |u(s_i) - \hat{u}(s_i)| \cdot |\hat{u}(t_i) - z_i| + |u(t_i) - \hat{u}(t_i)| \cdot |\hat{u}(s_i) - y_i|.$$

It follows that

$$|\hat{u}(t_i) - \hat{u}(s_i)|^2 \leq 2K(|u(s_i) - u(t_i)| + |\hat{u}(t_i) - z_i| + |\hat{u}(s_i) - y_i|).$$

But the right side has limit 0 as $i \to \infty$, since u is continuous, and by (2.10). We conclude that $|\hat{u}(t_i) - \hat{u}(s_i)| \to 0$ as $i \to \infty$. We have arrived at a contradiction of (2.8). $\hat{u}$ must therefore be continuous. $\square$

We refer to the "truncation" function $tr_\epsilon : R^n \to R^n$, defined to be

$$tr_\epsilon(\xi) := \begin{cases} \xi & \text{if } |\xi| \leq \epsilon \\ \epsilon|\xi|^{-1}\xi & \text{if } |\xi| > \epsilon \,. \end{cases} \tag{2.11}$$

Recall that $T(y, \epsilon)$ denotes the ϵ tube about the arc y:

$$T(y, \epsilon) := \{(t, x) \in [S, T] \times R^n : t \in [S, T], |x - y(t)| \leq \epsilon\}.$$

Theorem 2.4.3 (Generalized Filippov Existence Theorem) *Let Ω be a relatively open set in $[S, T] \times R^n$. Take a multifunction $F : \Omega \rightsquigarrow R^n$, an arc $y \in W^{1,1}([S, T]; R^n)$, a point $\xi \in R^n$, and $\epsilon \in (0, +\infty) \cup \{+\infty\}$ such that $T(y, \epsilon) \subset \Omega$. Assume that*

(i) *$F(t, x')$ is a closed nonempty set for all $(t, x') \in T(y, \epsilon)$, and F is $\mathcal{L} \times \mathcal{B}^n$ measurable;*

(ii) *there exists $k \in L^1$ such that*

$$F(t, x') \subset F(t, x'') + k(t)|x' - x''|B \tag{2.12}$$

for all $x', x'' \in y(t) + \epsilon B$, a.e. $t \in [S, T]$.

Assume further that

$$K\left(|\xi - y(S)| + \int_S^T \rho_F(t, y(t), \dot{y}(t))dt\right) \le \epsilon, \qquad (2.13)$$

where $K := \exp\left(\int_S^T k(t)dt\right)$.

Then there exists an F-trajectory x satisfying $x(S) = \xi$ such that

$$
\begin{aligned}
||x - y||_{L^\infty} &\le |x(S) - y(S)| + \int_S^T |\dot{x}(t) - \dot{y}(t)|\, dt \\
&\le K\left(|\xi - y(S)| + \int_S^T \rho_F(t, y(t), \dot{y}(t))dt\right). \quad (2.14)
\end{aligned}
$$

Now suppose that (i) and (ii) are replaced by the stronger hypotheses

(i)′ $F(t, x')$ is a nonempty, compact, convex set for all $(t, x') \in T(y, \epsilon)$;

(ii)′ there exists a function $o(.) : R^+ \to R^+$ and $k_\infty > 0$ such that $\lim_{\alpha \downarrow 0} o(\alpha) = 0$ and

$$F(s', x') \subset F(s'', x'') + k_\infty|x' - x''|B + o(|s' - s''|)B \qquad (2.15)$$

for all $(s', x'), (s'', x'') \in T(y, \epsilon)$.

Then, if y is continuously differentiable, x can be chosen also to be continuously differentiable.

(If $\epsilon = +\infty$ then in the above hypotheses $T(y, \epsilon)$ and ϵB are interpreted as $[S, T] \times R^n$ and R^n, respectively; the left side of Condition (2.13) is required to be finite.)

Remarks

(i) The hypotheses invoked in the first part of Theorem 2.4.3 do not require F to be convex-valued. For many developments in Optimal Control the requirement that F is convex is crucial; fortunately, proving this basic existence theorem is not one of them.

(ii) The proof of Theorem 2.4.3 is by construction. The iterative procedure used is a generalization to differential inclusions of the well-known Picard iteration scheme for obtaining a solution to the differential equation

$$\dot{x}(t) = f(t, x(t)), \ x(S) = \xi.$$

An initial guess y at a solution is made. It is then improved by "successive approximations" $x_0, x_1, x_2, \ldots$. These arcs are generated by the recursive equations

$$x_{i+1}(t) = \xi + \int_S^t f(s, x_i(s))ds$$

with starting condition $x_0(t) = y(t) + (\xi - y(S))$.

Proof. We may assume without loss of generality that $\epsilon = \infty$. Indeed if $\epsilon = \bar{\epsilon}$ for some finite $\bar{\epsilon}$ then we consider $\tilde{F}$ in place of F, where

$$\tilde{F}(t, x) := F(t, y(t) + tr_{\bar{\epsilon}}(x - y(t))).$$

(See (2.11) for the definition of $tr_{\bar{\epsilon}}$.)

$\tilde{F}$ satisfies the hypotheses (in relation to y) with $\epsilon = \infty$ and with the same $k \in L^1$ as before. Of course

$$\tilde{F}(t, x) = F(t, x) \quad \text{for } x \in y(t) + \bar{\epsilon}B.$$

Now apply the $\epsilon = +\infty$ case of the theorem to $\tilde{F}$. This gives an $\tilde{F}$-trajectory x such that $x(S) = \xi$ and (2.14) is satisfied (when $\tilde{F}$ replaces F). If, however,

$$K\left(|\xi - y(S)| + \int_S^T \rho_F(t, y(t), \dot{y}(t))dt\right) \le \bar{\epsilon},$$

then the theorem tell us that

$$||x - y||_{L^\infty} \le \bar{\epsilon},$$

and therefore x is an F-trajectory because $F(t, .)$ and $\tilde{F}(t, .)$ coincide on $y(t) + \bar{\epsilon}B$. This justifies setting $\epsilon = +\infty$.

It suffices to consider only the case $\xi = y(S)$. To show this, suppose that $\xi \neq y(S)$. Replace the underlying time interval $[S, T]$ by $[S - 1, T]$ and ξ by $\hat{\xi} = y(S)$. Replace also F by $\tilde{F} : [S - 1, T] \times R^n \rightsquigarrow R^n$ and y by $\tilde{y} : [S - 1, T] \to R^n$, defined as follows.

$$\tilde{F}(t, x) := \begin{cases} F(t, x) & \text{for } (t, x) \in [S, T] \times R^n \\ \{\xi - y(S)\} & \text{for } (t, x) \in [S - 1, S) \times R^n \end{cases}$$

and

$$\tilde{y}(t) := \begin{cases} y(t) & \text{for } t \ge S \\ y(S) & \text{for } t < S. \end{cases}$$

Now apply the special case of the theorem to find $\tilde{x} \in W^{1,1}([S-1, T]; R^n)$ such that $\tilde{x}(S - 1) = \tilde{y}(S - 1) = \hat{\xi} \equiv y(S)$. Take x to be the restriction of $\tilde{x}$ to $[S, T]$. We readily deduce that

$$||\dot{\tilde{x}} - \dot{\tilde{y}}||_{L^1([S-1; T]; R^n)} \le K \int_{S-1}^T \rho_{\tilde{F}}(t, \tilde{y}(t), \dot{\tilde{y}}(t))dt.$$

Now, to derive the desired estimate, we have merely to note that

$$||\dot{\tilde{x}} - \dot{\tilde{y}}||_{L^1([S-1, T]; R^n)} = |\xi - y(S)| + ||\dot{x} - \dot{y}||_{L^1([S, T]; R^n)}$$

and

$$\int_{S-1}^{T} \rho_{\tilde{F}}(t, \tilde{y}(t), \dot{\tilde{y}}(t))dt = |\xi - y(s)| + \int_{S}^{T} \rho_{F}(t, y(t), \dot{y}(t))dt.$$

Henceforth, then, we assume that $\epsilon = +\infty$ and $y(S) = \xi$; we must find an F-trajectory x such that $x(S) = \xi$ and

$$||\dot{x} - \dot{y}||_{L^1([S,T]:R^n)} \leq K \int_{S}^{T} \rho_F(t, y(t), \dot{y}(t))dt.$$

Write $x_0(t) = y(t)$. According to Theorem 2.3.11, we may choose a measurable function v_1 satisfying

$$v_1(t) \in F(t, x_0(t)) \quad \text{a.e. } t \in [S, T]$$

and

$$\rho_F(t, x_0(t), \dot{x}_0(t)) = |v_1(t) - \dot{x}_0(t)| \quad \text{a.e. } t \in [S, T].$$

This is because

$$t \rightsquigarrow G(t) := \{v \in F(t, x_0(t)) : \rho_F(t, x_0(t), \dot{x}_0(t)) = |v - \dot{x}_0(t)|\}$$

is a closed, nonempty, measurable multifunction. (We use Proposition 2.3.2 and the fact that

$$(t, v) \rightarrow g(t, v) := \rho_F(t, x_0(t), \dot{x}_0(t)) - |v - \dot{x}_0(t)|$$

is a Carathéodory function.) Under the hypotheses, $t \rightarrow \rho_F(t, x_0(t), \dot{x}_0(t))$ is integrable. Since $\dot{x}_0$ is integrable, v_1 is integrable too. We may therefore define x_1 according to

$$x_1(t) := y(S) + \int_{S}^{t} v_1(s)ds.$$

Note that

$$\rho_F(t, x_0(t), \dot{x}(t)) = 0 \quad \text{a.e.}$$

Again appealing to Theorem 2.3.11, we choose a measurable function v_2 satisfying

$$v_2(t) \in F(t, x_1(t)) \quad \text{a.e.}$$

and

$$|v_2(t) - \dot{x}_1(t)| = \rho_F(t, x_1(t), \dot{x}_1(t)) \quad \text{a.e.} \tag{2.16}$$

In view of the Lipschitz continuity properties of $\rho_F(t, ., v)$ and since $\dot{x}_1$ is integrable, we readily deduce from the integrability of $t \rightarrow \rho_F(t, x_0(t), \dot{x}_0(t))$

that $t \to \rho_F(t, x_1(t), \dot{x}_1(t))$ is also integrable. It then follows from (2.16) that v_2 is integrable and we may define

$$x_2(t) = y(S) + \int_S^t v_2(s)ds.$$

We proceed in this way to construct a sequence of absolutely continuous arcs $\{x_m\}$ satisfying

$$\rho_F(t, x_m(t), \dot{x}_{m+1}(t)) = 0 \quad \text{a.e.}$$

$$|\dot{x}_{m+1}(t) - \dot{x}_m(t)| = \rho_F(t, x_m(t), \dot{x}_m(t)) \quad \text{a.e.}$$

for $m = 0, 1, 2, \ldots$ and

$$\rho_F(t, x_0(t), \dot{x}_0(t)) = \rho_F(t, y(t), \dot{y}(t)) \quad \text{a.e.}$$

Notice that

$$\begin{aligned}
\|x_1 - x_0\|_{L^\infty} &\leq \int_S^T |\dot{x}_1(t) - \dot{x}_0(t)| dt \\
&= \int_S^T \rho_F(t, x_0(t), \dot{x}_0(t)) dt = \Lambda_F(y). \quad (2.17)
\end{aligned}$$

Applying Proposition 2.4.1, we deduce that, for $m \geq 1$ and a.e. t,

$$\begin{aligned}
|\dot{x}_{m+1}(t) - \dot{x}_m(t)| &\leq \rho_F(t, x_{m-1}(t), \dot{x}_m(t)) + k(t)|x_m(t) - x_{m-1}(t)| \\
&= k(t)|x_m(t) - x_{m-1}(t)|. \quad (2.18)
\end{aligned}$$

Since $x_{m+1}(S) = x_m(S)$, it follows that, for all $m \geq 1$ and a.e. $t \in [S, T]$,

$$\begin{aligned}
|x_{m+1}(t) &- x_m(t)| \\
&\leq \int_S^t k(t_1)|x_m(t_1) - x_{m-1}(t_1)| dt_1 \\
&\leq \int_S^t k(t_1) \int_S^{t_1} k(t_2) \ldots \int_S^{t_{m-1}} k(t_m)|x_1(t_m) - x_0(t_m)| dt_m \ldots dt_2 dt_1 \\
&\leq S_m(t)\Lambda_F(y), \quad (2.19)
\end{aligned}$$

in view of (2.17). Here

$$S_m(t) := \int_S^t k(t_1) \int_S^{t_1} k(t_2) \ldots \int_S^{t_{m-1}} k(t_m) dt_m \ldots dt_2 dt_1.$$

The right side can be reduced, one indefinite integral at a time, with the help of the integration by parts formula. There results:

$$S_m(t) = \frac{\left(\int_S^t k(t)dt\right)^m}{m!}.$$

It follows from (2.17) through (2.19) that, for any integers $M > N \geq 0$, we have

$$
\begin{aligned}
&\|\dot{x}_M - \dot{x}_N\|_{L^1} \\
&\quad \leq \ \|\dot{x}_M - \dot{x}_{M-1}\|_{L^1} + \ldots + \|\dot{x}_{N+1} - \dot{x}_N\|_{L^1} \\
&\quad \leq \ \left[\frac{\left(\int_S^T k(t)dt\right)^{M-1}}{(M-1)!} + \ldots + \frac{\left(\int_S^T k(t)dt\right)^{N}}{N!} \right] \Lambda_F(y). \quad (2.20)
\end{aligned}
$$

(Here $\left(\int_S^T k(t)dt\right)^m /m! := 1$ when $m = 0$.) It is clear from this inequality that $\{\dot{x}_m\}$ is a Cauchy sequence in L^1. It follows that

$$
\dot{x}_m \to v \text{ in } L^1,
$$

for some $v \in L^1$. Define $x \in W^{1,1}$ according to

$$
x(t) := \xi + \int_S^t v(s)ds.
$$

Since

$$
\|x - x_m\|_{L^\infty} \leq \int_S^T |v(s) - \dot{x}_m(s)|ds
$$

and $\dot{x}_m \to v$ in L^1, we know that

$$
x_m \to x \text{ uniformly.}
$$

By extracting a subsequence (we do not relabel), we can arrange that

$$
\dot{x}_m \to \dot{x} \text{ a.e.}
$$

Define $\mathcal{O}$ to be the subset of points $t \in [S,T]$ such that $\dot{x}_m(t) \in F(t, x_{m-1}(t))$ for all index values $m = 1, 2, \ldots$ and such that $\dot{x}_m(t) \to \dot{x}(t)$. Take any $t \in \mathcal{O}$. Then

$$
\dot{x}_m(t) \in F(t, x_{m-1}(t)).
$$

Since $F(t,.)$ has a closed graph, we obtain in the limit

$$
\dot{x}(t) \in F(t, x(t)).
$$

But $\mathcal{O}$ has full measure. It follows that x is an F-trajectory.

Next observe that by setting $N = 0$ in inequality (2.20) we arrive at

$$
\|\dot{x}_M - \dot{y}\|_{L^1} \leq \exp\left(\int_S^T k(t)dt\right) \Lambda_F(y)
$$

for $M = 1, 2, \ldots$. Since $\dot{x}_M \to \dot{x}$ in L^1 as $M \to \infty$ we deduce that

$$\|\dot{x} - \dot{y}\|_{L^1} \leq \exp\left(\int_S^T k(t)dt\right) \Lambda_F(y).$$

This is the required estimate.

It remains to prove that x can be chosen continuously differentiable when the "comparison function" y is continuously differentiable, under the additional hypotheses.

Construct a sequence $\{x_i\}$ as above. Under the additional hypotheses, $t \to F(t, x_0(t))$ is a continuous multifunction. In view of Proposition 2.4.2, $\dot{x}_1$ is a continuous function. ($\dot{x}_1(t)$ is the projection of $\dot{y}(t) = \dot{x}_0(t)$ onto $F(t, x_0(t))$.) Arguing inductively, we conclude that

$$t \to \dot{x}_i(t) \text{ is continuous for all } i \geq 0.$$

Fix any $i \geq 0$. For arbitrary t, $\dot{x}_{i+1}(t) \in F(t, x_i(t))$ and $\dot{x}_{i+2}(t)$ minimizes $v \to |\dot{x}_{i+1}(t) - v|$ over $v \in F(t, x_{i+1}(t))$, by construction. Under the hypotheses, we can find $w \in F(t, x_{i+1}(t))$ such that

$$|w - \dot{x}_{i+1}(t)| \leq k_\infty |x_{i+1}(t) - x_i(t)| \quad \text{for all } t.$$

Since $x_{i+2}(S) = x_{i+1}(S)$ and i was chosen arbitrarily, we conclude that

$$|\dot{x}_{i+2}(t) - \dot{x}_{i+1}(t)| \leq k_\infty |x_{i+1}(t) - x_i(t)| \quad \text{for all } t.$$

Now Hypothesis (ii)$'$ implies Hypothesis (ii) with $k(t) = k_\infty$ for all t. By (2.19) then, for any integers $M > N \geq 2$, we have

$$\|\dot{x}_M - \dot{x}_N\|_C \leq k_\infty \left[\frac{\left(\int_S^T k_\infty dt\right)^{M-2}}{(M-2)!} + \ldots + \frac{\left(\int_S^T k_\infty dt\right)^{N-1}}{(N-1)!}\right] \Lambda_F(y).$$

It follows that $\dot{x}_i$ is a Cauchy sequence in C. But C is complete, so the sequence has a strong C limit, some continuous function v. But v must coincide with $\dot{x}$, the strong L^1 limit of $\{\dot{x}_i\}$, following adjustment on a nullset. It follows that x is a continuously differentiable function. $\square$

Naturally, if we specialize to the point-valued case, we recover an existence theorem for differential equations. In this important special case the solution is unique. We require

Lemma 2.4.4 (Gronwall's Inequality) *Take an absolutely continuous function $z : [S, T] \to R^n$. Assume that there exist nonnegative integrable functions k and v such that*

$$|\frac{d}{dt}z(t)| \leq k(t)|z(t)| + v(t) \quad a.e. \quad t \in [S, T].$$

Then

$$|z(t)| \leq \exp\left(\int_S^t k(\sigma)d\sigma\right)\left[|z(S)| + \int_S^t \exp\left(-\int_S^\tau k(\sigma)d\sigma\right)v(\tau)d\tau\right]$$

for all $t \in [S,T]$.

Proof. Since z is absolutely continuous so too is $t \to |z(t)|$. Let $\mathcal{O} \subset [S,T]$ be the subset of points t such that $z(.)$ and $|z(.)|$ are both differentiable at t. $\mathcal{O}$ has full measure and, it is straightforward to show,

$$\frac{d}{dt}|z(t)| \leq |\dot{z}(t)| \quad \text{for all } t \in \mathcal{O}.$$

Now define the absolutely continuous function

$$\eta(t) := \exp\left(-\int_S^t k(\sigma)d\sigma\right)|z(t)| \quad \text{for all } t \in [S,T].$$

Then for every $t \in \mathcal{O}$ we have

$$\begin{aligned}
\dot{\eta}(t) &= \exp\left(-\int_S^t k(\sigma)d\sigma\right)\left[\frac{d}{dt}|z(t)| - k(t)|z(t)|\right] \\
&\leq \exp\left(-\int_S^t k(\sigma)d\sigma\right)[|\dot{z}(t)| - k(t)|z(t)|] \\
&\leq \exp\left(-\int_S^t k(\sigma)d\sigma\right)v(t).
\end{aligned}$$

It follows that for each $t \in [S,T]$,

$$\eta(t) \leq \eta(S) + \int_S^t \exp\left(-\int_S^\tau k(\sigma)d\sigma\right)v(\tau)d\tau.$$

Since $\eta(S) = |z(S)|$ and

$$|z(t)| = \exp\left(\int_S^t k(\sigma)d\sigma\right)\eta(t),$$

we deduce

$$|z(t)| \leq \exp\left(\int_S^t k(\sigma)d\sigma\right)\left[|z(S)| + \int_S^t \exp\left(-\int_S^\tau k(\sigma)d\sigma\right)v(\tau)d\tau\right].$$

$\square$

Corollary 2.4.5 (ODEs: Existence and Uniqueness of Solutions)
Take a function $f : [S,T] \times R^n \to R^n$, an arc $y \in W^{1,1}([S,T];R^n)$, $\epsilon \in (0,\infty) \cup \{+\infty\}$, and a point $\xi \in R^n$. (The case $\epsilon = +\infty$ is interpreted as in the statement of Theorem 2.4.3.) Assume that

(i) $f(.,x)$ is measurable for each $x \in R^n$;

(ii) there exists $k \in L^1$ such that

$$|f(t, x') - f(t, x'')| \leq k(t)|x' - x''|$$

for all $x', x'' \in y(t) + \epsilon B$, a.e. $t \in [S, T]$.

Assume further that

$$K\left(|\xi - y(S)| + \int_S^T |\dot{y}(t) - f(t, y(t))|dt\right) \leq \epsilon,$$

where $K := \exp\left(\int_S^T k(t)dt\right)$.

Then there exists a unique solution to the differential equation

$$\begin{aligned}
\dot{x}(t) &= f(t, x(t)) \quad a.e. \ t \in [S, T] \\
x(S) &= \xi
\end{aligned}$$

that satisfies

$$\begin{aligned}
||x - y||_{L^\infty} &\leq |x(S) - y(S)| + \int_S^T |\dot{x}(t) - \dot{y}(t)| \, dt \\
&\leq K\left(|\xi - y(S)| + \int_S^T |\dot{y}(t) - f(t, y(t))dt\right).
\end{aligned}$$

Proof. All the assertions of the corollary follow immediately from the Generalized Filippov Existence Theorem (Theorem 2.4.3), with the exception of "uniqueness." Suppose however that there are two solutions x' and x'' to the differential equation which satisfy $||x' - y||_{L^\infty} \leq \epsilon$ and $||x'' - y||_{L^\infty} \leq \epsilon$. Define $z(t) := x'(t) - x''(t)$. Then under the hypotheses, for almost every $t \in [S, T]$,

$$\begin{aligned}
|\dot{z}(t)| &= |f(t, x'(t)) - f(t, x''(t))| \\
&\leq k(t)|x'(t) - x''(t)| = k(t)|z(t)|.
\end{aligned}$$

Since $z(S) = 0$, it follows from Gronwall's Lemma that $z \equiv 0$. We conclude that $x' = x''$. Uniqueness is proved. $\square$

2.5   Perturbed Differential Inclusions

Consider a sequence of arcs whose elements satisfy perturbed versions of a "nominal" differential inclusion such that the perturbation terms in some sense tend to zero as we proceed along the sequence. When can we extract a subsequence with limit a solution to the nominal differential inclusion? As we show, this is possible under unrestrictive hypotheses on the differential inclusion and on the nature of the perturbations. Necessary conditions in Nonsmooth Optimal Control are usually obtained by deriving necessary conditions for simpler perturbed versions of the Optimal Control problem of interest and passing to the limit. The significance of the results of this section is that they justify the limit-taking procedures.

Use is made of a characterization of subsets of L^1 that are relatively sequentially compact.

Theorem 2.5.1 (Dunford–Pettis Theorem) *Let S be a bounded subset of $L^1([S,T]; R^n)$. Then the following conditions are equivalent.*

(i) *Every sequence in S has a subsequence converging to some L^1 function, with respect to the weak L^1 topology;*

(ii) *For every $\epsilon > 0$ there exists $\delta > 0$ such that for every measurable set $D \subset [S,T]$ and $x \in S$ satisfying meas $\{D\} < \delta$, we have $\int_D x(t)dt < \epsilon$.*

Proof. See [61]. □

When Condition (ii) above is satisfied, we say "the family of functions S is equi-integrable." A simple criterion for equi-integrability (a sufficient condition to be precise) is that the family of functions is "uniformly integrably bounded" in the sense that there exists an integrable function $\alpha \in L^1$ such that

$$|x(t)| \leq \alpha(t) \quad \text{a.e.} \ \ t \in [S,T]$$

for all $x \in S$.

We require also certain properties of the Hamiltonian $H(t,x,p)$ associated with a given multifunction $F : [S,T] \times R^n \rightsquigarrow R^n$, defined at points $(t,x,p) \in \text{dom}\, F \times R^n$.

$$H(t,x,p) := \sup_{v \in F(t,x)} p \cdot v. \tag{2.21}$$

Proposition 2.5.2 *Consider a multifunction $F : [S,T] \times R^n \rightsquigarrow R^n$, which has as values closed sets.*

(a) *Fix $(t,x) \in \text{dom}\, F$. Assume that there exists $c \geq 0$ such that $F(t,x) \subset cB$. Then*

$$|H(t,x,p)| \leq c|p| \quad \text{for every} \ \ p \in R^n,$$

and $H(t,x,.)$ is Lipschitz continuous with Lipschitz constant c.

(b) *Fix* $t \in [S,T]$. *Take convergent sequences* $x_i \to x$ *and* $p_i \to p$ *in* R^n. *Assume that there exists* $c \geq 0$ *such that*

$$(t_i, x_i) \in \text{dom } F \quad \text{and} \quad F(t, x_i) \subset cB \quad \text{for all } i,$$

and that $\text{Gr } F(t, .)$ *is closed. Then*

$$\text{limsup}_{i \to \infty} H(t, x_i, p_i) \leq H(t, x, p).$$

(c) *Take measurable functions* $x : [S,T] \to R^n$, $p : [S,T] \to R^n$ *such that* $\text{Gr } x \subset \text{dom } F$. *Assume that*

$$F \text{ is } \mathcal{L} \times \mathcal{B}^n \text{ measurable.}$$

Then $t \to H(t, x(t), p(t))$ *is a measurable function.*

Proof.

(a): Take $(t, x) \in \text{dom } F$. Choose any $p, p' \in R^n$. Since $F(t, x)$ is a compact, nonempty set, there exists $v \in F(t, x)$ such that

$$H(t, x, p) = p \cdot v.$$

Of course, $H(t, x, p') \geq p' \cdot v$. But then

$$H(t, x, p) - H(t, x, p') \leq (p - p') \cdot v \leq |v| |p - p'|.$$

Since $|v| \leq c$, and the roles of p and p' can be interchanged, we deduce that

$$H(t, x, p) - H(t, x, p') \leq c|p - p'|.$$

Notice, in particular, that $|H(t, x, p)| \leq c|p|$, since $H(t, x, 0) = 0$.

(b): Fix $t \in [S,T]$ and take any sequences $x_i \to x$ and $p_i \to p$ in R^n. $\limsup_i H(t, x_i, p_i)$ can be replaced by $\lim_i H(t, x_i, p_i)$, following extraction of a suitable subsequence (we do not relabel). For each i, $F(t, x_i)$ is a nonempty compact set (since $(t, x_i) \in \text{dom } F$); consequently, there exists $v_i \in F(t, x_i)$ such that $H(t, x_i, p_i) = p_i \cdot v_i$. But $|v_i| \leq c$ for $i = 1, 2, \ldots$. We can therefore arrange, by extracting another subsequence, that $v_i \to v$ for some $v \in R^n$. Since $\text{Gr } F(t, .)$ is closed, $v \in F(t, x)$ and so

$$\lim_i H(t, x_i, p_i) = \lim_i p_i \cdot v_i = p \cdot v \leq H(t, x, p).$$

(c): According to Proposition 2.3.2 and Theorem 2.3.11, we can find a measurable selection $v(.)$ of $t \rightsquigarrow F(t, x(t))$. Fix $k > 0$ and define for all $(t, x) \in [S,T] \times R^n$,

$$F_k'(t, x) := F(t, x) \cap (v(t) + kB)$$

and

$$F_k(t,x) \;=\; \begin{cases} F_k'(t,x) & \text{if } (t,x) \in \text{dom } F_k' \\ v(t) + kB & \text{otherwise .} \end{cases}$$

It is a straightforward exercise to show that F_k is $\mathcal{L} \times \mathcal{B}^n$ measurable. Since $\text{dom } F_k = [S,T] \times R^n$, we can define

$$H_k(t,x,p) \;:=\; \max\{p \cdot v \,:\, v \in F_k(t,x)\}$$

for all $(t,x,p) \in [S,T] \times R^n \times R^n$. Clearly

$$H_k(t,x(t),p(t)) \to H(t,x(t,p(t))) \text{ as } k \to \infty,$$

for all t. It therefore suffices to show that $t \to H_k(t,x(t),p(t))$ measurable.

However, by Corollary 2.3.3, this will be true if we can show that $(t,(x,p)) \to H_k(t,(x,p))$ is $\mathcal{L} \times \mathcal{B}^{2n}$ measurable.

Fix $r \in R$. We complete the proof by showing that

$$D \;:=\; \{(t,x,p) \in [S,T] \times R^n \times R^n \,:\, H_k(t,x,p) \geq r\}$$

is $\mathcal{L} \times \mathcal{B}^{2n}$ measurable. Let $\{v_i\}$ be a dense subset of R^n. Take a sequence $\epsilon \downarrow 0$. It is easy to verify that $D = \mathcal{D}$, where

$$\mathcal{D} \;:=\; \cap_{j=1}^{\infty} \cup_{i=1}^{\infty} (A_{i,j} \times B_{i,j}),$$

in which

$$A_{i,j} \;:=\; \{(t,x) \in [S,T] \times R^n \,:\, (v_i + \epsilon_j B) \cap F_k(t,x) \neq \emptyset\}$$

and

$$B_{i,j} \;:=\; \{p \in R^n \,:\, p \cdot v_i > r - \epsilon_j\}.$$

(That $D \subset \mathcal{D}$ is obvious; to show that $\mathcal{D} \subset D$, we exploit the fact that $\text{Gr } F_k(t,.)$ is a compact set.)

But D is obtainable from $\mathcal{L} \times \mathcal{B}^{2n}$ measurable sets, by means of a countable number of union and intersection operations. It follows that D is $\mathcal{L} \times \mathcal{B}^{2n}$ measurable, as claimed. $\square$

With these preliminaries behind us we are ready to answer the question posed at the beginning of the section.

Theorem 2.5.3 (Compactness of Trajectories) *Take a relatively open subset $\Omega \subset [S,T] \times R^n$ and a multifunction $F : \Omega \rightsquigarrow R^n$.*

Assume that, for some closed multifunction $X : [S,T] \rightsquigarrow R^n$ such that $\text{Gr } X \subset \Omega$, the following hypotheses are satisfied.

(i) F is a closed, convex, nonempty multifunction.

(ii) F is $\mathcal{L} \times \mathcal{B}^n$ measurable.

(iii) For each $t \in [S, T]$, the graph of $F(t, .)$ restricted to $X(t)$ is closed.

Consider a sequence $\{x_i\}$ of $W^{1,1}([S, T]; R^n)$ functions, a sequence $\{r_i\}$ in $L^1([S, T]; R)$ such that $\|r_i\|_{L^1} \to 0$ as $i \to \infty$, and a sequence $\{A_i\}$ of measurable subsets of $[S, T]$ such that $\operatorname{meas} A_i \to |T - S|$ as $i \to \infty$.
Suppose that:

(iv) $\operatorname{Gr} x_i \subset \operatorname{Gr} X$ for all i;

(v) $\{\dot{x}_i\}$ is a sequence of uniformly integrably bounded functions on $[S, T]$ and $\{x_i(S)\}$ is a bounded sequence;

(vi) there exists $c \in L^1$ such that

$$F(t, x_i(t)) \subset c(t)B$$

for a.e. $t \in A_i$ and for $i = 1, 2, \ldots$.

Suppose further that

$$\dot{x}_i(t) \in F(t, x_i(t)) + r_i(t)B \quad a.e. \ t \in A_i.$$

Then along some subsequence (we do not relabel)

$$x_i \to x \quad uniformly \quad and \quad \dot{x}_i \to \dot{x} \quad weakly \ in \ L^1$$

for some $x \in W^{1,1}([S, T]; R^n)$ satisfying

$$\dot{x}(t) \in F(t, x(t)) \quad a.e. \ t \in [S, T].$$

Proof. The $\dot{x}_i$s are uniformly integrably bounded on $[S, T]$. According to the Dunford–Pettis Theorem, we can arrange, by extracting a subsequence (we do not relabel), that $\dot{x}_i \to v$ weakly in L^1 for some L^1 function v. Since $\{x_i(S)\}$ is a bounded sequence we may arrange by further subsequence extraction that $x_i(S) \to \xi$ for some $\xi \in R^n$. Now define

$$x(t) := \xi + \int_S^t v(s)ds.$$

By weak convergence, $x_i(t) \to x(t)$ for every $t \in [S, T]$. Clearly too $\dot{x}_i \to \dot{x}$ weakly in L^1.

Now consider the Hamiltonian $H(t, \xi, p)$ defined by (2.21). Choose any p and any Lebesgue measurable subset $V \subset [S, T]$. For almost every $t \in V \cap A_i$, we have

$$H(t, x_i(t), p) \geq p \cdot \dot{x}_i(t) - r_i(t)|p|.$$

Since all terms in this inequality are integrable, we deduce

$$\int_{V \cap A_i} p \cdot \dot{x}_i(t)dt - \int_{V \cap A_i} r_i(t)|p|dt \leq \int_{V \cap A_i} H(t, x_i(t), p)dt.$$

Because the $\dot{x}_i$s are uniformly integrably bounded, meas $[A_i] \to |T - S|$, $\dot{x}_i \to \dot{x}$ weakly in L^1, and $||r_i||_{L^1} \to 0$, we see that the left side of this relationship has limit $\int_V p \cdot \dot{x}(t)$. It follows that

$$\int_V p \cdot \dot{x}(t)dt \leq \limsup_{i \to \infty} \int_V \chi_i(t)H(t, x_i(t), p)dt.$$

Here χ_i denotes the characteristic function of the set A_i.

Since $\chi_i(t)H(t, x_i(t), p)$ is bounded above by $c(t)|p|$, we deduce from Fatou's Lemma that

$$\int_V p \cdot \dot{x}(t) \leq \int_V \limsup_i \chi_i(t)H(t, x_i(t), p)dt.$$

From the upper semicontinuity properties of H then (see Proposition 2.5.2)

$$\int_V (H(t, x(t), p) - p \cdot \dot{x}(t))\, dt \geq 0.$$

Let $\{p_i\}$ be an ordering of the set of n-vectors having rational coefficients. Define $D \subset [S, T]$ to be the subset of points $t \in [S, T]$ such that t is a Lebesgue point of $t \to H(t, x(t), p_i) - p_i \cdot \dot{x}(t)$ for all i. D is a set of full measure. For any $t \in D \cap [S, T)$ and any i

$$H(t, x(t), p_i) - p_i \cdot \dot{x}(t) =$$

$$\lim_{\delta \downarrow 0} \frac{1}{\delta} \int_t^{t+\delta} [H(\sigma, x(\sigma), p_i) - p_i \cdot \dot{x}(\sigma)]\, d\sigma \geq 0\,.$$

Since $H(t, x(t), .)$ is continuous for each t, it follows that

$$\sup\{p \cdot e - p \cdot \dot{x}(t) : e \in F(t, x(t))\} \geq 0 \quad \text{for all} \quad p \in R^n \quad \text{a.e.} \quad t \in [S, T].$$

But $F(t, x(t))$ is closed and convex. We deduce with the help of the separation theorem that

$$\dot{x}(t) \in F(t, x(t)) \quad \text{a.e.} \quad t \in [S, T].$$

We have confirmed that x is an F-trajectory. $\square$

2.6 Existence of Minimizing F-Trajectories

Take a relatively open subset $\Omega \subset [S, T] \times R^n$, a multifunction $F : \Omega \rightsquigarrow R^n$, a closed multifunction $X : [S, T] \rightsquigarrow R^n$ with the property that $\text{Gr}\, X \subset \Omega$, and a closed set $C \subset R^n \times R^n$. We define the set of feasible F-trajectories (associated with the constraint sets $X(t)$, $S \leq t \leq T$, and C) to be

$$\mathcal{R}_F(X, C) := \{x \in C([S, T]; R^n) : x \text{ is an } F\text{-trajectory}$$
$$x(t) \in X(t) \text{ for all } t \in [S, T] \text{ and } (x(S), x(T)) \in C\}\,.$$

We deduce from the results of the previous section the following criteria for compactness of the set of feasible F-trajectories.

Proposition 2.6.1 *Take Ω, F, X, and C as above. Assume that*

(i) F is a closed, $\mathcal{L} \times \mathcal{B}^n$ measurable multifunction;

(ii) for each $t \in [S, T]$, the graph of $F(t, .)$ restricted to $X(t)$ is closed;

(iii) there exist $\alpha \in L^1$ and $\beta \in L^1$ such that

$$F(t, x) \subset (\alpha(t)|x| + \beta(t)) B \quad \text{for all } (t, x) \in \operatorname{Gr} X;$$

(iv) either $X(s)$ is bounded for some $s \in [S, T]$ or one of the following two sets
$$C_0 := \{x_0 \in R^n : (x_0, x_1) \in C \text{ for some } x_1 \in R^n\}$$
$$C_1 := \{x_1 \in R^n : (x_0, x_1) \in C \text{ for some } x_0 \in R^n\}$$

is bounded. Assume further that

(v) $F(t, x)$ is convex for all $(t, x) \in \operatorname{Gr} X$.

Then $\mathcal{R}_F(X, C)$ is compact with respect to the supremum norm topology.

Proof. Since the supremum norm topology is a metric topology, it suffices to prove sequential compactness. Accordingly, take any sequence of feasible F-trajectories $\{x_i\}$. We must show that there exists an F-trajectory x satisfying the constraints $x(t) \in X(t)$ for all $t \in [S, T]$ and $(x(S), x(T)) \in C$ such that

$$x_i \to x \quad \text{uniformly}$$

along some subsequence. But these conclusions can be drawn from Theorem 2.5.3 provided we can show that the set $\mathcal{R}_F(X, C)$ is bounded with respect to the supremum norm. By Hypothesis (iv) however there exists $k > 0$ and $\bar{s} \in [S, T]$ such that for any feasible F-trajectory y we have

$$|y(\bar{s})| \leq k.$$

By Hypothesis (iii),

$$|\dot{y}(t)| \leq \alpha(t)|y(t)| + \beta(t) \quad \text{a.e.}$$

It follows from the Gronwall Lemma (applied "backwards" in time on the interval $[S, \bar{s}]$ and "forwards" on $[\bar{s}, T]$) that

$$|y(t)| \leq K \quad \text{for all } t \in [S, T],$$

where

$$K = e^{||\alpha||_{L^1}} (k + ||\beta||_{L^1}).$$

We have confirmed that $\mathcal{R}(X, C)$ is bounded with respect to the supremum norm. $\square$

It is a simple step now to supply conditions for existence of solutions to the optimal control problem

$$(P) \begin{cases} \text{Minimize} g(x(S), x(T)) \text{ over } x \in W^{1,1}([S,T]; R^n) \\ \text{which satisfy} \\ \dot{x}(t) \in F(t, x(t)) \text{ a.e. } t \in [S,T], \\ x(t) \in X(t) \text{ for all } t \in [S,T], \\ (x(S), x(T)) \in C, \end{cases}$$

in which $g : R^n \times R^n \to R$ is a given lower semicontinuous function.

Indeed (P) can be equivalently formulated as a problem of seeking a minimizer of a lower semicontinuous function over a compact subset of $C([S,T]; R^n)$ equipped with the supremum norm, namely,

$$\text{minimize } \psi(y) \text{ over } y \in \mathcal{R}_F(X, C),$$

where

$$\psi(y) := g(y(S), y(T)).$$

(P) therefore has a minimizer provided $\mathcal{R}_F(X, C)$ is nonempty. We have proved:

Proposition 2.6.2 *Take Ω, F, X, C, and g as above. Assume that*

(i) F is a closed, $\mathcal{L} \times \mathcal{B}^n$ measurable multifunction;

(ii) for each $t \in [S,T]$, the graph of $F(t,.)$ restricted to $X(t)$ is closed;

(iii) there exist $\alpha \in L^1$ and $\beta \in L^1$ such that

$$F(t, x) \subset (\alpha(t)|x| + \beta(t)) B \quad \text{for all } (t, x) \in \mathrm{Gr}\, X;$$

(iv) either $X(s)$ is bounded for some $s \in [S,T]$ or one of the following two sets

$$C_0 := \{x_0 \in R^n : (x_0, x_1) \in C \text{ for some } x_1 \in R^n\}$$
$$C_1 := \{x_1 \in R^n : (x_0, x_1) \in C \text{ for some } x_0 \in R^n\}$$

is bounded.

Assume further that

(a) the set of feasible F-trajectories $\mathcal{R}_F(X, C)$ is nonempty;

(b) $F(t, x)$ is convex for each $(t, x) \in \mathrm{Gr}\, X$.

Then (P) has a minimizer.

2.7 Relaxation

Suppose that an optimization problem of interest fails to have a minimizer. "Relaxation" is the procedure of adding extra elements to the domain of the optimization problem to ensure existence of minimizers.

For a relaxation scheme to be of interest it usually needs to be accompanied by the information that an element $\bar{x}$ in the extended domain can be approximated by an element y in the original domain of the optimization problem (to the extent that we can arrange that the cost of y is arbitrarily close to that of $\bar{x}$). In these circumstances, we can find a suboptimal element for the original problem (i.e., one whose cost is arbitrarily close to the infimum cost) by finding a minimizer in the extended domain and approximating it.

Relaxation is now examined in connection with the optimization problem (P) of the preceding section, which for convenience we reproduce.

$$(P) \begin{cases} \text{Minimize } g(x(S), x(T)) \quad \text{over } x \in W^{1,1}([S,T]; R^n) \\ \text{which satisfy} \\ \dot{x}(t) \in F(t, x(t)) \quad \text{a.e. } t \in [S,T], \\ x(t) \in X(t) \quad \text{for all } t \in [S,T], \\ (x(S), x(T)) \in C . \end{cases}$$

We impose the hypotheses of Proposition 2.6.2 with the exception of the convexity hypothesis

$$F(t, x) \text{ is convex for all } (t, x) \in \operatorname{Gr} X .$$

In these circumstances (P) may fail to have a minimizer. The point is illustrated by the following example.

Example 2.1 Consider problem (P) when the state vector x is 2-dimensional. We write the components of the state vector $x = (y, z)$. Set

$$[S, T] = [0, 1], \ \Omega = [0, 1] \times R^2, \ X(t) = R^2,$$
$$g((y_0, z_0), (y_1, z_1)) = z_1,$$

$$C = \{((y_0, z_0), (y_1, z_1)) : z_0 = 0\},$$

$$F(y, z) := (\{-1\} \cup \{+1\}) \times \{|y|\}.$$

Notice that, if an arc (y, z) satisfies the constraints, then

$$z(t) = \int_0^t |y(s)| ds. \tag{2.22}$$

Evidently then this special case of (P) is a disguised version, arrived at through state augmentation, of the optimization problem

$$(P)' \begin{cases} \text{Minimize } J(y) = \int_0^1 |y(s)| ds \\ \text{over } y \in W^{1,1}([0,1];R) \text{ satisfying} \\ \dot{y}(t) \in \{-1\} \cup \{+1\} \text{ a.e.} \end{cases}$$

(If (y,z) is feasible for (P) then y and z are related by (2.22), y is feasible for $(P)'$, and the costs are the same; also, if y is feasible for $(P)'$ then (y,z), with z given by (2.22), is feasible for (P) and again the costs are the same.)

However, as we show, $(P)'$ has no minimizers. It follows that, in this case, (P) fails to have a solution. The fact that all the hypotheses of Proposition 2.6.2 are satisfied with the exception of (ii) confirms that we cannot dispense with the convexity hypothesis.

To show that $(P)'$ does not have a solution, notice first of all that

$$J(y) \geq 0 \tag{2.23}$$

for all arcs y satisfying the constraint of $(P)'$. Consider next the sequence of feasible arcs $\{y_i\}$,

$$y_i(t) = \int_0^t v_i(s) ds,$$

where

$$v_i(s) = \begin{cases} +1 & \text{for } s \in A_i \cap [0,1] \\ -1 & \text{for } s \notin A_i \cap [0,1] \end{cases}$$

and

$$A_i = \cup_{j=0}^{\infty} [(2i)^{-1} 2j, (2i)^{-1}(2j+1)].$$

An easy calculation yields:

$$J(y_i) = 2^{-(i+1)} \quad \text{for } i = 1, 2, \dots .$$

Since $J(y_i) \to 0$ as $i \to \infty$ we conclude from (2.23) that the infimum cost is 0. If there exists a minimizer $\bar{y}$ then

$$J(\bar{y}) = \int_0^1 |\bar{y}(s)| ds = 0.$$

This implies that $\bar{y} \equiv 0$. It follows that

$$\dot{\bar{y}}(t) = 0 \quad \text{a.e.}$$

But then $\bar{y}$ fails to satisfy the differential inclusion

$$\dot{\bar{y}}(t) \in (\{-1\} \cup \{+1\}) \quad \text{a.e.}$$

It follows that no minimizer exists.

The pathological feature of the above problem is that the limit point $\bar{y}$ of any minimizing sequence satisfies only the convexified differential inclusion

$$\dot{\bar{y}}(t) \in \text{co} \left(\{-1\} \cup \{+1\}\right) \quad \text{a.e.}$$

In light of this example, a natural relaxation procedure for us to adopt is to allow arcs that satisfy the convexified differential inclusion

$$\dot{x}(t) \in \text{co}\, F(t, x(t)) \quad \text{a.e.} \quad t \in [S, T]. \tag{2.24}$$

Accordingly, an arc $x \in W^{1,1}$ satisfying this dynamic constraint is called a "relaxed" F-trajectory. When it is necessary to emphasize the distinction with relaxed F-trajectories, we sometimes call F-trajectories "ordinary" F-trajectories.

As earlier discussed, for this concept of relaxed trajectory to be useful we need to know that arcs satisfying the "relaxed" dynamic constraint (2.24) can be adequately approximated by arcs satisfying the original, unconvexified, constraint. That this can be done is a consequence of the Generalized Filippov Existence Theorem and the following theorem of R. J. Aumann on integrals of multifunctions. (See [27].)

Theorem 2.7.1 (Aumann's Theorem) *Take a Lebesgue measurable multifunction* $\Gamma : [S, T] \rightsquigarrow R^n$ *that is closed and nonempty. Assume that there exists* $c \in L^1$ *such that*

$$\Gamma(t) \in c(t)B \quad \text{for all} \ \ t \in [S, T].$$

Then

$$\int_S^T \Gamma(s)ds = \int_S^T \text{co}\,\Gamma(s)ds,$$

where

$$\int_S^T A(s)ds := \left\{ \int_S^T \gamma(s)ds : \gamma \ \ \text{is a Lebesgue measurable selection of } A \right\}$$

for $A = \Gamma$ *and* $A = \text{co}\,\Gamma$.

The ground has now been prepared for:

Theorem 2.7.2 (Relaxation Theorem) *Take a relatively open set* $\Omega \subset [S, T] \times R^n$ *and an* $\mathcal{L} \times \mathcal{B}^n$ *measurable multifunction* $F : \Omega \rightsquigarrow R^n$ *that is closed and nonempty. Assume that there exist* $k \in L^1$ *and* $c \in L^1$ *such that*

$$F(t, x') \subset F(t, x'') + k(t)B \quad \text{for all} \ \ (t, x'), (t, x'') \in \Omega$$

and

$$F(t, x) \subset c(t)B \quad \text{for all} \ \ (t, x) \in \Omega.$$

Take any relaxed F-trajectory x with $\operatorname{Gr} x \subset \Omega$ *and any* $\delta > 0$. *Then there exists an ordinary F-trajectory y that satisfies* $y(S) = x(S)$ *and*

$$\max_{t\in[S,T]} |y(t) - x(t)| < \delta.$$

Proof. Choose $\epsilon > 0$ such that $T(x, 2\epsilon) \subset \Omega$ and let α be such that

$$0 < \alpha < \min\left\{\frac{\epsilon}{K \ln K}, \epsilon, \frac{\delta}{1 + K \ln K}\right\},$$

where $K := \exp(\|k\|_{L^1})$. Let $h > 0$ be such that

$$\int_I c(t)dt < \alpha/2$$

for any subinterval $I \subset [S, T]$ of length no greater than h.

Let $\{S = t_0, t_1, \ldots, t_k = T\}$ be a partition of $[S, T]$ such that $\operatorname{meas} I_i < h$ for $i = 1, 2, \ldots, k$, where $I_i := [t_{i-1}, t_i)$ for $i = 1, \ldots, k - 1$ and $I_k = [t_{k-1}, t_k]$. The multifunction $t \rightsquigarrow F(t, x(t))$ is Lebesgue measurable (see Proposition 2.3.2) and satisfies

$$F(t, x(t)) \subset c(t)B \quad \text{for all } t \in [S, T].$$

Recalling that x is a co F-trajectory, we deduce from Aumann's Theorem (Theorem 2.7.1) that there exist measurable functions $f_i : [S, T] \to R^n$ such that $f_i(t) \in F(t, x(t))$ a.e. $t \in I_i$ and

$$\int_{I_i} f_i(t)dt = \int_{I_i} \dot{x}(t)dt$$

for $i = 1, \ldots, k$. Define

$$f(t) := \sum_{i=1}^{k} f_i(t)\chi_{I_i}(t),$$

where χ_{I_i} denotes the indicator function of I_i and set

$$z(t) = x(S) + \int_S^t f(s)ds \quad \text{for } t \in [S, T].$$

Fix $t \in [S, T]$. Then for some $j \in \{1, \ldots, k\}$,

$$|z(t) - x(t)| = |\sum_{i=1}^{j-1} \int_{I_i} (f_i(\sigma) - \dot{x}(\sigma))d\sigma + \int_{I_j \cap [S,t]} (f_j(\sigma) - \dot{x}(\sigma))d\sigma|$$

$$\leq 0 + 2\int_{I_j} c(t)dt < \alpha.$$

It follows that

$$||z - x||_{L^\infty} < \alpha. \tag{2.25}$$

Since $\alpha < \epsilon$, we have

$$T(z, \epsilon) \subset \Omega.$$

Notice that, since $\dot{z}(t) \in F(t, x(t))$ a.e. and in view of Proposition 2.4.1,

$$\begin{aligned} \rho_F(t, z(t), \dot{z}(t)) &\leq \rho_F(t, x(t), \dot{z}(t)) + k(t)|x(t) - z(t)| \\ &< k(t)\alpha \quad \text{a.e.} \end{aligned}$$

We have

$$\Lambda_F(z) := \int_S^T \rho_F(t, z(t), \dot{z}(t))dt < \alpha \ln K.$$

Since $\alpha K \ln K < \epsilon$, we deduce from the Generalized Filippov Existence Theorem that there exists an F-trajectory y such that $y(S) = x(S)$ and

$$||y - z||_{L^\infty} \leq K\Lambda_F(z) < \alpha K \ln K.$$

By (2.25) then

$$||y - x||_{L^\infty} \leq ||y - z||_{L^\infty} + ||z - x||_{L^\infty} \leq \alpha(K \ln K + 1).$$

Since however

$$\alpha(K \ln K + 1) < \delta,$$

we conclude that the F-trajectory y satisfies

$$||y - x||_{L^\infty} < \delta.$$

$\square$

The optimization problem obtained when we replace the dynamic constraint $\dot{x} \in F$ in (P) by $\dot{x} \in \text{co}\, F$ is denoted $(P)_{\text{relaxed}}$. Minimizers for $(P)_{\text{relaxed}}$ are called relaxed minimizers for (P).

The following proposition provides information clarifying the relationship between (P) and $(P)_{\text{relaxed}}$.

Proposition 2.7.3 *Take Ω, F, X, C, and g as above. Assume that*

(i) F is a compact, $\mathcal{L} \times \mathcal{B}^n$ measurable multifunction;

(ii) for each $t \in [S, T]$, the graph of $F(t, .)$ restricted to $X(t)$ is closed;

(iii) there exist $\alpha \in L^1$ and $\beta \in L^1$ such that

$$F(t, x) \subset (\alpha(t)|x| + \beta(t))\, B \quad \text{for all } (t, x) \in \text{Gr}\, X;$$

(iv) either $X(s)$ is bounded for some $s \in [S,T]$ or one of the following two sets

$$C_0 := \{x_0 \in R^n : (x_0, x_1) \in C \text{ for some } x_1 \in R^n\}$$

$$C_1 := \{x_1 \in R^n : (x_0, x_1) \in C \text{ for some } x_0 \in R^n\}$$

is bounded.

Assume further that the set of feasible F-trajectories $\mathcal{R}_F(X,C)$ is nonempty. Then $(P)_{relaxed}$ has a minimizer.

If, in addition, we assume that

(a) there exists $k \in L^1$ such that

$$F(t,x) \subset F(t,x') + k(t)|x - x'|B \text{ for all } (t,x),(t,x') \in \Omega,$$

(b) g is continuous,

and, for some relaxed minimizer $\bar{x}$ and $\epsilon > 0$,

(c)

$$\bar{x}(t) + \epsilon B \subset X(t) \text{ for all } t \in [S,T], \tag{2.26}$$

(d)

$$\text{either } (\bar{x}(S) + \epsilon B) \times \{\bar{x}(T)\} \subset C \quad \text{or} \quad \{\bar{x}(S)\} \times (\bar{x}(T) + \epsilon B) \subset C, \tag{2.27}$$

then

$$\inf (P)_{relaxed} = \inf (P).$$

The right and left sides of the last relationship denote the infimum cost for (P) and $(P)_{relaxed}$, respectively.

Proof. Existence of a minimizer for $(P)_{relaxed}$ follows immediately from Proposition 2.6.2 applied to the modified version of (P) in which co F replaces F. (Notice that co F inherits the measurability properties of F according to Proposition 2.3.8, as well as the linear growth properties.)

Suppose that there exists a relaxed minimizer $\bar{x}$ and $\epsilon > 0$ such that

$$\bar{x}(t) + \epsilon B \subset X(t) \text{ for all } t \in [S,T]$$

and

$$\{\bar{x}(S)\} \times (\bar{x}(T) + \epsilon B) \subset C$$

(The case $(\bar{x}(S) + \epsilon B) \times \{\bar{x}(T)\} \subset C$. is treated by "reversing time".) Take any $\alpha > 0$. Then noting the continuity of the function g and applying the

Relaxation Theorem, Theorem 2.7.2, we can find an ordinary F-trajectory $\bar{y}$ such that

$$\bar{y}(S) = \bar{x}(S), \ \|\bar{y} - \bar{x}\|_{L^\infty} < \epsilon$$

and

$$g(\bar{y}(S), \bar{y}(T)) < g(\bar{x}(S), \bar{x}(T)) + \alpha.$$

Clearly $\bar{y}$ satisfies the constraints

$$(\bar{y}(S), \bar{y}(T)) \in C \ \text{and} \ \bar{y}(t) \in X(t) \ \text{for all} \ t \in [S, T].$$

In other words, $\bar{y}$ is a feasible (ordinary) F-trajectory. It follows that

$$\inf(P)_{\text{relaxed}} = g(\bar{x}(S), \bar{x}(T)) > g(\bar{y}(S), \bar{y}(T)) - \alpha \geq \inf(P) - \alpha.$$

Since $\alpha > 0$ is arbitrary and

$$\inf \ (P)_{\text{relaxed}} \leq \inf \ (P),$$

we conclude that

$$\inf \ (P)_{\text{relaxed}} = \inf \ (P).$$

$\square$

Notice the crucial role of the "interiority" hypotheses (2.26) and (2.27). If all relaxed minimizers violate these hypotheses then the Relaxation Theorem does not automatically imply that the infimum costs of (P) and $(P)_{\text{relaxed}}$ coincide; while it is true that any relaxed minimizer $\bar{x}$ can be uniformly approximated by an F-trajectory y, we cannot in general guarantee that y will satisfy the constraints for it to qualify as a feasible F-trajectory.

2.8 The Generalized Bolza Problem

In Section 2.6 we gave conditions for the existence of minimizers in the context of minimization problems over some class of arcs satisfying a given differential inclusion $\dot{x}(t) \in F(t, x(t))$. These conditions restricted attention to problems for which the velocity sets $F(t, x)$ are bounded. For traditional variational problems and also for many Optimal Control problems of interest there are no constraints on permitted velocities. To deal in a unified manner with problems with bounded and unbounded velocity sets, it is convenient to adopt a new framework for the optimization problems involved, namely, to regard them as special cases of the Generalized Bolza Problem:

$$(GBP) \left\{ \begin{array}{l} \text{Minimize } \Lambda(x) := l(x(S), x(T)) + \int_S^T L(t, x(t), \dot{x}(t))dt \\ \text{over arcs } x \in W^{1,1}([S, T]; R^n), \end{array} \right.$$

in which $[S, T]$ is a given interval, and $l : R^n \times R^n \to R \cup \{+\infty\}$ and $L : [S, T] \times R^n \times R^n \to R \cup \{+\infty\}$ are given extended-valued functions. Provided we arrange that $t \to L(t, x(t), \dot{x}(t))$ is measurable and minorized by an integrable function for every $x \in W^{1,1}$ (our hypotheses take care of this), Λ will be a well-defined $R \cup \{+\infty\}$ valued functional on $W^{1,1}$.

Notice that the functions l and L are permitted to takes values $+\infty$. So they can be used implicitly to take account of constraints. For example, the "differential inclusion" problem

$$\left\{ \begin{array}{l} \text{Minimize } g(x(S), x(T)) \\ \text{over arcs } x \in W^{1,1}([S, T]; R^n) \text{ satisfying} \\ \dot{x}(t) \in F(t, x(t)), \\ (x(S), x(T)) \in C \end{array} \right.$$

is a special case of the Generalized Bolza Problem in which

$$L(t, x, v) = \left\{ \begin{array}{ll} 0 & \text{if } v \in F(t, x) \\ +\infty & \text{otherwise} \end{array} \right.$$

and

$$l(x_0, x_1) = \left\{ \begin{array}{ll} g(x_0, x_1) & \text{if } (x_0, x_1) \in C \\ +\infty & \text{otherwise.} \end{array} \right.$$

In existence theorems covering problems with unbounded velocity sets, superlinear growth hypotheses on the cost integrand are typically invoked to compensate for unbounded velocity sets. The key advantage of the Generalized Bolza Problem as a vehicle for such theorems is that unrestrictive hypotheses ensuring existence of minimizers, which require coercivity of the cost integrand in precisely those "directions" in which velocities are unconstrained, can be economically expressed as conditions on the extended-valued function L.

Theorem 2.8.1 (Generalized Bolza Problem: Existence of Minimizers) *Assume that the data for (GBP) satisfy the following hypotheses:*

(H1): l is lower semicontinuous and there exist a lower semicontinuous function $l^0 : R^+ \to R$ satisfying

$$\lim_{r \uparrow +\infty} l^0(r) = +\infty$$

and either
$$l^0(|x_0|) \leq l(x_0, x_1) \quad \text{for all } x_0, x_1$$

or
$$l^0(|x_1|) \leq l(x_0, x_1) \quad \text{for all } x_0, x_1.$$

(H2): L is $\mathcal{L} \times \mathcal{B}^{n \times n}$ measurable.

(H3): $L(t, ., .)$ is lower semicontinuous for each $t \in [S, T]$.

(H4): For each $(t, x) \in [S, T] \times R^n$, $L(t, x, .)$ is convex and $\operatorname{dom} L(t, x, .) \neq \emptyset$.

(H5): For all $t \in [S, T]$, $x \in R^n$, and $v \in R^n$,

$$L(t, x, v) \geq \theta(|v|) - \alpha|x|,$$

for some $\alpha \geq 0$ and some lower semicontinuous convex function $\theta : R^+ \to R^+$ satisfying

$$\lim_{r \uparrow +\infty} \theta(r)/r = +\infty.$$

Then (GBP) has a minimizer. (We allow the possibility that $\Lambda(x) = +\infty$ for all $x \in W^{1,1}$. In this case all arcs x are regarded as minimizers.)

Comment

The proof of this theorem, which follows shortly, exploits properties of the Hamiltonian

$$H(t, x, p) := \sup_{v \in R^n} \{p \cdot v - L(t, x, v)\}.$$

To a large extent, the role of the growth condition (H5) is to ensure that the Hamiltonian has the required properties to furnish existence theorems. Loosely speaking, growth conditions on the Lagrangian translate into (one-sided) boundedness conditions on the Hamiltonian. One direction for generalizing this theorem is to replace (H5) by less restrictive conditions imposed directly on the Hamiltonian. Rockafellar has shown that the conclusions of Theorem 2.8.1 remain valid when (H5) is replaced by:

(H5)′ For all $t \in [S, T]$ and $x, p \in R^n$,

$$H(t, x, p) \leq \mu(t, p) + |x|(\sigma(t) + \rho(t)|p|),$$

for some integrable functions $\sigma(.)$, $\rho(.)$, and some function $\mu(. , .)$ such that $\mu(. , p)$ is integrable for each $p \in R^n$.

The ensuing analysis calls upon some properties of convex functions, which it is now convenient to summarize. We say that a convex function $f : R^n \to R \cup \{+\infty\}$ is *proper* if it is lower semicontinuous and $\operatorname{dom} f \neq \emptyset$. The *conjugate* of f is the function $f^* : R^n \to R \cup \{+\infty\}$, defined by the *Legendre–Fenchel transformation*:

$$f^*(y) = \sup_{x \in R^n} \{y \cdot x - f(x)\}.$$

Important facts are that, for any proper, convex function $f : R^n \to R \cup \{+\infty\}$, its conjugate f^*, too, is proper, convex. Furthermore, f can be recovered from f^* by means of a second application of the Legendre–Fenchel transformation:

$$f(x) = \sup_{y \in R^n} (y \cdot x - f^*(y)).$$

One consequence of these relationships is

Proposition 2.8.2 *(Jensen's Inequality) Take any proper, convex function $f : R^n \to R \cup \{+\infty\}$. Then, for any set $I \subset R$ of positive measure and any $v \in L^1(I; R^n)$, $t \to f(v(t))$ is a measurable function, minorized by an integrable function, and*

$$\int_I f(v(t))dt \geq |I|f\left(|I|^{-1}\int_I |v(t)|dt\right),$$

where $|I|$ denotes the Lebesgue measure of I.

Proof. Take any $y \in \text{dom } f^*$. Then, for each $t \in I$,

$$f(v(t)) \geq v(t) \cdot y - f^*(y).$$

It follows that the (measurable) function $t \to f(v(t))$ is minorized by an integrable function. Also,

$$\int_I f(v(t))dt \geq \left(\int_I v(t)dt\right) \cdot y - f^*(y)|I|$$

$$= |I|\left[\left(|I|^{-1}\int_I v(t)dt\right) \cdot y - f^*(y)\right].$$

This inequality is valid for all $y \in \text{dom } f^*$. Maximizing over $\text{dom } f^*$ and noting that f is obtained from f^* by applying the Legendre–Fenchel transformation, we obtain

$$\int_I f(v(t))dt \geq |I|f\left(|I|^{-1}\int_I |v(t)|dt\right),$$

as claimed. $\square$

Proof of Theorem 2.8.1 We assume that $l^0(x_0) \leq l(x_0, x_1)$ for all (x_0, x_1). (The case $l^0(x_1) \leq l(x_0, x_1)$ for all (x_0, x_1) is treated similarly.) Under the hypotheses, l^0 and θ are bounded below. By scaling and adding a constant to $\int L dt + l$ (this does not effect the minimizers) we can arrange that $l^0 \geq 0$ and $\theta \geq 0$. We can also arrange that the constant α is arbitrarily small.

Choose α such that $e^{\alpha|T-S|}|T - S|\alpha < 1$.

Since θ has superlinear growth, we can define $k : R^+ \to R^+$:

$$k(\beta) := \sup\{r \geq 0 : r = 0 \text{ or } \theta(r) \leq \beta r\}.$$

Step 1: Fix $M \geq 0$. We show that the level set

$$S_M := \{x \in W^{1,1} : \Lambda(x) \leq M\}$$

is weakly sequentially precompact; i.e., any sequence $\{x_i\}$ in $\mathcal{S}_M$ has a subsequence that converges, with respect to the weak $W^{1,1}$ topology, to some point in $W^{1,1}$.

Take any $x \in \mathcal{S}_M$ and define the L^1 function

$$q(t) := L(t, x(t), \dot{x}(t)).$$

Then

$$|\dot{x}(t)| \leq k(1) + \theta(|\dot{x}(t)|) \leq k(1) + q(t) + \alpha|x(t)| \quad \text{a.e.} \qquad (2.28)$$

But, for each $t \in [S, T]$,

$$
\begin{aligned}
\int_S^t q(s)ds &= \Lambda(x) - \int_t^T q(s)ds \leq M - \int_t^T \theta(|\dot{x}(s)|)ds + \alpha \int_t^T |x(s)|ds \\
&\leq M + \alpha|T - S|\,|x(t)| + \int_t^T (\alpha|T - S||\dot{x}(s)| - \theta(|\dot{x}(s)|))\,ds \\
&\leq M + \alpha|T - S|\,|x(t)| + \alpha|T - S|^2 k(\alpha|T - S|).
\end{aligned}
$$

It follows from (2.28) and Gronwall's Inequality that

$$
\begin{aligned}
|x(t)| &\leq e^{\alpha(t-S)}\left[|x(S)| + \int_S^T (k(1) + q(s))ds\right] \\
&\leq e^{\alpha(T-S)}\left[|x(S)| + k(1)|T - S| + M \right. \\
&\qquad \left. + \alpha|T - S|^2 k(\alpha|T - S|) + \alpha|T - S|\,|x(t)|\right].
\end{aligned}
$$

Therefore

$$|x(t)| \leq A|x(S)| + B, \qquad (2.29)$$

where the constants A and B (they do not depend on x) are

$$A := \left(1 - \alpha|T - S|e^{\alpha|T-S|}\right)^{-1} e^{\alpha|T-S|}$$

and

$$B := \left(1 - \alpha|T - S|e^{\alpha|T-S|}\right)^{-1}\left(k(1)|T - S| + M + \alpha|T - S|^2 k(\alpha|T - S|)\right).$$

We deduce from (2.29) and the fact that $l(x_0, x_1) \geq l^0(|x_0|)$ that

$$|x(S)| \leq K, \qquad (2.30)$$

where $K > 0$ is any constant (it can be chosen independent of x) such that

$$l^0(r) - \alpha[Ar + B] > M \quad \text{for all } r \geq K.$$

Now, for any set $I \subset [S,T]$ of positive measure, Jensen's inequality yields

$$
\begin{aligned}
\theta(|I|^{-1} \int_I |\dot{x}(t)|dt) &\leq |I|^{-1} \int_I \theta(|\dot{x}(t)|)dt \\
&\leq |I|^{-1} \int_S^T \theta(|\dot{x}(t)|)dt \\
&\leq |I|^{-1} \left(\int_S^T L(t, x(t), \dot{x}(t))dt + \alpha \int_S^T |x(t)|dt \right) \\
&\leq |I|^{-1}(M + \alpha AK + \alpha B).
\end{aligned}
$$

We conclude that, if $\int_I |\dot{x}(t)|dt > 0$,

$$
\frac{\theta(\int_I |\dot{x}(t)|dt / |I|)}{\int_I |\dot{x}(t)|dt / |I|} \leq \frac{(M + \alpha AK + \alpha B)}{\int_I |\dot{x}(t)|dt}.
$$

Since θ has superlinear growth, it follows from this inequality that there exists a function $\omega : R^+ \to R^+$ (which does not depend on x) such that $\lim_{\sigma \downarrow 0} \omega(\sigma) = 0$ and

$$
\int_I |\dot{x}(t)|dt \leq \omega(|I|) \quad \text{for all measurable } I \subset [S,T]. \tag{2.31}
$$

Take any sequence $\{x_i\}$ in $\mathcal{S}_M$. Then, by (2.30), $\{x_i(S)\}$ is a bounded sequence. On the other hand, $\{\dot{x}_i\}$ is an equicontinuous sequence, by (2.31). Invoking the Dunford–Pettis criterion for weak sequential compactness in L^1, we deduce that, along a subsequence,

$$
x_i(S) \to x(S) \quad \text{and} \quad \dot{x}_i \to \dot{x}_i \text{ weakly in } L^1.
$$

Otherwise expressed,

$$
x_i \to x \quad \text{weakly in } W^{1,1},
$$

for some $x \in W^{1,1}$. This is what we set out to prove.

Step 2: Take an $\mathcal{L} \times \mathcal{B}^n$ function $\phi : [S,T] \times R^n \to R \cup \{+\infty\}$ that satisfies the conditions:

(a): for each $t \in [S,T]$, $\phi(t, .)$ is lower semicontinuous and finite at some point, and

(b): for some $\tilde{p} \in L^\infty$, the function $t \to \phi(t, \tilde{p}(t))$ is minorized by an integrable function.

We show that

$$
\int_S^T \hat{\phi}(t)dt = \sup_{p(.) \in L^\infty} \int_S^T \phi(t, p(t))dt, \tag{2.32}
$$

where

$$\hat{\phi}(t) := \sup_{p \in R^n} \phi(x, p).$$

(Note that, under the hypotheses, $\hat{\phi}$ is measurable and minorized by an integrable function. So the left side of (2.32) is well defined. The right side is interpreted as the supremum of the specified integral over $p(.)$s such that the integrand is minorized by some integrable function.)

For any $p(.) \in L^\infty$, $\hat{\phi}(t) \geq \phi(t, p(t))$ for all t. It immediately follows that (2.32) holds, when "$\geq$" replaces "$=$".

It suffices then to validate (2.32) when "$\leq$" replaces "$=$". To this end, choose any $r \in R^n$ such that

$$\int_S^T \hat{\phi}(t) dt > r .$$

We can also choose $K > 0$ and $\epsilon > 0$ such that, writing

$$\hat{\hat{\phi}}(t) := \min\{\hat{\phi}(t), K\},$$

we have

$$\int_S^T (\hat{\hat{\phi}}(t) - \epsilon) dt > r.$$

Define the multifunction

$$\Gamma(t) := \{p \in R^n : \phi(t, p) > \hat{\hat{\phi}}(t) - \epsilon\}.$$

Under the hypotheses, Γ takes values nonempty (open) sets and $\operatorname{Gr}\Gamma$ is $\mathcal{L} \times \mathcal{B}^n$ measurable. According to Aumann's Measurable Selection Theorem then, Γ has a measurable selection, which we write $\bar{p}(.)$.

However, since $t \to \phi(t, \bar{p}(t))$ and $t \to \phi(t, \tilde{p}(t))$ are minorized by integrable functions, we can find a measurable set E such that $\bar{p}$ restricted to $[S, T] \setminus E$ is essentially bounded and

$$\int_{[S,T] \setminus E} \phi(t, \bar{p}(t)) dt + \int_E \phi(t, \tilde{p}(t)) dt > r.$$

It follows that

$$\int_S^T \phi(t, p(t)) dt > r,$$

in which $p(.)$ is the essentially bounded function

$$p(t) := \begin{cases} \tilde{p}(t) & \text{if } t \in I \\ \bar{p}(t) & \text{otherwise.} \end{cases}$$

Since r is an arbitrary strict lower bound on $\int_S^T \hat{\phi}(t) dt$, the desired inequality is confirmed.

Step 3: We show that $\Lambda(.)$ is weakly sequentially lower semicontinuous (w.r.t. the $W^{1,1}$ topology).

Since weak L^1 convergence implies uniform convergence, we deduce from the lower semicontinuity of l that $x \to l(x(S), x(T))$ is weakly sequentially lower semicontinuous. It remains therefore to show that

$$\tilde{\Lambda}(x) := \int_S^T L(t, x(t), \dot{x}(t))dt$$

is also weakly sequentially lower semicontinuous. Take any $x \in W^{1,1}$. Then, since $L(t, x(t), .)$ is a proper, convex function for each t,

$$\tilde{\Lambda}(x) = \int_S^T \sup_{p \in R^n} [p \cdot \dot{x}(t) - H(t, x(t), p)] \, dt$$

$$= \sup_{p(.) \in L^\infty} \int_S^T [p(t) \cdot \dot{x}(t) - H(t, x(t), p(t))] \, dt.$$

(We have used the results of Step 2 to justify the last equality. Note that the function $\phi(t, p) = p \cdot \dot{x}(t) - H(t, x(t), p)$ satisfies the relevant hypotheses. In particular, $\phi(t, \tilde{p}(t))$ is minorized by an integrable function for the choice $\tilde{p} \equiv 0$.)

For fixed $p \in L^\infty$, consider now the integral functional

$$\Lambda_p(x) := \int_S^T [p(t) \cdot \dot{x}(t) - H(t, x(t), p(t))] \, dt.$$

We claim that Λ_p is weakly sequentially lower semicontinuous.

To verify this assertion, take any weakly convergent sequence $y_i \to y$ in $W^{1,1}$. Then $\dot{y}_i \to \dot{y}$ weakly in L^1 and $y_i \to y$ uniformly. Since $H(t, ., p(t))$ is upper semicontinuous for each t and the $H(t, y_i(t), p(t))$s are minorized by a common integrable function (this last property follows from Hypothesis (H5)), we deduce from Fubini's Theorem that

$$\liminf_{i \to \infty} \Lambda_p(y_i) = \liminf_{i \to \infty} \int_S^T [p(t) \cdot \dot{y}_i(t) - H(t, y_i(t), p(t))] \, dt$$

$$\geq \int_S^T [p(t) \cdot \dot{y}(t) - H(t, y(t), p(t))] \, dt$$

$$= \Lambda_p(y).$$

Weak sequential lower semicontinuity of Λ_p is confirmed.

We have shown that, for each $x \in W^{1,1}$,

$$\tilde{\Lambda}(x) = \sup_{p(.) \in L^\infty} \Lambda_p(x).$$

But the upper envelope of a family of weakly sequentially lower semicontinuous functionals on $W^{1,1}$ is also weakly sequentially lower semicontinuous. It follows that $\tilde{\Lambda}$ is weakly sequentially lower semicontinuous.

Conclusion: We have shown in Steps 1 and 3 that Λ is sequentially lower semicontinuous and that the level sets of Λ are sequentially compact with respect to the weak $W^{1,1}$ topology. These properties guarantee existence of a minimizer. (We allow the possibility that $\Lambda(x) = +\infty$ for all x. In this case all xs are minimizers.) $\square$

2.9 Notes for Chapter 2

A standard reference for properties of measurable multifunctions is [27]. More recent texts covering the subject matter of Section 2.3 are [9], [12], and [125].

There is a substantial literature on differential inclusions. Material in Sections 2.4 through 2.6, which restricts attention largely to differential inclusions involving multifunctions that are Lipschitz continuous with respect to the state, only scratches the surface of this extensive field. (See [9] and [57].) The important existence theorem, Theorem 2.3.3, and accompanying estimates are essentially due to Filippov. The proof given here is an adaptation of that in [9], to allow for measurable time dependence. The Compactness of Trajectories Theorem (Theorem 2.4.3) is from [38]. Implicit in this theorem are early ideas for establishing existence of optimal controls under a convexity hypothesis on the "velocity set" associated with Tonelli, L. C. Young, Filippov, Gramkrelidze, Roxin, and others.

Relaxation is a much broader topic than is conveyed by Section 2.7, which focuses on relaxation schemes for Optimal Control problems formulated in terms of a differential inclusion. Relaxation schemes for many other classes of optimization problems have been proposed including, for example, variational problems in several independent variables [5] and Optimal Control problems with time delay [126]. Of particular importance are those introduced by Warga for Optimal Control problems with dynamic constraints in the form of a differential equation parameterized by control functions [147] and those introduced by Gamkrelidze [76].

Tonelli's techniques for proving existence of minimizers to variational problems with unbounded derivatives [135], based on establishing weak lower semicontinuity of integral functionals and weak compactness of level sets, underlie much research in this area to this day. Far-reaching extensions of Tonelli's existence theorems were achieved by Rockafellar. (See, for example, [122].) An accessible treatment of existence issues in the Calculus of Variations and Optimal Control is to be found in [88].

Chapter 3

Variational Principles

Since the fabric of the universe is most perfect and the work of a most wise Creator, nothing at all takes place in the universe in which some rule of the maximum or minimum does not appear.

– Leonhard Euler

3.1 Introduction

The name "variational principle" is traditionally attached to a law of nature asserting that some quantity is minimized. Examples are Dirichlet's Principle (the spatial distribution of an electrostatic field minimizes some quadratic functional), Fermat's Principle (a light ray follows a shortest path), or Hamilton's Principle of Least Action (the evolution of a dynamical system is an "extremum" for the action functional). These principles are called "variational" because working through their detailed implications entails solving problems in the Calculus of Variations.

Nowadays the term "variational principle" is used to describe any procedure in which a property of interest is shown to imply that some quantity is minimized. Variational principles in this broader sense are at the heart of Nonsmooth Analysis and its applications. It is no exaggeration to say that they have as significant a role in a nonsmooth setting as, say, the inverse function theorem and fixed point theorems do in traditional real analysis. Variational principles, "some function is minimized", make sense even if the function is not differentiable, and the part they play in Nonsmooth Analysis is not therefore all that surprising.

This section brings together a number of important variational principles from the point of view of applications to Nonsmooth Analysis. The Exact Penalization Theorem gives conditions under which a minimizer for a constrained optimization problem is also a minimizer for (an often more tractable) unconstrained problem when the data are Lipschitz continuous. Ekeland's Theorem tells us that if a point $\bar{x}$ is an "ϵ-minimizer" of a function then some neighboring point to $\bar{x}$ is a minimizer for a perturbed problem. The section concludes with a Mini-Max Theorem that gives conditions under which interchanging the order in which "min" and "max" are applied in multilevel optimization problems is permissible.

3.2   Exact Penalization

Take a metric space $(X, m(.,.))$, a nonempty subset $C \subset X$, and a function $f : X \to R$. We can use the metric $m(.,.)$ to define a "distance function" d_C on X, with respect to the set C:

$$d_C(x) := \inf_{x' \in C} m(x, x') \quad \text{for each } x \in X.$$

It is easy to show that, for any $x \in C$, "$x \in C$" implies "$d_C(x) = 0$" and, if C is a closed subset, the converse is also true.

We say that the function f satisfies a Lipschitz condition on X (with Lipschitz constant K) if

$$|f(x) - f(x')| \leq K m(x, x') \quad \text{for all } x, x' \in X.$$

Consider the problem of minimizing the function $f : X \to R$ over the set $C \subset X$. We would like to replace this problem (or at least approximate it) by a more amenable one involving no constraint; an approach of long standing is to drop the constraint, but to compensate for its absence by adding a "penalty term" $K g(x)$ to the cost (K is the penalty parameter). The function g is chosen to be zero on C and positive outside C. The larger K, the more severe the penalty for violating the constraint, so one would expect that solving the "penalized problem,"

$$\text{minimize } f(x) + K g(x) \text{ over } x \in X,$$

for large K would yield a point x that approximately minimizes the cost for the original problem and approximately satisfies the constraint. Now suppose that C is closed and f satisfies a Lipschitz condition on X. A remarkable feature of the distance function is that, if it is adopted as the penalty function, then penalization is "exact," in the sense that a minimizer for the original problem is also a minimizer for the penalized problem.

Justification of this assertion is provided by the following proposition.

Theorem 3.2.1 (Exact Penalization Theorem) *Let $(X, m(.,.))$ be a metric space. Take a set $C \subset X$ and a function $f : X \to R$. Assume that f satisfies a Lipschitz condition on X with Lipschitz constant K. Let $\bar{x}$ be a minimizer for the constrained minimization problem,*

$$\text{minimize } f(x) \text{ over points } x \in X \text{ satisfying } x \in C. \tag{3.1}$$

Choose any $\hat{K} \geq K$. Then $\bar{x}$ is a minimizer also for the unconstrained minimization problem,

$$\text{minimize } f(x) + \hat{K} d_C(x) \text{ over points } x \in X. \tag{3.2}$$

If $\hat{K} > K$ and C is a closed set, then the converse assertion is also true: any minimizer $\bar{x}$ for the unconstrained problem (3.2) is also a minimizer for the constrained problem (3.1) and so, in particular, $\bar{x} \in C$.

Proof. Let $\bar{x}$ be a minimizer for (3.1) and let $\hat{K} \geq K$. Suppose that, contrary to the claims of the theorem, $\bar{x}$ fails to be a minimizer for (3.2). Then there exist a point $y \in X$ and $\epsilon > 0$ such that $f(y) + \hat{K}d_C(y) < f(\bar{x}) - \hat{K}\epsilon$. Choose a point $z \in C$ such that $m(y, z) \leq d_C(y) + \epsilon$. Since $\hat{K}$ is a Lipschitz constant for f on X,

$$f(z) \leq f(y) + \hat{K}m(y, z) \leq f(y) + \hat{K}(d_C(y) + \epsilon) < f(\bar{x}).$$

This is not possible since $\bar{x}$ minimizes f over C.

Suppose next that $\hat{K} > K$ and C is closed. Let $\bar{x}$ be a minimizer for (3.2). Choose any $\epsilon > 0$. Then we can find a point $z \in C$ such that

$$d_C(\bar{x}) > m(\bar{x}, z) - \hat{K}^{-1}\epsilon.$$

We have

$$
\begin{aligned}
f(z) &\leq f(\bar{x}) + Km(\bar{x}, z) \\
&\leq f(\bar{x}) + Kd_C(\bar{x}) + (K/\hat{K})\epsilon \\
&< f(\bar{x}) + \hat{K}d_C(\bar{x}) - (\hat{K} - K)d_C(\bar{x}) + \epsilon \\
&\leq f(z) - (\hat{K} - K)d_C(\bar{x}) + \epsilon.
\end{aligned}
$$

It follows that $(\hat{K} - K)d_C(\bar{x}) < \epsilon$. Since $\epsilon > 0$ is arbitrary, $d_C(\bar{x}) = 0$. But then $\bar{x} \in C$, because C is closed. We deduce that $f(c) \geq f(\bar{x})$ for all $c \in C$. In other words, $\bar{x}$ is a minimizer for (3.1). $\square$

3.3   Ekeland's Theorem

Take a complete metric space $(X, d(., .))$, a lower semicontinuous function $f : X \to R \cup \{+\infty\}$, a point $x_0 \in \text{dom } f$, and some $\epsilon > 0$.

Suppose that x_0 is an ϵ-minimizer for f. This means

$$f(x_0) \leq \inf_{x \in X} f(x) + \epsilon.$$

In these circumstances, as we show, there exists some $\bar{x} \in \text{dom } f$ satisfying

$$d(\bar{x}, x_0) \leq \epsilon^{1/2}$$

which is a minimizer for the perturbed function

$$f_\epsilon(x) = f(x) + \epsilon^{1/2}d(x, \bar{x});$$

i.e.,

$$f_\epsilon(\bar{x}) = \inf_{x \in X} f_\epsilon(x).$$

This is a version of Ekeland's Variational Principle. It tells us that we can perturb f in such a way as to ensure that a minimizer $\bar{x}$ for the perturbed

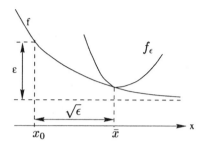

FIGURE 3.1. Ekeland's Variational Principle.

problem exists. Furthermore we can arrange that both the distance of the minimizer $\bar{x}$ for the perturbed function from x_0 and also the perturbation term are small, if ϵ is small.

The essential idea is captured by Figure 3.1.

Here X is R with the metric induced by the Euclidean norm, and f is a strictly monotone decreasing function. The function f has no minimizer "close" to the ϵ-minimizer x_0. (In fact f has no minimizers at all!) However an appropriately chosen point $\bar{x}$, close to x_0, is a minimizer for the perturbed function

$$x \to f(x) + \epsilon^{1/2}d(x, \bar{x}).$$

The perturbation term penalizes deviation from $\bar{x}$; by raising the graph of the function away from $\bar{x}$, we force $\bar{x}$ to be a minimizer. In this example it is essential that the perturbation term be a nonsmooth function and also that we allow $\bar{x}$ to be different from x_0.

Widespread use of this variational principle has been made in the field of Optimization and Nonlinear Analysis generally. Its main role in Optimization has been to justify techniques for deriving necessary conditions for a minimization problem based on applying available necessary conditions to a simpler perturbed problem and passage to the limit. The fact that it allows $(X, d(.,.))$ to be an arbitrary complete metric space adds greatly to it flexibility.

The variational principle summarized above is a special case of a slightly more general theorem in which, by making different choices of parameter $\alpha > 0$ and $\lambda > 0$, we can trade off the size of the perturbation term and the distance of $\bar{x}$ from x_0. (The preceding version involves the choices $\lambda = \epsilon^{1/2}$ and $\alpha = \epsilon^{1/2}$.)

Theorem 3.3.1 (Ekeland's Variational Principle) *Take a complete metric space* $(X, d(.,.))$, *a lower semicontinuous function* $f : X \to R \cup \{+\infty\}$, *a point* $x_0 \in \text{dom } f$, *and numbers* $\alpha > 0$ *and* $\lambda > 0$. *Assume that*

$$f(x_0) \leq \inf_{x \in X} f(x) + \lambda\alpha. \tag{3.3}$$

Then there exists $\bar{x} \in X$ such that

(i) $f(\bar{x}) \leq f(x_0)$,

(ii) $d(x_0, \bar{x}) \leq \lambda$,

(iii) $f(\bar{x}) \leq f(x) + \alpha d(x, \bar{x})$ *for all $x \in X$.*

Proof. It suffices to find some $\bar{x}$ such that

(a): $f(\bar{x}) + \alpha d(x_0, \bar{x}) \leq f(x_0)$
and
(b): $f(\bar{x}) < f(x) + \alpha d(x, \bar{x})$ for all $x \neq \bar{x}$.
Indeed (a) implies (i) and (b) implies (iii). Notice also that (3.3) and (a) imply

$$f(\bar{x}) + \alpha d(x_0, \bar{x}) \leq f(x_0) \leq \inf_{x \in X} f(x) + \lambda \alpha \leq f(\bar{x}) + \lambda \alpha.$$

Since $f(\bar{x})$ is finite (from (a)) and $\alpha > 0$ we conclude that $d(x_0, \bar{x}) \leq \lambda$. The remaining assertion (ii) is also confirmed.

The validity of (a) and (b) is proved as follows.
Define a multifunction $T : X \rightsquigarrow X$,

$$T(x) := \{y \in X : f(y) + \alpha d(x, y) \leq f(x)\}.$$

Notice that for each $x \in X$, $T(x)$ is a closed set. It is clear that

$$x \in T(x) \quad \text{for all } x \in X. \tag{3.4}$$

A further significant property of the multifunction T is

$$y \in T(x) \quad \text{implies} \quad T(y) \subset T(x). \tag{3.5}$$

This is obviously true if $x \notin \text{dom} f$ (for then $T(x) = X$). So assume $x \in \text{dom} f$. Since $y \in T(x)$,

$$f(y) + \alpha d(x, y) \leq f(x).$$

Take any $z \in T(y)$. Then

$$f(z) + \alpha d(z, y) \leq f(y).$$

These inequalities, together with the triangle inequality, imply

$$f(y) + f(z) + \alpha d(z, x) \leq f(x) + f(y).$$

However $f(y) < \infty$, since $f(x) < \infty$. We conclude that

$$f(z) + \alpha d(z, x) \leq f(x).$$

This inequality implies $z \in T(x)$, which is the desired condition.

Now define $\xi : \text{dom} f \to R \cup \{+\infty\}$,

$$\xi(x) := \inf_{y \in T(x)} f(y). \tag{3.6}$$

We see that

$$y \in T(x) \quad \text{implies} \quad \xi(x) \le f(x) - \alpha d(x, y).$$

It follows that

$$\text{diam } T(x) \le 2\alpha^{-1}(f(x) - \xi(x)), \tag{3.7}$$

where

$$\text{diam } T(x) = \sup\{d(y, y') : y \in T(x), y' \in T(x)\}.$$

Now construct a sequence $x_0, x_1, \ldots$, starting with the x_0 of the theorem statement, as follows: for $k = 0, 1, \ldots$ choose $x_{k+1} \in T(x_k)$ to satisfy

$$f(x_{k+1}) \le \xi(x_k) + 2^{-k}.$$

This is possible since, by definition,

$$\xi(x_k) = \inf_{x \in T(x_k)} f(x).$$

Fix $k \ge 0$. Since $x_{k+1} \in T(x_k)$, we deduce from (3.5) that

$$\xi(x_k) \le \xi(x_{k+1}).$$

On the other hand, Property (3.4) implies

$$\xi(x) \le f(x) \quad \text{for all } x \in X.$$

It follows that

$$\xi(x_{k+1}) \le f(x_{k+1}) \le \xi(x_k) + 2^{-k} \le \xi(x_{k+1}) + 2^{-k}.$$

From the preceding inequalities we deduce that

$$0 \le f(x_{k+1}) - \xi(x_{k+1}) \le 2^{-k}.$$

Recalling (3.7) we see that

$$\text{diam } T(x_{k+1}) \le 2^{-k} \cdot 2\alpha^{-1}.$$

We have shown that $\{T(x_k)\}$ is a nested sequence of nonempty closed sets in X whose diameters tend to zero as $k \to \infty$. $\{x_k\}$ is a Cauchy sequence then and, since $(X, d(., .))$ is complete, $x_k \to \bar{x}$ for some $\bar{x} \in X$. Furthermore we have

$$\cap_{k=0}^{\infty} T(x_k) = \{\bar{x}\}. \tag{3.8}$$

It remains to show that, for this choice of $\bar{x}$, properties (a) and (b) are satisfied. But (3.8) implies $\bar{x} \in T(x_0)$ which simply means

$$f(\bar{x}) + \epsilon d(x_0, \bar{x}) \leq f(x_0).$$

We have verified (a). On the other hand (3.5) and (3.8) imply

$$T(\bar{x}) \subset \cap_{k=0}^{\infty} T(x_k) = \{\bar{x}\}.$$

It follows that if $x \neq \bar{x}$ then $x \notin T(\bar{x})$ and so

$$f(\bar{x}) < f(x) + \alpha d(x, \bar{x}).$$

We have verified (b). $\square$

3.4   Mini-Max Theorems

Take nonempty sets X and Y and a function $F : X \times Y \to R$. The central question addressed in this section is: under what circumstances is the relationship

$$\inf_{x \in X} \sup_{y \in Y} F(x, y) = \sup_{y \in Y} \inf_{x \in X} F(x, y) \tag{3.9}$$

valid? In other words, when do the operations of taking the supremum over Y and taking the infimum over X commute?

Notice that, without the imposition of any additional hypotheses whatsoever, we are assured that the two sides of (3.9) are related by inequality:

Proposition 3.4.1 *Let X, Y, and F be as above. Then*

$$\inf_{x \in X} \sup_{y \in Y} F(x, y) \geq \sup_{y \in Y} \inf_{x \in X} F(x, y).$$

Proof. For any fixed $x' \in X$ and $y' \in Y$,

$$\sup_{y \in Y} F(x', y) \geq F(x', y') \geq \inf_{x \in X} F(x, y').$$

It follows that

$$\inf_{x \in X} \sup_{y \in Y} F(x, y) \geq \inf_{x \in X} F(x, y')$$

and consequently

$$\inf_{x \in X} \sup_{y \in Y} F(x, y) \geq \sup_{y \in Y} \inf_{x \in X} F(x, y).$$

$\square$

The challenge then is to establish when the reverse inequality holds:

$$\inf_{x \in X} \sup_{y \in Y} F(x,y) \leq \sup_{y \in Y} \inf_{x \in X} F(x,y)$$

for this combines with the assertion of Proposition 3.4.1 to give (3.9).

A related question, and one of great independent interest, is whether there exists a point $(x^*, y^*) \in X \times Y$ satisfying

$$\sup_{y \in Y} F(x^*,y) = F(x^*,y^*) = \inf_{x \in X} F(x,y^*).$$

A pair (x^*, y^*) having this property is called a *saddlepoint*.

The connection is that existence of a saddlepoint is a sufficient condition for the commutability condition (3.9).

Proposition 3.4.2 *Suppose that* (x^*, y^*) *is a saddlepoint. Then*

$$\inf_{x \in X} \sup_{y \in Y} F(x,y) = F(x^*,y^*) = \sup_{y \in Y} \inf_{x \in X} F(x,y).$$

Furthermore,

$$\sup_{y \in Y} F(x^*,y) = \sup_{y \in Y} \inf_{x \in X} F(x,y) \qquad (3.10)$$

and

$$\inf_{x \in X} F(x,y^*) = \inf_{x \in X} \sup_{y \in Y} F(x,y). \qquad (3.11)$$

Proof. By definition of a saddlepoint and in view of Proposition 3.4.1 we have

$$F(x^*,y^*) = \sup_{y \in Y} F(x^*,y) \geq \inf_{x \in X} \sup_{y \in Y} F(x,y)$$

$$\geq \sup_{y \in Y} \inf_{x \in X} F(x,y) \geq \inf_{x \in X} F(x,y^*) = F(x^*,y^*).$$

These relationships therefore hold with equality. The assertions of the proposition follow immediately. □

Relationships (3.10) and (3.11) do in fact fully characterize a saddlepoint:

Proposition 3.4.3 *There exists a saddlepoint if and only if the following conditions both hold.*

(a) There exists $x^* \in X$ *such that*

$$\sup_{y \in Y} F(x^*,y) = \sup_{y \in Y} \inf_{x \in X} F(x,y).$$

(b) There exists $y^* \in Y$ such that

$$\inf_{x \in X} F(x, y^*) = \inf_{x \in X} \sup_{y \in Y} F(x, y).$$

Furthermore, if (a) and (b) both hold, then (x^*, y^*) is a saddlepoint.

Proof. In view of the preceding proposition all we have to show is that, if (x^*, y^*) satisfies Conditions (a) and (b), then (x^*, y^*) is a saddlepoint. However (a) implies

$$\inf_{x \in X} \sup_{y \in Y} F(x, y) \le \sup_{y \in Y} F(x^*, y) = \sup_{y \in Y} \inf_{x \in X} F(x, y).$$

This combines with the assertions of Proposition 3.4.1 to give

$$\inf_{x \in X} \sup_{y \in Y} F(x, y) = \sup_{y \in Y} \inf_{x \in X} F(x, y).$$

So if (b) also holds, then

$$F(x^*, y^*) \le \sup_{y \in Y} F(x^*, y) =$$
$$\sup_{y \in Y} \inf_{x \in X} F(x, y) = \inf_{x \in X} \sup_{y \in Y} F(x, y) = \inf_{x \in X} F(x, y^*) \le F(x^*, y^*).$$

It follows that

$$\sup_{y \in Y} F(x^*, y) = F(x^*, y^*) = \inf_{x \in X} F(x, y^*).$$

This is the saddlepoint condition. $\square$

Establishing existence of a saddlepoint is by no means straightforward and requires the imposition of stringent hypotheses (see Von Neumann's Mini-Max Theorem below). It turns out however that these hypotheses can be relaxed significantly if we are willing to settle for just one of the two "one-sided" properties (a) and (b) characterizing existence of a saddlepoint. The One-Sided Mini-Max Theorem, giving hypotheses under which Property (a) of Proposition 3.4.3 holds, has come to be recognized as a powerful analytic tool with important implications for Optimization. Numerous applications are made in future chapters.

We make one final observation. It is that, while Property (a) falls somewhat short of guaranteeing existence of a saddlepoint, it nonetheless implies the commutability of the infimum and supremum operations. Validity of this assertion is a byproduct of the proof of Proposition 3.4.3. (Replacing F by $-F$ so that "inf" becomes "sup" and vice versa, we arrive at an analogous statement in relation to Property (b).)

Proposition 3.4.4 *Suppose that either of the following conditions holds.*

(a) There exists x^* such that

$$\sup_{y \in Y} F(x^*, y) = \sup_{y \in Y} \inf_{x \in X} F(x, y);$$

(b) there exists y^* such that

$$\inf_{x \in X} F(x, y^*) = \inf_{x \in X} \sup_{y \in Y} F(x, y).$$

Then

$$\inf_{x \in X} \sup_{y \in Y} F(x, y) = \sup_{y \in Y} \inf_{x \in X} F(x, y).$$

The main theorems of the section now follow.

Theorem 3.4.5 (Aubin One-Sided Mini-Max Theorem) *Consider a function $F : X \times Y \to R$ in which X is a subset of a linear space and Y is a subset of a topological linear space. Assume that*

(i) X and Y are convex sets,

(ii) $F(.\,, y)$ is convex for every $y \in Y$,

(iii) $F(x, .)$ is concave and upper semicontinuous for every $x \in X$, and

(iv) Y is compact.

Then there exists $y^* \in Y$ such that

$$\inf_{x \in X} F(x, y^*) = \inf_{x \in X} \sup_{y \in Y} F(x, y).$$

This implies in particular (see Proposition 3.4.4) that

$$\inf_{x \in X} \sup_{y \in Y} F(x, y) = \sup_{y \in Y} \inf_{x \in X} F(x, y).$$

Proof. Define

$$\eta^- := \sup_{y \in Y} \inf_{x \in X} F(x, y)$$

and

$$\eta^+ := \inf_{x \in X} \sup_{y \in Y} F(x, y).$$

For any subset $K \subset X$ write

$$\eta_K^- := \sup_{y \in Y} \inf_{x \in K} F(x, y).$$

Now let S denote the class of subsets of X comprising only a finite number of points and define:

$$\tilde{\eta}^- := \inf_{K \in S} \sup_{y \in Y} \inf_{x \in K} F(x, y) = \inf_{K \in S} \eta_K^-.$$

Step 1: We show that

$$\eta^- \leq \tilde{\eta}^- \leq \eta^+.$$

Take any $K \in S$. Then

$$\sup_{y \in Y} \inf_{x \in K} F(x, y) \geq \sup_{y \in Y} \inf_{x \in X} F(x, y) = \eta^-.$$

Taking the infimum of the left side over $K \in S$ we obtain

$$\tilde{\eta}^- \geq \eta^+.$$

On the other hand, for any x we have $\{x\} \subset S$ whence

$$\sup_{y \in Y} F(x, y) \geq \inf_{K \in S} \sup_{y \in Y} \inf_{x \in K} F(x, y) = \tilde{\eta}^-.$$

Taking the infimum of the left side over x yields

$$\eta^+ \geq \tilde{\eta}^-,$$

as required.

Step 2: We show that there exists $y^* \in Y$ such that

$$\inf_{x \in X} F(x, y^*) \geq \tilde{\eta}^-.$$

For each $x \in X$, define $E_x \subset X$ to be the set

$$E_x := \{y \in Y : F(x, y) \geq \tilde{\eta}^-\}.$$

We must show that

$$\cap_{x \in X} E_x \neq \emptyset. \tag{3.12}$$

Notice however that, because $F(x, .)$ is an upper semicontinuous function, the "level set" E_x is compact for every $x \in X$. Property (3.12) will follow then if we can show that, for an arbitrary finite set $\{x_1, \ldots, x_m\}$ of points in X,

$$\cap_{i=1}^m E_{x_i} \neq \emptyset.$$

Take any arbitrary finite subset $K = \{x_1, \ldots, x_m\} \subset X$. Then

$$\cap_{i=1}^m E_{x_i} = \{y \in Y : \min_{i=1,\ldots,m} F(x_i, y) \geq \tilde{\eta}^-\}.$$

But $\min_{i=1,\ldots,m} F(x_i,.)$ is upper semicontinuous and its maximum value is therefore achieved over the compact set Y at some $y^* \in Y$. We have

$$\inf_{x \in K} F(x,y^*) = \sup_{y \in Y} \inf_{x \in K} F(x,y) \geq \tilde{\eta}^-.$$

It follows that $\cap_{i=1}^m E_{x_i}$ contains the point y^* and is therefore nonempty.

Step 3: For a finite subset $K = \{x_1,\ldots,x_m\} \subset X$, define

$$\zeta_K := \inf_{(\lambda_1,\ldots,\lambda_m) \in \Sigma^m} \sup_{y \in Y} \sum_i^m \lambda_i F(x_i,y),$$

in which

$$\Sigma^m := \{(\lambda_1,\ldots,\lambda_m) : \lambda_i \geq 0 \quad \text{for each } i \text{ and } \sum_{i=1}^m \lambda_i = 1\}.$$

We show that

$$\zeta_K \leq \eta_K^-.$$

(Recall that

$$\eta_K^- := \sup_{y \in Y} \inf_{x \in K} F(x,y) .)$$

Indeed suppose, contrary to the claim, that there exists $\alpha > 0$ such that

$$\zeta_K - \alpha > \sup_{y \in Y} \inf_{x \in K} F(x,y).$$

This implies

$$(\zeta_K - \alpha)(1,\ldots,1) \notin C,$$

where $C \subset R^m$ is the subset

$$C := \{(\xi_1,\ldots,\xi_m) : \quad \text{there exists } y \in Y$$
$$\text{such that } \xi_i \leq F(x_i,y) \text{ for } i = 1,\ldots,m\}.$$

However it may be deduced from the concavity of the functions $F(x_i,.)$ and the convexity of the set Y, that C is a convex set.

By the separation theorem there exists a nonzero vector $\lambda \in R^m$ (we may assume that $\sum_{i=1}^m |\lambda_i| = 1$) such that

$$(\zeta_K - \alpha)(1,\ldots,1) \cdot \lambda \geq c \cdot \lambda \quad \text{for all } c \in C.$$

This inequality can be satisfied only if $\lambda_i \geq 0$ for all i. It follows that $\lambda \in \sum^m$ and $(1,\ldots,1) \cdot \lambda = 1$. Inserting

$$c := (F(x_1,y),\ldots,F(x_m,y))$$

into this relationship for arbitrary $y \in Y$ gives

$$\zeta_K - \alpha \geq \sum_{i=1}^{m} \lambda_i F(x, y).$$

Taking the supremum of the right side over $y \in Y$, we arrive at

$$\zeta_K - \alpha \geq \sup_{y \in Y} \sum_{i=1}^{m} \lambda_i F(x_i, y)$$

$$\geq \inf_{\lambda \in \Sigma^m} \sup_{y \in Y} \sum_{i=1}^{m} \lambda_i F(x_i, y) = \zeta_K.$$

From this contradiction we conclude that

$$\zeta_K \leq \eta_K^-.$$

Step 4: We show that

$$\tilde{\eta}^- \geq \eta^+.$$

Take any finite set $K = \{x_1, \ldots, x_m\}$ and $\lambda \in \Sigma^m$. Define

$$x_\lambda := \Sigma_{i=1}^{m} \lambda_i x_i.$$

Since X is convex and $F(., y)$ is a convex function for each $y \in Y$, we have

$$\sup_{y \in Y} \sum_{i=1}^{m} \lambda_i F(x_i, y) \geq \sup_{y \in Y} F(x_\lambda, y) \geq \inf_{x \in X} \sup_{y \in Y} F(x, y) = \eta^+.$$

Taking the infimum on the left over $\lambda \in \Sigma^m$ we deduce that

$$\zeta_K \geq \eta^+.$$

But then, by Step 3,

$$\eta_K^- \geq \eta^+.$$

Taking the infimum of the left side over $K \subset S$ we obtain

$$\tilde{\eta}^- \geq \eta^+.$$

Conclusion

Steps 1, 2, and 4 give

$$\eta^- \leq \tilde{\eta}^- \leq \eta^+,$$

$$\eta^- \left(= \sup_{y \in Y} \inf_{x \in X} F(x, y) \right) \geq \inf_{x \in X} F(x, y^*) \geq \tilde{\eta}^-$$

for some $y^* \in Y$ and

$$\tilde{\eta}^- \geq \eta^+ = \left(\inf_{x \in X} \sup_{y \in Y} F(x, y) \right).$$

We conclude that

$$\inf_{x \in X} F(x, y^*) = \inf_{x \in X} \sup_{y \in Y} F(x, y).$$

This is what we set out to prove. □

Theorem 3.4.6 (Von Neumann Mini-Max Theorem) *Consider a function $F : X \times Y \to R$ in which X and Y are subsets of topological linear spaces. Assume that*

(i) X and Y are convex compact sets,

(ii) $F(.,y)$ is convex and lower semicontinuous for every $y \in Y$, and

(iii) $F(x,.)$ is concave and upper semicontinuous for every $x \in X$.

Then there exists an element $(x^*, y^*) \in X \times Y$ that is a saddlepoint for F; i.e.,

$$\sup_{y \in Y} F(x^*, y) = F(x^*, y^*) = \inf_{x \in X} F(x, y^*). \tag{3.13}$$

This implies in particular (see Proposition 3.4.2)

$$\inf_{x \in X} \sup_{y \in Y} F(x, y) = \sup_{y \in Y} \inf_{x \in X} F(x, y).$$

Proof. Apply Theorem 3.4.5 to $F(x, y)$ and also to $-F(x, y)$ (in the latter case interchanging the roles of X and Y). We deduce existence of some $(x^*, y^*) \in X \times Y$ such that

$$\inf_{x \in X} F(x^*, y^*) = \inf_{x \in X} \sup_{y \in Y} F(x, y)$$

and

$$\sup_{y \in Y} \inf_{x \in X} F(x, y) = \sup_{y \in Y} F(x^*, y).$$

In view of Proposition 3.4.5 it follows that

$$F(x^*, y^*) \leq \sup_{y \in Y} F(x^*, y) = \inf_{x \in X} F(x, y^*) \leq F(x^*, y^*).$$

This implies the saddlepoint condition (3.13). □

Compactness of the underlying sets X and Y (or at least of one of them) plays an essential part in proving the above Mini-Max Theorems. The compactness hypotheses on X and Y can be replaced in certain circumstances by coercivity hypotheses on the "objective function" F; here compactness of the level sets of certain derived functions in some sense substitutes for that of X and Y.

We illustrate the point by proving existence of a saddlepoint for an objective function $F(x,y)$ neither of whose variables x or y is confined to compact sets. This objective function is of interest because of its relevance to studying the relationship between different necessary conditions in Optimal Control.

Proposition 3.4.7 *Take a lower semicontinuous, convex function $h : R^n \to R \cup \{+\infty\}$ such that $\operatorname{dom} h \neq \emptyset$, a number $\sigma > 0$, and vectors $\bar{x} \in R^n$ and $\bar{y} \in R^n$.*
Define
$$F(x,y) := x \cdot (y - \bar{y}) + \sigma |x - \bar{x}|^2 - h(y).$$
Then there exists a point $(x^*, y^*) \in R^n \times \operatorname{dom} h$ such that
$$F(x^*, y) \leq F(x^*, y^*) \leq F(x, y^*) \quad \text{for all } x, y \in R^n.$$

Furthermore
$$x^* = \bar{x} - \frac{1}{2}\sigma^{-1}(y^* - \bar{y}).$$

Proof. Translating $(\bar{x}, \bar{y})$ to the origin and replacing the function h by $y \to \bar{x} \cdot y + h(y + \bar{y})$, we reduce consideration to the case when $\bar{x} = \bar{y} = 0$.

Define $D := \operatorname{dom} h$. By a basic property of lower semicontinuous $R \cup \{+\infty\}$-valued convex functions, h is minorized by an affine function. It follows that there exists $\alpha > 0$ and $\beta > 0$ such that
$$h(y) \geq -\alpha - \beta|y| \quad \text{for all } y \in R^n. \tag{3.14}$$
Define $\psi : R^n \to R \cup \{-\infty\}$ to be the function
$$\psi(y) := \inf_{x \in R^n} F(x, y).$$
We explicity calculate
$$\psi(y) = -\frac{1}{4}\sigma^{-1}|y|^2 - h(y) \quad \text{for all } y \in R^n.$$
Notice that $\psi(y') > -\infty$ for some $y' \in R^n$ (since $D \neq \emptyset$). Also, by (3.14),
$$\limsup_{|y|\to\infty} \psi(y) \leq \limsup_{|y|\to\infty}\{-\frac{1}{4}\sigma^{-1}|y|^2 + \alpha + \beta|y|\} = -\infty.$$

It follows that the number d defined by

$$d := \sup_{y \in R^n} \psi(y)$$

is finite and there exists $K > 0$ such that

$$\sup_{y \in R^n} \psi(y) = \sup_{y \in KB \cap D} \psi(y).$$

The right side is expressible as

$$\sup_{y \in KB \cap D} \inf_{x \in R^n} F(x,y).$$

Applying the One-Sided Mini-Max Theorem (Theorem 3.4.5), we obtain a point $y^* \in D \cap KB$ such that

$$\inf_{x \in R^n} F(x,y^*) = \sup_{y \in KB \cap D} \inf_{x \in R^n} F(x,y) = \sup_{y \in D} \inf_{x \in R^n} F(x,y). \tag{3.15}$$

Now choose $k > 0$ such that

$$-(k - 2\beta - \frac{\alpha}{\sigma k})\sigma k < d - 1 \text{ and } \beta < k, \tag{3.16}$$

and define, for each $y \in R^n$,

$$\psi_k(y) := \inf_{x \in kB} F(x,y).$$

We easily calculate

$$\psi_k(y) = \begin{cases} \psi(y) & \text{if } |y| < 2\sigma k \\ -k|y| + \sigma k^2 - h(y) & \text{if } |y| \geq 2\sigma k. \end{cases}$$

Note however that, in view of (3.14) and (3.16),

$$\psi_k(y) \leq d - 1 = \sup_{y \in D} \psi(y) - 1 \text{ if } |y| \geq 2\sigma k.$$

Since ψ_k majorizes ψ and $\psi_k = \psi$ on $2\sigma kB$, this last relation can be true only if

$$\sup_{y \in D} \psi(y) = \sup_{y \in D} \inf_{x \in kB} F(x,y). \tag{3.17}$$

Now apply the One-Sided Mini-Max Theorem (Theorem 3.4.5) to

$$\inf_{x \in kB} \sup_{y \in D} F(x,y).$$

This gives the existence of x^* such that

$$\sup_{y \in D} F(x^*,y) = \sup_{y \in D} \inf_{x \in kB} F(x,y) = \sup_{y \in D} \inf_{x \in R^n} F(x,y) \tag{3.18}$$

by (3.17).

But then, by Proposition 3.4.4,

$$\sup_{y \in D} \inf_{x \in R^n} F(x, y) = \inf_{x \in R^n} \sup_{y \in D} F(x, y).$$

We conclude from (3.15) that

$$\inf_{x \in R^n} F(x, y^*) = \inf_{x \in R^n} \sup_{y \in D} F(x, y). \tag{3.19}$$

According to Proposition 3.4.3, Assertions (3.18) and (3.19) imply that

$$\sup_{y \in D} F(x^*, y) = F(x^*, y^*) = \inf_{x \in R^n} F(x, y^*).$$

This is the saddlepoint property.

Notice that, since $y^* \in D$, x^* minimizes the function

$$x \to F(x, y^*) = x \cdot y^* + \sigma |x|^2 - h(y^*)$$

over R^n. It follows that the gradient of $F(., y^*)$ vanishes at $x = x^*$. We conclude that

$$x^* = -\frac{1}{2\sigma} y^*.$$

This is the final property to be verified and the proof is complete. $\square$

3.5   Notes for Chapter 3

The important role of exact penalization in the derivation of necessary conditions for constrained optimization problems was early recognized by Clarke [33]. The proof of the Exact Penalization Theorem, Theorem 3.2.1, is taken from [38], adapted to allow for an underlying space that is a (possibly incomplete) metric space in place of a Banach space.

Ekeland's Variational Principle was initially devised to show that approximate minimizers approximately satisfy necessary conditions of optimality [62]. Our proof is taken from [11]. An early application to Optimal Control, which provided a pattern for later research into constrained optimization, was Clarke's derivation of necessary conditions for nonsmooth optimal control problems with endpoint constraints [32]. Since then the principle has been put to many and diverse uses in nonlinear analysis, some of which are described in [63].

It turns out that Ekeland's Variational Principle is just one of a number of related principles, now available, concerning properties of approximate minimizers. Of particular interest is the Variational Principle of Borwein and Preiss [23] which, like that of Ekeland, asserts the existence of a nearby point minimizing a perturbed functional. The difference is that Borwein and Preiss substitute a *smooth* perturbation term in place of Ekeland's nonsmooth one. This is an important feature for certain applications. Discussions of alternative variational principles and transparent proofs are to be found in [54].

The father of Mini-Max is von Neumann. Mini-Max theorems appear, at first sight, rather specialized affairs, because they concern saddle properties of (from some perspectives) the rather narrow class of convex–concave functions. However they are powerful tools in nonconvex optimization, because they can be used to convert the nonnegativity property of the first variation (which can be interpreted as a one-sided mini-max property) into a multiplier rule. The centerpiece of Section 3.4 is Aubin's One-Sided Mini-Max Theorem, which retains part of Von Neumann's Mini-Max Theorem under reduced hypotheses and which is well suited to such applications. We reproduce the proof in [6].

Chapter 4

Nonsmooth Analysis

> *To many mathematicians I became the man of the functions without derivatives ... whenever I tried to take part in a mathematical discussion there would always be an analyst who would say, "This won't interest you; we are discussing functions having derivatives" ... or a geometer would say ... "We're discussing surfaces that have tangent planes."*
>
> – Henri Lebesgue

4.1   Introduction

Let $\bar{x} \in R^k$ be a point in the manifold

$$C := \{x : g_i(x) = 0 \quad \text{for } i = 1, \ldots, m\}$$

in which $g_i : R^k \to R, i = 1, \ldots, m$ are given continuously differentiable functions such that $\nabla g_1(\bar{x}), \ldots, \nabla g_m(\bar{x})$ are linearly independent. Then the set of normal vectors to C at $\bar{x}$ is

$$\left\{ \sum_{i=1}^m \lambda_i \nabla g_i(\bar{x}) : \lambda_1, \ldots, \lambda_m \in R \right\},$$

and its orthogonal complement (translated to $\bar{x}$),

$$\bar{x} + \{y : y \cdot \nabla g_i(\bar{x}) = 0 \text{ for } i = 1, 2, \ldots, m\},$$

is an affine subspace of R^k that provides a local approximation to C "near" $\bar{x}$.

If, on the other hand, we are given a continuously differentiable function $f : R^k \to R$ and a point $\bar{x}$, then

$$x \to f(\bar{x}) + \nabla f(\bar{x}) \cdot (x - \bar{x})$$

is an affine function that approximates f near $\bar{x}$.

We see here how approximations to smooth manifolds and functions are traditionally constructed: they take the form of affine subspaces for smooth manifolds and of affine functions for smooth functions. The importance of these approximations is that it is often possible to predict qualitative properties of smooth manifolds and functions from properties of their "affine" approximations, which are in almost all cases simpler to investigate. A case in point is the Inverse Function Theorem, which tells us that a continuously differentiable function $f : R^k \to R$ is invertible on a neighborhood of a point $\bar{x} \in R^k$ if its affine approximation $x \to f(\bar{x}) + \nabla f(\bar{x}) \cdot (x - \bar{x})$ is invertible.

What general principles and techniques may be developed, governing the approximation of sets and functions in topological vector spaces when the regularity and differentiability hypotheses underpinning traditional methods are violated? These are precisely the questions that Nonsmooth Analysis aims to answer.

The key idea is to abandon the notion of affine approximation. In Nonsmooth Analysis, closed sets (and in particular manifolds) are approximated, not by affine subspaces, but by *cones* and functions are approximated, not by a single affine function, but by a *family* of affine functions.

Once this idea is accepted, it is possible to extend to a nonsmooth setting many principles of traditional nonlinear analysis and to assemble a calculus of rules for estimating approximations to many specific sets and functions of interest. There has been a great deal of research activity in Nonsmooth Analysis over the past 25 years. A sophisticated and far-reaching theory has now been put together, the significance of which is confirmed by a growing body of applications in Mathematical Programming, Variational Analysis, Optimal Control, and other fields of applied nonlinear analysis.

Our aim in this and the next chapter is to provide a self-contained treatment of those aspects of Nonsmooth Analysis of particular relevance to Optimal Control. In this chapter, basic constructs of Nonsmooth Analysis are defined and relationships between them are explored. In the next, a "generalized calculus" is developed.

4.2   Normal Cones

Take a closed set $C \subset R^k$ and a point $x \in R^k$. We wish to give meaning to "normals" to C at x, that is vectors which in some sense point out of C from the "basepoint" x. We introduce three notions of normals: proximal normals, strict normals, and limiting normals.

Proximal normals are vectors that satisfy an outward-pointing condition with quadratic error term. Proximal normals have a simple geometric interpretation and, for this reason, are often easy to characterize in specific applications. Strict normals are similarly defined, except an "order" term replaces the quadratic error term in the defining condition. Limiting normals, as their name suggests, are limits of proximal normals at neighboring basepoints; they can be equivalently defined as limits of strict normals at neighboring basepoints, as we show.

Limiting normals feature most prominently in the applications to optimization explored in this book, owing to their superior analytic properties. Many of these properties are consequences of the fact that the cone of limiting normals has a closed graph, regarded as a set-valued function of the basepoint. Proximal normals have the role of building blocks of limiting normals. Strict normals provide an alternative route to generating limiting normals via limit-taking.

Definition 4.2.1 Given a closed set $C \subset R^k$ and a point $x \in C$, the *proximal normal cone* to C at x, written $N_C^P(x)$, is the set

$$N_C^P(x) := \{p \in R^k : \exists\, M > 0 \text{ such that Condition (4.1) is satisfied }\}$$

$$p \cdot (y - x) \leq M|y - x|^2 \quad \text{for all } y \in C. \tag{4.1}$$

Elements in $N_C^P(x)$ are called *proximal normals* to C at x.

The defining condition (4.1) for a proximal normal, which is referred to as the "proximal normal inequality (for sets)," can be expressed as follows: there exists $M > 0$ such that

$$|(x + (2M)^{-1}p) - y| \geq (2M)^{-1}|p| \quad \text{for all } y \in C. \tag{4.2}$$

In terms of $r = 2M$ and $z = x + (2M)^{-1}p$, the condition amounts to:

$$|z - x| = \min\{|z - y| : y \in C\} \quad \text{and} \quad p = r(z - x).$$

These observations lead to the following geometric interpretation of proximal normals.

Proposition 4.2.2 *Take a closed set $C \subset R^k$ and points $x \in C$ and $p \in R^k$. Then p is a proximal normal to C at x if and only if there exist a point $z \in R^k$ and a scaling factor $r > 0$ such that*

$$|z - x| = \min\{|z - y| : y \in C\} \quad \text{and} \quad p = r(z - x).$$

This proposition tells us that p is a nonzero proximal normal to C at x precisely when there is some $z \in R^k$ lying outside C such that x is the closest point in C to z (with respect to the Euclidean distance function) and the vector p points in the direction $(z - x)$. See Figure 1.9.

The derivation of many useful relationships involving normal cones to a set C at x involves limit-taking with respect to the basepoint x. This is often not possible if the cones in question are proximal normal cones, because membership of proximal normal cones is not in general preserved under such operations. We can construct a normal cone from the proximal normal cone that does have the desired stability with respect to changes of the basepoint, by adding extra normals which are limits of proximal normals as nearby basepoints. In this way we arrive at the limiting normal cone. See Figure 1.10.

Definition 4.2.3 Given a closed set $C \subset R^k$ and a point $x \in C$, the *limiting normal cone* to C at x, written $N_C(x)$, is the set

$$N_C(x) := \left\{ p : \text{ there exists } x_i \overset{C}{\to} x, p_i \to p \text{ such that } p_i \in N_C^P(x_i) \text{ for all } i \right\}.$$

Elements in $N_C(x)$ are called *limiting normals* to C at x.

Another normal cone of interest is the strict normal cone.

Definition 4.2.4 *Given a closed set $C \subset R^k$ and a point $x \in C$, the strict normal cone to C at x, written $\hat{N}_C(x)$, is the set*

$$\hat{N}_C(x) := \left\{ p : \limsup_{y \overset{C}{\to} x} \frac{p \cdot (y - x)}{|y - x|} \leq 0 \right\}.$$

Elements in $\hat{N}_C(x)$ are called strict normals to C at x.

Otherwise expressed, a strict normal p to C at x is a vector satisfying

$$p \cdot (y - x) \leq o(|y - x|) \quad \text{for all } y \in C, \tag{4.3}$$

for some function $o : R^+ \to R^+$ such that $\lim_{\epsilon \downarrow 0} o(\epsilon)/\epsilon = 0$; in other words p satisfies (to "within first-order") conditions asserting that it is an outward normal to a hyperplane supporting C at x. Notice that the defining relationship (4.3) for strict normals is weaker than the defining relationship (4.1) for proximal normals: for a strict normal p also to be a proximal normal, we stipulate that a special form of error modulus $o(\cdot)$ can be used, namely, a quadratic function. It follows that

$$N_C^P(x) \subset \hat{N}_C(x).$$

Proximal normal cones and strict normal cones are distinct concepts: it is not difficult to supply examples in which

$$N_C^P(x) \overset{\text{strict}}{\subset} \hat{N}_C(x).$$

However limits of strict normals at basepoints converging to x generate the *same* set $(N_C(x))$ as limits of proximal normals. It is for this reason that strict normals, in place of proximal normals, provide an alternative foundation on which to define and establish properties of limiting normal cones.

Proposition 4.2.5 *Take a closed set $C \subset R^k$ and points $x \in C$ and $p \in R^k$. Then the following assertions are equivalent:*

(i) $p \in N_C(x)$;

(ii) there exist sequences $x_i \xrightarrow{C} x$ and $p_i \to p$ such that $p_i \in \hat{N}_C(x_i)$ for all i.

Proof. (i) implies (ii), by the definition of $N_C(x)$ and since $N_C^P(x) \subset \hat{N}_C(x)$.

To establish that (ii) implies (i), we show that strict normals can be suitably approximated by proximal normals. Fix $q \in \hat{N}_C(y)$ for some $y \in C$. It suffices to demonstrate that sequences $y_i \xrightarrow{C} y$ and $q_i \to q$ may be found such that $q_i \in N_C^P(y_i)$ for all i. For then we can show, by constructing a suitable diagonal sequence, that if a vector p is the limit of a sequence of strict normals at neighboring points, then it is also the limit of proximal normals; that is, it lies in the limiting normal cone.

Take any $\epsilon_i \downarrow 0$. Then for each i, we can find a point y_i that is a closest point to $(y + \epsilon_i q)$ in C. Since $\epsilon_i q \to 0$, we have that $y_i \to y$. For each i define $q_i := \epsilon_i^{-1}(y + \epsilon_i q - y_i)$. Then $q_i \in N_C^P(y_i)$. We see that $q_i = q + \epsilon_i^{-1}(y - y_i)$, so the proof will be complete if we can show that $\epsilon_i^{-1}(y - y_i) \to 0$. However by the "closest point" property of y_i,

$$|(y + \epsilon_i q) - y_i|^2 \le |(y + \epsilon_i q) - z|^2 \quad \text{for all } z \in C.$$

Choosing $z = y$ we arrive at

$$0 \ge |y_i - (y + \epsilon_i q)|^2 - \epsilon_i^2 |q|^2 = |y_i - y|^2 - 2\epsilon_i(y_i - y) \cdot q.$$

But $q \in \hat{N}_C(y)$. It follows that

$$|y_i - y|^2 \le 2\epsilon_i(y_i - y) \cdot q \le 2\epsilon_i o(|y_i - y|)$$

for some function $o(\cdot) : R^+ \to R^+$ such that $\lim_{\epsilon \downarrow 0} o(\epsilon)/\epsilon = 0$. We conclude that $\epsilon_i^{-1}|y_i - y| \to 0$. $\square$

Some elementary properties of the cones that have been introduced are now listed without proof.

Proposition 4.2.6 *Take a closed set $C \subset R^k$ and a point $x \in C$. Then:*

(i) $N_C^P(x)$, $\hat{N}_C(x)$, and $N_C(x)$ are all cones in R^k, containing $\{0\}$ and

$$N_C^P(x) \subset \hat{N}_C(x) \subset N_C(x);$$

(ii) $N_C^P(x)$ is convex (but possibly not closed);

(iii) $\hat{N}_C(x)$ is closed and convex;

(iv) the set-valued mapping $y \to N_C(y) : C \to R^k$ has a closed graph, in the sense that, for any sequences $y_i \xrightarrow{C} y$ and $p_i \to p$ such that $p_i \in N_C(y_i)$ for all i, we have $p \in N_C(x)$.

We note also

Proposition 4.2.7 *Take a closed set $C \subset R^k$ and a point $x \in C$. Then:*

(i) $x \in \text{int}\{C\}$ *implies* $N_C(x) = \{0\}$ *(and hence*

$$N_C^P(x) = \hat{N}_C(x) = \{0\}).$$

(ii) $x \in \text{bdy}\{C\}$ *implies* $N_C(x)$ *contains nonzero elements.*

Proof. (i) Suppose $x \in \text{int}\{C\}$. It follows from Proposition 4.2.2 that all proximal normals at points in C near to x are zero. So $N_C(x) = \{0\}$.

(ii) Assume $x \in \text{bdy}\{C\}$. Then there exists a sequence $x_i \to x$ such that $x_i \notin C$ for all i. For each i select a closest point c_i to x_i in C. Then $x_i - c_i \neq 0$ and $\xi_i := |x_i - c_i|^{-1}(x_i - c_i)$ is a proximal normal to C at c_i, of unit norm. But $|x - c_i| \leq |x - x_i| + |x_i - c_i| \leq 2|x - x_i|$. So $c_i \overset{C}{\to} x$. Along a subsequence then $\xi_i \to \xi$. The vector ξ has unit norm and belongs to $\xi \in N_C(x)$. $\square$

We note also

Proposition 4.2.8 *Take closed subsets $C_1 \subset R^m$ and $C_2 \subset R^n$, and a point $(x_1, x_2) \in C_1 \times C_2$. Then*

$$\begin{aligned}
N_{C_1 \times C_2}^P(x_1, x_2) &= N_{C_1}^P(x_1) \times N_{C_2}^P(x_2) \\
\hat{N}_{C_1 \times C_2}(x_1, x_2) &= \hat{N}_{C_1}(x_1) \times \hat{N}_{C_2}(x_2) \\
N_{C_1 \times C_2}(x_1, x_2) &= N_{C_1}(x_1) \times N_{C_2}(x_2).
\end{aligned}$$

Proof. The first two relationships are direct consequences of the definitions of proximal normals and strict normals. The last relationship is a consequence of the first relationship and the manner in which limiting normals are obtained as limits of proximal normals. $\square$

In the convex case, the normal cones above all coincide with the normal cone in the sense of Convex Analysis

Proposition 4.2.9 *Take a closed convex set $C \subset R^k$ and a point $\bar{x} \in C$. Then*

$$\begin{aligned}
N_C^P(\bar{x}) &= \hat{N}_C(\bar{x}) = N_C(\bar{x}) \\
&= \{\xi : \xi \cdot (x - \bar{x}) \leq 0 \text{ for all } x \in C\}.
\end{aligned}$$

Proof. For $y \in C$ write

$$S(y) := \{\xi : \xi \cdot (y' - y) \leq 0 \text{ for all } y' \in C\}.$$

Since the conditions for membership of $S(\bar{x})$ are more severe than those for membership of $N_C^P(\bar{x})$, we know that

$$S(\bar{x}) \subset N_C^P(\bar{x}).$$

Take an arbitrary element $\xi \in N_C^P(\bar{x})$. By definition, there exists $M \geq 0$ such that

$$\xi \cdot (x' - \bar{x}) \leq M|x' - \bar{x}|^2,$$

for all $x' \in C$. Choose any $x \in C$. Since C is convex, $x' = \epsilon x + (1 - \epsilon)\bar{x} \in C$ for any $\epsilon \in (0, 1)$. Inserting this choice of x' into the above inequality, dividing across by ϵ, and passing to the limit as $\epsilon \downarrow 0$ gives

$$\xi \cdot (x - \bar{x}) \leq 0.$$

It follows that $\xi \in S(\bar{x})$. This establishes that $S(\bar{x}) \subset N_C^P(\bar{x})$. We conclude that

$$S(\bar{x}) = N_C^P(\bar{x}).$$

However it is easy to deduce from the lower semicontinuity of f that, for any $y \in \operatorname{dom} f$, sequences $y_i \xrightarrow{C} y$ and $\xi_i \to \xi$ such that $\xi_i \in S(y_i)$ for all i, we have $\xi \in S(y)$. It follows from the definition of $N_C^P(\bar{x})$ that

$$S(\bar{x}) = N_C(\bar{x}).$$

The remaining relationships in the proposition to be proved follow from Proposition 4.2.6 (i). $\square$

4.3   Subdifferentials

Consider a function $f : R^k \to R$ and a point $x \in R^k$. Suppose that f is C^2 (twice continuously differentiable). Then the gradient $\nabla f(x)$ of f at x is the unique vector $\xi \in R^k$ such that, for some $M > 0$ and $\epsilon > 0$,

$$|f(y) - f(x) - \xi \cdot (y - x)| \leq M|y - x|^2 \quad \text{for all } y \in x + \epsilon B. \tag{4.4}$$

This description evokes the familiar "difference quotient" characterization of the gradient. The fact that we assume f is a C^2 function permits us to use a quadratic error term on the right side.

An alternative description of $\nabla f(x)$ is in terms of the epigraph of f, epi f,

$$\operatorname{epi} f := \{(x, \alpha) \in R^k \times R : \alpha \geq f(x)\}.$$

It can be shown that $\nabla f(x)$ is the unique vector $\xi \in R^k$ such that, for some $M \geq 0$,

$$(\xi, -1) \cdot ((y, \alpha) - (x, f(x))) \leq M(|y - x|^2 + |\alpha - f(x)|^2) \qquad (4.5)$$

for all $(y, \alpha) \in \text{epi } f$. (It is always possible to find some M satisfying this "global" inequality when $\xi = \nabla f(x)$; the "extra" error term $|\alpha - f(x)|^2$ ensures that the inequality holds for all y, not just ys near x). To see that this last condition does indeed identify $\nabla f(x)$, we note that $(y, f(y)) \in \text{epi } f$. Hence

$$\begin{aligned}
0 \;\leq\; & \liminf_{y \to x} \frac{f(y) - f(x) - \xi \cdot (y - x)}{|y - x|} \\
=\; & \liminf_{y \to x} \left[\frac{f(y) - f(x) - \nabla f(x) \cdot (y - x)}{|y - x|} + \frac{(\nabla f(x) - \xi) \cdot (y - x)}{|y - x|} \right] \\
=\; & 0 + \liminf_{y \to x} \frac{(\nabla f(x) - \xi) \cdot (y - x)}{|y - x|} \\
=\; & -|\nabla f(x) - \xi|.
\end{aligned}$$

It follows that $\xi = \nabla f(x)$.

The reasons for introducing relationship (4.5) may not be immediately apparent. All becomes clear however when it is interpreted in terms of normal cones: (4.5) can be equivalently stated

$$(\xi, -1) \in N^P_{\text{epi } f}(x, f(x)), \qquad (4.6)$$

where, we recall, $N^P_{\text{epi } f}(x, f(x))$ denotes the proximal normal cone.

Up to this point, we have discussed different ways of looking at the gradient of a C^2 function f. Our ulterior motive of course has been to come up with some description that makes sense when f is no longer "differentiable" in the conventional sense.

Two routes are open to us, to generalize either (4.4) (the analytical approach) or (4.6) (the geometric approach). As they stand, Conditions (4.4) and (4.6) are often rather restrictive: even for Lipschitz continuous functions f there will be many points x at which the sets of ξs satisfying either (4.4) or (4.6) will be empty. It is therefore often helpful to play variations on these themes, namely, to consider limits of ξ_is satisfying (4.4) along a sequence of points $x_i \to x$, or to replace the quadratic error term $M|y - x|^2$ by an error modulus $o(|y - x|)$ (if we are going down the analytical road), or to insert different kinds of normal cones in place of $N^P_{\text{epi } f}(x, f(x))$ (if we want to take a more geometric approach).

Different choices are available regarding what we deem to be definitions and what to be consequences of the definitions. The theory would appear to unfold most simply when the basic definitions are given in terms of normal cones to epigraphs, and this therefore is our chosen approach.

To what class of functions should we aim to generalize the traditional notion of derivatives? The class of lower semicontinuous functions $f : R^k \to R \cup \{+\infty\}$ is a natural choice here. This is because generalized derivatives are defined in terms of normal cones to the epigraph set epi f, so we would like to arrange that the epigraph set is closed. The class of lower semicontinuous functions is precisely the class of functions with closed epigraph sets.

Definition 4.3.1 Take a lower semicontinuous function $f : R^k \to R \cup \{+\infty\}$ and a point $x \in \text{dom } f$.

(i) The *proximal subdifferential* of f at x, written $\partial^P f(x)$, is the set

$$\partial^P f(x) := \left\{ \xi : (\xi, -1) \in N_{\text{epi } f}^P (x, f(x)) \right\}.$$

Elements in $\partial^P f(x)$ are called *proximal subgradients*.

(ii) The *strict subdifferential* of f at x, written $\hat{\partial} f(x)$, is the set

$$\hat{\partial} f(x) := \{ \xi : (\xi, -1) \in \hat{N}_{\text{epi } f}(x, f(x)) \}.$$

Elements in $\hat{\partial} f(x)$ are called *strict subgradients*.

(iii) The *limiting subdifferential* of f at x, written $\partial f(x)$, is the set

$$\partial f(x) := \{ \xi : (\xi, -1) \in N_{\text{epi } f}(x, f(x)) \}.$$

Elements in $\partial f(x)$ are called *limiting subgradients*.

Proximal subgradients and limiting subdifferentials are illustrated in Figures 1.11 and 1.12 respectively.

Notice that, since

$$N_C^P(x) \subset \hat{N}_C(x) \subset N_C(x),$$

we have

$$\partial^P f(x) \subset \hat{\partial} f(x) \subset \partial f(x).$$

These different subdifferentials provide local information about a function f near a point x. The most noteworthy departure from classical real analysis is that, for lower semicontinuous functions, the subdifferentials are in general *set-valued*.

Example 4.1   (In each of the following cases, f is a function from R to R.)

(i) $f(y) = |y|$ and $x = 0$. Then

$$\partial^P f(x) = \hat{\partial} f(x) = \partial f(x) = [-1, +1].$$

(ii) $f(y) = -|y|$ and $x = 0$. Then

$$\partial^P f(x) = \hat{\partial} f(x) = \emptyset \quad \text{and} \quad \partial f(x) = \{-1\} \cup \{+1\}.$$

(iii) $f(y) = |y|^{1/2}$ and $x = 0$. Then

$$\partial^P f(x) = \hat{\partial} f(x) = \partial f(x) = (-\infty, +\infty).$$

(iv) $f(y) = \text{sgn}\{y\}|y|^{1/2}$ and $x = 0$. Then

$$\partial^P f(x) = \hat{\partial} f(x) = \partial f(x) = \emptyset.$$

Unbounded or empty subdifferentials (as occur in (iii) and (iv)) give warning that the slopes of the function on an arbitrary small ϵ-ball about the basepoint are unbounded. In the presence of such pathologies, we would like to know the *direction* of these arbitrarily large slopes. Information of this nature is conveyed by some new constructs; they are the "asymptotic" analogues of the subdifferentials already defined.

Definition 4.3.2 Take a lower semicontinuous function $f : R^k \to R \cup \{+\infty\}$ and a point $x \in \text{dom } f$.

(i) The *asymptotic proximal subdifferential* of f at x, written $\partial_P^\infty f(x)$, is

$$\partial_P^\infty f(x) := \{\xi : (\xi, 0) \in N_{\text{epi } f}^P(x, f(x))\}.$$

Elements in $\partial_P^\infty f(x)$ are called *asymptotic proximal subgradients*.

(ii) The *asymptotic strict subdifferential* of f at x, written $\hat{\partial}^\infty f(x)$, is

$$\hat{\partial}^\infty f(x) := \{\xi : (\xi, 0) \in \hat{N}_{\text{epi } f}(x, f(x))\}.$$

Elements in $\hat{\partial}^\infty f(x)$ are called *asymptotic strict subgradients*.

(iii) The *asymptotic limiting subdifferential* of f at x, written $\partial^\infty f(x)$, is

$$\partial^\infty f(x) := \{\xi : (\xi, 0) \in N_{\text{epi } f}(x, f(x))\}.$$

Elements in $\partial^\infty f(x)$ are called *asymptotic limiting subgradients*.

Example 4.2   Consider the function $f : R \to R$ given by $f(y) = \text{sgn}\{y\}|y|^{1/2}$ and the point $x = 0$. As we have noted $\partial f(x) = \emptyset$. However $\partial^\infty f(x) = [0, \infty)$. The calculation of $\partial^\infty f(x)$ has revealed to us that the "infinite" slopes near x are positive.

The asymptotic limiting subdifferential $\partial^\infty f(x)$ is of interest also because of the information that nonzero asymptotic limiting subgradients give about "non-Lipschitz behavior" of a function f near a point $x \in \text{dom } f$. It is to be expected then that the asymptotic limiting subdifferential of a Lipschitz continuous function is the trivial set $\{0\}$.

Proposition 4.3.3 *Take a lower semicontinuous function $f : R^k \to R \cup$ $\{+\infty\}$ and a point $\bar{x} \in R^k$. Assume that f is Lipschitz continuous on a neighborhood of $\bar{x}$ with Lipschitz constant K. Then:*

(i) $\partial f(\bar{x}) \subset KB$;

(ii) $\partial^\infty f(\bar{x}) = \{\ 0\ \}$.

Proof. Take a point

$$(\xi, -\lambda) \in N_{\text{epi}\,f}(\bar{x}, f(\bar{x})).$$

Then there exist convergent sequences

$$(x_i, \alpha_i) \overset{\text{epi}\,f}{\to} (\bar{x}, f(\bar{x})) \quad \text{and} \quad (\xi_i, -\lambda_i) \to (\xi, -\lambda)$$

such that

$$(\xi_i, -\lambda_i) \in N^P_{\text{epi}\,f}(x_i, \alpha_i) \quad \text{for all } i \text{ sufficiently large,}$$

and such that f is Lipschitz continuous on a neighborhood of x_i with Lipschitz constant K.

We know that $\alpha_i = f(x_i) + \beta_i$ for some $\beta_i \geq 0$. It follows now from the definition of proximal subgradients that

$$\xi_i \cdot (x - x_i) - \lambda_i(\beta + f(x) - \beta_i - f(x_i))$$
$$\leq M(|x - x_i|^2 + |\beta - \beta_i + f(x) - f(x_i)|^2)$$

for all $\beta \geq 0$ and $x \in R^n$.

Setting $x = x_i$ we deduce

$$-\lambda_i(\beta - \beta_i) \leq M|\beta - \beta_i|^2 \quad \text{for all } \beta \geq 0.$$

This implies $\lambda_i \geq 0$. Passing to the limit as $i \to \infty$, we obtain $\lambda \geq 0$.

Setting $\beta = \beta_i$ we deduce from the Lipschitz continuity of f that

$$\xi_i \cdot (x - x_i) \leq \lambda_i K|x - x_i| + (M + MK^2)|x - x_i|^2$$

for all x in a neighborhood of x_i. Choosing $x = x_i + r\xi_i$ for $r > 0$ sufficiently small gives

$$r|\xi_i|^2 \leq rK\lambda_i|\xi_i| + M(1 + K^2)r^2|\xi_i|^2.$$

Dividing across by r and passing to the limit as $r \downarrow 0$ gives

$$|\xi_i|^2 \leq K\lambda_i|\xi_i|.$$

We deduce that

$$|\xi_i| \leq K\lambda_i \quad \text{for each } i.$$

Passing to the limit as $i \to \infty$, we arrive at

$$|\xi| \leq K\lambda.$$

It follows that $\partial^\infty f(\bar{x}) = \{0\}$ and $\partial f(\bar{x}) \subset KB$. $\square$

A converse of the above proposition is proved later in the chapter (Theorem 4.9.1 and Corollary 4.9.2).

Limiting subgradients and asymptotic limiting subgradients are defined in terms of the limiting normal cone to the epigraph set. Taking a reverse view, we can represent the limiting normal cone to the epigraph of a function by means of the subgradients of the function.

Proposition 4.3.4 *Take a lower semicontinuous function $f : R^k \to R \cup \{+\infty\}$ and a point $x \in \mathrm{dom}\, f$. Then:*

$$N_{\mathrm{epi}\, f}(x, f(x)) = \{(\lambda\xi, -\lambda) : \lambda > 0, \ \xi \in \partial f(x)\} \cup (\partial^\infty f(x) \times \{0\}).$$

Either $\partial f(x)$ is nonempty or $\partial^\infty f(x)$ contains nonzero elements.

Proof. The first part of the proposition merely describes how the limiting subdifferential and asymptotic limiting subdifferential are constructed from the limiting normal cone. Since $(x, f(x))$ is a boundary point of epi f, $N_{\mathrm{epi}\, f}(x, f(x))$ contains nonzero elements (see Proposition 4.2.7). The fact that either $\partial f(x)$ is nonempty or $\partial^\infty f(x)$ contains nonzero elements is an immediate consequence. $\square$

The following "closure" properties of the limiting subdifferential and the asymptotic limiting subdifferential follow from their representation in terms of $N_{\mathrm{epi}\, f}(x, f(x))$ and from the fact that the set-valued function $z \to N_{\mathrm{epi}\, f}(z) : \mathrm{epi}\, f \to R^k \times R$ has a closed graph.

Proposition 4.3.5 *Take a lower semicontinuous function $f : R^k \to R \cup \{+\infty\}$ and a point $x \in \mathrm{dom}\, f$. Then:*

(i) *$\partial f(x)$ is a closed set. Given sequences $x_i \xrightarrow{f} x$ and $\xi_i \to \xi$ such that $\xi_i \in \partial f(x_i)$ for all i, then $\xi \in \partial f(x)$;*

(ii) *$\partial^\infty f(x)$ is a closed cone. Given sequences $x_i \xrightarrow{f} x$ and $\xi_i \to \xi$ such that $\xi_i \in \partial^\infty f(x_i)$ for all i, then $\xi \in \partial^\infty f(x)$.*

Finally we note that, for convex functions, the definitions of proximal, strict, and limiting subdifferential all coincide with the subdifferential in the sense of Convex Analysis.

Proposition 4.3.6 *Take a lower semicontinuous, convex function $f : R^k \to R \cup \{+\infty\}$ and a point $\bar{x} \in \operatorname{dom} f$. Then*

$$\partial^P f(\bar{x}) = \hat{\partial} f(\bar{x}) = \partial f(\bar{x})$$
$$= \{\xi : \xi \cdot (x - \bar{x}) \leq f(x) - f(\bar{x}) \text{ for all } x \in R^k\}.$$

Proof. Define

$$D(\bar{x}) = \{\xi : \xi \cdot (x - \bar{x}) \leq f(x) - f(\bar{x}) \text{ for all } x \in R^k\}.$$

It is easy to see that $D(\bar{x})$ can be alternatively expressed in terms of the normal cone (in the sense of convex analysis) to $\operatorname{epi} f$ at $(\bar{x}, f(\bar{x}))$:

$$D(\bar{x}) = \{\xi : (\xi, -1) \cdot ((x, \alpha) - (\bar{x}, f(\bar{x}))) \leq 0 \text{ for all } (x, \alpha) \in \operatorname{epi} f\}.$$

All the assertions of the proposition now follow from Proposition 4.2.9 and the definitions of the proximal, strict, and limiting normal cones. $\square$

4.4 Difference Quotient Representations

Subgradients have been defined, somewhat indirectly, via normals to epigraph sets. This approach has certain advantages from the point of view of unifying the treatment of normal vectors and subgradients and of deriving a number of important properties of subgradients. This section provides alternative conditions for vectors to be proximal subgradients (and strict subgradients), in terms of limits of difference quotients. These often simplify the task of investigating the detailed properties of subgradients in specific cases.

We start with a very useful characterization of proximal subgradients.

Proposition 4.4.1 *Take a lower semicontinuous function $f : R^k \to R \cup \{+\infty\}$ and points $x \in \operatorname{dom} f$ and $\xi \in R^k$. Then the following two statements are equivalent.*

(i) $\xi \in \partial^P f(x)$;

(ii) there exist $M > 0$ and $\epsilon > 0$ such that

$$\xi \cdot (y - x) \leq f(y) - f(x) + M|y - x|^2 \tag{4.7}$$

for all $y \in x + \epsilon B$.

The significance of Condition (ii), which is referred to as the proximal normal inequality for functions, is the absence of the term $M|f(y) - f(x)|^2$ from the right side of inequality (4.7), a term which you would expect to be required to provide a "difference quotient" characterization of proximal subgradient. In situations when f is not continuous this troublesome error term complicates the analysis of limits of proximal normal vectors. Notice that (4.7) is required to hold only locally.

Proof. Assume (ii). Inequality (4.7) implies

$$\xi \cdot (y - x) \le \alpha - f(x) + M[|y - x|^2 + |\alpha - f(x)|^2]$$

for any $(y, \alpha) \in \text{epi } f \cap [(x, f(x)) + \epsilon B]$. The inequality can be written

$$(\xi, -1) \cdot ((y, \alpha) - (x, f(x))) \le M|(y, \alpha) - (x, f(x))|^2. \qquad (4.8)$$

The geometric interpretation of this relation is that the open ball, with center $(x, f(x)) + (2M)^{-1}(\xi, -1)$ and of radius $(2M)^{-1}|(\xi, -1)|$, is disjoint from epi $f \cap [(x, f(x)) + \epsilon B]$, and the closure of this ball contains $(x, f(x))$. (See Proposition 4.2.2 and preceding comments). So for $\tilde{M} \ge M$ sufficiently large, when we substitute $\tilde{M}$ in place of M, this open ball is contained in $(x, f(x)) + \epsilon \text{ int } B$, and is consequently disjoint from epi f. Inequality (4.8), which is valid when M is replaced by the larger $\tilde{M}$, then tells us that

$$\left((x, f(x)) + (2\tilde{M})^{-1}(\xi, -1) + (2\tilde{M})^{-1}|(\xi, -1)|B\right) \cap \text{epi } f = \emptyset.$$

Again appealing to the geometric interpretation of proximal normals, we deduce that $(\xi, -1) \in N^P_{\text{epi } f}(x, f(x))$. We have shown that (i) is true.

Assume (i). We may suppose that $(x, f(x)) = (0, 0)$ (this amounts simply to translating the origin in $R^k \times R$ to $(x, f(x))$).

By assumption there exists $M > 0$ such that

$$(\xi, -1) \cdot (y, \alpha) \le M(|y|^2 + |\alpha|^2) \quad \text{for all } (y, \alpha) \in \text{epi } f.$$

Set $\lambda = (2M)^{-1}$. Denote by int $B^k(y, r)$ the open ball in R^k with center y and of radius r. According to Proposition 4.2.2 and the preceding discussion $(0, 0) + \text{int } B^{k+1}(s, \rho)$ is disjoint from epi f for

$$s = (s_1, s_2) := (\lambda \xi, -\lambda) \quad \text{and} \quad \rho = \lambda\sqrt{1 + |\xi|^2}.$$

For each $x \in \text{int } B^k(s_1, \rho)$ consider the vertical line $\{x\} \times R$. This penetrates the ball int $B^{k+1}(s, \rho)$. Define the functionals $\phi^+(x)$ and $\phi^-(x)$ to take as values the vertical coordinates of the points at which this line intercepts the upper and lower hemispheres, respectively.

For each $x \in \text{int } B^k(s_1, \rho)$, $\phi^+(x) > \phi^-(x)$ and consequently

$$\{x\} \times (\phi^-(x), \phi^+(x))$$

is a nonempty subset of int $B^{k+1}(s, \rho)$. But $\{x\} \times [f(x), \infty) \subset \text{epi } f$. Since epi f and int $B^{k+1}(s, \rho)$ are disjoint, we conclude that

$$f(x) \ge \phi^+(x) \quad \text{for all } x \in \text{int } B^k(s_1, \rho). \qquad (4.9)$$

In particular this inequality holds for all x in the neighborhood int $B^k(0, \epsilon')$ of the origin, in which ϵ' is the positive number

$$\epsilon' := \frac{1}{2}(\lambda\sqrt{1 + |\xi|^2} - \lambda|\xi|).$$

A simple calculation yields the following formula for $\phi^+(x)$,

$$\phi^+(x) = \sqrt{\lambda^2 + \lambda^2|\xi|^2 - |x - \lambda\xi|^2} - \lambda$$

for $x \in$ int $B^k(0, \epsilon')$. $\phi^+(x)$ is analytic on int $B^k(0, \epsilon')$. The gradient $\nabla\phi^+(0)$ and Hessian $\nabla^2\phi^+(0)$ at the origin are

$$\begin{aligned}\nabla\phi^+(0) &= \xi \\ \nabla^2\phi^+(0) &= -\lambda^{-1}(I + \xi\xi^T).\end{aligned}$$

Since $\lambda = (2M)^{-1}$ and $\phi^+(0) = 0$, we have

$$\begin{aligned}\phi^+(x) &= \phi^+(0) + \nabla\phi^+(0) \cdot x + \frac{1}{2}x \cdot \left(\nabla^2\phi^+(0)\right)(x) + o(x) \\ &= \xi \cdot x - M|x|^2 - M|\xi \cdot x|^2 + o(x). \end{aligned} \qquad (4.10)$$

Here, $o(\cdot)$ is some function that satisfies $\lim_{x \to 0} |o(x)|/|x|^2 = 0$. Now take any $\epsilon \in (0, \epsilon')$ such that

$$|o(x)| < M(1 + |\xi|^2)|x|^2 \quad \text{for all} x \in \text{int } B^k(0, \epsilon).$$

It follows then from (4.9) and (4.10) that

$$\begin{aligned}\xi \cdot x &\leq f(x) + M|x|^2 + M|\xi \cdot x|^2 - o(x) \\ &\leq f(x) + M(1 + |\xi|^2)|x|^2 + M(1 + |\xi|^2)|x|^2 \\ &= f(x) + \tilde{M}|x|^2 \quad \text{for all } x \in \text{int } B^k(0, \epsilon),\end{aligned}$$

where $\tilde{M} := 2M(1 + |\xi|^2)$. This is what we set out to prove. $\square$

We note for future use a variant on the preceding characterization of proximal normals, valid in the Lipschitz case.

Proposition 4.4.2 *Take a Lipschitz continuous function $f : R^k \to R$ and points $x \in R^k$ and $\xi \in R^k$. Then the following two statements are equivalent.*

(i) $\xi \in \partial^P f(x)$;

(ii) there exists $M > 0$ such that

$$\xi \cdot (y - x) \leq f(y) - f(x) + M|y - x|^2 \qquad (4.11)$$

for all $y \in R^k$.

The difference is of course that, now, the "proximal inequality" (4.11) must be satisfied for *all* ys in R^k, not just ys in a neighborhood of x.

Proof. Assume (ii). Then (i) follows from Proposition 4.4.1.

Assume (i). Then we know from Proposition 4.4.1 that there exist $M > 0$ and $\epsilon > 0$ such that (4.11) is valid for all $y \in x + \epsilon B$. Let K be a Lipschitz constant for f. Define $N := \epsilon^{-1}(|\xi| + K)$. Then

$$\xi \cdot (y - x) - (f(y) - f(x)) \leq |\xi||y - x| + K|y - x| \leq N|y - x|^2 \quad \text{if } y \notin x + \epsilon B.$$

It follows that (4.11) is satisfied for all y when M is replaced by $\tilde{M} := \max\{M, N\}$. □

The next result provides a "one-sided difference quotient" characterization of the strict subdifferential.

Proposition 4.4.3 *Take a lower semicontinuous function $f : R^k \to R \cup \{+\infty\}$ and points $x \in \operatorname{dom} f$ and $\xi \in R^k$. Then the following two statements are equivalent.*

(i) $\xi \in \hat{\partial} f(x)$;

(ii)

$$\limsup_{y \to x} \frac{\xi \cdot (y - x) - (f(y) - f(x))}{|y - x|} \leq 0. \tag{4.12}$$

Reference is made in the proof to the Bouligand tangent cone $T_C(x)$ to a closed set $C \subset R^k$ at the point $x \in C$, defined as

$$T_C(x) := \{\eta : \text{ there exists } y_i \overset{C}{\to} y \text{ and } t_i \downarrow 0 \text{ such that } t_i^{-1}(y_i - x) \to \eta\},$$

whose properties are explored later in the chapter.

Proof. Assume (ii). Condition (4.12) can be expressed:

$$\xi \cdot (y - x) - (f(y) - f(x)) \leq o(|y - x|) \quad \text{for all } y \in R^k,$$

in which $o(\cdot) : R^+ \to R^+$ is a function satisfying $o(\epsilon)/\epsilon \to 0$ as $\epsilon \downarrow 0$. It follows that

$$(\xi, -1) \cdot (z - (x, f(x))) \leq o(|z - (x, f(x))|) \quad \text{for all } z \in \operatorname{epi} f.$$

Otherwise expressed

$$(\xi, -1) \in \hat{N}_{\operatorname{epi} f}(x, f(x)),$$

which implies (i).

Assume now (i). Suppose that Condition (4.12) is not satisfied. This means that there exists a number $\alpha > 0$ and a sequence $y_i \to x$ such that $y_i \neq x_i$ and

$$\xi \cdot (y_i - x) - (f(y_i) - f(x)) \geq \alpha |y_i - x|, \quad \text{for all } i. \tag{4.13}$$

This inequality can be rearranged to give

$$f(y_i) \leq f(x) + \xi \cdot (y_i - x) - \alpha |y_i - x| \quad \text{for all } i. \tag{4.14}$$

We see that $\limsup_{i \to \infty} f(y_i) \leq f(x)$. Since f is lower semicontinuous, it follows that $f(y_i) \to f(x)$ as $i \to \infty$.

Define $z_i := (y_i, f(y_i))$, $t_i := |z_i - (x, f(x))|$ and $\eta_i := t_i^{-1}(z_i - (x, f(x)))$. Observe that, for each i, $|\eta_i| = 1$. Following extraction of subsequences we have that $\eta_i \to \eta$ for some nonzero vector η. Notice that $z_i \overset{\text{epi} f}{\to} (x, f(x))$ and $t_i \to 0$. By definition of the tangent cone then $\eta \in T_{\text{epi} f}(x, f(x))$.

We claim that

$$(\xi, -1) \cdot \eta > 0.$$

If this is true the proof is complete because, according to Theorem 4.10.4 below, the strict normal cone $\hat{N}_{\text{epi} f}(x, f(x))$ is contained in the polar set of the tangent cone $T_{\text{epi} f}(x, f(x))$. This implies $(\xi, -1) \cdot \eta \leq 0$, a contradiction.

To verify the claim, we make use of the following identity, which follows from the definition of η_i.

$$(\xi, -1) \cdot \eta_i = t_i^{-1}(\xi \cdot (y_i - x) - (f(y_i) - f(x))). \tag{4.15}$$

Now (4.14) implies that

$$\limsup_{i \to \infty} |y_i - x|^{-1}(f(y_i) - f(x)) < \infty.$$

There are therefore two possibilities to be considered:

(a) there exists $K > 0$ such that

$$f(y_i) - f(x) \leq K|y_i - x| \quad \text{for all } i; \tag{4.16}$$

(b) along a subsequence

$$\lim_{i \to \infty} |y_i - x|^{-1}(f(y_i) - f(x)) = -\infty.$$

Assume (a) is true. By (4.14) through (4.16), we have:

$$(\xi, -1) \cdot \eta_i \geq t_i^{-1} \alpha |y_i - x| = \frac{\alpha |y_i - x|}{|(y_i, f(y_i)) - (x, f(x))|} \geq (1 + K^2)^{-1/2}\alpha > 0.$$

Since $\eta_i \to \eta$, we must have $(\xi, -1) \cdot \eta > 0$.

Finally assume (b) is true. (Along the subsequence) $t_i^{-1}|y_i - x| \to 0$ and $t_i^{-1}(f(y_i) - f(x)) \to -1$. But then by (4.15), $(\xi, -1) \cdot \eta = \lim_i (\xi, -1) \cdot \eta_i = +1$. This confirms the claim in Case (b) also. $\square$

4.5   Nonsmooth Mean Value Inequalities

We pause to establish some consequences of the preceding theory, namely, generalizations of the Mean Value Theorem. The results of this section are used to carry out a deeper investigation of subdifferentials, normal cones, etc., and also to study solutions to optimization problems.

It will be convenient to use the following notation. Given a point $\bar{x} \in R^k$ and a subset $Y \subset R^k$, we denote by $[\bar{x}, Y]$ the set

$$[\bar{x}, Y] := \{x : x = \lambda \bar{x} + (1 - \lambda)y \text{ for some } y \in Y \text{ and } \lambda \in [0, 1]\}.$$

$(\bar{x}, Y)$ denotes the related set, in which the qualifier $\lambda \in [0, 1]$ is replaced by $\lambda \in (0, 1)$.

If $Y = \{\bar{y}\}$ we write $[\bar{x}, \bar{y}]$ in place of $[\bar{x}, Y]$.

The classical Mean Value Theorem tells us that if $f : R^n \to R$ is a C^1 function on an open set containing the line segment $[\bar{x}, \bar{y}]$ then, for some $\bar{z} \in [\bar{x}, \bar{y}]$,

$$\nabla f(\bar{z}) \cdot (\bar{y} - \bar{x}) = f(\bar{y}) - f(\bar{x}).$$

We might ask whether there is an analogue of this result covering situations where, say, f is a Lipschitz continuous function, expressed in terms of a limiting subgradient. The answer to this question is no, as is evident from the following example.

Example 4.3   Define the function $f : R \to R$ to be

$$f(x) = 1 - |x|$$

and set $\bar{x} = -1$ and $\bar{y} = +1$. A simple calculation shows that

$$\partial f(x) \subset \{-1\} \cup \{+1\} \quad \text{for all } x \in R.$$

It follows that, for any $z \in R$ and any $\xi \in \partial f(z)$,

$$0 = f(\bar{y}) - f(\bar{x}) \notin \xi \cdot (\bar{y} - \bar{x}).$$

The above example rules out a direct generalization of the Mean Value Theorem involving limiting subgradients, to allow for the function f to be nonsmooth. We do notice however in this example that the inequality

$$f(\bar{y}) - f(\bar{x}) \le \xi \cdot (\bar{y} - \bar{x})$$

is satisfied for some $\xi \in \partial f(\bar{z})$ with $\bar{z} \in [\bar{x}, \bar{y}]$. Possible choices are $\bar{z} = -0.5$ and $\xi = +1$.

It is the "inequality" version of the Mean Value Theorem that may be extended to a nonsmooth setting. Nonsmooth Mean Value Inequalities have many uses as an aid to proving fundamental relationships in nonsmooth analysis and in the analysis of solutions to optimization problems. The role of these theorems is usually to provide some upper bound and the fact that they take the form only of an inequality is seldom a significant shortcoming.

The Mean Value Inequality proved here is due to Clarke and Ledyaev. Besides allowing the function concerned to be nonsmooth, it departs from the classical Mean Value Theorem in one other respect. It is to assert that the inequality involved holds in some *uniform* sense.

In the setup for this Mean Value Inequality, the line segment $[\bar{x}, \bar{y}]$ is replaced by the "generalized" line segment joining a point $\bar{x}$ to a compact, convex set $Y \subset R^k$, namely, $[\bar{x}, Y]$. A natural extension of the smooth Mean Value Inequality would be: for each $y \in Y$ there exists $\bar{z} \in [\bar{x}, Y]$ and $\xi \in \partial f(\bar{z})$ such that

$$f(y) - f(\bar{x}) \leq \xi \cdot (y - \bar{x}).$$

What comes as a surprise is that, if we replace the left side by

$$\min_{y \in Y} f(y) - f(\bar{x}),$$

then the inequality is valid whatever choice is made of y in Y for the *same* ξ (a limiting subgradient of f evaluated at some point in $[\bar{x}, Y]$). This result, when first proved, was new even for smooth functions.

The main step is to prove the following "approximate" Mean Value Inequality involving proximal subgradients.

Theorem 4.5.1 (The Proximal Mean Value Inequality) *Take a lower semicontinuous function* $f : R^n \to R \cup \{+\infty\}$, *a point* $\bar{x} \in \operatorname{dom} f$, *and a compact convex set* $Y \subset \operatorname{dom} f$. *Define* $\hat{r} \in R \cup \{+\infty\}$ *to be*

$$\hat{r} := \inf_{y \in Y} f(y) - f(\bar{x}).$$

Then for any finite $r < \hat{r}$ *and* $\epsilon > 0$, *we can choose* $\bar{z} \in [\bar{x}, Y] + \epsilon B$ *and* $\xi \in \partial^P f(\bar{z})$ *such that*

$$r < \xi \cdot (y - \bar{x}) \quad \text{for all } y \in Y$$

and

$$f(\bar{z}) - f(\bar{x}) \quad \leq \quad \max\{r, 0\}.$$

Proof. By translation of the origin in $R^n \times R$ we can arrange that $\bar{x} = 0$ and $f(\bar{x}) = 0$. Fix $r < \hat{r}$ and $\epsilon > 0$. We deduce from the compactness of Y and the lower semicontinuity of f that positive numbers δ, M, and k may be chosen such that $0 < \delta < \epsilon$ and

$$f(z) \quad \geq \quad -M \text{ for all } z \in [0, Y] + \delta B$$
$$\min_{y \in Y + \delta B} f(y) \quad \geq \quad (r + \hat{r})/2$$

and
$$k > (M + |r| + 1)/\delta^2.$$

The proof hinges on examining the properties of the function $g : [0,1] \times R^n \times R^n \to R \cup \{+\infty\}$:

$$g(t, y, z) := f(z) + k|ty - z|^2 - tr.$$

Write
$$S := Y \times ([0, Y] + \delta B).$$

Notice that, for any $y \in Y$,

$$(0, y, 0) \in [0, 1] \times S \quad \text{and} \quad g(0, y, 0) = 0.$$

Since S is compact and g is lower semicontinuous, the minimization problem

$$\text{Minimize } g(t, y, z) \text{ over } (t, y, z) \in [0, 1] \times S$$

has a solution $(\bar{t}, \bar{y}, \bar{z})$ and

$$\min_{(t,y,z) \in [0,1] \times S} g(t, y, z) \leq 0.$$

This relationship implies that

$$g(\bar{t}, \bar{y}, \bar{z}) \leq 0, \tag{4.17}$$

and so
$$f(\bar{z}) - \max\{0, r\} \leq 0.$$

In particular, $f(\bar{z}) < +\infty$. Let us show that

$$\bar{z} \in [0, Y] + \delta \text{ int } B.$$

If $\bar{z}$ does not satisfy this condition then $|\bar{z} - \bar{t}\bar{y}| \geq \delta$, because $\bar{t}\bar{y} \in [0, Y]$. But then

$$g(\bar{t}, \bar{y}, \bar{z}) = f(\bar{z}) + k|\bar{t}\bar{y} - \bar{z}|^2 - \bar{t}r > -M + (M + |r| + 1) - |r| > 1.$$

We have contradicted (4.17).

We note next that

$$\bar{t} < 1. \tag{4.18}$$

This is because if $\bar{t} = 1$, it would follow that

$$
\begin{aligned}
g(\bar{t}, \bar{y}, \bar{z}) &= f(\bar{z}) + k|\bar{y} - \bar{z}|^2 - r \\
&\geq \begin{cases} f(\bar{z}) + k|\bar{y} - \bar{z}|^2 - r & \text{if } |\bar{y} - \bar{z}| \leq \delta \\ f(\bar{z}) + k\delta^2 - r & \text{if } |\bar{y} - \bar{z}| > \delta \end{cases} \\
&\geq \min\{(r + \hat{r})/2 - r, -M + M + |r| + 1 - r\} > 0.
\end{aligned}
$$

But then, in view of (4.17), $(\bar{t}, \bar{y}, \bar{z})$ cannot be a minimizer. We deduce from this contradiction the validity of (4.18).

We know that, for any z in the open set $[0, Y] + \delta \text{int } B$

$$g(\bar{t}, \bar{y}, z) \geq g(\bar{t}, \bar{y}, \bar{z}).$$

Using the identity

$$|\bar{t}\bar{y} - z|^2 - |\bar{t}\bar{y} - \bar{z}|^2 = -2(\bar{t}\bar{y} - \bar{z}) \cdot (z - \bar{z}) + |z - \bar{z}|^2,$$

we can write the preceding inequality as

$$2k(\bar{t}\bar{y} - \bar{z}) \cdot (z - \bar{z}) \leq f(z) - f(\bar{z}) + k|z - \bar{z}|^2.$$

This holds, in particular, for all points z in some neighborhood of $\bar{z}$. Define

$$\xi = 2k(\bar{t}\bar{y} - \bar{z}).$$

According to Proposition 4.4.1,

$$\xi \in \partial^P f(\bar{z}).$$

Two cases need now to be considered.

Case (a): $\bar{t} = 0$. Here $\xi = -2k\bar{z}$. Choose $t \in (0, 1]$. For any $y \in Y$ we have

$$\begin{aligned}
0 &\leq g(t, y, \bar{z}) - g(0, \bar{y}, \bar{z}) \\
&= f(\bar{z}) + k|ty - \bar{z}|^2 - tr - f(\bar{z}) - k|\bar{z}|^2 \\
&= t^2 k|y|^2 - 2kt\bar{z} \cdot y - tr.
\end{aligned}$$

Dividing across the inequality by t and passing to the limit as $t \downarrow 0$ we obtain

$$\xi \cdot y = -2k\bar{z} \cdot y \geq r.$$

This is the required relationship.

Case (b): $\bar{t} \in (0, 1)$. The fact that $\bar{y}$ minimizes the quadratic convex function $y \to g(\bar{t}, y, \bar{z})$ over the convex set Y implies

$$2\bar{t}k(\bar{t}\bar{y} - \bar{z}) \cdot (y - \bar{y}) \geq 0, \quad \text{for all } y \in Y.$$

Since $\bar{t} > 0$, it follows that

$$\xi \cdot (y - \bar{y}) \geq 0.$$

But $\bar{t}$ minimizes the quadratic function $t \to g(t, \bar{y}, \bar{z})$ over $(0, 1)$ so

$$2k(\bar{t}\bar{y} - \bar{z}) \cdot \bar{y} - r = 0.$$

It follows that $\xi \cdot \bar{y} = r$. We conclude that

$$\xi \cdot y \geq r.$$

The theorem is proved. $\square$

Performing a simple exercise in limit–taking now yields an "exact" Mean Value Inequality for Lipschitz functions in terms of limiting subdifferentials.

Theorem 4.5.2 (The Lipschitz Case) *Take a lower semicontinuous function $f : R^n \to R \cup \{+\infty\}$, a point $\bar{x} \in R^n$, and a compact convex set $Y \subset R^n$. Assume that, for some $\delta > 0$, f is Lipschitz continuous on $[\bar{x}, Y] + \delta B$.*

Then there exists $\bar{z} \in [\bar{x}, Y]$ and $\xi \in \partial f(\bar{z})$ such that

$$\min_{y \in Y} f(y) - f(\bar{x}) \leq \xi \cdot (y - \bar{x}) \text{ for all } y \in Y$$

and

$$f(\bar{z}) - f(\bar{x}) \leq \max\{\min_{y \in Y} f(y) - f(\bar{x}), 0\}.$$

Proof. Choose sequences $\epsilon_i \downarrow 0$ and $r_i \uparrow \hat{r}$, where

$$\hat{r} := \min_{y \in Y} f(y) - f(\bar{x}).$$

Now apply Theorem 4.5.1 with ϵ and r taken to be ϵ_i and r_i, respectively, for $i = 1, 2, \ldots$. For each i, we deduce the existence of some $z_i \in [\bar{x}, Y] + \epsilon_i B$ and $\xi_i \in \partial^P f(z_i)$ such that

$$r_i < \xi_i \cdot (y - \bar{x}) \quad \text{for all } y \in Y \tag{4.19}$$

and

$$f(z_i) - f(\bar{x}) \leq \max\{r_i, 0\}. \tag{4.20}$$

$\{z_i\}$ and $\{\xi_i\}$ are bounded sequences because Y is compact and f is Lipschitz continuous on $[\bar{x}, Y] + \delta B$ (see Proposition 4.3.3). We may arrange then by extracting subsequences that $z_i \to \bar{z}$ and $\xi_i \to \xi$ for some $\bar{z} \in [\bar{x}, Y]$ and some $\xi \in R^n$. It follows now from Proposition 4.3.5 that $\xi \in \partial f(\bar{z})$. The property that ξ and $\bar{z}$ satisfy the desired inequalities is now recovered in the limit as $i \to \infty$ from (4.19) and (4.20). $\square$

If we specialize to the case $Y = \{\bar{y}\}$ and substitute the convexified limiting subdifferential co∂f in place of ∂f, then a full (two-sided) Mean Value Theorem is valid. This result, due to Lebourg, dates back to 1975. To prove it, we make use of properties of limiting subgradients, derived in the next chapter.

Theorem 4.5.3 (A Two-Sided Mean Value Theorem) *Take a locally Lipschitz continuous function* $f : R^n \to R$ *and a line segment* $[\bar{x}, \bar{y}] \subset R^n$. *Then there exist* $z \in \{\epsilon\bar{x} + (1 - \epsilon)\bar{y} : 0 < \epsilon < 1\}$ *and* $\xi \in co\,\partial f(z)$ *such that*

$$f(\bar{y}) - f(\bar{x}) = \xi \cdot (\bar{y} - \bar{x}).$$

Proof. Consider the function $g : R \to R \cup \{+\infty\}$ defined by

$$g(t) := f(\bar{x} + t(\bar{y} - \bar{x})) - t(f(\bar{y}) - f(\bar{x})) + \Psi_{[0,1]}(t),$$

in which $\Psi_{[0,1]}(t)$ denotes the indicator function of the set $[0, 1]$. Notice that $g(0) = g(1) = f(\bar{x})$. So either of the following conditions is satisfied.

(a): g has a minimizer at some "interior" point $\tau \in (\bar{x}, \bar{y})$,

(b): $-g$ has a minimizer at some "interior" point $\sigma \in (\bar{x}, \bar{y})$.

Here $(\bar{x}, \bar{y})$ denotes the relative interior of $[\bar{x}, \bar{y}]$.

Consider Case (a). We deduce from the proximal normal inequality that $0 \in \partial^P g(\tau)$. It follows from the Chain Rule (see Chapter 5) that

$$0 = \xi \cdot (\bar{y} - \bar{x}) - (f(\bar{y}) - f(\bar{x}))$$

for some $\xi \in \partial f(\bar{x} + \tau(\bar{y} - \bar{x}))$ as required.

Consider Case (b). The Chain Rule applied to $-g$ now gives

$$0 = \xi \cdot (\bar{y} - \bar{x}) + (f(\bar{y}) - f(\bar{x}))$$

for some $\xi \in \partial(-f)(\sigma)$. Since however

$$co\,\partial f(x) = -co\,\partial(-f(x))$$

(see Theorem 4.7.5) we find that

$$0 = \xi' \cdot (\bar{y} - \bar{x}) - (f(\bar{y}) - f(\bar{x}))$$

for some $\xi' \in \partial f(\sigma)$ in this case also. $\square$

4.6 Characterization of Limiting Subgradients

We now seek an analytic description of limiting subgradients (and their asymptotic relatives). Our goal is to describe these objects in terms of limits of proximal subgradients, for which an analytic characterization is already available.

That limiting subgradients can be expressed as limits of proximal subgradients at neighboring points is a straightforward consequence of the definition of limiting subgradients in terms of limiting normal vectors to the epigraph set and the property that limiting normal vectors are limits of proximal normal vectors. This is Part (a) of Theorem 4.6.2 below.

It is easy also to show that asymptotic limiting subgradients are expressible as limits of vectors that (apart from a positive scale factor) are either proximal subgradients or asymptotic proximal subgradients. *However there is a more convenient description of asymptotic limiting subgradients that involves sequences of proximal subgradients alone.* This is Part (b) of Theorem 4.6.2. It has important implications because the validity of various subdifferential calculus rules centers on the properties of asymptotic subdifferentials of the functions involved, and any results that restrict the possible ways in which asymptotic limiting subdifferentials can arise simplify the analysis of pathological phenomena.

As a first step we prove a lemma, whose conclusions may be summarized "proximal normals to an epigraph set can be approximated by nonhorizontal proximal normals."

Lemma 4.6.1 *Take a lower semicontinuous function $f : R^k \to R \cup \{+\infty\}$, a point $\bar{x} \in \operatorname{dom} f$, and a nonzero point $\xi \in R^k$ such that*

$$(\xi, 0) \in N^P_{\operatorname{epi} f}(\bar{x}, f(\bar{x})).$$

Then there exist convergent sequences $x_i \xrightarrow{f} \bar{x}$, $\xi_i \to \xi$, and a sequence of positive numbers $\lambda_i \downarrow 0$ such that

$$(\xi_i, -\lambda_i) \in N^P_{\operatorname{epi} f}(x_i, f(x_i)) \text{ for all } i.$$

Use is made in the proof of properties of the distance function derived later in the chapter. (See Lemmas 4.8.1 and 4.8.2 and Theorem 4.8.4.)

Proof. Since $(\xi, 0)$ is a nonzero "horizontal" proximal normal, there exists a point $(x, f(\bar{x})) \notin \operatorname{epi} f$ and a number $\sigma > 0$ such that $(\bar{x}, f(\bar{x}))$ is a closest point in $\operatorname{epi} f$ to $(x, f(\bar{x}))$ and $\xi = \sigma(x - \bar{x})$. We can arrange that $(\bar{x}, f(\bar{x}))$ is the *unique* closest point; also that there exists some $\alpha > 0$ such that $(\bar{x}, f(\bar{x}))$ is the closest point in $\operatorname{epi} f$ to the point

$$(\bar{x} + (1 + \alpha)(x - \bar{x}), f(\bar{x})).$$

This can be achieved by replacing x by a point on the "open" line segment joining x and $\bar{x}$ and by increasing σ appropriately.

We may now deduce from Lemmas 4.8.3 and 4.8.4 and Theorem 4.8.4 the following properties.

First,

$$(|\xi|^{-1}\xi, 0) \in \partial^P d_{\operatorname{epi} f}(x, f(\bar{x})).$$

Second, if for some sequences $(y_i, r_i) \to (x, f(\bar{x}))$ and $\{(\xi_i', -\lambda_i')\}$ in $R^n \times R$,

$$(\xi_i', -\lambda_i') \in \partial^P d_{\text{epi } f}(y_i, r_i) \quad \text{for all } i, \tag{4.21}$$

then there exists (x_i, s_i), a unique closest point to (y_i, r_i) in epi f, such that

$$(\xi_i', -\lambda_i') \in \partial^P d_{\text{epi } f}(x_i, s_i) \quad \text{for all } i \tag{4.22}$$

$$(x_i, s_i) \to (\bar{x}, f(\bar{x})) \tag{4.23}$$

and

$$(\xi_i', -\lambda_i') \to (|\xi|^{-1}\xi, 0). \tag{4.24}$$

We show presently that

$$d_{\text{epi } f}(x, f(\bar{x}) - t) > d_{\text{epi } f}(x, f(\bar{x})) \quad \text{for all } t > 0. \tag{4.25}$$

Assuming this to be the case, we take a sequence $t_i \downarrow 0$ and, for each i apply the Mean Value Inequality (Theorem 4.5.1) to the function $d_{\text{epi } f}$ with base point $(x, f(\bar{x}) - t_i)$ and "Y set" $\{(x, f(\bar{x}))\}$. The theorem tells us that there exist sequences $(y_i, r_i) \to (x, f(\bar{x}))$ and $\{(\xi_i', -\lambda_i')\}$ in $R^n \times R$ satisfying (4.21) and, for all i,

$$(\xi_i', -\lambda_i') \cdot ((x, f(\bar{x})) - (x, f(\bar{x}) - t_i)) \le$$
$$d_{\text{epi } f}(x, f(\bar{x})) - d_{\text{epi } f}(x, f(\bar{x}) - t_i).$$

These inequalities and (4.25) imply that $\lambda_i' > 0$ for all i. It follows from (4.22) that in fact $s_i = f(x_i)$. (4.23) may therefore be replaced by $x_i \overset{f}{\to} \bar{x}$. Now define

$$(\xi_i, -\lambda_i) := (|\xi|\xi_i', -|\xi|\lambda_i') \quad \text{for each } i.$$

Note that, since $\xi \ne 0$, $\lambda_i > 0$ for each i. We have from (4.22) and Lemma 4.8.1 that

$$(\xi_i, -\lambda_i) \in N^P_{\text{epi } f}(x_i, f(x_i)) \quad \text{for each } i$$

and, in view of (4.24),

$$(\xi_i, -\lambda_i) \to (\xi, 0) \quad \text{as } i \to \infty.$$

The sequences $\{x_i\}, \{\xi_i\}$, and $\{\lambda_i\}$ therefore have the properties asserted in the lemma.

It remains to confirm (4.25). If it is untrue, there exists $t > 0$ such that

$$d_{\text{epi } f}(x, f(\bar{x}) - t) \le d_{\text{epi } f}(x, f(\bar{x})).$$

We may find $(y, r) \in R^n \times R$ with $r \ge f(y)$ such that

$$d_{\text{epi } f}(x, f(\bar{x}) - t) = |(x, f(\bar{x}) - t) - (y, r)|.$$

Notice that $y \neq \bar{x}$. Indeed if this were not the case then

$$|(x, f(\bar{x}) - t) - (\bar{x}, r)| = |(x - \bar{x}, (t + r - f(\bar{x})))| > |x - \bar{x}| = d_{\text{epi } f}(x, f(\bar{x})),$$

since $t > 0$ and $r \geq f(\bar{x})$, in contradiction of our premise. It follows that

$$\begin{aligned}
d_{\text{epi } f}(x, f(\bar{x})) &\geq d_{\text{epi } f}(x, f(\bar{x}) - t) \\
&= |(x, f(\bar{x}) - t) - (y, r)| = |(x, f(\bar{x})) - (y, r + t)|.
\end{aligned}$$

But $(y, r + t) \in \text{epi } f$ since $t \geq 0$. It follows that $(y, r + t)$ is a closest point in $\text{epi } f$ to $(x, f(\bar{x}))$ with $y \neq \bar{x}$. This is impossible because $(\bar{x}, f(\bar{x}))$ is the unique closest point. $\square$

It is now a simple matter to prove:

Theorem 4.6.2 (Asymptotic Description of Limiting Subgradients)
Take a lower semicontinuous function $f : R^k \to R \cup \{+\infty\}$ and points $x \in \text{dom } f$ and $\xi \in R^k$.

(a) The following conditions are equivalent.

(i) $\xi \in \partial f(x)$;

(ii) there exist $x_i \xrightarrow{f} x$ and $\xi_i \to \xi$ such that $\xi_i \in \partial^P f(x_i)$ for all i.

(b) The following conditions are equivalent.

(iii) $\xi \in \partial^\infty f(x)$;

(iv) there exist $x_i \xrightarrow{f} x$, $t_i \downarrow 0$ and $\xi_i \to \xi$ such that $t_i^{-1}\xi_i \in \partial^P f(x_i)$ for all i.

Proof.

(a): Assume (i). Then there exist sequences $(\xi_i', -t_i) \to (\xi, -1)$ and $z_i \xrightarrow{\text{epi } f} (x, f(x))$ such that $(\xi_i', -t_i) \in N_{\text{epi } f}^P(z_i)$ for all i. We may assume $t_i > 0$ for all i. We deduce from the proximal inequality that $z_i = (x_i, f(x_i))$ for all i and hence $x_i \xrightarrow{f} x$. Define $\xi_i := t_i^{-1}\xi_i'$. Then $\xi_i \in \partial^P f(x_i)$ for each i. Since $\xi_i \to \xi$ as $i \to \infty$, (ii) follows.

Assume (ii). For the given sequences $\{x_i\}$ and $\{\xi_i\}$, we have that $(\xi_i, -1) \in N_{\text{epi } f}^P(x_i, f(x_i))$ for each i and $(x_i, f(x_i)) \to (x, f(x))$ as $i \to \infty$. It follows from the closure properties of the limiting normal cone that $(\xi, -1) \in N_{\text{epi } f}(x, f(x))$. But then $\xi \in \partial f(x)$.

(b): Assume (iii). Suppose to begin with that $\xi \neq 0$. By Lemma 4.6.1, there exist sequences $(\xi_i', -t_i) \to (\xi, 0)$ and $x_i \xrightarrow{f} x$ such that $t_i > 0$ and $(\xi_i', -t_i) \in N_{\text{epi } f}^P(x_i, f(x_i))$ for each i. Define $\xi_i := t_i^{-1}\xi_i'$ for each i. Then

$(\xi_i, -1) \in N^P_{\text{epi } f}(x_i, f(x_i))$, whence $\xi_i \in \partial f(x_i)$ for each i. We have shown (iv).

Suppose on the other hand that $\xi = 0$. Since $x \in \text{dom } f$, it follows from Proposition 4.2.7 that there exists $(\xi', -\beta) \neq (0,0)$ such that

$$(\xi', -\beta) \in N_{\text{epi } f}(x, f(x)).$$

Since limiting normals are limits of proximal normals and proximal normals to epigraph sets have as second coordinate a nonnegative number, we know that $\beta \geq 0$. Whether or not $\beta = 0$, we deduce from Part (a) of Theorem 4.6.2 or Part (b) (in the case $\xi' \neq 0$) that there exist sequences $x_i \xrightarrow{f} x$, $\xi'_i \to \xi'$, and $\beta_i \to \beta$ such that

$$\beta_i > 0 \quad \text{and} \quad \beta_i^{-1}\xi'_i \in \partial^P f(x_i) \quad \text{for all } i.$$

Choose a sequence $\alpha_i \downarrow 0$ such that

$$\alpha_i \beta_i \to 0 \quad \text{as} \quad i \to \infty.$$

Now set $\xi_i = \alpha_i \xi'_i$, $t_i = \alpha_i \beta_i$ for each i.

We see that

$$\xi_i \to \xi(=0), \quad t_i \downarrow 0 \quad \text{and} \quad t_i^{-1}\xi_i \in \partial^P f(x_i) \quad \text{for all } i.$$

We have shown (iv) in this case also.

Suppose (iv). For the given sequences $\{x_i\}$, $\{t_i\}$, and $\{\xi_i\}$, we have $(\xi_i, -t_i) \in N^P_{\text{epi } f}(x_i, f(x_i))$ for all i. Since $(\xi_i, -t_i) \to (\xi, 0)$ and $(x_i, f(x_i)) \xrightarrow{\text{epi } f} (x, f(x))$, it follows $(\xi, 0) \in N^P_{\text{epi } f}(x, f(x))$. This is (iii). $\square$

A companion piece to the above theorem is a characterization of limiting subgradients and asymptotic limiting subgradients in terms of limits of strict (rather than proximal) subgradients.

Theorem 4.6.3 *Take a lower semicontinuous function $f : R^k \to R \cup \{+\infty\}$ and points $x \in \text{dom } f$ and $\xi \in R^k$.*

(a) The following conditions are equivalent.

(i) $\xi \in \partial f(x)$;

(ii) there exist $x_i \xrightarrow{f} x$ and $\xi_i \to \xi$ such that $\xi_i \in \hat{\partial} f(x_i)$, for all i.

(b) The following conditions are equivalent.

(iii) $\xi \in \partial^\infty f(x)$;

(iv) there exist $x_i \xrightarrow{f} x$, $t_i \downarrow 0$ and $\xi_i \to \xi$ such that $t_i^{-1}\xi_i \in \hat{\partial} f(x_i)$ for all i.

The most significant assertions in this theorem are the implications (i)
$\Rightarrow$ (ii) and (iii) $\Rightarrow$ (iv); but these follow directly from Theorem 4.6.2 since
proximal normals are certainly strict normals. The reverse implications are
routine consequences of the fact that limiting normal vectors to epi f are
limits of neighboring strict normal vectors (see Proposition 4.2.5).

4.7   Subgradients of Lipschitz Continuous Functions

The techniques so far assembled provide local descriptions of Lipschitz
continuous functions, since they are special cases of lower semicontinous
functions. But Lipschitz continuous functions are encountered so frequently
in applications of Nonsmooth Analysis that it is important to exploit the
extra structure they introduce and alternative approaches to local approximation.

The Lipschitz continuity hypothesis is essentially a hypothesis that the
function in question has a bounded slope. It is hardly surprising then that
the subdifferentials of Lipschitz continuous functions are bounded sets, a
fact that follows easily from the preceding analysis.

Proposition 4.7.1 *Take a lower semicontinuous function $f : R^k \to R \cup \{+\infty\}$ and a point $x \in R^k$. Assume that f is Lipschitz continuous on a neighborhood of x with Lipschitz constant K. Then:*

(i) $\partial f(x)$ is nonempty and $\partial f(x) \subset KB$;

(ii) $\partial^\infty f(x) = \{0\}$.

Proof. Take ξ and y such that $\xi \in \partial^P f(y)$ and f is Lipschitz continuous
on a neighborhood of y. Then there exist $M > 0$ and $\epsilon > 0$ such that

$$\xi \cdot (z - y) \le f(z) - f(y) + M|z - y|^2 \quad \text{for all } z \in y + \epsilon B.$$

It follows that, for all $\lambda > 0$, sufficiently small,

$$\lambda|\xi|^2 \le f(y + \lambda\xi) - f(y) + M\lambda^2|\xi|^2 \le \lambda K|\xi| + M\lambda^2|\xi|^2.$$

Dividing across by λ and passing to the limit gives

$$|\xi|^2 \le K|\xi|.$$

We conclude that $|\xi| \le K$. This shows that for all y sufficiently close to x,

$$\xi \in \partial^P f(y) \text{ implies } |\xi| \le K.$$

The assertions of the proposition now follow from Theorem 4.6.2 and Proposition 4.3.4. □

The above proposition is a special case of more far-reaching results, covered by Corollary 4.9.2 below.

We now examine an alternative approach to defining subdifferentials of Lipschitz continuous functions, based on convex approximations and duality ideas. It is important partly because of the insights it gives into the relationship between subdifferentials of nonsmooth functions and their counterparts in Convex Analysis, but also because the approach will provide new representations of subdifferentials that are extremely useful in applications.

The starting point is the definition of the (Clarke) generalized directional derivative of a locally Lipschitz continuous function.

Definition 4.7.2 Take a function $f : R^k \to R$ and points $x \in R^k$ and $v \in R^k$. Assume that f is Lipschitz continuous on a neighborhood of x. The *generalized directional derivative of f at x in the direction v*, written $f^0(x, v)$, is the number

$$f^0(x, v) := \limsup_{y \to x, t \downarrow 0} t^{-1}[f(y + tv) - f(y)].$$

Notice that, because f is assumed merely to be Lipschitz continuous, the right side would not make sense if "lim" replaced "lim sup."

We list some salient properties of $f^0(x, v)$.

Proposition 4.7.3 *Take a function $f : R^k \to R$ and a point $x \in R^k$. Assume that f is Lipschitz continuous on a neighborhood of x with Lipschitz constant K. Then the function $v \to f^0(x, v)$ with domain R^k has the following properties.*

(i) It is finite-valued, Lipschitz continuous with Lipschitz constant K, and positively homogeneous, in the sense that

$$f^0(x, \alpha v) = \alpha f^0(x, v) \quad \text{for all } v \in R^k \text{ and } \alpha \geq 0;$$

(ii) it is convex.

Proof.

(i): These properties are straightforward consequences of the definition of f^0 and the assumed Lipschitz continuity of f (near x).

(ii): To show convexity of $f^0(x, \cdot)$, take any $v, w \in R^k$. Since $f^0(x, \cdot)$ is positively homogeneous, it suffices to show "sub-additivity," namely,

$f^0(x, v + w) \leq f^0(x, v) + f^0(x, w)$. We know that there exist sequences $y_i \to x$ and $t_i \downarrow 0$ such that

$$f^0(x, v + w) = \lim_i \{t_i^{-1}(f(y_i + t_i(v + w)) - f(y_i))\}.$$

But the term between braces on the right can be expressed as

$$t_i^{-1}(f(z_i + t_i v) - f(z_i)) + t_i^{-1}(f(y_i + t_i w) - f(y_i)),$$

in which $z_i = y_i + t_i w$. Since $z_i \to x$, we conclude

$$
\begin{aligned}
f^0(x, v + w) &\leq \limsup_{z \to x, t \downarrow 0} \{t^{-1}(f(z + tv) - f(z))\} \\
&\quad + \limsup_{y \to x, t \downarrow 0} \{t^{-1}(f(y + tw) - f(y))\} \\
&= f^0(x, v) + f^0(x, w),
\end{aligned}
$$

as claimed. $\square$

We have constructed a convex function $f^0(x, \cdot)$ that approximates f "near" x. What would be more natural then than to introduce a "subdifferential," written $\bar{\partial} f(x)$, which is the subdifferential in the sense of convex analysis of the convex function $v \to f^0(x, v)$ at $v = 0$? Since $f^0(x, 0) = 0$, this approach gives

$$\bar{\partial} f(x) := \{\xi : f^0(x, v) \geq \xi \cdot v \quad \text{for all } v \in R^k\}.$$

$\bar{\partial} f(x)$ is called the *Clarke subdifferential*[1] of f at x. $\bar{\partial} f(x)$ is a nonempty, compact, convex set (for fixed x). Elements in $\bar{\partial} f(x)$ are uniformly bounded in the Euclidean norm by the Lipschitz constant of f on a neighborhood of x.

The generalized directional derivative can be interpreted as the support function of $\bar{\partial} f(x)$:

Proposition 4.7.4 *Take a function $f : R^k \to R$ that is Lipschitz continuous on a neighborhood of a point $x \in R^k$. Then*

$$f^0(x, v) = \max\{v \cdot \xi : \xi \in \bar{\partial} f(x)\} \quad \text{for all } v \in R^k.$$

---

[1]While we have defined this subdifferential only for Lipschitz continuous functions, Clarke early provided an extension to lower semicontinuous functions [29]. This was achieved by defining normal cones to closed sets in R^k via the Clarke subdifferential of the distance function, which is Lipschitz continuous, and then defining subdifferentials to lower semicontinuous functions in terms of normal cones to epigraph sets, as in Section 4.3.

Proof. We have $f^0(x, v) \geq \max\{v \cdot \xi : \xi \in \bar{\partial}f(x)\}$ for all v, by definition of $\bar{\partial}f(x)$. Fix any v. It remains to show equality for this v. Choose ξ to be a subgradient (in the sense of convex analysis) to $f^0(x, \cdot)$ at v. Then

$$f^0(x, w) - f^0(x, v) \geq \xi \cdot (w - v) \quad \text{for all } w \in R^k.$$

Take $w = \alpha\bar{w}$ for arbitrary $\bar{w} \in R^k$ and $\alpha > 0$. Since $f^0(x, w) = \alpha f^0(x, \bar{w})$, we may divide across the inequality by α and pass to the limit as $\alpha \to \infty$. This gives $f^0(x, \bar{w}) \geq \xi \cdot \bar{w}$. We conclude that $\xi \in \bar{\partial}f(x)$. On the other hand, setting $w = 0$, we obtain $\xi \cdot v \geq f^0(x, v)$. It follows that $f^0(x, v) = \max\{v \cdot \xi : \xi \in \bar{\partial}f(x)\}$. $\square$

The Clarke subdifferential commutes with -1.

Proposition 4.7.5 *Take a function $f : R^k \to R$ that is Lipschitz continuous on a neighborhood of a point $x \in R^k$. Then*

$$\bar{\partial}(-f)(x) = -\bar{\partial}f(x).$$

Proof. It suffices to show that the two sets have the same support function, which (as we now know) are $(-f)^0(x, v)$ and $f^0(x, -v)$ for any v. But

$$(-f)^0(x, v) = \limsup_{x' \to x, t \downarrow 0} \frac{(-f)(x' + tv) - (-f)(x')}{t},$$

which, following the substitution $y' = x' + tv$, becomes

$$(-f)^0(x, v) = \limsup_{y' \to x, t \downarrow 0} \frac{f(y' - tv) - f(y')}{t} = f^0(x, -v).$$

This is the required relation. $\square$

Note that assertions of Proposition 4.7.5 are in general false if the limiting subdifferential is substituted for the Clarke subdifferential. This point is illustrated by functions (i) and (ii) in Example 4.1.

How does the Clarke subdifferential $\bar{\partial}f(x)$ fit in with our other nonsmooth constructs? It relates very neatly to the limiting subdifferential $\partial f(x)$.

Proposition 4.7.6 *Take a lower semicontinuous function $f : R^k \to R$ that is Lipschitz continuous on a neighborhood of some point $x \in R^k$. Then*

$$\bar{\partial}f(x) = \text{co}\,\partial f(x).$$

Proof. Since the two sets are closed convex sets, it suffices to show that they have the same support function, i.e., to show $f^0(x, v) = \max\{v \cdot \xi : \xi \in \partial f(x)\}$ for all $v \in R^k$.

(a): We show $f^0(x, v) \geq \max\{v \cdot \xi : \xi \in \partial f(x)\}$ for all $v \in R^k$.

Choose any $\xi \in \partial f(x)$ and any $v \in R^k$. Then there exist sequences $x_i \xrightarrow{f} x$ and $\xi_i \to \xi$ such that $\xi_i \in \partial^P f(x_i)$ for each i, by Theorem 4.6.2. We know that for each i there exist $M_i > 0$ and $\delta_i > 0$ such that

$$f(x) - f(x_i) \geq \xi_i \cdot (x - x_i) - M_i|x - x_i|^2 \quad \text{for all } x \in x_i + \delta_i B.$$

It follows that we can find a sequence $t_i \downarrow 0$ such that

$$f(x_i + t_i v) - f(x_i) \geq t_i \xi_i \cdot v - i^{-1} t_i \quad \text{for all } i.$$

Then

$$f^0(x, v) \geq \limsup_{i \to \infty} t_i^{-1}(f(x_i + t_i v) - f(x_i)) \geq \xi \cdot v.$$

Since ξ and v were arbitrary, we have

$$f^0(x, v) \geq \max\{v \cdot \xi : \xi \in \partial f(x)\} \quad \text{for all } v \in R^k.$$

(b): We show $f^0(x, v) \leq \max\{v \cdot \xi : \xi \in \partial f(x)\}$ for all $v \in R^k$.

Pick an arbitrary $v \in R^k$. Then sequences $x_i \to x$ and $t_i \downarrow 0$ can be found such that

$$f^0(x, v) = \lim_{i \to \infty} t_i^{-1}(f(x_i + t_i v) - f(x_i)).$$

The Mean Value Inequality (Theorem 4.5.2) applied to the function f with basepoint x_i and "Y set" $\{x_i + t_i v\}$ for each i yields the following information: for each i there exists $z_i \in x_i + t_i|v|B$ and $\xi_i \in \partial f(z_i)$ such that

$$f(x_i + t_i v) - f(x_i) \leq t_i \xi_i \cdot v. \tag{4.26}$$

Clearly $z_i \to x$. Because f is Lipschitz continuous on a neighborhood of x, the ξ_is are uniformly bounded. Along a subsequence then $\xi_i \to \xi$, for some $\xi \in R^k$. By Theorem 4.6.2 then $\xi \in \partial f(x)$. Dividing across (4.26) by t_i and passing to the limit gives

$$f^0(x, v) \leq \xi \cdot v \quad \text{for some } \xi \in \partial f(x).$$

We conclude that

$$f^0(x, v) \leq \max\{v \cdot \xi : \xi \in \partial f(x)\} \quad \text{for all } v.$$

$\square$

Consider now a function $f : R^k \to R$ and a point $x \in R^k$ such that f is Lipschitz continuous on a neighborhood of x. According to Rademacher's Theorem, f is differentiable almost everywhere on this neighborhood (with respect to k-dimensional Lebesgue measure). Can we construct limiting subgradients of f at x as limits of neighboring derivatives? The answer to this question, supplied by the following theorem, is a qualified yes: the convex hull of the set of limits of neighboring derivatives coincides with the convex hull of the limiting subdifferential at x. The same is true, even if we exclude neighboring derivatives on some specified subset of measure zero. This important theorem is of intrinsic interest (it relates classical and modern concepts of derivatives in a very concrete way), but also provides a computational tool for the convex hull of the limiting subdifferential of great power. Often the simplest approach to calculating $\operatorname{co} \partial f(x)$, by far, is to look at limits of neighboring derivatives.

When a function $f : R^k \to R$ is Lipschitz continuous on a neighborhood of x, the concepts of Gâteaux, Hadamard, and Fréchet differentiability all coincide. For convenience, when we say that a Lipschitz continuous function f is differentiable at x, we mean Gâteaux differentiable; i.e., there is a unique vector $\xi \in R^k$, the Gâteaux derivative $\nabla f(x)$ of f at x, such that

$$\lim_{h \downarrow 0} h^{-1}(f(x + hv) - f(x) - h\,\xi \cdot v) = 0 \quad \text{for all } v \in R^k. \tag{4.27}$$

Notice that if f is differentiable at x (and Lipschitz continuous on a neighborhood of x) then

$$\nabla f(x) \in \operatorname{co} \partial f(x).$$

This is true since (4.27) (with $\nabla f(x)$ substituted in place of ξ) implies $f^0(x, v) \geq \xi \cdot v$ for every v, whence (by Proposition 4.7.6) $\nabla f(x) \in \operatorname{co} \partial f(x)$.

Theorem 4.7.7 *Take a function $f : R^k \to R$, a point $x \in R^k$ and any subset $\Omega \subset R^k$ having Lebesgue measure zero. Assume that f is Lipschitz continuous on a neighborhood of x. Then*

$$\operatorname{co} \partial f(x) = \operatorname{co}\{\xi : \exists\, x_i \to x, x_i \notin \Omega, \nabla f(x_i) \text{ exists and } \nabla f(x_i) \to \xi\}.$$

Proof. We know that $y \rightsquigarrow \partial f(y)$ has a closed graph on a neighborhood of x (see Proposition 4.3.5). However the vectors in $\partial f(y)$ are uniformly bounded as y ranges over this neighborhood, as may be deduced from Theorem 4.7.1. It follows from Carathéodory's Theorem that the set-valued function $y \rightsquigarrow \operatorname{co} \partial f(y)$ has a closed graph on a neighborhood of x. But, we have observed, if f is differentiable at a point y near x, that $\nabla f(y) \in \operatorname{co}\partial f(y)$. It follows that $S \subset \operatorname{co}\partial f(x)$, where S is the set on the right side in the theorem statement.

Both the set S and $\operatorname{co}\partial f(x)$ are compact convex sets and, as we have shown, $S \subset \operatorname{co}\partial f(x)$. So it remains to establish that the support functions of the two sets satisfy the inequality: for any $v \neq 0$,

$$\sup\{\xi \cdot v : \xi \in \partial f(x)\} \leq \sup\{\xi \cdot v : \xi \in S\}.$$

Choose an arbitrary vector $v \neq 0$. In view of Proposition 4.7.4, it suffices to show that

$$f^0(x, v) \leq \sup\{\xi \cdot v : \xi \in S\}.$$

In fact we need only demonstrate that

$$f^0(x, v) \leq \alpha, \tag{4.28}$$

where the number α is defined to be

$$\alpha := \limsup\{\nabla f(y) \cdot v : f \text{ is differentiable at } y, y \notin \Omega, y \to x\}.$$

Choose any ϵ. Denote by S the subset of R^k on which f fails to be differentiable. By definition of α, there exists $\delta > 0$ such that

$$y \in x + \delta B \text{ and } y \notin \Omega \cup S \text{ implies } \nabla f(y) \cdot v \leq \alpha + \epsilon. \tag{4.29}$$

We suppose that δ has been chosen small enough that $x + \delta B$ lies in the neighborhood on which f is Lipschitz continuous. By Rademacher's Theorem then, $\Omega \cup S$ has k-dimensional Lebesgue measure zero in $x + \delta B$. For each $y \in R^k$, consider the line segments $L_y := \{y + tv : 0 < t < \delta/(2|v|)\}$. Since $\Omega \cup S$ has k-dimensional Lebesgue measure 0 in $x + \delta B$, it follows from Fubini's Theorem that, for almost every y in $x + (\delta/2)B$, the line segment L_y intersects $\Omega \cup S$ on a set of zero one-dimensional Lebesgue measure. Take y to be any point in $x + (\delta/2)B$ having this property, and take any $t \in (0, \delta/(2|v|))$. Then

$$f(y + tv) - f(y) = \int_0^t \nabla f(y + sv) \cdot v \, ds.$$

This formula is valid because f is Lipschitz continuous on

$$\{y + sv : 0 \leq s \leq t\}.$$

Since $y + sv \in x + \delta B$ and $y + sv \notin \Omega \cup S$ for almost all $s \in (0, t)$, we conclude from (4.29) that

$$f(y + tv) - f(y) \leq t(\alpha + \epsilon).$$

This inequality holds for all $y \in x + (\delta/2)B$, excluding a subset of measure zero and for all $t \in (0, \delta/(2|v|))$. But then, since f is continuous, it is true for all $y \in x + (\delta/2)B$, $t \in (0, \delta/(2|v|))$. This implies that

$$f^0(x, v) = \limsup_{y \to x, t \downarrow 0} t^{-1}[f(y + tv) - f(y)] \leq \alpha + \epsilon.$$

Since $\epsilon > 0$ was arbitrary, we conclude (4.28). The proof is complete. $\square$

4.8   The Distance Function

Take a set $C \subset R^k$. The distance function $d_C : R^k \to R$ is

$$d_C(x) := \inf\{|x - y| : y \in C\}.$$

The distance function features prominently in nonsmooth analysis. This is not surprising since a key building block is the proximal normal, whose very definition revolves around the concept of "closest point" (a point that minimizes the distance function).

Chapter 3 has already provided a foretaste of the distance function's significance. Take a set A in R^n and a Lipschitz continuous function $f : A \to R$. Take also a subset $C \subset A$. Then, for the "exact penalty" parameter K chosen sufficiently large, a minimizer for the constrained optimization problem

minimize $f(x)$ over points $x \in A$ satisfying $x \in C$,

is a minimizer also for the unconstrained problem,

minimize $f(x) + K d_C(x)$ over points $x \in A$.

This is a consequence of the Exact Penalization Theorem (Proposition 3.2.1), which permits us to trade off the complications associated with a constraint against an extra nonsmooth term in the cost.

Another important property of d_C is that it provides a description of a set C in terms of a Lipschitz continuous function.

Proposition 4.8.1 *For any set $C \subset R^k$, the distance function $d_C : R^k \to R$ is Lipschitz continuous with Lipschitz constant 1.*

Proof. Take any $x, y \in R^k$. Choose $\epsilon > 0$. By definition of d_C, there exists $z \in C$ such that $d_C(y) \geq |y - z| - \epsilon$. Then

$$d_C(x) \leq |x - z| \leq |x - y| + |y - z| \leq |x - y| + d_C(y) + \epsilon.$$

But $\epsilon > 0$ was arbitrary. So $d_C(x) - d_C(y) \leq |x - y|$. Since the roles of x and y are interchangeable, we conclude $|d_C(x) - d_C(y)| \leq |x - y|$, which implies d_C is Lipschitz continuous with Lipschitz constant 1. $\square$

Our goal now is to describe the limiting subgradients of d_C at a point in C. As usual, we start by studying proximal normals.

Proposition 4.8.2 *Take a closed set $C \subset R^k$ and a point $x \in C$. Then*

$$\partial^P d_C(x) = N_C^P(x) \cap B.$$

Proof.

(a): We show $\partial^P d_C(x) \subset N_C^P(x) \cap B$.

Suppose $\xi \in \partial^P d_C(x)$. Then, since d_C is Lipschitz continuous with Lipschitz constant 1, by Proposition 4.4.2, there exists $M > 0$ such that

$$\xi \cdot (y - x) \leq d_C(y) - d_C(x) + M|y - x|^2 \quad \text{for all } y \in R^k.$$

Since d_C vanishes on C, we have

$$\xi \cdot (y - x) \leq M|y - x|^2 \quad \text{for all } y \in C.$$

But this implies $\xi \in N_C^P(x)$. Hence $|\xi| \leq 1$, by Proposition 4.3.3, so $\xi \in N_C^P(x) \cap B$.

(b): We show $\partial^P d_C(x) \supset N_C^P(x) \cap B$.

Take any $\xi \in N_C^P(x) \cap B$. We need to show that $\xi \in \partial^P d_C(x)$. This is certainly true for $\xi = 0$ so we may assume $\xi \neq 0$. We know that there exists $M > 0$ such that

$$\xi \cdot (y - x) \leq M|y - x|^2 \quad \text{for all } y \in C.$$

This fact can be expressed:

$$C \cap Q = \emptyset,$$

where Q is the open sphere

$$Q := \{y \in R^k : |x + (2M)^{-1}\xi - y| < (2M)^{-1}|\xi|\}.$$

It follows that, for any $y \in Q$,

$$d_C(y) \geq g(y) \tag{4.30}$$

in which

$$g(v) := (2M)^{-1}|\xi| - |v - x - (2M)^{-1}\xi|.$$

($g(v)$ will be recognized as the distance of v from the surface of the sphere.) However since d_C is nonnegative, and g is nonpositive on the complement of Q, (4.30) is actually valid for all y.

The function g is analytic on a neighborhood of x (since $\xi \neq 0$). We calculate $\nabla g(x) = |\xi|^{-1}\xi$. It follows that there exist some $\alpha > 0$ and $\epsilon > 0$ such that

$$g(y) - g(x) \geq |\xi|^{-1}\xi \cdot (y - x) - \alpha|y - x|^2 \quad \text{for all } y \in x + \epsilon B.$$

Since $g(x) = d_C(x) = 0$, we deduce from (4.30) that

$$d_C(y) - d_C(x) \geq g(y) - g(x) \geq |\xi|^{-1}\xi \cdot (y - x) - \alpha|y - x|^2$$

for all $y \in x + \epsilon B$. Noting however that $|\xi| \leq 1$ and $d_C(x) = 0$, we deduce that

$$d_C(y) - d_C(x) \geq |\xi|(d_C(y) - d_C(x)) \geq \xi \cdot (y - x) - \alpha|\xi||y - x|^2$$

for all $y \in x + \epsilon B$. From this we conclude (via Proposition 4.4.1) that $\xi \in \partial^P d_C(x)$. $\square$

Our next objective is to derive a similar description of limiting normal cones in terms of limiting subdifferentials of the distance function. First of all however we need to assemble more information about proximal normals $\xi \in \partial^P d_C(x)$ at basepoints $x \notin C$. The following lemma tells us that $\xi \in \partial^P d_C(\bar{x})$ for some $\bar{x} \in C$, and a little bit more.

Lemma 4.8.3 *Take a closed set $C \subset R^k$, a point $x \notin C$, and a vector $\xi \in \partial^P d_C(x)$. Then x has a unique closest point $\bar{x}$ in C and*

$$\xi = |x - \bar{x}|^{-1}(x - \bar{x}).$$

We conclude, in particular, that $\xi \in \partial^P d_C(\bar{x})$.

Proof. By Proposition 4.4.2, there exists $M > 0$ such that

$$\xi \cdot (y - x) \leq d_C(y) - d_C(x) + M|y - x|^2 \quad \text{for all } y \in R^k. \qquad (4.31)$$

Let $\bar{x}$ be a closest point to x in C. Then

$$|x - \bar{x}| = d_C(x) \ (\neq 0).$$

Evidently then $\bar{x}$ is the closest point to $\bar{x} + (1 - \alpha)(x - \bar{x})$ for all $\alpha \in (0, 1)$, a property that can be expressed analytically as

$$d_C(\bar{x} + (1 - \alpha)(x - \bar{x})) = (1 - \alpha)|x - \bar{x}| = (1 - \alpha)d_C(x).$$

Combining this relationship with the proximal inequality (4.31), we deduce

$$\begin{aligned} -\alpha|x - \bar{x}| &= d_C(\bar{x} + (1 - \alpha)(x - \bar{x})) - d_C(x) \\ &\geq -\alpha\xi \cdot (x - \bar{x}) - \alpha^2 M|x - \bar{x}|^2 \quad \text{for all } \alpha \in (0, 1). \end{aligned}$$

Divide across this inequality by α and pass to the limit as $\alpha \downarrow 0$. This gives

$$|x - \bar{x}| \leq \xi \cdot (x - \bar{x}).$$

But, by Proposition 4.8.2, $|\xi| \leq 1$. Since $x - \bar{x} \neq 0$, we deduce that

$$\xi = |x - \bar{x}|^{-1}(x - \bar{x}).$$

If $\bar{y}$ is any other closest point to x in C, the above reasoning gives $d_C(x) = |x - \bar{y}| = |x - \bar{x}| \neq 0$ and $\xi = |x - \bar{y}|^{-1}(x - \bar{y}) = |x - \bar{x}|^{-1}(x - \bar{x})$, from which we conclude $\bar{x} = \bar{y}$.

Of course, the fact that $\xi \in \partial^P d_C(\bar{x})$ follows from Proposition 4.8.1. $\square$

We note also the following partial converse.

Lemma 4.8.4 *Take a closed set $C \subset R^k$ and points $x \notin C$ and $\bar{x} \in C$. Assume that, for some $\alpha > 0$, $y = \bar{x}$ minimizes the function*

$$y \rightarrow |\bar{x} + (1 + \alpha)(x - \bar{x}) - y|$$

over C. Define

$$\xi := |x - \bar{x}|^{-1}(x - \bar{x}).$$

We have:

(i) $\xi \in \partial^P d_C(x)$,

(ii) if $\{x_i\}$ and $\{\xi_i\}$ are sequences such that $x_i \rightarrow x$ as $i \rightarrow \infty$ and $\xi_i \in \partial^P d_C(x_i)$ for each i, then there exists a sequence of points $\{y_i\}$ in C such that $\xi_i = |x_i - y_i|^{-1}(x_i - y_i)$ for all i, and $y_i \rightarrow \bar{x}$ and $\xi_i \rightarrow \xi$ as $i \rightarrow \infty$.

Proof. We arrange, by translation of the origin, that $\bar{x} = 0$.

(i): The minimizing property of $\bar{x}$ can be expressed geometrically: the open ball $(1 + \alpha)x + (1 + \alpha)|x|B$ is disjoint from C. It follows that for any point $v \in R^k$ with $|v - \alpha x| \leq |x|(1 + \alpha)$,

$$d_C(x + v) \geq \psi(v),$$

where

$$\psi(v) := (1 + \alpha)|x| - |x + v - (1 + \alpha)x| = (1 + \alpha)|x| - |v - \alpha x|.$$

(The function ψ measures the distance of the point $x + v$ in the ball $(1 + \alpha)x + (1 + \alpha)|x|B$ to its boundary.)

Notice that $d_C(x) = |x| = \psi(0)$, so

$$d_C(x + v) - d_C(x) \geq \psi(v) - \psi(0).$$

But ψ is analytic on $\alpha|x|\text{int } B$. (Notice $x \neq 0$ since $x \notin C$.) We calculate its gradient at $v = 0$ to be

$$\nabla\psi(0) = |x|^{-1}x.$$

It follows that, for some $M > 0$ and $\epsilon > 0$,

$$d_C(x + v) - d_C(x) \geq |x|^{-1}v \cdot x - M|v|^2 \quad \text{for all } v \in \epsilon B.$$

By Proposition 4.4.1, then,

$$|x|^{-1}x \in \partial^P d_C(x).$$

(ii): We begin by showing that $\bar{x}(= 0)$ is the unique minimizer of $y \to |x - y|$ over C. Let y' be an arbitrary such minimizer. Since $0 \in C$, $|x-y'|^2 \leq |x|^2$. Hence

$$-2x \cdot y' + |y'|^2 \leq 0.$$

But the minimizing property of $\bar{x}(= 0)$ tells us that

$$|(1+\alpha)x - y'|^2 \geq (1+\alpha)^2|x|^2,$$

and hence that

$$-2(1+\alpha)x \cdot y' + |y'|^2 \geq 0.$$

These relationships combine to give

$$|y'|^2 - (1+\alpha)|y'|^2 \geq 0.$$

This is possible only if $y' = 0$. So $\bar{x}(= 0)$ is indeed the unique minimizer.

Now take sequences $\{x_i\}$ and $\{\xi_i\}$ with the stated properties. Since $x \notin C$ and C is closed, we may assume $x_i \notin C$ for all i. According to Lemma 4.8.3 then there exists a sequence $\{y_i\}$ in C such that

$$d_C(x_i) = |x_i - y_i| \tag{4.32}$$

and

$$\xi_i = |x_i - y_i|^{-1}(x_i - y_i). \tag{4.33}$$

Extract an arbitrary subsequence. $\{y_i\}$ is bounded. Along a further subsequence then we have $y_i \to \bar{y}$ for some $\bar{y} \in C$. By the continuity of the distance function, we deduce from (4.32) that $d_C(x) = |x - \bar{y}|$. In view of our earlier observations then, $\bar{y} = 0$. Noting (4.33), we see that

$$y_i \to 0 \quad \text{and} \quad \xi_i \to |x|^{-1}x = \xi. \tag{4.34}$$

Since the limits here are independent of the subsequence initially selected, (4.34) is in fact true for the original sequence. $\square$

Simple limit-taking procedures now permit us to relate the limiting normal cone to a set and the limiting subdifferential of the distance function.

Theorem 4.8.5 *Take a closed set $C \subset R^k$ and a point $x \in C$. Then*

$$\partial d_C(x) = N_C(x) \cap B.$$

Proof.

(a): We show $\partial d_C(x) \subset N_C(x) \cap B$.

Take a vector $\xi \in \partial d_C(x)$. Then there exist $x_i \to x$ and $\xi_i \to \xi$ such that $\xi_i \in \partial^P d_C(x_i)$ for all i (by Proposition 4.6.3). In consequence of Proposition 4.8.2 and Lemma 4.8.3, there exists a sequence of points $\{z_i\}$ in C such that $z_i \overset{C}{\to} x$, and $\xi_i \in N_C^P(z_i)$ and $|\xi_i| \leq 1$ for each i. But then by definition of $N_C(x)$, $\xi \in N_C(x) \cap B$.

(b): We show $\partial d_C(x) \supset N_C(x) \cap B$.

Since $N_C(x)$ is a cone, generated by taking limits of proximal normals and $x \rightsquigarrow \partial d_C(x)$ has a closed graph, it suffices to show that $N_C^P(x) \cap B \subset \partial^P d_C(x)$. But this property is supplied by Proposition 4.8.2. $\square$

4.9   Criteria for Lipschitz Continuity

On occasion, we wish to establish Lipschitz continuity of some specific function. In the case when an explicit formula is available for evaluation of the function, we can expect to be able to assess whether this property holds directly. There are, too, certain elementary criteria for Lipschitz continuity that are sometimes of use: "the pointwise supremum of a uniformly bounded family of Lipschitz continuous functions with common Lipschitz constant is Lipschitz continuous," for example. But for certain implicitly defined functions, most notably for many of the "value" functions considered in optimization, the task of establishing Lipschitz continuity on a neighborhood of a basepoint is a challenging one. Here conditions involving boundedness of subdifferentials are often useful. The following theorem makes precise the intuitive notion that lack of Lipschitz continuity should be reflected in the presence of "unbounded" neighboring derivatives.

Theorem 4.9.1 *Take a lower semicontinuous function $f : R^k \to R \cup \{+\infty\}$, a point $x \in \mathrm{dom}\, f$, and a constant $K > 0$. Then the following conditions are equivalent.*

(i) f is Lipschitz continuous with Lipschitz constant K on some neighborhood of x;

(ii) there exists $\epsilon > 0$ such that for every $x \in R^k$ and $\xi \in \partial^P f(x)$ satisfying

$$|x - \bar{x}| \leq \epsilon \quad and \quad |f(x) - f(\bar{x})| \leq \epsilon,$$

we have

$$|\xi| \leq K.$$

The theorem tells us that, in order to check Lipschitz continuity, we have only to test boundedness of neighboring proximal subgradients, at points where they are defined.

This theorem and earlier derived properties of subdifferentials lead to an alternative criterion for Lipschitz continuity involving the asymptotic subdifferential.

Corollary 4.9.2 *Take a lower semicontinuous function $f : R^k \to R \cup \{+\infty\}$ and a point $\bar{x} \in dom\ f$. Then the following conditions are equivalent:*

(i) f is Lipschitz continuous on a neighborhood of $\bar{x}$;

(ii) $\partial^\infty f(\bar{x}) = \{0\}$.

Proof. Condition (i) implies Condition (ii), by Proposition 4.7.1. To show the reverse implication, assume that (i) is not true. Then, by Theorem 4.9.1, there exist sequences $x_i \xrightarrow{f} \bar{x}$ and $\{\xi_i\}$ such that

$$\xi_i \in \partial^P f(x_i) \text{ for each } i$$

and

$$|\xi_i| \to \infty \text{ as } i \to \infty.$$

For each i, we have

$$(|\xi_i|^{-1}\xi_i, -|\xi_i|^{-1}) \in N^P_{\text{epi}f}(x_i, f(x_i)).$$

Along a subsequence however

$$(|\xi_i|^{-1}\xi_i, -|\xi_i|^{-1}) \to (\bar{\xi}, 0)$$

for some $\bar{\xi}$ with $|\bar{\xi}| = 1$. In view of the closure properties of the limiting normal cone

$$(\bar{\xi}, 0) \in N_{\text{epi}f}(\bar{x}, f(\bar{x})).$$

This means that $\bar{\xi}$ is a nonzero vector in $\partial^\infty f(\bar{x}) = \{0\}$. We have shown that (ii) is not true.
□

As a preliminary step in the proof of Theorem 4.9.1, we establish some properties of proximal subgradients of a special "min" function.

Lemma 4.9.3 *Take a lower semicontinuous function $f : R^k \to R \cup \{+\infty\}$, $x \in R^k$, and $\beta \in R$. Define*

$$g(y) := \min\{f(y), \beta\}.$$

Suppose that ξ is a nonzero vector such that

$$\xi \in \partial^P g(x).$$

Then

$$f(x) \le \beta \quad \text{and} \quad \xi \in \partial^P f(x).$$

Proof. Since $\xi \in \partial^P g(x)$, there exists $\alpha > 0$ such that

$$\min\{f(x'), \beta\} - \min\{f(x), \beta\} \ \geq \ \xi \cdot (x' - x) - M|x' - x|^2$$

for all $x' \in x + \alpha B$.

Suppose that $f(x) > \beta$. Then

$$0 = \beta - \beta \geq \min\{f(x'), \beta\} - \min\{f(x), \beta\} \geq \xi \cdot (x' - x) - M|x' - x|^2$$

for all $x' \in x + \alpha B$. It follows that $\xi = 0$. But this is impossible since we have assumed $\xi \neq 0$. We have shown that

$$f(x) \leq \beta.$$

Notice also that

$$
\begin{aligned}
f(x') - f(x) &\geq \min\{f(x'), \beta\} - \min\{f(x), \beta\} \\
&\geq \xi \cdot (x' - x) - M|x' - x|^2
\end{aligned}
$$

for all $x' \in x + \alpha B$. It follows that $\xi \in \partial^P f(x)$. $\square$

Proof of Theorem 4.9.1. That Condition (i) implies Condition (ii) follows from Theorem 4.7.1. To prove the reverse implication, assume that f is not Lipschitz continuous with Lipschitz constant K on any neighborhood of $\bar{x}$. We show that condition (ii) cannot be satisfied.

Choose any $\epsilon > 0$. Since f is lower semicontinuous, we can find $\delta > 0$, $0 < \delta < \epsilon$, such that

$$z \in \bar{x} + \delta B \quad \text{implies} \quad f(z) > f(\bar{x}) - \epsilon. \tag{4.35}$$

Under the hypothesis there exist sequences

$$x_i \to \bar{x} , \ y_i \to \bar{y}$$

such that

$$x_i \neq y_i$$

and

$$f(y_i) - f(x_i) > K|y_i - x_i| \tag{4.36}$$

for all i. (The case $f(y_i) = +\infty$ is permitted.) We can assume that for some subsequence (we do not relabel)

$$f(x_i) \to f(\bar{x}) \quad \text{as} \quad i \to \infty. \tag{4.37}$$

This is because if this property fails to hold then, in view of the lower semicontinuity of f, we have

$$\liminf_{i\to\infty} f(x_i) > f(\bar{x}).$$

But $|x_i - \bar{x}| \to 0$. It follows that (4.36) remains valid (for all i sufficiently large) when we redefine $x_i := \bar{x}$. Of course (4.37) is automatically satisfied.

In view of these properties, we can choose a sequence $\epsilon_i \downarrow 0$ such that

$$f(x_i) - f(\bar{x}) \;\leq\; \epsilon_i \tag{4.38}$$

and

$$K|y_i - x_i| \;<\; f(\bar{x}) - f(x_i) + \epsilon_i. \tag{4.39}$$

Fix a value of the index i such that

$$|x_i - \bar{x}| < \tfrac{1}{2}\delta \;,\; |y_i - \bar{x}| < \tfrac{1}{2}\delta \;\text{ and }\; \epsilon_i < \epsilon. \tag{4.40}$$

Define the function

$$g_i(y) := \min\{f(\bar{x}) + \epsilon_i, f(y)\}.$$

It follows from (4.36) and (4.39) that

$$g_i(y_i) - g_i(x_i)$$
$$= \min\{f(\bar{x}) - f(x_i) + \epsilon_i, f(y_i) - f(x_i)\} \;>\; K|y_i - x_i|.$$

Since g_i is a lower semicontinuous function that is finite at x_i and y_i, the following conclusions can be drawn from this last relationship and the Mean Value Inequality (Theorem 4.5.1). There exist

$$z \in \bar{x} + \delta B \;\text{ and }\; \xi \in \partial^P g_i(z)$$

such that

$$K|y_i - x_i| \;<\; \xi \cdot (y_i - x_i).$$

But $y_i \neq x_i$. It follows that
$$|\xi| > K.$$

Recall however that that $\delta < \epsilon$. So $|z - \bar{x}| < \epsilon$. From (4.35) we deduce that

$$f(z) - f(\bar{x}) > -\epsilon.$$

Since the proximal subgradient ξ is nonzero, we deduce, from Lemma 4.9.3 and the preceding relationships, that

$$|z - \bar{x}| < \epsilon, \; |f(z) - f(\bar{x})| < \epsilon, \quad\text{and}\quad \xi \in \partial^P f(z).$$

But ϵ is an arbitrary positive number. It follows that Condition (ii) is not satisfied. $\square$

4.10   Relationships Between Normal and Tangent Cones

Up to this point, tangent vectors have put in only an occasional appearance. In this optimization oriented treatment of the theory, it is natural to emphasize the significance of normal vectors, since it is in terms of these objects (and their close relatives, subgradients) that Lagrange multiplier rules and other important principles of nonsmooth optimization are expressed. There is however a rich web of relationships involving tangent vectors and normal vectors, as we now reveal.

Take a closed set $C \subset R^k$ and a point $x \in C$. With the help of the set limit notation of Section 2.1, we now review the various concepts of normal cones and their interrelation.

The *proximal normal cone* to C at x, $N_C^P(x)$, is

$$N_C^P(x) := \{\xi : \exists\, M > 0 \text{ such that } \xi \cdot (y - x) \le M|y - x|^2 \text{ for all } y \in C\}.$$

The *strict normal cone* to C at x, $\hat{N}_C(x)$, is

$$\hat{N}_C(x) := \{\xi : \limsup_{y \xrightarrow{C} x} |y - x|^{-1}\xi \cdot (y - x) \le 0\}.$$

The *limiting normal cone* is expressible in terms of our new notation as

$$N_C(x) := \limsup_{y \xrightarrow{C} x} N_C^P(y).$$

Proposition 4.2.5 provides us also with the following representation of $N_C(x)$:

$$N_C(x) := \limsup_{y \xrightarrow{C} x} \hat{N}_C(y).$$

We recall $N_C^P(x)$, $\hat{N}_C(x)$, and $N_C(x)$ are all cones containing $\{0\}$. $N_C^P(x)$ is convex, $\hat{N}_C(x)$ is closed and convex, $N_C(x)$ is closed, and

$$N_C^P(x) \subset \hat{N}_C(x) \subset N_C(x).$$

The tangent cones of primary interest here are the *Bouligand* and the *Clarke* tangent cones defined as follows.

Definition 4.10.1 *Take a closed set $C \subset R^k$ and a point $x \in C$.*
The Bouligand Tangent Cone *to C at $x \in C$, written $T_C(x)$, is the set*

$$T_C(x) := \limsup_{t \downarrow 0} t^{-1}(C - x).$$

The Clarke tangent cone *to C at x, written $\bar{T}_C(x)$, is the set*

$$\bar{T}_C(x) := \liminf_{t \downarrow 0,\, y \xrightarrow{C} x} t^{-1}(C - y).$$

Equivalent "sequential" definitions are as follows.

$T_C(x)$ comprises vectors ξ corresponding to which there exist some sequence $\{c_i\}$ in C and some sequence $t_i \downarrow 0$ such that $t_i^{-1}(c_i - x) \to \xi$.

$\bar{T}_C(x)$ comprises vectors ξ such that for any sequences $x_i \overset{C}{\to} x$ and $t_i \downarrow 0$ there exists a sequence $\{c_i\}$ in C such that $t_i^{-1}(c_i - x_i) \to \xi$.

We state without proof the following elementary properties.

Proposition 4.10.2 *Take a closed set $C \subset R^k$ and a point $x \in C$. Then $T_C(x)$ and $\bar{T}_C(x)$ are closed cones containing the origin and $\bar{T}_C(x) \subset T_C(x)$.*

A less obvious property is:

Proposition 4.10.3 *Take a closed set $C \subset R^k$ and a point $x \in C$. Then $\bar{T}_C(x)$ is a convex set.*

Proof. Take any u and v in $\bar{T}_C(x)$. Since $\bar{T}_C(x)$ is a cone, to establish convexity we must show that $u + v \in \bar{T}_C(x)$. Take any sequence $x_i \overset{C}{\to} x$ and $t_i \downarrow 0$. Because $u \in \bar{T}_C(x)$, there exists a sequence of points $u_i \to u$ such that $x_i + t_i u_i \in C$ for each i. But $x_i + t_i u_i \to x$, so, since $v \in \bar{T}_C(x)$ also, there exists a sequence of points $v_i \to v$ such that $(x_i + t_i u_i) + t_i v_i \in C$ for each i. A rearrangement of this inclusion gives $x_i + t_i(u_i + v_i) \in C$. But this implies $u + v \in \bar{T}_C(x)$. $\square$

Denote by S^* the polar cone of a set $S \subset R^k$; namely,

$$S^* := \{\xi : \xi \cdot x \leq 0 \text{ for all } x \in S\}.$$

Theorem 4.10.4 *Take a closed set $C \subset R^k$ and a point $x \in C$. Then the strict normal cone $\hat{N}_C(x)$ and the Bouligand tangent cone $T_C(x)$ are related according to*

$$\hat{N}_C(x) = T_C(x)^*.$$

Proof.

(a): We show that $\hat{N}_C(x) \subset T_C(x)^*$.
Take any $\xi \in \hat{N}_C(x)$. Then

$$\xi \cdot (y - x) \leq o(|y - x|) \quad \text{for all } y \in C \tag{4.41}$$

for some $o(\cdot) : R^+ \to R^+$ that satisfies $o(s)/s \to 0$ as $s \downarrow 0$. Choose any $v \in T_C(x)$. Then there exists $x_i \overset{C}{\to} x$ and $t_i \downarrow 0$ such that, if we define $v_i := t_i^{-1}(x_i - x)$ for each i, then $v_i \to v$. It is claimed that $\xi \cdot v \leq 0$; it will follow that $\xi \in T_C(x)^*$. The claim is certainly valid when $v = 0$. So assume $v \neq 0$. For each i then

$$\xi \cdot v_i = t_i^{-1}\xi \cdot (x_i - x) \leq |v_i||x_i - x|^{-1}o(|x_i - x|)$$

by (4.41). Passing to the limit as $i \to \infty$, we obtain $\xi \cdot v \leq 0$.

(b): We show that $\hat{N}_C(x) \supset T_C(x)^*$.
Suppose that $\xi \notin \hat{N}_C(x)$. Then there exist $\epsilon > 0$ and $x_i \overset{C}{\to} x$ such that

$$\xi \cdot (x_i - x) > \epsilon|x_i - x| \text{ for all } i.$$

Note that $x_i \neq x$ for each i. Set $t_i = |x_i - x|$ and $v_i := t_i^{-1}(x_i - x)$. The v_is all have unit length. Along a subsequence then, $v_i \to v$ for some $v \in T_C(x)$ with $|v| = 1$. Along the subsequence we have

$$\xi \cdot v_i = t_i^{-1}\xi \cdot (x_i - x) > \epsilon t_i^{-1}|x_i - x| = \epsilon|v_i| = \epsilon.$$

In the limit, we obtain $\xi \cdot v \geq \epsilon$. It follows that $\xi \notin T_C(x)^*$. $\square$

The next theorem tells us (among other things) that we get the Clarke tangent cone by applying the lim inf operation (with respect to the base-point) to the Bouligand tangent cone.

Theorem 4.10.5 *Take a closed set $C \subset R^k$ and a point $x \in C$. Then the Bouligand tangent cone $T_C(x)$, its closed convex hull, and the Clarke tangent cone $\bar{T}_C(x)$ are related as follows.*

$$\liminf_{y \overset{C}{\to} x} \overline{co}T_C(y) = \liminf_{y \overset{C}{\to} x} T_C(y) = \bar{T}_C(x).$$

Proof. Since $T_C(y) \subset \overline{co}T_C(y)$ for each y, it suffices to show

(i) $\liminf_{y \overset{C}{\to} x} \overline{co}T_C(y) \subset \bar{T}_C(x)$,

(ii) $\bar{T}_C(x) \subset \liminf_{y \overset{C}{\to} x} T_C(y)$,

since these inclusions imply

$$\liminf_{y \overset{C}{\to} x} \overline{co}T_C(y) \subset \bar{T}_C(x) \subset \liminf_{y \overset{C}{\to} x} T_C(y) \subset \liminf_{y \overset{C}{\to} x} \overline{co}T_C(y)$$

from which the required relationships follow.

We show

$$\liminf_{y \overset{C}{\to} x} \overline{co}T_C(y) \subset \bar{T}_C(x).$$

Take any $v \in \liminf_{y \xrightarrow{C} x} \overline{\mathrm{co}} T_C(y)$. Let $x_i \xrightarrow{C} x$ and $t_i \downarrow 0$ be arbitrary sequences. Suppose we are able to find a sequence $\epsilon_i \downarrow 0$ such that

$$t_i^{-1} d_C(x_i + t_i v) \le \epsilon_i. \tag{4.42}$$

Then we have

$$\limsup_i t_i^{-1} d_C(x_i + t_i v) = 0.$$

This implies $v \in \bar{T}_C(x)$ which is the required relationship.

It remains to show (4.42). For each i, we consider the function

$$g_i(t) := d_C(x_i + tv), \quad 0 \le t \le t_i.$$

Choose $z_i(t) \in \arg\min\{|x_i + tv - z| : z \in C\}$. Notice that, since $x_i \in C$,

$$|x - z_i(t)| \le |x_i + tv - z_i(t)| + |x_i - x + tv| \le 2t_i|v| + |x_i - x|$$

for $0 \le t \le t_i$ and $i = 1, 2, \dots$.

Evidently then

$$\sup_{0 \le t \le t_i} |x - z_i(t)| \to 0 \text{ as } i \to \infty.$$

Since $v \in \liminf_{y \xrightarrow{C} x} \overline{\mathrm{co}} T_C(y)$, there exist $\epsilon_i \downarrow 0$ and functions $w_i(\cdot)$: $[0, t_i] \to R^n$ such that

$$w_i(t) \in \overline{\mathrm{co}} T_C(z_i(t)) \quad \text{and} \quad |w_i(t) - v| < \epsilon_i \quad \text{for all } t \in [0, t_i].$$

Fix i. We verify (4.42) for the above choice of ϵ_i.

Let us first investigate the properties of a point $t \in [0, t_i)$ at which $g_i > 0$ and where g_i is differentiable.

To simplify the notation, write z for $z_i(t)$ and set $p = x_i + tv$. We have for sufficiently small $h > 0$,

$$\begin{aligned} g_i(t+h) - g_i(t) &\le |x_i + tv + hv - z| - |x_i + tv - z| \\ &= |p - z + hv| - |p - z|. \end{aligned}$$

Since $p \ne z$ ($g_i(t) > 0$, remember) the function $u \to |p - z + u|$ is differentiable at any point sufficiently close to the origin, in particular at hv (for $h > 0$ sufficiently small). In view of the convexity of the norm then the "subgradient inequality" for convex functions gives

$$g_i(t+h) - g_i(t) \le h|p - z + hv|^{-1}(p - z + hv) \cdot v.$$

We conclude that

$$\begin{aligned} h^{-1}(g_i(t+h) - g_i(t)) &\le \left(\frac{p - z + hv}{|p - z + hv|} - \frac{p - z}{|p - z|} \right) \cdot v + \frac{p - z}{|p - z|} \cdot v \\ &\le \left(\frac{p - z + hv}{|p - z + hv|} - \frac{p - z}{|p - z|} \right) \cdot v \\ &\qquad + \frac{p - z}{|p - z|} \cdot w_i + \epsilon_i. \tag{4.43} \end{aligned}$$

Since $p - z \in N_C^P(z)$ and $w_i(t) \in \bar{co}T_C(z)$ and in view of the fact that

$$\hat{N}_C(z) = T_C^*(z) = (\bar{co}T_C(z))^*,$$

it follows that

$$(p - z) \cdot w_i(t) \leq 0.$$

We therefore retain inequality (4.43) when we drop the second term on the right. Passing to the limit as $h \downarrow 0$, we obtain

$$\frac{d}{dt}g_i(t) \leq \epsilon_i.$$

Since g_i is Lipschitz continuous and therefore almost everywhere differentiable, this inequality is valid for all points $t \in \{s \in [0, t_i] : g_i(s) > 0\}$ excluding a nullset.

Condition (4.42) is automatically satisfied if $g_i(t_i) = 0$. Suppose then that $g_i(t_i) > 0$. Define

$$\tau :=; \inf \{t \in [0, t_i] : g_i(s) > 0 \quad \text{for all } s \in [t, t_i]\}.$$

Since g_i is continuous and $g_i(0) = 0$, we conclude $g_i(\tau) = 0$. It follows that

$$g_i(t_i) = g_i(\tau) + \int_\tau^{t_i} \dot{g}_i(\sigma)d\sigma \leq 0 + t_i\epsilon_i.$$

But then

$$t_i^{-1}d_C(x_i + t_iv)(= t_i^{-1}g_i(t_i)) \leq \epsilon_i.$$

Condition (4.42) is therefore satisfied in this case also. The proof is complete.

We show that

$$\bar{T}_C(x) \subset \liminf_{y \xrightarrow{C} x} T_C(y).$$

Take $v \in \bar{T}_C(x)$. Consider an arbitrary sequence $x_i \xrightarrow{C} x$. It follows from the definition of $\bar{T}_C(x)$ that there exist $\epsilon_i \downarrow 0$, $t_i \downarrow 0$, and $N_i \uparrow \infty$ such that

$$t^{-1}d_C(x_k + tv) \leq \epsilon_i \quad \text{for all } 0 \leq t \leq t_i, \text{ and } k \geq N_i.$$

Relabel $\{x_{N_i}\}_{i=1}^\infty$ as $\{x_i\}$ and $\{t_{N_i}\}_{i=1}^\infty$ as $\{t_i\}$. Then for each $i = 1, 2, \ldots$ and for all $t \in [0, t_i]$, we have in particular

$$t^{-1}d_C(x_i + tv) \leq \epsilon_i.$$

Fix i and $t \in [0, t_i]$, choose $z_i(t) \in C$ such that

$$d_C(x_i + tv) = |x_i + tv - z_i(t)|.$$

Define

$$v_i(t) := t^{-1}(z_i(t) - x_i), \quad \text{for all } t \in (0, t_i], i = 1, 2, \dots . \tag{4.44}$$

Then

$$|v_i(t) - v| \le \epsilon_i \text{ for all } t \in (0, t_i], \quad \text{for } i = 1, 2, \dots .$$

Fix i. Choose $\bar{v}_i$ to be a cluster point of $\{v_i(t)\}_{t>0}$ at $t = 0$ (such $\bar{v}_i$ exists since $v_i(\cdot)$ is bounded on $(0, t_i]$). We have $|\bar{v}_i - v| \le \epsilon_i$. By (4.44) and in view of the definition of $T_C(x_i)$, we have that

$$\bar{v}_i \in T_C(x_i) \quad \text{and} \quad \bar{v}_i \to v.$$

Starting with an arbitrary sequence $\{x_i\}$ in C converging to x, we have found a subsequence $\{x_{i_n}\}_{n=1}^{\infty}$ and a sequence $\{\bar{v}_n\}$ such that

$$\bar{v}_n \in T_C(x_{i_n}) \text{ for all } n \quad \text{and} \quad \bar{v}_n \to v \text{ as } n \to \infty.$$

It follows that $v \in \liminf_{\substack{y \xrightarrow{C} x}} T_C(y)$. □

Our next objective is to relate the Clarke tangent cone and the limiting normal cone. First a lemma is required on the interaction between limit-taking and the construction of polar sets.

Lemma 4.10.6 *Take a set-valued function* $S : R^k \rightsquigarrow R^n$, *a set* $C \subset R^k$, *and a point* $x \in R^k$. *Assume that* $S(y)$ *is a closed convex cone for all* y. *Then*

$$\liminf_{\substack{y \xrightarrow{C} x}} S(y) = [\limsup_{\substack{y \xrightarrow{C} x}} S(y)^*]^*.$$

Proof.

(a): We show that

$$\liminf_{\substack{y \xrightarrow{C} x}} S(y) \subset [\limsup_{\substack{y \xrightarrow{C} x}} S(y)^*]^*.$$

Take $v \in \liminf_{\substack{y \xrightarrow{C} x}} S(y)$. Choose any $\xi \in \limsup_{\substack{y \xrightarrow{C} x}} S(y)^*$. Then there exist sequences $x_i \xrightarrow{C} x$ and $\{\xi_i\}$ in R^n such that

$$\xi_i \in S(x_i)^* \quad \text{for all } i \quad \text{and} \quad \xi_i \to \xi.$$

We may also choose a sequence $v_i \to v$ such that $v_i \in S(x_i)$ for each i. For each i then we have $\xi_i \cdot v_i \le 0$. In the limit we get $\xi \cdot v \le 0$. It follows that

$$v \in [\limsup_{\substack{y \xrightarrow{C} x}} S(y)^*]^*.$$

(b): We show that

$$\liminf_{y \xrightarrow{C} x} S(y) \supset [\limsup_{y \xrightarrow{C} x} S(y)^*]^*.$$

Take

$$v \in [\limsup_{y \xrightarrow{C} x} S(y)^*]^*.$$

Assume, contrary to the assertions of the lemma that

$$v \notin \liminf_{y \xrightarrow{C} x} S(y).$$

Then there is some $\epsilon > 0$ and a sequence $x_i \xrightarrow{C} x$ such that

$$(v + \epsilon B) \cap S(x_i) = \emptyset \quad \text{for each } i.$$

Applying the separation theorem we obtain, for each i, a vector ξ_i such that $|\xi_i| = 1$ and

$$\sup_{w \in S(x_i)} w \cdot \xi_i \leq \inf_{w \in v + \epsilon B} w \cdot \xi_i = v \cdot \xi_i - \epsilon .$$

Since $S(x_i)$ is a cone, we deduce that $\sup_{w \in S(x_i)} w \cdot \xi_i = 0$ and so $\xi_i \in S(x_i)^*$. Along a subsequence $\xi_i \to \xi$ for some ξ with $|\xi| = 1$. We have that $\xi \in \limsup_{y \xrightarrow{C} x} S(y)^*$. We also have $0 \leq v \cdot \xi_i - \epsilon$. In the limit we arrive at $v \cdot \xi \geq \epsilon$, from which we conclude that $v \notin [\limsup_{y \xrightarrow{C} x} S(y)^*]^*$. This contradiction concludes the proof. $\square$

Theorem 4.10.7 *Take a closed set $C \subset R^k$ and $x \in C$. The Clarke tangent cone $\bar{T}_C(x)$ and the limiting normal cone $N_C(x)$ are related according to*

$$\bar{T}_C(x) = N_C(x)^*.$$

Proof. Apply the preceding lemma to $S(y) := \overline{\text{co}}T_C(y)$. This gives

$$\liminf_{y \xrightarrow{C} x} \overline{\text{co}}T_C(y) = (\limsup_{y \xrightarrow{C} x} \overline{\text{co}}T_C(y)^*)^*.$$

But $\liminf_{y \xrightarrow{C} x} \overline{\text{co}}T_C(y) = \bar{T}_C(x)$ by Theorem 4.10.5. Also, for each y we have $(\overline{\text{co}}T_C(y))^* = T_C(y)^* = \hat{N}_C(y)$ by Theorem 4.10.4. But then, by Proposition 4.2.5, $N_C(x) = \limsup_{y \xrightarrow{C} x} (\overline{\text{co}}T_C(y))^*$. Assembling those relationships, we get $\bar{T}_C(x) = N_C(x)^*$. $\square$

The relationships we have established between $N_C^P(x)$, $\hat{N}_C(x)$, $N_C(x)$, $\bar{T}_C(x)$, and $T_C(x)$ associated with a closed set $C \in R^k$ and point $x \in C$, are summarized in the following diagram, in which " $*$" denotes "take the polar cone":

$$
\begin{array}{ccc}
T_C(x) & \overset{\text{lim inf}}{\longrightarrow} & \bar{T}_C(x) \\
\downarrow * & & \uparrow * \\
\hat{N}_C(x) & \overset{\text{lim sup}}{\longrightarrow} \; N_C(x) & \overset{\text{lim sup}}{\longleftarrow} \; N_C^P(x) \, .
\end{array}
$$

Notes for Chapter 4

Notes on Chapters 4 and 5 are provided at the end of Chapter 5.

Chapter 5

Subdifferential Calculus

If Newton had thought that continuous functions do not necessarily have a derivative – and this is the general case – the differential calculus would never have been created.

– Emile Picard

5.1 Introduction

In this chapter, we assemble a number of useful rules for calculating and estimating the limiting subdifferentials of composite functions in terms of their constituent mappings. A typical rule is the "Sum Rule"

$$\partial(f + g)(x) \subset \partial f(x) + \partial g(x). \tag{5.1}$$

(The right side denotes, of course, the set of all vectors ξ expressible as $\xi = \eta_1 + \eta_2$ for some $\eta_1 \in \partial f(x)$ and $\eta_2 \in \partial g(x)$.)

We notice at once that the rule is "one sided": it takes the form of a set inclusion not a set equivalence. Fortunately the inclusion that comes naturally is in the helpful direction: we want an estimate of $\partial(f + g)$ in terms of the subgradients of the (usually simpler) functions f and g of which $f + g$ is constituted. It is inevitable that inclusions feature in these rules if they are to handle functions which are locally Lipschitz continuous (or even less regular). This point is illustrated in the following example.

Example 5.1 Take $f : R \to R$ and $g : R \to R$ to be the Lipschitz continuous functions

$$f(x) = \min\{0, x\} \quad \text{and} \quad g(x) = -\min\{0, x\}.$$

Then

$$\partial(f + g)(0) = \{0\}.$$

Yet

$$\partial f(0) = \{0\} \cup \{1\} \quad \text{and} \quad \partial g(0) = [-1, 0].$$

We see that

$$\partial(f + g)(0) \overset{\text{strict}}{\subset} \partial f(0) + \partial g(0).$$

It is desirable to have calculus rules that apply to general lower semi-continuous functions. But even a "one-sided" rule such as (5.1) may fail unless some nondegeneracy hypothesis is imposed. The precise nature of the hypotheses required will differ from rule to rule, but in each case they will eliminate certain kinds of interaction of the relevant asymptotic limiting subdifferentials. These nondegeneracy hypotheses are automatically satisfied when the functions involved are Lipschitz continuous.

Example 5.2   Take the function $s : R \to R$ to be

$$s(x) := \text{sgn} \ \{x\}|x|^{1/2}$$

and define the functions $f : R \to R$ and $g : R \to R$ according to

$$f(x) = s(x) \quad \text{and} \quad g(x) = -s(x).$$

We find that $\partial f(0) = \partial g(0) = \emptyset$ and $\partial(f + g)(0) = \{0\}$. So

$$\{0\} = \partial(f + g)(0) \not\subset \partial f(0) + \partial g(0) = \emptyset.$$

Notice that $\partial^\infty f(0) = [0, \infty)$ and $\partial^\infty g(0) = (-\infty, 0]$. The pathological aspect of this example is that there exist nonzero numbers a and b, such that $a \in \partial^\infty f(0)$, $b \in \partial^\infty g(0)$, and $a + b = 0$. (Take $a = 1$ and $b = -1$, for example.) The data then fail to satisfy the condition:

$$a \in \partial^\infty f(0), b \in \partial^\infty g(0), a + b = 0 \quad \text{implies} \quad a = b = 0,$$

which will turn out to be precisely the nondegeneracy hypothesis appropriate to the Sum Rule.

For a concept of subdifferential to be useful in the field of optimization, we require it, at the very least, to have the property: "If f achieves its minimum value at $\bar{x} \in \text{dom} f$, then $\{0\}$ is contained in the subdifferential of f at $\bar{x}$." Fortunately this is true even of the "tightest" subdifferential we have introduced, the proximal subdifferential.

Proposition 5.1.1 *Take a lower semicontinuous function $f : R^k \to R \cup \{+\infty\}$ and a point $x \in \text{dom} f$. Assume that x achieves the minimum value of f over a neighborhood of x; then*

$$0 \in \partial^P f(x).$$

Proof. The fact that x is a local minimum means that there exists $\epsilon > 0$ such that

$$\xi \cdot (y - x) \leq f(y) - f(x) + M|y - x|^2 \quad \text{for all } y \in x + \epsilon B$$

when we take $\xi = 0$ and any $M \geq 0$. We conclude from Proposition 4.4.1 that $0 \in \partial^P f(x)$. $\square$

To make a start, we need the following rudimentary Sum Rule.

Lemma 5.1.2 *Take functions $f : R^k \rightarrow R \cup \{+\infty\}$ and $g : R^k \rightarrow R$, a closed set $C \subset R^k$, and a point $x \in (\text{int } C) \cap (\text{dom } f)$. Assume that f is lower semicontinuous and g is of class C^2 on a neighbourhood of x.*

$$\partial^P (f + g + \Psi_C)(x) = \partial^P f(x) + \{\nabla g(x)\},$$
$$\partial(f + g + \Psi_C)(x) = \partial f(x) + \{\nabla g(x)\},$$
$$\partial^\infty (f + g + \Psi_C)(x) = \partial^\infty f(x).$$

(As usual, Ψ_C denotes the indicator function of the set C.)

Proof. For some $\epsilon > 0$ such that $x + \epsilon B \subset \text{int } C$, there exists $m > 0$ such that

$$|g(x') - g(x) - \nabla g(x) \cdot (x' - x)| \leq m|x' - x|^2$$

for all $x' \in x + \epsilon B$, since g is assumed C^2 on a neighborhood of x. Take any $y \in x + \epsilon B$, $\xi \in R^k$. Then, in consequence of this observation, there exist $M > 0$ and $\delta > 0$ such that

$$(\xi + \nabla g(x)) \cdot (x' - x) \leq (f + g + \Psi_C)(x') - (f + g + \Psi_C)(x) + M|x' - x|^2$$

for all $x' \in x + \delta B$ if and only if there exist $M' > 0$ and $\delta' > 0$ such that

$$\xi \cdot (x' - x) \leq f(x') - f(x) + M'|x' - x|^2$$

for all $x' \in x + \delta' B$.

It follows that $\xi + \nabla g(x) \in \partial^P (f + g + \Psi_C)(x)$ if and only if $\xi \in \partial^P f(x)$. We have confirmed the first assertion of the lemma.

The remaining assertions follow from Thmeorem 4.6.2 , upon limit-taking, when we note that $x' \xrightarrow{f} x$ if and only if $x' \xrightarrow{f+g+\Psi_C} x$. $\square$

5.2 A Marginal Function Principle

Take a function $F : R^k \times R^l \rightarrow R \cup \{+\infty\}$. Let $f : R^k \rightarrow R \cup \{+\infty\}$ be the marginal function:

$$f(x) := F(x, 0) \quad \text{for all } x \in R^k$$

and let $\bar{x}$ be a minimizer for f:

$$f(\bar{x}) = \min \{f(x) : x \in R^k\}.$$

In convex optimization there is a standard procedure for constructing a function $g : R^l \to R \cup \{-\infty\}$ such that

$$f(\bar{x}) \geq \sup\{g(v) : v \in R^l\}. \tag{5.2}$$

It is to introduce the Fenchel conjugate functional $F^* : R^k \times R^l \to R \cup \{+\infty\}$:

$$F^*(y, v) := \sup\{x \cdot y + u \cdot v - F(x, u) : (x, u) \in R^k \times R^l\}$$

and set

$$g(v) := -F^*(0, v).$$

The validity of the "weak duality" condition (5.2) follows directly from the definition of F^*. If F is jointly convex in its arguments and satisfies an appropriate nondegeneracy hypothesis then Condition (5.2) can be replaced by the stronger condition: there exists $\bar{v} \in R^l$ such that

$$f(\bar{x}) = \sup\{g(v) : v \in R^l\} = g(\bar{v}). \tag{5.3}$$

Assertions of this nature ("Duality Principles") have an important role in the derivation of optimality conditions in convex optimization and can be used as a starting point for developing a subdifferential calculus for convex functions.

Now suppose that F is no longer convex. Then it will be possible to find $\bar{v}$ such that (5.3) is true only under very special circumstances. Surprisingly however we can still salvage some ideas from the above constructions for analyzing nonconvex functions.

The key observation is that Condition (5.3) can be expressed

$$F(\bar{x}, 0) + F^*(0, \bar{v}) = \bar{x} \cdot 0 + 0 \cdot \bar{v}$$

or, alternatively (in terms of the subdifferential of F in the sense of convex analysis),

$$(0, \bar{v}) \in \partial F(\bar{x}, 0). \tag{5.4}$$

The above duality principle can therefore be formulated as: there exists $\bar{v} \in R^l$ such that (5.4) is satisfied. The advantage of writing the condition in this way is that it makes no reference to conjugate functionals and is suitable for generalization to situations where F is no longer convex.

We now prove a theorem, a corollary of which asserts the following: suppose $\bar{x}$ is a minimizer for $x \to F(x, 0)$. Then, under a mild nondegeneracy hypothesis on F, there exists $\bar{v}$ such that (5.4) is true, when ∂F is interpreted as a limiting subdifferential. As we have discussed, this "marginal function" principle has a similar role in the derivation of nonsmooth calculus rules to that of duality principles in convex analysis.

Theorem 5.2.1 (Marginal Function Principle) *Take a lower semicontinuous function* $F : R^n \times R^m \to R \cup \{+\infty\}$ *and a point* $(\bar{x}, \bar{u}) \in \operatorname{dom} F$. *Assume that*

(i) there exists a bounded set $K \subset R^n$ *such that*

$$\operatorname{dom} F(.\,, u) \subset K \quad \text{for all } u \in R^m;$$

(ii) $\bar{x}$ *is the unique minimizer of* $x \to F(x, \bar{u})$.

Define $V : R^m \to R \cup \{+\infty\}$ *to be*

$$V(u) := \min_{x \in R^n} F(x, u).$$

Then V *is a lower semicontinuous function and*

$$
\begin{aligned}
\partial V(\bar{u}) &\subset \{\xi : (0, \xi) \in \partial F(\bar{x}, \bar{u})\}, \\
\partial^\infty V(\bar{u}) &\subset \{\xi : (0, \xi) \in \partial^\infty F(\bar{x}, \bar{u})\}.
\end{aligned}
$$

Proof. We begin by establishing the following properties of F and V.

(a): For every $u \in R^m$, $x \to F(x, u)$ has a minimizer (with infinite cost if $\operatorname{dom} F(.\,, u) = \emptyset$).

(b): V is lower semicontinuous.

(c): Given any sequences $u_i \overset{V}{\to} \bar{u}$ and $\{x_i\}$ such that $V(u_i) = F(x_i, u_i)$ for each i, then $(x_i, u_i) \overset{F}{\to} (\bar{x}, \bar{u})$.

Take any $u \in R^m$. If $\operatorname{dom} F(.\,, u) = \emptyset$, then any x is a minimizer. If $\operatorname{dom} F(.\,, u) \neq \emptyset$, then the search for the minimizer of $F(.\,, u)$ must be carried out over the nonempty set $\operatorname{dom} F(.\,, u)$. But this set is compact, since F is lower semicontinuous and $\operatorname{dom} F(.\,, u)$ is assumed bounded. A minimizer exists then in this case too. We have shown (a).

Take any sequence $u_i \to u$. We wish to show $\liminf_i V(u_i) \geq V(u)$. If all but a finite number of the $V(u_i)$s are infinite there is nothing to prove. Otherwise we can replace the sequence by a subsequence along which $V(u_i)$ is finite. For each i choose a minimizer x_i of the function $x \to F(x, u_i)$. By (i), we can arrange by extracting a further subsequence that $x_i \to x$ for some $x \in R^m$. We know then that $(x_i, u_i) \to (x, u)$. It follows now from the lower semicontinuity of F that

$$\liminf_i V(u_i) = \liminf_i F(x_i, u_i) \geq F(x, u) \geq V(u).$$

We have confirmed property (b).

Take any sequence $u_i \xrightarrow{V} \bar{u}$ and $\{x_i\}$ such that $V(u_i) = F(x_i, u_i)$ for each i. Then $V(u_i) < +\infty$ (for i sufficiently large), and the x_is are confined to a compact set. Along a subsequence then, $x_i \to x$ for some $x \in R^n$. By lower semicontinuity of F,

$$V(\bar{u}) = \lim_i V(u_i) = \lim F(x_i, u_i) \geq F(x, \bar{u}) \geq V(\bar{u}).$$

It follows that

$$F(x_i, u_i) \to F(x, \bar{u}).$$

We see also that $x = \bar{x}$ since $F(., \bar{u})$ has a unique minimizer. From the fact that the limits are independent of the subsequence extracted, we conclude that, for the original sequence,

$$(x_i, u_i) \xrightarrow{F} (\bar{x}, \bar{u}).$$

Property (c) is confirmed.

We are now ready to estimate subgradients. Let F, V and $(\bar{x}, \bar{u})$ be as in the theorem. Take any $\xi \in \partial V(\bar{u})$. According to Theorem 4.6.2 there exist sequences $u_i \xrightarrow{V} \bar{u}$ and $\xi_i \to \xi$ such that $\xi_i \in \partial^P V(u_i)$ for all i. For each i, there exist $\epsilon_i > 0$ and $M_i > 0$ such that

$$\xi_i \cdot (u - u_i) \leq V(u) - V(u_i) + M_i |u - u_i|^2 \quad \text{for all } u \in u_i + \epsilon_i B. \quad (5.5)$$

Let x_i be a minimizer for $x \to F(x, u_i)$. Then, since $V(u) \leq F(x, u)$ for all x and u, we deduce from (5.5) that

$$(0, \xi_i) \cdot (x - x_i, u - u_i) \leq F(x, u) - F(x_i, u_i) + M_i |(x, u) - (x_i, u_i)|^2$$

for all $(x, u) \in (x_i, u_i) + \epsilon_i B$. It follows that

$$(0, \xi_i) \in \partial^P F(x_i, u_i).$$

In view of Property (c) above, $(x_i, u_i) \xrightarrow{F} (\bar{x}, \bar{u})$. By Theorem 4.6.2 then

$$(0, \xi) \in \partial F(\bar{x}, \bar{u}).$$

Next take $\xi \in \partial^\infty V(\bar{u})$. We know that there exist sequences $u_i \xrightarrow{V} \bar{u}$, $\xi_i \to \xi$, and $t_i \downarrow 0$ such that $t_i^{-1} \xi_i \in \partial^P V(u_i)$ for all i. The preceding arguments yield

$$t_i^{-1}(0, \xi_i) \in \partial^P F(x_i, u_i).$$

Here x_i is a minimizer for $x \to F(x, u_i)$. Recalling that $(x_i, u_i) \xrightarrow{F} (\bar{x}, \bar{u})$, we conclude that $(0, \xi) \in \partial^\infty F(\bar{x}, \bar{u})$. $\square$

Frequently the theorem is used in the form of the following corollary:

Corollary 5.2.2 *Take a lower semicontinuous function* $F : R^n \times R^m \to R \cup \{+\infty\}$ *and a point* $(\bar{x}, \bar{u}) \in \operatorname{dom} F$. *Assume that* $\bar{x}$ *minimizes* $x \to F(x, \bar{u})$ *over some neighborhood of* $\bar{x}$. *Suppose that*

$$\{\eta : (0, \eta) \in \partial^\infty F(\bar{x}, \bar{u})\} = \{0\}.$$

Then there exists a point $\xi \in R^m$ *such that*

$$(0, \xi) \in \partial F(\bar{x}, \bar{u}).$$

Proof. Since $\bar{x}$ is a local minimizer of $F(., \bar{u})$, there exists $\epsilon > 0$ such that $F(\bar{x}, \bar{u}) = \min\{F(x, \bar{u}) : x \in \bar{x} + \epsilon B\}$. Set $C := \bar{x} + \epsilon B$. Choose any $r > 0$. Define $\tilde{F} : R^n \times R^m \to R \cup \{+\infty\}$:

$$\tilde{F}(x, u) := F(x, u) + r|x - \bar{x}|^2 + \Psi_C(x).$$

Define also

$$V(u) := \inf_x \tilde{F}(x, u).$$

The function $\tilde{F}$ satisfies the hypotheses of Theorem 5.2.1. Notice in particular that the "penalty term" $r|x - \bar{x}|^2$ ensures that $\tilde{F}(., \bar{u})$ has a unique minimizer. Bearing in mind that the limiting subgradients and asymptotic limiting subgradients of F and $\tilde{F}$ at $(\bar{x}, \bar{u})$ coincide (Lemma 5.1.2), we deduce that

$$\partial V(\bar{u}) \subset \{\xi : (0, \xi) \in \partial F(\bar{x}, \bar{u})\} \quad \text{and} \quad \partial^\infty V(\bar{u}) \subset \{\xi : (0, \xi) \in \partial^\infty F(\bar{x}, \bar{u})\}.$$

But under the nondegeneracy hypothesis, $\partial^\infty V(\bar{u}) = \{0\}$. We know from Corollary 4.9.2 and Proposition 4.7.1 that $\partial V(\bar{u})$ is nonempty. It follows that there exists some $\xi \in R^m$ such that $(0, \xi) \in \partial F(\bar{x}, \bar{u})$. $\square$

5.3 Partial Limiting Subgradients

The first calculus rule relates the partial limiting subdifferential of a function of two variables and the projection of the "total" limiting subdifferential onto the relevant coordinate.

Theorem 5.3.1 (Partial Limiting Subgradients) *Take a lower semicontinuous function* $f : R^n \times R^m \to R \cup \{+\infty\}$ *and a point* $(\bar{x}, \bar{y}) \in \operatorname{dom} f$. *Assume that*

$$(0, \eta) \in \partial^\infty f(\bar{x}, \bar{y}) \text{ implies } \eta = 0.$$

Then:

$$\partial_x f(\bar{x}, \bar{y}) \subset \{\xi : \text{ there exists } \eta \text{ such that } (\xi, \eta) \in \partial f(\bar{x}, \bar{y})\}$$
$$\partial_x^\infty f(\bar{x}, \bar{y}) \subset \{\xi : \text{ there exists } \eta \text{ such that } (\xi, \eta) \in \partial^\infty f(\bar{x}, \bar{y})\}.$$

Proof. Take any $\xi \in \partial_x f(\bar{x}, \bar{y})$. We know then that there exist $\xi_i \to \xi$ and $x_i \stackrel{f(.,\bar{y})}{\to} \bar{x}$ such that $\xi_i \in \hat{\partial}_x f(x_i, \bar{y})$ for all i. For each i then we may choose $\epsilon_i > 0$ and $M_i > 0$ such that x_i is the unique minimizer of $F_i(., \bar{y})$, where

$$F_i(x, y) := f(x, y) - \xi_i \cdot (x - x_i) + M_i |x - x_i|^2 + \Psi_{x_i + \epsilon_i B}(x).$$

Corollary 5.2.2 and Lemma 5.1.2 applied to F_i yield: either there exists $\eta' \in R^m$ such that $(\xi_i, \eta') \in \partial f(x_i, \bar{y})$ or there exists a nonzero $\eta'' \in R^m$ such that $(0, \eta'') \in \partial^\infty f(x_i, \bar{y})$. There are two possible situations:

(a): for an infinite number of is there exists $\eta_i \in R^m$ such that

$$(\xi_i, \eta_i) \in \partial f(x_i, \bar{y});$$

(b): for an infinite number of is there exists $\eta_i \in R^m$ such that $|\eta_i| = 1$ and

$$(0, \eta_i) \in \partial^\infty f(x_i, \bar{y})$$

(we can arrange by scaling that $|\eta_i| = 1$ since $\partial^\infty f$ is a cone).

Case (b) however is never encountered. This is because, in Case (b), the η_is that satisfy $(0, \eta_i) \in \partial^\infty f(x_i, \bar{y})$ have an accumulation point η satisfying $(0, \eta) \in \partial^\infty f(\bar{x}, \bar{y})$ and $|\eta| = 1$, in view of the fact that $(x_i, \bar{y}) \stackrel{f}{\to} (\bar{x}, \bar{y})$. This is in violation of the nondegeneracy hypothesis.

It remains then to attend to (a). We restrict attention to a subsequence, thereby ensuring that η_i, with the stated properties, exists for each i.

It may be assumed that $\{\eta_i\}$ is a bounded sequence, for otherwise we are in a situation where, following a further subsequence extraction, $|\eta_i| \to \infty$ and $|\eta_i|^{-1}\eta_i \to v$ for some v with $|v| = 1$. But then, for $t_i := |\eta_i|^{-1}$, we have

$$t_i \downarrow 0 \quad \text{and} \quad t_i(\xi_i, \eta_i) \to (0, v).$$

Since $(x_i, \bar{y}) \stackrel{f}{\to} (\bar{x}, \bar{y})$, we conclude

$$(0, v) \in \partial^\infty f(\bar{x}, \bar{y}).$$

This is ruled out by the nondegeneracy hypothesis. We have confirmed that $\{\eta_i\}$ is a bounded sequence.

By extracting a subsequence we can arrange that $\{\eta_i\} \to \eta$ for some $\eta \in R^m$. Along the subsequence $(\xi_i, \eta_i) \in \partial f(x_i, \bar{y})$ and $(x_i, \bar{y}) \stackrel{f}{\to} (\bar{x}, \bar{y})$. It follows that $(\xi, \eta) \in \partial f(\bar{x}, \bar{y})$. We have shown that

$$\partial_x f(\bar{x}, \bar{y}) \subset \{\xi : \text{there exists } \eta \text{ such that } (\xi, \eta) \in \partial f(\bar{x}, \bar{y})\}.$$

Verification of the estimate governing the asymptotic partial limiting subdifferential

$$\partial_x^\infty f(\bar{x}, \bar{y}) \subset \{\xi : \text{there exists } \eta \text{ such that } (\xi, \eta) \in \partial^\infty f(\bar{x}, \bar{y})\}$$

is along similar lines. A sketch of the arguments involved is as follows. Take any $\xi \in \partial_x^\infty f(\bar{x}, \bar{y})$. We deduce from Theorem 4.6.2 that there exist $\xi_i \to \xi$, $x_i \overset{f(\cdot, \bar{y})}{\to} x$, and $t_i \downarrow 0$ such that

$$\xi_i \in t_i \partial_x^P f(x_i, \bar{y}).$$

Once again there are two possibilities to be considered:

(a)′: for an infinite number of is there exists $\eta_i \in R^m$ such that $(\xi_i, \eta_i) \in t_i \partial f(x_i, \bar{y})$,

(b)′: for an infinite number of is there exists $\eta_i \in R^m$ such that $|\eta_i| = 1$ and

$$(0, \eta_i) \in t_i \partial^\infty f(x_i, \bar{y}).$$

(b)′ is analogous to (b), a possibility which, as we have noted, cannot occur. So we may assume (a)′. We deduce from the nondegeneracy hypothesis that the η_is are bounded and so have an accumulation point η. We deduce (as before) the required relationship:

$$(\xi, \eta) \in \partial^\infty f(\bar{x}, \bar{y}).$$

□

5.4 A Sum Rule

A general Sum Rule is another consequence of the marginal function principle.

Theorem 5.4.1 (Sum Rule) *Take lower semicontinuous functions* f_i : $R^n \to R \cup \{+\infty\}$, $i = 1, \ldots, m$, *and a point* $\bar{x} \in \cap_i \mathrm{dom}\, f_i$. *Define* $f = f_1 + \ldots + f_m$. *Assume that*

$$v_i \in \partial^\infty f_i(\bar{x}) \ \text{for} \ i = 1, \ldots, m \ \text{and} \ \sum_i v_i = 0 \ \text{imply} \ v_i = 0 \ \text{for all} \ i.$$

Then

$$\partial f(\bar{x}) \subset \partial f_1(\bar{x}) + \ldots + \partial f_m(\bar{x})$$

and

$$\partial^\infty f(\bar{x}) \subset \partial^\infty f_1(\bar{x}) + \ldots + \partial^\infty f_m(\bar{x}).$$

Proof. Consider the function $F : R^n \times (R^n)^m \to R \cup \{+\infty\}$ defined by

$$F(x, y) := \sum_{i=1}^m f_i(x + y_i) \ \text{for} \ y = (y_1, \ldots, y_m).$$

Evidently $\partial f(\bar{x}) = \partial_x F(\bar{x}, 0)$. This formula relating limiting subgradients of f and partial limiting subgradients of F provides the link with the marginal function principle.

Take $(\xi, \eta) \in \partial^P F(x, y)$ at some point $(x, y) \in \text{dom } F$. Then there exists $M > 0$ such that

$$F(x', y') - F(x, y) \geq (\xi, \eta) \cdot ((x', y') - (x, y)) - M|(x', y') - (x, y)|^2$$

for all (x', y') in some neighborhood of (x, y). Expressed in terms of the f_is this inequality informs us that

$$\sum_i [f_i(x' + y_i') - f_i(x + y_i)] \geq \xi \cdot (x' - x) + \sum_i \eta_i \cdot (y_i' - y_i) - M|(x', y') - (x, y)|^2.$$

Fix an index value $j \in \{1, \ldots, m\}$. Set $x' = x$ and $y_i' = y_i$ for $i \neq j$ in the above inequality. There results

$$f_j(x + y_j') - f_j(x + y_j) \geq \eta_j \cdot (y_j' - y_j) - M|y_j' - y_j|^2$$

for all y_j's close to y_j. It follows that $\eta_j \in \partial^P f_j(x + y_j)$.

Next, for any x' close to x, set $y_i' = x + y_i - x'$ for $i = 1, \ldots, m$. This gives

$$0 \geq \xi \cdot (x' - x) - \left(\sum_i \eta_i\right) \cdot (x' - x) - (m + 1)M|x' - x|^2.$$

We conclude that $\xi = \sum_i \eta_i$. To summarize:

$$\partial^P F(x, y) \subset \{(\eta_1 + \ldots + \eta_m, \eta) : \eta = (\eta_1, \ldots, \eta_m) \text{ and}$$
$$\eta_i \in \partial^P f_i(x + y_i) \text{ for } i \in \{1, \ldots, m\}\}.$$

Now take any $(\xi, \eta) \in \partial F(\bar{x}, 0)$. Then $(\xi, \eta) = \lim_k (\xi_k, \eta^k)$ for some sequences $\{(\xi_k, \eta^k)\}$ and $\{(x_k, y^k)\}$ such that $(\xi_k, \eta^k) \in \partial^P F(x_k, y^k)$ for each k, and $(x_k, y^k) \xrightarrow{F} (\bar{x}, 0)$. As we have shown

$$\eta_i^k \in \partial^P f_i(x_k + y_i^k) \quad \text{for } i = 1, \ldots, m$$

and $\xi_k = \sum_{i=1}^m \eta_i^k$. We claim also that

$$f_i(x_k + y_i^k) \to f_i(\bar{x}) \quad \text{for each } i.$$

To confirm this relationship, consider

$$\gamma_i := \limsup_{k \to \infty} (f_i(x_k + y_i^k) - f_i(\bar{x})) \quad \text{for } i = 1, \ldots, m.$$

Since the f_is are lower semicontinuous, we must show that $\gamma_i = 0$ for each i. By extracting subsequences if necessary, we can arrange that

$$\gamma_i = \lim_{k \to \infty} (f_i(x_k + y_i^k) - f_i(\bar{x})) \quad \text{for } i = 1, \ldots, m.$$

("lim" has replaced "lim sup" here.) But since $(x_k, y^k) \xrightarrow{F} (\bar{x}, 0)$,

$$\lim_k \sum_{i=1}^{m} f_i(x_k + y_i^k) = \sum_{i=1}^{m} f_i(\bar{x}),$$

whence

$$\sum_i \liminf_{k \to \infty} (f_i(x_k + y_i^k) - f_i(\bar{x})) \leq 0.$$

By lower semicontinuity, each term in the summation is nonnegative. It follows that, for each i,

$$0 = \liminf_{k \to \infty} (f_i(x_k + y_i^k) - f_i(\bar{x})) = \lim_{k \to \infty} (f_i(x_k + y_i^k) - f_i(\bar{x})) = \gamma_i.$$

Our claim then is justified.

Since

$$x_k + y_i^k \xrightarrow{f} \bar{x}, \; \eta_i^k \to \eta_i \quad \text{as } k \to \infty,$$
$$\eta_i^k \in \partial^P f_i(x_k + y_i^k) \quad \text{for all } k,$$

we have

$$\eta_i \in \partial f_i(\bar{x}) \quad \text{for each } i.$$

We have shown that

$$\partial F(\bar{x}, 0) \subset \{(\eta_1 + \ldots + \eta_m, \eta) :$$
$$\eta = (\eta_i, \ldots, \eta_m) \text{ and } \eta_i \in \partial f_i(\bar{x}) \text{ for } i \in \{1, \ldots, m\}\}.$$

Similar arguments based on the representation of $\partial^\infty F$ in terms of scaled proximal subgradients at neighboring points yield

$$\partial^\infty F(\bar{x}, 0) \subset \{(\eta_1 + \ldots + \eta_m, \eta) :$$
$$\eta = (\eta_i, \ldots, \eta_m) \text{ and } \eta_i \in \partial^\infty f_i(\bar{x}) \text{ for each } i\}.$$

Now apply Theorem 5.3.1 to F. Note that the nondegeneracy hypothesis of Theorem 5.3.1 is satisfied since, if $(0, \eta) \in \partial^\infty F(\bar{x}, 0)$ and $\eta \neq 0$, then $\sum_i \eta_i = 0$ and $\eta_i \in \partial^\infty f_i(\bar{x})$ for each i, a possibility excluded by our hypotheses. Recalling that $f(x) = F(x, 0)$ for all $x \in R^n$, we deduce that

$$\partial f(\bar{x}) = \partial_x F(\bar{x}, 0) \subset \partial f_1(\bar{x}) + \ldots + \partial f_m(\bar{x})$$

and

$$\partial^\infty f(\bar{x}) = \partial_x^\infty F(\bar{x}, 0) \subset \partial^\infty f_1(\bar{x}) + \ldots + \partial^\infty f_m(\bar{x}).$$

$\square$

As a first application of the Sum Rule, we estimate the normal cone to the graph of a Lipschitz continuous map.

Proposition 5.4.2 *Take a lower semicontinuous map $G : R^n \to R^m$ and a point $u \in R^n$. Assume that G is Lipschitz continuous on a neighborhood of u. Then*

$$N_{\mathrm{Gr}\,G}(u, G(u)) \subset \{(\xi, -\eta) : \xi \in \partial(\eta \cdot G)(u), \eta \in R^m\}.$$

Proof. Take any $(\xi, -\eta) \in N_{\mathrm{Gr}\,G}(u, G(u))$. Then $(\xi, -\eta) = \lim_i (\xi_i, -\eta_i)$ for sequences $\{(\xi_i, -\eta_i)\}$ and $u_i \overset{G}{\to} u$ such that

$$(\xi_i, -\eta_i) \in N^P_{\mathrm{Gr}\,G}(u_i, G(u_i)) \quad \text{for all } i.$$

For each i, there exists $M_i > 0$ such that

$$(\xi_i, -\eta_i) \cdot (u' - u_i, G(u') - G(u_i)) \leq M_i |u' - u_i|^2$$

for all u' close to u_i. This inequality can be rearranged to give

$$\eta_i \cdot G(u') \geq \eta_i \cdot G(u_i) + \xi_i \cdot (u' - u_i) - M_i |u' - u_i|^2,$$

from which we deduce that

$$\xi_i \in \partial^P(\eta_i \cdot G)(u_i) .$$

If follows from Proposition 4.7.1 and the Sum Rule (Theorem 5.4.1) that

$$\xi_i \in \partial(\eta \cdot G)(u_i) + \partial((\eta_i - \eta) \cdot G)(u_i)$$

for each i. Also by Proposition 4.7.1., then, for sufficient large i,

$$\xi_i \in \partial(\eta \cdot G)(u_i) + K|\eta_i - \eta|B.$$

Here K is a Lipschitz constant for G on a neighborhood of u. Since the limiting normal cone is "robust" under limit-taking, we obtain $\xi \in \partial(\eta \cdot G)(u)$ in the limit as $i \to \infty$. $\square$

5.5   A Nonsmooth Chain Rule

We now derive a far-reaching Chain Rule.

Theorem 5.5.1 (A Chain Rule) *Take a locally Lipschitz continuous function $G : R^n \to R^m$, a lower semicontinuous function $g : R^m \to R \cup \{+\infty\}$, and a point $\bar{u} \in R^n$ such that $G(\bar{u}) \in \mathrm{dom}\, g$. Define the lower semicontinuous function $f(u) := g \circ G(u)$. Assume that:*

The only vector $\eta \in \partial^\infty g(G(\bar{u}))$ such that $0 \in \partial(\eta \cdot G)(\bar{u})$ is $\eta = 0$.

Then

$$\partial f(\bar{u}) \subset \{\xi : \text{ there exists } \eta \in \partial g(G(\bar{u})) \text{ such that } \xi \in \partial(\eta \cdot G)(\bar{u})\} \quad (5.6)$$

and

$$\partial^\infty f(\bar{u}) \subset \{\xi : \text{ there exists } \eta \in \partial^\infty g(G(\bar{u})) \text{ such that } \xi \in \partial(\eta \cdot G)(\bar{u})\}. \quad (5.7)$$

Otherwise expressed, inclusion (5.6) tells us that, given some $\xi \in \partial f(\bar{u})$, we may find some $\eta \in \partial g$, evaluated at $G(\bar{u})$, such that $\xi \in \partial d(\bar{u})$, where d is the scalar-valued function $d(u) := (\eta \cdot G)(u)$. ((5.7) can be likewise interpreted.)

Proof. Define the lower semicontinuous function

$$F(x, u) := g(x) + \Psi_{\mathrm{Gr}G}(u, x) + \Psi_{\{\bar{u}, G(\bar{u})\} + B}(u, x). \quad (5.8)$$

We see that $f(u) = \min_x F(x, u)$. Notice that $(G(\bar{u}), \bar{u}) \in \operatorname{dom} F$, $\operatorname{dom} F$ is bounded, and $G(\bar{u})$ is the unique minimizer of $F(., \bar{u})$. The scene is set then for applying Theorem 5.2.1. This tells us that

$$\partial f(\bar{u}) \subset \{\xi : (0, \xi) \in \partial F(G(\bar{u}), \bar{u})\}$$

and

$$\partial^\infty f(\bar{u}) \subset \{\xi : (0, \xi) \in \partial^\infty F(G(\bar{u}), \bar{u})\}.$$

Now the Sum Rule (Theorem 5.4.1) is applied to estimate the limiting subdifferential and asymptotic limiting subdifferential of F given by (5.8). The last term in (5.8) makes no contribution to the subdifferentials and can be ignored.

It is necessary to check the Sum Rule nondegeneracy condition. Now, as is easily shown, a general element v_1 in the asymptotic limiting subdifferential of $(x, u) \to g(x)$ at $(G(\bar{u}), \bar{u})$ is expressible as $v_1 = (\gamma, 0)$ where $\gamma \in \partial^\infty g(G(\bar{u}))$. On the other hand, by Proposition 5.4.2,

$$\begin{aligned}
\partial\{(x', u') \to \Psi_{\mathrm{Gr}\,G}(u', x')\}(x, u) \\
= \ \{(-\eta, \xi) : (\xi, -\eta) \in \partial\Psi_{\mathrm{Gr}G}(u, x)\} \\
\subset \ \{(-\eta, \xi) : \xi \in \partial(\eta \cdot G)(u)\}.
\end{aligned}$$

So an element in $\partial\{(x, u) \to \Psi_{\mathrm{Gr}G}(u, x)\}(G(\bar{u}), \bar{u})$ is expressible as $v_2 = (-\eta, \xi)$ for some ξ and η that satisfy $\xi \in \partial(\eta \cdot G)(\bar{u})$. We must examine the consequences of $v_1 + v_2 = 0$. They are that $\xi = 0$ and $0 \in \partial(\eta \cdot G)(\bar{u})$ for some $\eta \in \partial^\infty g(G(\bar{u}))$. But then $\eta = 0$ under the present hypotheses. So $v_1 + v_2 = 0$ implies $v_1 = v_2 = 0$. The nondegeneracy hypothesis is therefore satisfied. The Sum Rule (Theorem 5.4.1) and previous considerations now give

$$
\begin{aligned}
\partial f(\bar{u}) &\subset \{\xi : (0,\xi) = (\nu,0) + (-\eta,\xi), \nu \in \partial g(G(\bar{u})), \xi \in \partial(\eta \cdot G)(\bar{u})\} \\
&= \{\xi : \text{there exists } \eta \in \partial g(G(\bar{u})) \text{ such that } \xi \in \partial(\eta \cdot G)(\bar{u})\}.
\end{aligned}
$$

Similarly

$$\partial^\infty f(\bar{u}) \subset \{\xi : \text{there exists } \eta \in \partial^\infty g(G(\bar{u})) \text{ such that } \xi \in \partial(\eta \cdot G)(\bar{u})\}.$$

These are the relationships we set out to prove. $\square$

A number of important calculus rules are now obtainable as corollaries of the Nonsmooth Chain Rule.

Theorem 5.5.2 (Max Rule) *Take locally Lipschitz continuous functions* $f_i : R^n \to R, i = 1,\ldots,m$, *and a point* $\bar{x} \in R^n$. *Define* $f(x) = \max_i f_i(x)$ *and* $\Lambda := \{\lambda = (\lambda_1,\ldots,\lambda_m) \in R^m : \lambda_i \geq 0, \sum_i \lambda_i = 1\}$. *Then*

$$\partial f(\bar{x}) \subset \{\partial(\sum_{i=1}^m \lambda_i f_i)(\bar{x}) : \lambda \in \Lambda, \text{ and } \lambda_i = 0 \text{ if } f_i(\bar{x}) < f(\bar{x})\}.$$

Let us be clear about what this theorem tells us: given $\xi \in \partial f(\bar{x})$, there exists a "convex combination" $\{\lambda_i\}$ of "active" f_is such that

$$\xi \in \partial(\sum_i \lambda_i f_i)(\bar{x}).$$

This implies (via the Sum Rule and in view of the positive homogeneity properties of the subdifferential)

$$\xi \in \sum_i \lambda_i \partial f_i(\bar{x}),$$

another, slighly weaker, version of the Max Rule.

Proof. We readily calculate the limiting subgradient of

$$g(y) := \max\{y_1,\ldots,y_m\}$$

for $y = (y_1,\ldots,y_m)$. It is

$$\partial g(y) = \{\lambda = (\lambda_1,\ldots,\lambda_m) \in \Lambda : \lambda_i = 0 \text{ if } y_i < \max_j y_j\}.$$

Now we apply the Chain Rule (Theorem 5.5.1) with this g and

$$G(x) := (f_1,\ldots,f_m)(x),$$

noting that the nondegeneracy hypothesis is satisfied in view of Proposition 4.7.1. $\square$

Theorem 5.5.3 (Product Rule) *Take locally Lipschitz continuous functions* $f_i : R^n \to R, i = 1, \ldots, m$, *and a point* $\bar{x} \in R^n$. *Define* $f(x) = f_1(x)f_2(x)\ldots f_m(x)$. *Then*

$$\partial f(\bar{x}) \subset \partial(\sum_i^m \Pi_{j \neq i} f_j(\bar{x}) f_i)(\bar{x}).$$

This theorem tells us for example that, in the case $m = 2$,

$$\partial f(\bar{x}) \subset \partial(f_1(\bar{x})f_2 + f_2(\bar{x})f_1)(\bar{x}).$$

In the event that $f_1(\bar{x}) \geq 0$, $f_2(\bar{x}) \geq 0$, this condition implies

$$\partial f(\bar{x}) \subset f_1(\bar{x})\partial f_2(\bar{x}) + f_2(\bar{x})\partial f_1(\bar{x}).$$

Proof. Apply the Chain Rule with $g(y) := \Pi_i y_i$ for $y = (y_1, \ldots, y_m)$ and $G(x) := (f_1, \ldots, f_m)(x)$ for $x \in R^n$. $\square$

5.6 Lagrange Multiplier Rules

Finally we make contact with optimization. A general theorem is proved, providing necessary conditions for a point to be a minimizer of a composite function. Choosing different ingredients for this function supplies a variety of Lagrange multiplier rules, one example of which we investigate in detail.

Theorem 5.6.1 (Generalized Multiplier Rule) *Take Lipschitz continuous functions* $f : R^n \to R$ *and* $F : R^n \to R^m$, *a lower semicontinuous function* $h : R^m \to R \cup \{+\infty\}$, *a closed set* $C \subset R^n$, *and a point* $\bar{x} \in C$ *such that* $h(F(\bar{x})) < \infty$. *Define the lower semicontinuous function* $l : R^n \to R \cup \{+\infty\}$ *to be*

$$l(x) := f(x) + h(F(x)) + \Psi_C(x).$$

Let $\bar{x}$ *attain the minimum of* $l(.)$. *Assume that*

$$0 \in \partial(\eta \cdot F)(\bar{x}) + N_C(\bar{x}) \text{ for some } \eta \in \partial^\infty h(F(\bar{x})) \text{ implies } \eta = 0.$$

Then

$$0 \in \partial f(\bar{x}) + \partial(\eta \cdot F)(\bar{x}) + N_C(\bar{x}) \text{ for some } \eta \in \partial h(F(\bar{x})).$$

Proof. Choose $g(x, y) := f(x) + h(y) + \Psi_C(x)$ and $G(x) = (x, F(x))$. Then $\bar{x}$ minimizes $l(x) := g(G(x))$, and so

$$0 \in \partial^P l(\bar{x}) \subset \partial(g \circ G)(\bar{x}).$$

Set $\bar{y} = F(\bar{x})$. Now apply the Chain Rule (Theorem 5.5.1). To begin with, we need to check the nondegeneracy hypothesis. Take any

$$\eta = (\eta_1, \eta_2) \in \partial^\infty g(\bar{x}, \bar{y}) = \partial^\infty (f(x) + h(y) + \Psi_C(x))|_{(x,y)=(\bar{x},\bar{y})}.$$

It is a straightforward exercise to check that the conditions under which the Sum Rule (Theorem 5.4.1) supplies an estimate for (η_1, η_2) are satisfied. Consequently there exist $\eta_1 \in N_C(\bar{x})$ and $\eta_2 \in \partial^\infty h(F(\bar{x}))$. We see that if $0 \in \partial(\eta \cdot G)(\bar{x})$ then $0 \in \eta_1 + \partial(\eta_2 \cdot F)(\bar{x})$. Under current hypotheses then η_2 (and therefore also η_1) is zero, so the hypotheses under which we may apply the chain rule are satisfied.

A further application of the Sum Rule tells us that if $(\xi, \eta) \in \partial g(G(\bar{x}))$ then $\xi \in \partial f(\bar{x}) + N_C(\bar{x})$ and $\eta \in \partial h(\bar{y})$. From the Chain Rule then

$$0 \in \partial(g \circ G)(\bar{x}) \subset \partial f(\bar{x}) + \partial(\eta \cdot F)(\bar{x}) + N_C(\bar{x})$$

for some $\eta \in \partial h(F(\bar{x}))$. This is what we set out to prove. □

In applications the function h in the Generalized Multiplier Theorem is usually taken to be the indicator function of some closed set E (comprising allowable values for $F(x)$). In this case a minimizer $\bar{x}$ for l is a minimizer for the constrained optimization problem

$$\text{minimize } \{f(x) : F(x) \in E \text{ and } x \in C\}.$$

Now

$$\partial \psi_E(F(\bar{x})) = \partial^\infty \psi_E(F(\bar{x})) = N_E(F(\bar{x})).$$

It follows from Theorem 5.6.1 that, if

$$0 \in \partial(\eta \cdot F)(\bar{x}) + N_C(\bar{x}) \text{ and } \eta \in N_E(F(\bar{x})) \text{ implies } \eta = 0, \qquad (5.9)$$

then there exists

$$\eta \in N_E(F(\bar{x}))$$

such that

$$0 \in \partial f(\bar{x}) + \partial(\eta \cdot F)(\bar{x}) + N_C(\bar{x}).$$

The vector η is a Lagrange multiplier, which must be directed into the normal cone $N_E(F(\bar{x}))$ of the constraint set E at $F(\bar{x})$. The nondegeneracy condition, Condition (5.9), under which this Lagrange Multiplier Rule is valid requires, when F is a smooth function, that there does not exist a nonzero vector

$$\eta = \{\eta_1, \ldots, \eta_n\}$$

such that $-\eta$ is a limiting normal to E at $F(\bar{x})$ and the linear combination $\sum_i \eta_i \nabla F_i(\bar{x})$ of gradients of the components of F at $\bar{x}$ is a limiting normal to C at $\bar{x}$. This condition is a generalization of the Mangasarian–Fromovitz type constraint qualifications invoked in the Mathematical Programming literature.

A widely studied optimization problem to which the above theorem is applicable is

$$\text{(NLP)} \quad \text{minimize } f_0(x) \text{ over } x \in X \cap C,$$

where

$$X := \{x \in R^n : f_1(x) \le 0, \ldots, f_p(x) \le 0, g_1(x) = 0, \ldots, g_q(x) = 0\}.$$

Here $f_0, f_1, \ldots, f_p$, and $g_1, \ldots, g_q$ are all R-valued functions with domain R^n, and C is a closed subset of R^n.

Theorem 5.6.2 (Lagrange Multiplier Rule) *Let $\bar{x}$ be a local minimizer for (NLP). Assume that $f_0, \ldots, f_p$ and $g_1, \ldots, g_q$ are locally Lipschitz continuous functions. Then there exist $\lambda_0 \ge 0, \lambda_1 \ge 0, \ldots, \lambda_p \ge 0$ (with $\lambda_0 = 0$ or 1), and real numbers $\gamma_1, \ldots, \gamma_q$ such that*

$$\lambda_i = 0 \quad \text{if} \quad f_i(\bar{x}) < 0 \quad \text{for} \quad i = 1, \ldots, p, \tag{5.10}$$

$$\sum_{i=0}^{p} \lambda_i + \sum_{i=1}^{q} |\gamma_i| \ne 0$$

and

$$0 \in \partial[\sum_{i=0}^{p} \lambda_i f_i + \sum_{i=1}^{q} \gamma_i g_i](\bar{x}) + N_C(\bar{x}).$$

Proof. By replacing C by $C \cap (\bar{x} + \epsilon B)$, for $\epsilon > 0$ sufficiently small, we can arrange that $\bar{x}$ is a minimizer of f_0 over $X \cap C$ ($N_C(\bar{x})$ is unaffected). We can arrange also that, if $f_i(\bar{x}) < 0$, for some $i \in \{1, \ldots, p\}$ then $f_i(x) < 0$ for all $x \in \bar{x} + \epsilon B$. The constraint $f_i(x) < 0$ is thereby rendered irrelevant and we can ignore it. Let us assume that index values corresponding to inequality constraints inactive at $\bar{x}$ have been removed and we have labeled the remaining inequality constraint functions by indices in some new (possibly reduced) index set. The assertions of the theorem for the new index set (excluding the complementary slackness Condition (5.10)) imply those for the original index set (including this last condition). Complementary slackness therefore takes care of itself, if we attend to the other assertions of the theorem.

It is clear that $(\bar{\alpha}, \bar{x})$ (where $\bar{\alpha} = f_0(\bar{x})$) is a minimizer for the optimization problem:

$$\text{minimize } \{\alpha : f_0(x) - \alpha \le 0 \text{ and } x \in X \cap C\}.$$

Set

$$f(\alpha, x) = \alpha, \ F(\alpha, x) = (f_0(x) - \alpha, f_1(x), \ldots, f_p(x), g_1(x), \ldots, g_q(x))$$

and

$$E = (-\infty, 0]^{p+1} \times \{0\}^q.$$

We find $N_E(0) = M$ where

$$M = \{(\lambda_0, \lambda_1, \ldots, \lambda_p, \gamma_1, \ldots, \gamma_q) : \lambda_i \geq 0 \text{ for } i \in \{0, \ldots, p\}\}.$$

Let us now examine the nondegeneracy hypothesis of Theorem 5.6.1 for the above identifications of f and F and for $h = \Psi_E$. If $\eta \in \partial^\infty h(0)$, then

$$\eta = (\lambda_0, \lambda_1, \ldots, \lambda_p, \gamma_1, \ldots, \gamma_q) \in M.$$

It follows that the relationship

$$0 \in \partial(\eta \cdot F)(\bar{\alpha}, \bar{x}) + \{0\} \times N_C(\bar{x})$$

can be written

$$0 \in \{-\lambda_0\} \times \partial(\sum_{i=0}^{p} \lambda_i f_i + \sum_{i=1}^{q} \gamma_i g_i)(\bar{x}) + \{0\} \times N_C(\bar{x})$$

which implies

$$0 \in \{0\} \times \partial(\sum_{i=1}^{p} \lambda_i f_i + \sum_{i=1}^{q} \gamma_i g_i)(\bar{x}) + \{0\} \times N_C(\bar{x}).$$

It follows that if the nondegeneracy hypothesis is violated, then

$$0 \in \partial(\sum_{i=1}^{p} \lambda_i f_i + \sum_{i=1}^{q} \gamma_i g_i)(\bar{x}) + N_C(\bar{x})$$

for some nonzero $(\lambda_0, \lambda_1, \ldots, \lambda_p, \gamma_1, \ldots, \gamma_q) \in M$ and $\lambda_0 = 0$. The assertions of the theorem are valid then (albeit only in a degenerate sense).

On the other hand, if the nondegeneracy hypothesis is satisfied then there exists $\eta = (\lambda_0, \ldots, \lambda_p, \gamma_1, \ldots, \gamma_q) \in M$ such that

$$0 \in \partial f(\bar{\alpha}, \bar{x}) + \partial(\eta \cdot F)(\bar{\alpha}, \bar{x}) + \{0\} \times N_C(\bar{x}).$$

Since $\partial f(\bar{\alpha}, \bar{x}) = \{(1, 0, \ldots, 0)\}$, we conclude

$$0 \in \partial(\sum_{i=0}^{p} \lambda_i f_i + \sum_{i=1}^{q} \gamma_i g_i)(\bar{x}) + N_C(\bar{x}),$$

for some $(\lambda_0, \lambda_1, \ldots, \lambda_p, \gamma_1, \ldots, \gamma_q) \in M$ with $\lambda_0 = 1$. $\square$

5.7 Notes for Chapters 4 and 5

Developments in Convex Analysis in the 1960s, centered substantially on Rockafellar's contributions, revealed that, for purposes of characterizing minimizers to convex optimization problems with possibly nondifferentiable data, tangent cones, normal cones, and set-valued subdifferentials can in many ways do the work of tangent spaces, cotangent spaces and derivatives, respectively, in traditional analysis.

A desire to reproduce some of these successes in a nonconvex setting was the impetus behind the field of Nonsmooth Analysis, initiated in the following decade. The key advance was the introduction by Clarke in his 1973 thesis [29] of various "robust" nonsmooth constructs, including the generalized gradient of a lower semicontinuous function and what are now widely referred to as the Clarke normal cone and Clarke tangent cone to a closed set [29],[30], [38]. Numerous earlier definitions of "derivative" of a nondifferentiable function had previously been proposed, but the generalized gradient was the first to stand out for its generality, its extensive calculus, geometric interpretations, and the breadth of its applications, notably in the derivation of necessary conditions of optimality.

Once the idea of local approximation of nonsmooth functions and sets with nonsmooth boundaries by means of cones and set-valued "gradients" was out of the bag, there was an explosion of definitions. The book of Aubin and Frankowska [12], which limits itself largely to a systematic study of the "menagerie" of tangent cones, obtained by attaching different qualifiers to the limit operations in their definitions, and derivatives which they induce by consideration of tangent cones to graphs of functions, is evidence of this.

Chapters 4 and 5 concern those aspects of nonsmooth analysis required to support future chapters on necessary conditions and Dynamic Programming in Optimal Control. Here, fortunately, we need consider only relatively few constructs. Much of the material in these chapters focuses on the limiting normal cone (and the related limiting subdifferential of a lower semicontinuous function, defined via limiting normals to the epigraph set). But these chapters also feature a limited repertoire of tangent cones of relevance to our analysis. All material is more or less standard. Proofs of basic properties of limiting normals and subdifferentials follows, in many respects, those in unpublished notes of Rockafellar [123] and [94], and ideas implicit in [38]. Proofs of the relationships among the various normal and tangent cones considered here in many respects follow those in [12]. Material on generalized gradients in Section 4.7 is taken from [38].

The limiting normal cone was introduced by Mordukhovich in his 1976 paper [104], where it was used to formulate generalized transversality conditions in Optimal Control. The limiting normal cone is also referred to as the approximate normal cone by Ioffe [82] and simply as the normal cone (by Mordukhovich [107] and, recently, by Rockafellar and Wets [125]). The names "coderivative" and "subdifferential" are also used for the limiting subdifferential, in [107] and [125], respectively. These constructs were studied in detail in papers of Mordukhovich, Kruger, and Ioffe, including [106], [91], and [82]. More recent expository treatments are provided in [107], [125], and [94]. Proximal normal vectors, the generators of the limiting normal cone, were used by Clarke to prove properties of (Clarke) normal cones and generalized gradients [30]. But focusing attention on limiting normal cones as interesting objects in their own right was a significant departure.

The proximal normal cone, limiting normal cone, and its convex hull the Clarke normal cone, together with their associated subdifferentials, namely, the proximal subdifferential, limiting subdifferential, and the generalized gradient, respectively, all figure prominently in Nonsmooth Optimal Control. Of these constructs, the generalized gradient is well suited to the formulation of adjoint inclusions for nonsmooth dynamics, by virtue of its convexity properties. On the other hand, the limiting normal cone is a natural choice of normal cone to express transversality conditions for general endpoint constraint sets. The proximal subdifferential is a convenient vehicle for the interpretation of generalized solutions to the Hamilton–Jacobi Equation of Dynamic Programming.

Lebourg [92] proved the first nonsmooth Mean Value Theorem (the Two-Sided Mean Value Theorem for Lipschitz functions). Clarke and Ledyaev's approximate Mean Value Inequality [40], the uniform nature of which was novel even for smooth functions, has found numerous applications – in fixed point theorems, the interpretation of generalized solutions to partial differential inequalities, and other areas. (See also [41] and [54].) Approximate generalizations of the Mean Value Theorem were earlier investigated by Zagrodny [154] and Loewen [95].

We stick exclusively to Nonsmooth Analysis in finite-dimensional spaces. This suffices for topics in Optimal Control covered in this book. However optimal control problems are inherently infinite-dimensional. The investigation of certain issues in Optimal Control, such as sensitivity of the minimum cost to "infinite-dimensional perturbations" of the dynamics [36], as well as some alternative proofs of results in this book, involve the application of Nonsmooth Analysis in infinite-dimensional spaces. Clarke's theory of generalized gradients can be developed, with the accompanying calculus largely intact, in the context of Lipschitz continuous functions on Banach spaces [38]. A number of researchers have been involved in building alternative frameworks for Nonsmooth Analysis in infinite dimensions, including Clarke, Borwein, Ioffe, Loewen, and Mordukhovich. Borwein and Zhu have provided a helpful detailed review [24]. An appealing approach for its simplicity and broad applicability, followed by Clarke and his collaborators [54], centers on proximal normals in Hilbert space. Proximal normal cones and proximal subdifferentials do not have a satisfactory *exact* calculus. They do however admit a rich *fuzzy* calculus. Fuzzy calculus, initiated by Ioffe (see [84]), provides rules, involving ϵ error terms that can be made arbitrarily small, for the estimation of proximal subdifferentials of composite functions. The idea is to retain ϵ terms in the general theory and to attempt to dispose of them only at the applications stage, using special features of the problem at hand.

Chapter 6

The Maximum Principle

The solution of a whole range of technical problems, which are important in contemporary technology, is outside the classical calculus of variations.... The solution presented here is unified in one general mathematical method, which we call the maximum principle.

– Lev S. Pontryagin et al., *The Mathematical Theory of Optimal Processes*

6.1 Introduction

This chapter focuses on a set of optimality conditions known as the Maximum Principle. Many competing sets of optimality conditions are now available, but the Maximum Principle retains a special significance. An early version of the Maximum Principle due to Pontryagin et al. was after all the breakthrough marking the emergence of Optimal Control as a distinct field of research. Also, whatever additional information about minimizers is provided by Dynamic Programming, higher-order conditions, and the analysis of the geometry of state trajectories, first-order necessary conditions akin to the Maximum Principle remain the principal vehicles for the solution of specific optimal control problems (either directly or indirectly via the computational procedures they inspire), or at least for generating "suspects" for their solution.

A number of first-order necessary conditions are derived in this book. We attend to the Maximum Principle at the outset not just because it was the first necessary condition to handle in a satisfactory way constraints typically encountered in dynamic optimization, but because of the key role it performs in the derivation of all the others.

The most general version of the Maximum Principle derived in this chapter, namely, Theorem 6.2.1, applies to optimal control problems with general endpoint constraints and dynamic constraints expressed in terms of a "nonsmooth" differential equation parameterized by a control variable. Generalizations to allow for pathwise state constraints are provided in Chapter 9.

The chain of arguments used to derive the "nonsmooth" Maximum Principle is as follows. The Maximum Principle is first proved for optimal control problems with "smooth" dynamic constraints. This "smooth" Maximum

Principle is used to derive the Extended Euler–Lagrange Condition (a link which is provided in the next chapter). The Extended Euler–Lagrange Condition is then itself invoked in the derivation we give of the "nonsmooth" Maximum Principle.

The most important step taken in this chapter is a self-contained proof of a version of the Maximum Principle applicable to problems with smooth dynamics. This "smooth" Maximum Principle, besides playing a key part in the proof of more general necessary conditions, meets the requirements of many applications of optimal control and is therefore of independent interest.

The optimal control problem studied here is

$$(P) \quad \begin{cases} \text{Minimize } g(x(S), x(T)) \\ \text{over } x \in W^{1,1}([S,T]; R^n) \text{ and measurable} \\ \qquad \text{functions } u : [S,T] \to R^m \text{ satisfying} \\ \dot{x}(t) = f(t, x(t), u(t)) \quad \text{a.e.,} \\ u(t) \in U(t) \quad \text{a.e.,} \\ (x(S), x(T)) \in C, \end{cases}$$

the data for which comprise an interval $[S,T]$, functions $g : R^n \times R^n \to R$ and $f : [S,T] \times R^n \times R^m \to R^n$, a nonempty multifunction $U : [S,T] \rightsquigarrow R^m$, and a closed set $C \subset R^n \times R^n$.

A measurable function $u : [S,T] \to R^m$ that satisfies

$$u(t) \in U(t) \quad \text{a.e.}$$

is called a *control function*. The set of all control functions is written $\mathcal{U}$.

A *process* (x, u) comprises a control function u together with an arc $x \in W^{1,1}([S,T]; R^n)$ which is a solution to the differential equation

$$\dot{x}(t) = f(t, x(t), u(t)) \quad \text{a.e.}$$

A *state trajectory* x is the first component of some process (x, u). A process (x, u) is said to be *feasible* for (P) if the state trajectory x satisfies the endpoint constraint

$$(x(S), x(T)) \in C.$$

The Maximum Principle and related optimality conditions are satisfied by all minimizers. As we show, these conditions are satisfied also by processes that are merely "local" minimizers for (P). Different choices of topology on the set of processes give rise to different notions of local minimizer. The one adopted in this chapter is that of the $W^{1,1}$ local minimizer, although we make reference in the proofs to strong local minimizers.

Definition 6.1.1 *Take a feasible process* $(\bar{x}, \bar{u})$.

$(\bar{x}, \bar{u})$ *is a* $W^{1,1}$ *local minimizer if there exists* $\delta > 0$ *such that*

$$g(x(S), x(T)) \geq g(\bar{x}(S), \bar{x}(T)),$$

for all feasible processes (x, u) *which satisfy*

$$\|x - \bar{x}\|_{W^{1,1}} \leq \delta.$$

$(\bar{x}, \bar{u})$ *is a* strong local minimizer *if there exists* $\delta > 0$ *such that*

$$g(x(S), x(T)) \geq g(\bar{x}(S), \bar{x}(T)),$$

for all feasible processes (x, u) *which satisfy*

$$\|x - \bar{x}\|_{L^\infty} \leq \delta.$$

The point in showing that the optimality conditions are valid for local minimizers is to focus attention on the limitations of these conditions: they may lead us to a local minimizer in place of a global minimizer of primary interest. The $W^{1,1}$ norm is stronger than the L^∞ norm and therefore the class of $W^{1,1}$ local minimizers is larger than the class of L^∞ minimizers. It follows that, by choosing to work with $W^{1,1}$ local minimizers, we are carrying out a sharper analysis of the local nature of the Maximum Principle than would be the case if we chose to derive conditions satisfied by strong local minimizers.

6.2   The Maximum Principle

Denote by $\mathcal{H} : [S, T] \times R^n \times R^n \times R^m \to R$ the Unmaximized Hamiltonian function

$$\mathcal{H}(t, x, p, u) := p \cdot f(t, x, u).$$

Theorem 6.2.1 (The Maximum Principle) *Let* $(\bar{x}, \bar{u})$ *be a* $W^{1,1}$ *local minimizer for* (P). *Assume that, for some* $\delta > 0$, *the following hypotheses are satisfied.*

(H1) *For fixed* x, $f(., x, .)$ *is* $\mathcal{L} \times \mathcal{B}^m$ *measurable. There exists an* $\mathcal{L} \times \mathcal{B}$ *measurable function* $k : [S, T] \times R^m \to R$ *such that* $t \to k(t, \bar{u}(t))$ *is integrable and, for a.e.* $t \in [S, T]$,

$$|f(t, x, u) - f(t, x', u)| \leq k(t, u)|x - x'|$$

for all $x, x' \in \bar{x}(t) + \delta B$ *and* $u \in U(t)$;

(H2) $\operatorname{Gr} U$ *is an* $\mathcal{L} \times \mathcal{B}^m$ *measurable set;*

(H3) g *is locally Lipschitz continuous.*

Then there exist $p \in W^{1,1}([S, T]; R^n)$ *and* $\lambda \geq 0$ *such that*

(i) $(p, \lambda) \neq (0, 0)$;

(ii) $-\dot{p}(t) \in co\, \partial_x \mathcal{H}(t, \bar{x}(t), p(t), \bar{u}(t))$ *a.e.;*

(iii) $\mathcal{H}(t, \bar{x}(t), p(t), \bar{u}(t)) = \max_{u \in U(t)} \mathcal{H}(t, \bar{x}(t), p(t), u)$ *a.e.;*

(iv) $(p(S), -p(T)) \in \lambda \partial g(\bar{x}(S), \bar{x}(T)) + N_C(\bar{x}(S), \bar{x}(T))$.

Now assume, also, that

$$f(t, x, u) \text{ and } U(t) \text{ are independent of } t.$$

Then, in addition to the above conditions, there exists a constant r such that

(v) $\mathcal{H}(t, \bar{x}(t), p(t), \bar{u}(t)) = r$ *a.e.*

($\partial_x \mathcal{H}$ denotes the limiting subdifferential of $\mathcal{H}(t, ., p, u)$ for fixed (t, p, u).)

The proof of the theorem is deferred to the next section.

Elements (λ, p) whose existence is asserted in the Maximum Principle are called *multipliers* for (P). The components λ and p are referred to as the *cost multiplier* and *adjoint arc*, respectively.

Remarks

(a): The adjoint inclusion (Condition (ii) in the theorem statement) is often stated in terms of the Clarke's *Generalized Jacobian*:

Definition 6.2.2 *Take a point $y \in R^n$ and a function $L : R^n \to R^m$ that is Lipschitz continuous on a neighborhood of y. Then the Generalized Jacobian $DL(y)$ of L at y is the set of $m \times n$ matrices:*

$$DL(y) := co\{\eta : \exists y_i \to y \text{ such that}$$
$$\nabla L(y_i) \text{ exists } \forall i \text{ and } \nabla L(y_i) \to \eta\}.$$

A noteworthy property of the generalized Jacobian $DL(y)$ of a function $L : R^n \to R^m$ at a point y is that, for any row vector $r \in R^m$,

$$r\, DL(y) = co\, \partial(r\, L)(y)$$

(see [38], Proposition 2.6.4). Here, $\partial(r\, L)(y)$ is the limiting subdifferential of the function $y \to r\, L(y)$. It follows immediately that the adjoint inclusion can be equivalently written

$$-\dot{p}(t) \in p\, D_x f(t, \bar{x}(t).\dot{\bar{x}}(t)),$$

in which $D_x f(t, x, u)$ denotes the generalized Jacobian with respect to the x variable.

(b): The Maximum Principle (as stated above) extends to cover problems with an integral term in the cost function, i.e., problems in which we seek to minimize a functional of the form:

$$\int_S^T L(t, x(t), \dot{x}(t))dt + g(x(S), x(T)).$$

This is accomplished by introducing a new state variable z that satisfies the relationships

$$\begin{cases} \dot{z}(t) = L(t, x(t), \dot{x}(t)) \\ z(S) = 0. \end{cases}$$

The "mixed" cost can then be replaced by the pure endpoint cost

$$z(T) + g(x(S), x(T)).$$

Under appropriate conditions on L, there results a problem to which Theorem 6.2.1 is applicable. Applying the theorem and interpreting the conclusions in terms of the data for the original problem give a Maximum Principle for problems with mixed integral and endpoint cost terms, in which the Unmaximized Hamiltonian now takes the form

$$\mathcal{H}_\lambda(t, x, u) := p \cdot f(t, s, x) - \lambda L(t, x, u).$$

(In other respects the necessary conditions remain the same.) This procedure for eliminating integral cost terms, which is used frequently in this book, is referred to as *state augmentation*.

(c): We observe finally that the necessary conditions of Theorem 6.2.1 are homogeneous with respect to the multipliers (p, λ). This means that if (p, λ) serves as a set of multipliers then, for any $\alpha > 0$, $(\alpha p, \alpha \lambda)$ also serves. Since $(p, \lambda) \neq 0$, we can always arrange by choosing an appropriate α that

$$\|p\|_{L^\infty} + \lambda = 1.$$

Scaling multipliers in this way is often carried out to assist convergence analysis.

Examples of the following optimal control problem, in which the endpoint constraints are specified as functional constraints, are often encountered.

$$(FEC) \begin{cases} \text{Minimize } g(x(S), x(T)) \\ \text{over } x \in W^{1,1}([S, T]; R^n) \text{ and measurable} \\ \qquad \text{functions } u : [S, T] \to R^m \text{ satisfying} \\ \dot{x}(t) = f(t, x(t), u(t)) \quad \text{a.e.,} \\ u(t) \in U(t) \quad \text{a.e.,} \\ \phi_i(x(S), x(T)) \leq 0 \text{ for } i = 1, 2, \dots, k_1, \\ \psi_i(x(S), x(T)) = 0 \text{ for } i = 1, 2, \dots, k_2. \end{cases}$$

The new ingredients here are the constraint functions $\phi_i : R^n \times R^n \to R$, $i = 1, \ldots, k_1$ and $\psi_i : R^n \times R^n \to R$, $i = 1, \ldots, k_2$. We permit the cases $k_1 = 0$ (no inequality constraints) and $k_2 = 0$ (no equality constraints).

Of course the necessary conditions of Theorem 6.2.1 allow for such constraints, but they are rather cumbersome to apply, since this involves construction of a normal cone to an endpoint constraint set defined implicitly as a feasible region for a collection of functional inequality and equality constraints. It is convenient then to have at hand necessary conditions expressed directly in terms of the constraint functionals themselves by means of additional Lagrange multipliers. Such conditions are provided by the next theorem.

Theorem 6.2.3 (The Maximum Principle for Functional Endpoint Constraints) *Suppose that $(\bar{x}, \bar{u})$ is a $W^{1,1}$ local minimizer for (FEC). Assume that Hypotheses (H1) and (H2) of Theorem 6.2.1 are satisfied and that $g, \phi_1, \ldots \phi_{k_1}, \psi_1, \ldots, \psi_{k_2}$ are locally Lipschitz continuous.*

Then there exist $p \in W^{1,1}$, $\lambda \geq 0$, $\alpha_i \geq 0$ for $i = 1, \ldots, k_1$ and numbers β_i for $i = 1, \ldots, k_2$ such that

(i) $\lambda + \|p\|_{L^\infty} + \sum_{i=1}^{k_1} \alpha_i + \sum_{i=1}^{k_2} |\beta_i| \neq 0$,

(ii) $-\dot{p}(t) \in \mathrm{co}\, \partial_x \mathcal{H}(t, \bar{x}(t), p(t), \bar{u}(t))$ *a.e.*,

(iii) $\mathcal{H}(t, \bar{x}(t), p(t), \bar{u}(t)) = \max_{u \in U(t)} \mathcal{H}(t, \bar{x}(t), p(t), u)$ *a.e.*,

(iv) $(p(S), -p(T)) \in \partial(\lambda g(\bar{x}(S), \bar{x}(T))$
$\qquad + \sum_{i=1}^{k_1} \alpha_i \phi_i(\bar{x}(S), \bar{x}(T)) + \sum_{i=1}^{k_2} \beta_i \psi_i(\bar{x}(S), \bar{x}(T)))$,

and

$$\alpha_i = 0 \quad \text{for all } i \in \{1, \ldots, k_1\} \text{ such that } \phi_i(\bar{x}(S), \bar{x}(T)) < 0.$$

(In the preceding relationships, $\sum_{i=1}^{k}$ is interpreted as 0 if $k = 0$.)

If, furthermore,

$$f(t, x, u) \text{ and } U(t) \text{ are independent of } t,$$

then there exists a constant r such that

(v) $\mathcal{H}(t, \bar{x}(t), p(t), \bar{u}(t)) = r$ *a.e.*

Note that if the functions

$$g, \phi_1, \ldots, \phi_{k_1}, \psi_1, \ldots, \psi_{k_2}$$

are continously differentiable and if $f(t, x, u)$ is continuously differentiable in the x variable, then the "nonsmooth" conditions (ii) and (iv) can be replaced by relationships involving classical (Fréchet) derivatives

$$-\dot{p}(t) = \mathcal{H}_x(t, \bar{x}(t), p(t), \bar{u}(t)) \quad \text{a.e.}$$

and

$$(p(S), -p(T)) = \lambda \nabla g(\bar{x}(S), \bar{x}(T))$$
$$+ \sum_{i=1}^{k_1} \alpha_i \nabla \phi_i(\bar{x}(S), \bar{x}(T))$$
$$+ \sum_{i=1}^{k_2} \beta_i \nabla \psi_i(\bar{x}(S), \bar{x}(T)).$$

This version of the Maximum Principle (addressing problems with functional endpoint constraints and smooth data) does not require, for its formulation, the modern apparatus of subdifferentials, etc. Many of the more recent alternative necessary conditions of optimality, such as the Extended Euler–Lagrange and Extended Hamilton Conditions which are the subject matter of Chapter 7, are, by contrast, inherently nonsmooth: even if the dynamic constraint originates as a differential equation parameterized by a control variable, with smooth right side, the statement of these necessary conditions involves constructs of Nonsmooth Analysis (outside rather special cases).

Proof of Theorem 6.2.3. We assume that $k_1 \geq 1$ and $k_2 \geq 1$ and that all the endpoint inequality constraints are active. There is no loss of generality in so doing. To see this, note that $(\bar{x}, \bar{z}_0 \equiv 0, \bar{z}_1 \equiv 0, \bar{u})$ is a $W^{1,1}$ local minimizer for a modified version of problem (FEC) obtained by deleting the inactive endpoint inequality constraints, replacing the state vector x by $(x, z_0, z_1) \in R^n \times R \times R$, and by requiring the new state variables z_0 and z_1 to satisfy the differential equations

$$\dot{z}_0(t) = 0, \dot{z}_1(t) = 0$$

and the endpoint constraints

$$z_0(T) \leq 0, z_1(T) = 0.$$

The additional hypotheses (nonemptiness of the sets of equality and inequality constraints and all inequality constraints are active) are met. If the assertions of the theorem were true in these circumstances, we would be able to find an adjoint arc which we write (p, q_0, q_1) satisfying the conditions of the Maximum Principle with functional endpoint constraints. But because $z_0(.)$ and $z_1(.)$ are unconstrained at $t = S$, we conclude from the

Transversality Conditions that $q_0 \equiv 0$ and $q_1 \equiv 0$. The resulting conditions may be interpreted as the assertions of the theorem for the original problem when we associate with each inactive inequality endpoint constraint a zero multiplier.

Define $G : R^n \times R^n \to R \times R^{k_1} \times R^{k_2}$ according to

$$G(x_0, x_1) := (g(x_0, x_1) - g(\bar{x}(S), \bar{x}(T)),$$
$$\phi_1(x_0, x_1), \ldots, \phi_{k_1}(x_0, x_1), \psi_1(x_0, x_1), \ldots, \psi_{k_2}(x_0, x_1)).$$

Now observe that $(\bar{x}, \bar{y} \equiv 0, \bar{u})$ is a $W^{1,1}$ local minimizer for the optimal control problem with state vector $(x, y = (y_0, \ldots, y_{k_1+k_2})) \in R^n \times R^{1+k_1+k_2}$:

$$\left\{ \begin{array}{l} \text{Minimize } y_0(T) \\ \text{over } x \in W^{1,1}, (y_0, \ldots, y_{k_1+k_2}) \in W^{1,1} \\ \quad \text{and measurable functions } u \text{ satisfying} \\ \dot{x} = f, \dot{y}_0 = 0, \ldots, \dot{y}_{k_1+k_2} = 0, \quad \text{a.e.,} \\ u(t) \in U(t) \quad \text{a.e.,} \\ (x(S), x(T), y(S)) \in \operatorname{Gr} G, \\ y_i(T) \leq 0 \text{ for } i = 1, \ldots, k_1, \\ y_i(T) = 0 \text{ for } i = k_1 + 1, \ldots, k_1 + k_2. \end{array} \right.$$

The assertions of Theorem 6.2.1, which are valid for this problem, are expressed in terms of an adjoint arc, which we write

$$(p, q = (-\alpha_0, \ldots, -\alpha_{k_1}, -\beta_1, \ldots, -\beta_{k_2})),$$

and a cost multiplier $\lambda \geq 0$, not both zero. From the Transversality Condition of Theorem 6.2.1, we know

$$(p(S), -p(T), -(\alpha_0, \ldots, \alpha_{k_1}, \beta_1, \ldots, \beta_{k_2})) \in N_{\operatorname{Gr} G}(\bar{x}(S), \bar{x}(T), 0, \ldots, 0)$$

$$\alpha_0 = \lambda, \quad \alpha_1, \ldots, \alpha_{k_1} \geq 0.$$

With the help of Proposition 5.4.2, however, we deduce that

$$(p(S), -p(T)) \in \partial \left(\lambda g + \sum_{i=1}^{k_1} \alpha_i \phi_i + \sum_{i=1}^{k_2} \beta_i \psi_i \right) (\bar{x}(S), \bar{x}(T)).$$

This is the Transversality Condition as it appears in the theorem statement. Clearly all the other assertions of the theorem are confirmed (with "multipliers" $p, \alpha_1, \ldots, \alpha_{k_1}, \beta_1, \ldots, \beta_{k_2}$ and λ as above). $\square$

6.3   Derivation of the Maximum Principle from the Extended Euler Condition

In this section, the Maximum Principle (Theorem 6.2.1) is deduced from a general set of necessary conditions centered on the Extended Euler–Lagrange inclusion, the proof of which is given in Chapter 7.

We omit all mention of the constancy of the Hamiltonian condition (v), a proof of which is given in Section 8.7, in the more general context of Free Time optimal contol problems.

The following lemma justifies paying attention only to a special case of (P) in which some additional hypotheses are imposed.

Lemma 6.3.1 *Suppose it can be proved that there exists $\lambda \geq 0$ and $p \in W^{1,1}$ satisfying conditions (i) to (iv) of Theorem 6.2.1 under Hypotheses (H1) to (H3) and also the supplementary hypotheses:*

(H4)' $(\bar{x}, \bar{u})$ is a strong local minimizer (not merely a $W^{1,1}$ local minimizer);

(H5)' for all $t \in [S, T]$, $U(t)$ is a finite set and

$$f(t, \bar{x}(t), u) \neq f(t, \bar{x}(t), v) \text{ for all } u, v \in U(t) \text{ such that } u \neq v;$$

(H6)' there exist $k_f \in L^1$ and $c_f \in L^1$ such that

$$|f(t, x, u) - f(t, x', u)| \leq k_f(t)|x - x'| \text{ and } |f(t, x, u)| \leq c_f(t)$$

for all $x, x' \in \bar{x}(t) + \delta B$, $u \in U(t)$ a.e.

(δ is the constant of hypotheses (H1) to (H3) in the theorem statement.)

Then there exist $\lambda \geq 0$ and p satisfying conditions (i) to (iv), if merely Hypotheses (H1) to (H3) are satisfied.

Proof. Assume that the assertions of the theorem are valid in the special case when, besides (H1) to (H3), (H4)' to (H6)' are satisfied. Let us suppose that, in addition to (H1) to (H3), (H4'), and (H6') are satisfied but that hypothesis (H5)' is possibly violated. Set

$$L := \exp\left\{\int_S^T k(t, \bar{u}(t))dt\right\},$$

in which $k(.,.)$ is the function whose existence is hypothesized in (H1). Note that, for any $p \in W^{1,1}([S, T]; R^n)$ satisfying

$$-\dot{p}(t) \in \text{co } \partial_x \mathcal{H}(t, \bar{x}(t), p(t), \bar{u}(t))   \text{a.e.},$$

we have

$$\partial_x \mathcal{H}(t, \bar{x}(t), p(t), \bar{u}(t)) \subset k(t, \bar{u}(t))|p(t)|B.$$

It follows from Gronwall's Inequality (Lemma 2.4.4) that if, additionally, $|p(T)| \leq 1$ then

$$|p(t)| \leq L   \text{for all } t \in [S, T].$$

Choose $\{S_j\}$ an increasing family of finite subsets of LB such that $LB \subset S_j + j^{-1}B$ for all j. Fix j. Let N be the number of elements in S_j.

For each $s \in S_j$, we may select a measurable function $w_s(.)$ such that $w_s(t) \in U(t)$ a.e. and

$$H(t, \bar{x}(t), s) \leq s \cdot f(t, \bar{x}(t), w_s(t)) + j^{-1} \quad \text{a.e.}$$

Here

$$H(t, x, s) := \sup\{s \cdot f(t, x, u) : u \in U(t)\}.$$

Order the finite collection of functions $\{\bar{u}\} \cup \{w_s : s \in S_j\}$ as $\{v_k\}_{k=0}^N$, taking $v_0(t) = \bar{u}(t)$. We modify the values of the v_ks as follows: for each t, replace (without relabeling) $v_k(t)$ by $v_0(t)$, for all $k \geq 1$ such that

$$f(t, \bar{x}(t), v_k(t)) = f(t, \bar{x}(t), v_0(t)).$$

Then replace (again without relabeling) $v_k(t)$ by $v_1(t)$ for all $k \geq 2$ such that

$$f(t, \bar{x}(t), v_k(t)) = f(t, \bar{x}(t), v_1(t)).$$

Continue in this way, redefining function values at time t, finally replacing $v_N(t)$ by $v_{N-1}(t)$ if

$$f(t, \bar{x}(t), v_N(t)) = f(t, \bar{x}(t), v_{N-1}(t)).$$

Now consider a new problem in which the multifunction U is replaced by U_j,

$$U_j(t) := \cup_{j=0}^N \{v_j(t)\}.$$

It can be shown that U_j is a measurable multifunction. The above constructions ensure that, for each t, $U_j(t)$ is a finite set, $\bar{u}(t) \in U_j(t)$ and

$$w_1, w_2 \in U_j, \ w_1 \neq w_2$$
$$\text{implies} \quad f(t, \bar{x}(t), w_1) \neq f(t, \bar{x}(t), w_2).$$

Furthermore, for each $s \in S_j$ we can choose a selector $\tilde{w}_s$ of U_j such that

$$H(t, \bar{x}(t), s) \leq s \cdot f(t, \bar{x}(t), \tilde{w}_s(t)) + j^{-1} \quad \text{a.e.} \tag{6.1}$$

For the new problem, $(\bar{x}, \bar{u})$ is a strong local minimizer because $(\bar{x}, \bar{u})$ remains feasible and the new control constraint set $U_j(t)$ is a subset of $U(t)$, a.e. Now however Hypothesis (H5)$'$ is satisfied.

But we have postulated that, if (H5)$'$ is satisfied (in addition to the other hypotheses), then the assertions of Theorem 6.2.1 are valid. We deduce then existence of multipliers p and λ with the properties listed in Theorem 6.2.1, except that the Weierstrass Condition here takes the form

$$p(t) \cdot \dot{\bar{x}}(t) = H_j(t, \bar{x}(t), p(t)) \quad \text{a.e.,} \tag{6.2}$$

in which
$$H_j(t, x, s) := \max\{s \cdot f(t, x, u) : u \in U_j(t)\}.$$

We know $(p, \lambda) \neq (0, 0)$. However the adjoint inclusion (ii) implies

$$|\dot{p}(t)| \leq k(t, \bar{u}(t))|p(t)|.$$

By Gronwall's Inequality (Lemma 2.4.4)

$$p(S) = 0 \quad \text{implies} \quad p(.) \equiv 0.$$

It follows that $(p(S), \lambda) \neq (0, 0)$. We can arrange then, by scaling the multipliers, that

$$|p(S)| + \lambda = 1. \tag{6.3}$$

For each t at which (6.2) is valid, choose $s \in S_j$ such that

$$p(t) \in \{s\} + j^{-1}B.$$

Since $c_f(t)$ (the Lipschitz constant of Hypothesis (H6)$'$) is a Lipschitz constant for $H(t, \bar{x}(t), .)$ and $H_j(t, \bar{x}(t), .)$, it follows that

$$
\begin{aligned}
H_j(t, \bar{x}(t), p(t)) &\geq H_j(t, \bar{x}(t), s) - j^{-1}c_f(t) \\
&\geq s \cdot f(t, \bar{x}(t), w_s(t)) - j^{-1}c_f(t) \\
&\geq H(t, \bar{x}(t), s) - j^{-1} - j^{-1}c_f(t) \\
&\geq H(t, \bar{x}(t), p(t)) - j^{-1} - 2j^{-1}c_f(t).
\end{aligned}
$$

But then, from (6.2),

$$-p(t) \cdot \dot{\bar{x}}(t) + H(t, \bar{x}(t), p(t)) \leq j^{-1}(1 + 2c_f(t)) \quad \text{a.e.} \tag{6.4}$$

Relabel (p, λ) as (p_j, λ_j) and carry out the above construction for $j = 1, 2, \ldots$.

We deduce from the normalization condition (6.3) and the adjoint inclusion (ii) that, as j ranges over the positive integers, the λ_js are uniformly bounded, the p_js are uniformly bounded with respect to the supremum norm, and the $\dot{p}_j$s are dominated by a common integrable function.

The hypotheses are satisfied under which the Compactness of Trajectories Theorem (Theorem 2.5.3) is applicable to

$$-\dot{p}_j \in \text{co } \partial_x H.$$

We conclude that, along some subsequence,

$$p_j \to p \quad \text{uniformly}$$

for some $p \in W^{1,1}$ which satisfies

$$-\dot{p}(t) \in \text{co } \partial_x \mathcal{H}(t, \bar{x}(t), p(t), \bar{u}(t)) \quad \text{a.e.}$$

We obtain also in the limit

$$(p(S), -p(T)) \in \lambda \partial g(\bar{x}(S), \bar{x}(T)) + N_C(\bar{x}(S), \bar{x}(T)),$$

since the right side of this relationship is a closed set. The continuity of $H(t, \bar{x}(t), .)$ also permits us to deduce from (6.4) (in which p_j replaces p) that

$$p(t) \cdot \dot{\bar{x}}(t) = H(t, \bar{x}(t), p(t)) \quad \text{a.e.}$$

This is the Weierstrass Condition.

We have verified the necessary conditions under Hypotheses (H1) to (H3), (H4)$'$, and (H6)$'$ alone.

Next suppose that $(\bar{x}, \bar{u})$ is a strong local minimizer, hypotheses (H1) to (H3) and (H4)$'$ are satisfied, but Hypothesis (H6)$'$ is possibly violated.

For each j define

$$U_j(t) := \{u \in U(t) : k(t, u) \leq k(t, \bar{u}(t)) + j, |f(t, \bar{x}(t), u)| \leq |\dot{\bar{x}}(t)| + j\}$$

and consider the problem obtained by replacing $U(t)$ in problem (P) by $U_j(t)$. (Gr U_j is $\mathcal{L} \times \mathcal{B}^m$ measurable; this fact may be deduced from (H1), Proposition 2.2.4, and from the hypothesized measurability properties of the functions k and f.) For each j, the data for this new problem satisfy (H6)$'$ and so the special case of the necessary conditions just proved applies. We are assured then of the existence of multipliers p and λ (they both depend on j) with the properties listed in Theorem 6.2.1, except that the Weierstrass Condition now takes the form

$$p(t) \cdot \bar{x}(t) \geq p(t) \cdot f(t, \bar{x}(t), u) \quad \text{for all } u \in U_j(t), \qquad \text{a.e. } t \in [S, T].$$

A convergence analysis along the lines just undertaken, which makes use of the fact for any $t \in [S, T]$

$$u \in U(t) \quad \text{implies} \quad u \in U_j(t) \text{ for } j \text{ sufficiently large},$$

yields multipliers in the limit, in relation to which the conditions are valid. The assertions of the theorem have been confirmed under the hypotheses (H1) to (H3) and (H4)$'$.

To complete the proof, assume that $(\bar{x}, \bar{u})$ is merely a $W^{1,1}$ local minimizer. In this case, there exists $\alpha > 0$ such that $(\bar{x}, \bar{y} \equiv 0, \bar{u})$ is a strong local minimizer for

$$\begin{cases} \text{Minimize } g(x(S), x(T)) \\ \text{over } x \in W^{1,1}, \ y \in W^{1,1}, \text{ and measurable functions } u \text{ satisfying} \\ \dot{x}(t) = f(t, x(t), u(t)) \text{ and } \dot{y}(t) = |f(t, x(t), u(t)) - \dot{\bar{x}}(t)| \quad \text{a.e.}, \\ u(t) \in U(t) \quad \text{a.e.}, \\ ((x(S), x(T)), y(S), y(T)) \in C \times \{0\} \times (\alpha B). \end{cases}$$

Apply the preceding special case of the Maximum Principle with reference to the strong local minimizer $(\bar{x}, \bar{u})$. We find that the adjoint function

r associated with the y variable is constant, because the right sides of the differential equations governing the state are independent of y. But then $r = 0$ since the endpoint constraint

$$y(T) \in \alpha B$$

is inactive.

The other relationships we obtain are the conditions of the Maximum Principle for the original problem. (Notice that all reference to the extra state variable y disappears.) We have shown then that the assertions of the Maximum Principle are valid merely under Hypotheses (H1) to (H3). This is what we set out to prove. $\square$

After reducing the hypotheses in this way, derivation of the Maximum Principle from the Extended Euler–Lagrange Conditions of the next chapter is relatively straightforward.

According to the lemma, we may assume without loss of generality that Hypotheses (H5)$'$ and (H6)$'$) are in force, in addition to (H1) to (H3). (The fact that we can also arrange that (H4)$'$ is satisfied is required only at a later stage of our investigations.)

Define the multifunction $F : [S,T] \times R^n \rightsquigarrow R^n$ to be

$$F(t,x) := \{f(t,x,u) : u \in U(t)\}.$$

We note that, since $U(t)$ is here assumed to be a finite set, F takes as values compact nonempty sets. Also, F is $\mathcal{L} \times \mathcal{B}^n$ measurable, Gr $F(t,.)$ is closed for each t, and $F(.,x)$ is measurable for each $x \in R^n$. It follows also from the hypotheses on f that

$$F(t,x) \subset F(t,x') + k_f(t)|x - x'|B \qquad \text{for all } x, x' \in \bar{x}(t) + \delta B \quad \text{a.e.}$$

The state trajectory $\bar{x}$ is a $W^{1,1}$ local minimizer for

$$\left\{ \begin{array}{l} \text{Minimize } g(x(S), x(T)) \\ \text{over } x \in W^{1,1}([S,T]; R^n) \text{ satisfying} \\ \dot{x}(t) \in F(t,x(t)) \quad \text{a.e.,} \\ (x(S), x(T)) \in C. \end{array} \right.$$

The hypotheses hold under which Theorem 7.4.1 applies. It asserts the existence of $p \in W^{1,1}$ and $\lambda \geq 0$ such that all the conditions of Theorem 6.2.1 are satisfied, except that, in place of the adjoint inclusion (ii), we have

$$\dot{p}(t) \in \text{co } \{\xi : (\xi, p(t)) \in N_{\text{Gr }F(t,.)}(\bar{x}(t), \dot{\bar{x}}(t))\} \quad \text{a.e.} \tag{6.5}$$

But in view of Hypothesis (H5)$'$, for each t, there exists $\alpha > 0$ (α will depend on t) such that

$$\text{Gr }F(t,.) \cap ((\bar{x}(t), \dot{\bar{x}}(t)) + \alpha B) = \text{Gr }f(t,.,\bar{u}(t)) \cap ((\bar{x}(t), \dot{\bar{x}}(t)) + \alpha B).$$

It follows that

$$N_{\mathrm{Gr}\,F(t,.)}(\bar{x}(t), \dot{\bar{x}}(t)) = N_{\mathrm{Gr}\,f(t,.,\bar{u}(t))}(\bar{x}(t), \dot{\bar{x}}(t)) \qquad \text{for all } t.$$

Now use the fact that for a function $d : R^n \to R^n$ and a point $y \in R^n$ such that d is Lipschitz continuous on a neighborhood of y we have

$$(q', q) \in N_{\mathrm{Gr}\,d(.)}(y, d(y)) \qquad \text{implies} \qquad -q' \in \partial[q \cdot d(x)]|_{x=y}.$$

(See Proposition 5.4.2.)

We conclude now from (6.5) that

$$
\begin{aligned}
-\dot{p}(t) \quad &\in \quad -\mathrm{co}\{-\xi : \xi \in \partial_x \mathcal{H}(t, \bar{x}(t), p(t), \bar{u}(t))\} \\
&= \quad \mathrm{co}\ \partial_x \mathcal{H}(t, \bar{x}(t), p(t), \bar{u}(t)) \quad \text{a.e.}
\end{aligned}
$$

This is the required adjoint inclusion. The proof of the theorem is complete.
□

6.4   A Smooth Maximum Principle

This section, the technical core of the chapter, provides a self-contained proof of Conditions (i) through (iv) of the Maximum Principle, Theorem 6.2.1, in the case when the dynamics constraint is "smooth" with respect to the state variable. The problem of interest remains (P):

$$(P) \quad \begin{cases} \text{Minimize } g(x(S), x(T)) \\ \text{over } x \in W^{1,1}([S,T]; R^n) \text{ and measurable} \\ \qquad \text{functions } u : [S,T] \to R^m \text{ satisfying} \\ \dot{x}(t) = f(t, x(t), u(t)) \quad \text{a.e.,} \\ u(t) \in U(t) \quad \text{a.e.,} \\ (x(S), x(T)) \in C, \end{cases}$$

with data an interval $[S,T]$, functions $g : R^n \times R^n \to R$ and $f : [S,T] \times R^n \times R^m \to R^n$, a nonempty multifunction $U : [S,T] \rightsquigarrow R^m$, and a closed set $C \subset R^n \times R^n$.

Theorem 6.4.1 (A Smooth Maximum Principle) *Let $(\bar{x}, \bar{u})$ be a $W^{1,1}$ local minimizer for (P). Assume that, in addition to hypotheses of Theorem 6.2.1, namely, there exists $\delta > 0$ such that*

(H1): $f(.,x,.)$ is $\mathcal{L} \times \mathcal{B}^m$ measurable for fixed x. There exists a Borel measurable function $k : [S,T] \times R^m \to R$ such that $t \to k(t, \bar{u}(t))$ is integrable and, for a.e. $t \in [S,T]$,

$$|f(t, x, u) - f(t, x', u)| \le k(t, u)|x - x'|$$

for all $x, x' \in \bar{x}(t) + \delta B, \ u \in U(t),$

(H2): Gr U is $\mathcal{L} \times \mathcal{B}^m$ measurable,

(H3): g is locally Lipschitz continuous,

and the following hypothesis is satisfied.

(S1): $f(t, . , u)$ is continuously differentiable on $\bar{x}(t) + \delta \, int B$ for all $u \in U(t)$, a.e. $t \in [S, T]$.

Then there exist $p \in W^{1,1}([S,T]; R^n)$ and $\lambda \geq 0$ such that

(i) $(p, \lambda) \neq (0,0)$,

(ii) $-\dot{p}(t) = \mathcal{H}_x(t, \bar{x}(t), p(t), \bar{u}(t))$ a.e.,

(iii) $(p(S), -p(T)) \in \lambda \partial g(\bar{x}(S), \bar{x}(T)) + N_C(\bar{x}(S), \bar{x}(T))$,

(iv) $\mathcal{H}(t, \bar{x}(t), p(t), \bar{u}(t)) = \max_{u \in U(t)} \mathcal{H}(t, \bar{x}(t), p(t), u)$ a.e.

The Smooth Maximum Principle is built up in stages, in which the optimality conditions are proved under hypotheses that are progressively less restrictive. The results obtained at completion of each stage are summarized as a proposition.

The first case treated is that when the velocity set is compact and convex, the cost function is smooth, there are no endpoint constraints, and, finally, $(\bar{x}, \bar{u})$ is a strong local minimizer.

Proposition 6.4.2 *Let $(\bar{x}, \bar{u})$ be a strong local minimizer for (P). Then the assertions of Theorem 6.4.1 are valid with $\lambda = 1$ when we assume that, in addition to (H1) to (H3) and (S1), the following hypotheses are satisfied:*

(S2): There exist $k_f \in L^1$ and $c_f \in L^1$ such that, for a.e. $t \in [S,T]$,

$$|f(t, x, u) - f(t, x', u)| \leq k_f(t)|x - x'| \quad and \quad |f(t, x, u)| \leq c_f(t)$$

for all $x, x' \in \bar{x}(t) + \delta B$, $u \in U(t)$;

(S3): $f(t, x, U(t))$ is a compact set for all $x \in \bar{x}(t) + \delta B$, a.e. $t \in [S,T]$;

(S4): $f(t, x, U(t))$ is a convex set for all $x \in \bar{x}(t) + \delta B$, a.e. $t \in [S,T]$;

(S5): g is continuously differentiable;

(S6): $C = R^n \times R^n$.

Proof. By reducing $\delta > 0$ if necessary, we arrange that $(\bar{x}, \bar{u})$ is a minimizer with respect to all (x, u)s satisfying

$$\|x - \bar{x}\|_{L^\infty} \leq \delta.$$

Step 1 (Linearization): Set

$$\mathcal{U} := \{u : [S,T] \to R^m : u(.) \text{ is measurable and } u(t) \in U(t) \quad \text{a.e.}\}.$$

We show that for any $u \in \mathcal{U}$ and $y \in W^{1,1}$ such that

$$\dot{y}(t) = f_x(t, \bar{x}(t), \bar{u}(t))y(t) + (f(t, \bar{x}(t), u(t)) - f(t, \bar{x}(t), \bar{u}(t))) \quad \text{a.e.,}$$

we have

$$\nabla g(\bar{x}(S), \bar{x}(T)) \cdot (y(S), y(T)) \geq 0.$$

Fix $\xi \in R^n$ and $u \in \mathcal{U}$. Take a sequence $\epsilon_i \downarrow 0$ and, for each i, consider the differential equation

$$\begin{aligned}
\dot{x}(t) &= f(t, x(t), \bar{u}(t)) + \epsilon_i \Delta f(t, x(t), u(t)) \\
x(S) &= \bar{x}(S) + \epsilon_i \xi
\end{aligned}$$

in which

$$\Delta f(t, x, u) := f(t, x, u) - f(t, x, \bar{u}(t)).$$

We deduce from the Generalized Filippov Existence Theorem (Theorem 2.4.3) that, for all i sufficiently large, there exists a solution $x_i \in W^{1,1}$ satisfying $\|x_i - \bar{x}\|_{L^\infty} < \delta$. Also, if we define

$$y_i(t) := \epsilon_i^{-1}(x_i(t) - \bar{x}(t)),$$

then the y_is are uniformly bounded and the $\dot{y}_i$s are uniformly integrably bounded.

For a subsequence then,

$$y_i(t) \to y(t) \quad \text{uniformly},$$

for some $y \in W^{1,1}$. Subtracting the differential equation for $\bar{x}$ from that for x_i, dividing across by ϵ_i and integrating gives

$$\epsilon_i^{-1}(x_i - \bar{x})(t) =$$
$$\xi + \int_S^t \epsilon_i^{-1}\left(f(s, x_i(s), \bar{u}(s)) - f(s, \bar{x}(s), \bar{u}(s))\right) ds$$
$$+ \int_S^t \Delta f(s, x_i(s), u(s)) ds.$$

With the aid of the Dominated Convergence Theorem and making use of Hypothesis (S2), we obtain in the limit

$$y(t) = \xi + \int_S^t f_x(s, \bar{x}(s), \bar{u}(s))y(s) ds + \int_S^t \Delta f(s, \bar{x}(s), u(s)) ds.$$

The arc y may be interpreted as the solution to the differential equation

$$\begin{aligned}
\dot{y}(t) &= f_x(t, \bar{x}(t), \bar{u}(t))y(t) + \Delta f(t, \bar{x}(t), u(t)), \\
y(S) &= \xi.
\end{aligned} \tag{6.6}$$

Under the convexity hypothesis (S4) and in view of the Generalized Filippov Selection Theorem (we refer to the discussion following the statement of Theorem 2.3.13), there exists, for each i, a control function $\tilde{u}_i \in \mathcal{U}$ such that

$$\dot{x}_i(t) = f(t, x_i(t), \tilde{u}_i(t)) \quad \text{a.e.}$$

Since $(\bar{x}, \bar{u})$ is a strong local minimizer and $(x_i = \bar{x} + \epsilon_i y_i, \tilde{u}_i)$ is a feasible process for (P) satisfying

$$\|x_i - \bar{x}\|_{L^\infty} \leq \delta$$

(for sufficiently large i), we must have

$$\epsilon_i^{-1} \left(g((\bar{x}(S), \bar{x}(T)) + \epsilon_i(y_i(S), y_i(T))) - g(\bar{x}(S), \bar{x}(T)) \right) \geq 0.$$

But g is continuously differentiable at $(\bar{x}(S), \bar{x}(T))$. In the limit as $i \to \infty$, we obtain

$$\nabla g(\bar{x}(S), \bar{x}(T)) \cdot (y(S), y(T)) \geq 0.$$

Step 2 (Dualization): Let $p \in W^{1,1}$ be the solution of the differential equation

$$
\begin{aligned}
-\dot{p}(t) &= p(t) f_x(t, \bar{x}(t), \bar{u}(t)) \quad \text{a.e.} \\
-p(T) &= \nabla_{x_1} g(\bar{x}(S), \bar{x}(T)).
\end{aligned}
\tag{6.7}
$$

(We denote by $\nabla_{x_1} g$ and $\nabla_{x_2} g$ the partial gradients of $g(x_1, x_2)$ with respect to x_1 and x_2, respectively.)

Choose any y, u, and ξ satisfying (6.6). Add the term

$$\int_S^T p(t) \cdot (\dot{y}(t) - f_x(t, \bar{x}(t), \bar{u}(t))y(t) - \Delta f(t, \bar{x}(t), u(t)))dt$$

(which is zero) to the left side of the inequality of Step 1. This gives

$$\nabla g(\bar{x}(S), \bar{x}(T)) \cdot (y(S), y(T))$$
$$+ \int_S^T p(t) \cdot (\dot{y}(t) - f_x(t, \bar{x}(t), \bar{u}(t))y(t) - \Delta f(t, \bar{x}(t), u(t)))dt \geq 0.$$

Integration by parts yields

$$\int_S^T p(t) \cdot \dot{y}(t)dt = -\int_S^T \dot{p}(t) \cdot y(t)dt + p(T) \cdot y(T) - p(S) \cdot \xi.$$

These last two identities combine with (6.7) to give

$$-\int_S^T (\dot{p}(t) + p(t)f_x(t, \bar{x}(t), \bar{u}(t))) \cdot y(t)dt$$
$$-\int_S^T p(t) \cdot \Delta f(t, \bar{x}(t), u(t))dt$$
$$+ (\nabla_{x_0} g(\bar{x}(S), \bar{x}(T)) - p(S)) \cdot \xi \geq 0.$$

Since this inequality is valid for all $\xi \in R^n$ and $u \in \mathcal{U}$, it follows

$$p(S) = \nabla_{x_0} g(\bar{x}(S), \bar{x}(T)).$$

We conclude from this equation and (6.7) that

$$\int_S^T p(t) \cdot (f(t, \bar{x}(t), u(t)) - f(t, \bar{x}(t), \bar{u}(t)))dt \leq 0 \qquad (6.8)$$

for all $u \in \mathcal{U}$. We claim that

$$p(t) \cdot f(t, \bar{x}(t), \bar{u}(t)) = \max_{u \in U(t)} p(t) \cdot f(t, \bar{x}(t), u) \quad \text{a.e. } t \in [S, T]. \qquad (6.9)$$

If this is not the case, there exists $\alpha > 0$ and a subset A of $[S, T]$ having positive measure such that

$$p(t) \cdot f(t, \bar{x}(t), \bar{u}(t)) < \sup_{u \in U(t)} p(t) \cdot f(t, \bar{x}(t), u) - \alpha.$$

for all $t \in A$. We may select $\tilde{u} \in \mathcal{U}$ such that

$$p(t) \cdot f(t, \bar{x}(t), \bar{u}(t)) < p(t) \cdot f(t, \bar{x}(t), \tilde{u}(t)) - \alpha. \qquad (6.10)$$

for almost every $t \in A$.

Since A has positive measure, we may choose a point $\bar{t} \in (S, T)$ such that $\bar{t} \in A$, $\tilde{u}(t) \in U(\bar{t})$, (6.10) is satisfied, and $\bar{t}$ is a Lebesgue point for

$$t \to p(t) \cdot (f(t, \bar{x}(t), \tilde{u}(t)) - f(t, \bar{x}(t), \bar{u}(t))).$$

Now for each $\epsilon \in (0, T - \bar{t})$ substitute the control

$$u(t) := \begin{cases} \tilde{u}(t) & \text{for } t \in [\bar{t}, \bar{t} + \epsilon] \\ \bar{u}(t) & \text{otherwise} \end{cases}$$

into (6.8). Dividing across the inequality by ϵ we arrive at

$$\epsilon^{-1} \int_{\bar{t}}^{\bar{t}+\epsilon} p(s) \cdot (f(s, \bar{x}(s), \tilde{u}(s)) - f(s, \bar{x}(s), \bar{u}(s)))ds \leq 0.$$

This relationship is valid for all $\epsilon > 0$ sufficiently small. Since $\bar{t}$ is a Lebesgue point of the integrand, we obtain in the limit as $\epsilon \downarrow 0$

$$p(\bar{t}) \cdot f(\bar{t}, \bar{x}(\bar{t}), \tilde{u}(\bar{t})) \leq p(t) \cdot f(\bar{t}, \bar{x}(\bar{t}), \bar{u}(\bar{t})).$$

This contradiction of (6.10) confirms the Weierstrass Condition (6.9). The proof of the proposition is complete. $\square$

Our next task is to remove Hypothesis (S6), that is, to allow general endpoint constraints.

Proposition 6.4.3 *The assertions of Proposition 6.4.2 are valid when, in addition to (H1) through (H3) and (S1), we impose merely (S2) through (S5).*

Proof. Fix $\epsilon \in (0,1]$. Reduce the size of $\delta > 0$, if necessary, to ensure that $(\bar{x}, \bar{u})$ is minimizing with respect to all feasible processes (x, u) for (P) satisfying $\|x - \bar{x}\|_{L^\infty} \leq \delta$. Embed (P) (augmented by the constraint $\|x - \bar{x}\|_{L^\infty} \leq \delta$) in a family of problems $\{P(a) : a \in R^n \times R^n\}$,

$$P(a) \begin{cases} \text{Minimize } g(x(S), x(T)) \\ \text{over } x \in W^{1,1} \text{ and measurable functions } u \text{ satisfying} \\ \dot{x}(t) = (1 - \epsilon)f(t, x(t), \bar{u}(t)) + \epsilon f(t, x(t), u(t)), \\ u(t) \in U(t) \quad \text{a.e.,} \\ (x(S), x(T)) \in C + a, \\ \|x - \bar{x}\|_{L^\infty} \leq \delta. \end{cases}$$

Since $f(t, x, U(t))$ is convex and in view of the Generalized Filippov Selection Theorem (Theorem 2.3.13), $(\bar{x}, \bar{u})$ is a minimizer for $P(0)$.

We impose an interim hypothesis,

(HS): If (x, u) is a minimizer for $P(0)$ then $x = \bar{x}$.

(It is discarded later in the proof.)

Denote by $V(a)$ the infimum cost of $P(a)$. (Set $V(a) = +\infty$ if there exist no (x, u)s satisfying the constraints of $P(a)$.)

Note the following properties of V.

(i) $V(a) > -\infty$ for all $a \in R^n \times R^n$ and if $V(a) < +\infty$ then $P(a)$ has a minimizer.

(ii) V is a lower semicontinuous function on $R^n \times R^n$.

(iii) If $a_i \to 0$ and $V(a_i) \to V(0)$ and if (x_i, u_i) is a minimizer for $P(a_i)$ for each i, then $x_i \to \bar{x}$ uniformly and $\dot{x}_i \to \dot{\bar{x}}$ weakly in L^1 as $i \to \infty$.

These are straightforward consequences of the Compactness of Trajectories Theorem (Theorem 2.5.3), results of Section 2.6 governing "convex" differential inclusions applied to the multifunction

$$F(t, x) := \{(1 - \epsilon)f(t, x, \bar{u}(t)) + \epsilon f(t, x, u) : u \in U(t)\},$$

and of the Generalized Filippov Selection Theorem (Theorem 2.3.13), which tells that if $x \in W^{1,1}$ satisfies the differential inclusion

$$\dot{x}(t) \in F(t, x(t)) \quad \text{a.e.,}$$

then there is a $\tilde{u} \in \mathcal{U}$ such that

$$\dot{x}(t) = (1 - \epsilon)f(t, x(t), \bar{u}(t)) + \epsilon f(t, x(t), \tilde{u}(t)) \quad \text{a.e.}$$

Since V is lower semicontinuous and $V(0) < +\infty$, there exists a sequence $a_i \to 0$ such that $V(a_i) \to V(0)$ as $i \to \infty$ and V has a proximal subdifferential ξ_i at a_i for each i. This means that, for each i, there exists $\alpha_i > 0$ and $M_i > 0$ such that

$$V(a) - V(a_i) \geq \xi_i \cdot (a - a_i) - M_i |a - a_i|^2 \qquad (6.11)$$

for all $a \in \{a_i\} + \alpha_i B$.

In view of the above properties of V, $P(a_i)$ has a minimizer (x_i, u_i) for each i and $x_i \to \bar{x}$ uniformly. By eliminating initial terms in the sequence we may arrange that

$$\|x_i - \bar{x}\|_{L^\infty} < \delta/4$$

for all i.

Fix i. Take any (x, u) such that $u \in \mathcal{U}$, x satisfies the differential equation constraint of $P(a)$, and also

$$\|x - x_i\|_{L^\infty} < \delta/2.$$

Choose an arbitrary point $c \in C$. Notice that

$$(x(S), x(T)) \in C + ((x(S), x(T)) - c).$$

This means that (x, u) is feasible process for $P((x(S), x(T)) - c)$. The cost of (x, u) cannot be smaller than the infimum cost $V((x(S), x(T)) - c)$. It follows that

$$g(x(S), x(T)) \geq V((x(S), x(T)) - c). \qquad (6.12)$$

Define

$$c_i := (x_i(S), x_i(T)) - a_i.$$

Since (x_i, u_i) solves $P(a_i) (= P((x_i(S), x_i(T)) - c_i))$, we have

$$g(x_i(S), x_i(T)) = V((x_i(S), x_i(T)) - c_i). \qquad (6.13)$$

Now define the function

$$\begin{aligned} J_i((x, u), c) \quad := \quad & g(x(S), x(T)) - \xi_i \cdot ((x(S), x(T)) - c) \\ & + M_i \left(|(x(S), x(T)) - (x_i(S), x_i(T)) - (c - c_i)|^2 \right). \end{aligned}$$

From (6.11) through (6.13), we deduce that

$$J_i((x, u), c) \geq J_i((x_i, u_i), c_i) \qquad (6.14)$$

for all $c \in C$ and all (x, u)s satisfying

$$\begin{aligned} &\dot{x}(t) = (1 - \epsilon) f(t, x(t), \bar{u}(t)) + \epsilon f(t, x(t), u(t)) \quad \text{a.e.,} \\ &u(t) \in U(t) \quad \text{a.e.,} \\ &\|x - \bar{x}\|_{L^\infty} \leq \delta/2. \end{aligned}$$

Set $(x, u) = (x_i, u_i)$ in (6.14). The inequality implies

$$-\xi_i \cdot (c - c_i) \leq M_i |c - c_i|^2 \text{ for all } c \in C.$$

We conclude that

$$-\xi_i \in N_C^P((x_i(S), x_i(T)) - a_i).$$

Next set $c = c_i$. We see that (x_i, u_i) is a strong local minimizer for

$$\begin{cases} \text{Minimize } g(x(S), x(T)) + M_i |(x(S), x(T)) - (x_i(S), x_i(T))|^2 \\ \qquad\qquad\qquad\qquad\qquad - \xi_i \cdot (x(S), x(T)) \\ \text{over } x \in W^{1,1} \text{ and measurable functions } u \text{ satisfying} \\ \dot{x}(t) = (1 - \epsilon)f(t, x(t), \bar{u}(t)) + \epsilon f(t, x(t), u(t)) \quad \text{a.e.,} \\ u(t) \in U(t) \quad \text{a.e.} \end{cases}$$

This is an "endpoint constraint-free" optimal control problem to which the necessary conditions of the special case of the Maximum Principle (Proposition 6.4.2), already proved, are applicable. We deduce the existence of an adjoint arc $p_i \in W^{1,1}$ such that

$$-\dot{p}_i(t) =$$
$$p_i(t)((1 - \epsilon)f_x(t, x_i(t), \bar{u}(t)) + \epsilon f_x(t, x_i(t), u_i(t))), \qquad (6.15)$$
$$p_i(t) \cdot \dot{x}_i(t) \geq p_i(t) \cdot ((1 - \epsilon)f(t, x_i(t), \bar{u}(t))$$
$$+ \epsilon f(t, x_i(t), u)) \text{ for all } u \in U(t), \qquad (6.16)$$
$$(p_i(S), -p_i(T)) \, (= \lambda_i \nabla g(x_i(S), x_i(T)) - \lambda_i \xi_i)$$
$$\in \lambda_i \nabla g(x_i(S), x_i(T)) + N_C((x_i(S), x_i(T)) - a_i), \quad (6.17)$$

in which $\lambda_i = 1$. Now scale p_i and λ_i (we do not relabel) so that

$$|p_i(S)| + \lambda_i = 1. \qquad (6.18)$$

Recall that

$$x_i \to \bar{x} \text{ uniformly.}$$

Since $\{p_i(S)\}$ is a bounded sequence, we deduce from (6.15) that the p_is are uniformly bounded and $\dot{p}_i$s are uniformly integrably bounded. Along a subsequence then $p_i \to p$ uniformly for some $p \in W^{1,1}$. Since $\{\lambda_i\}$ is a bounded sequence, we may arrange by yet another subsequence extraction that $\lambda_i \to \lambda$. We deduce from (6.15) with the help of the Compactness of Trajectories Theorem (Theorem 2.5.3) that p satisfies

$$-\dot{p}(t) \in p(t)f_x(t, \bar{x}(t), \bar{u}(t)) + 2\epsilon |p(t)| k_f(t)B. \qquad (6.19)$$

From (6.16) we see that, for arbitrary $u \in \mathcal{U}$,

$$\int_S^T p_i(t) \cdot \dot{x}_i(t)dt \geq$$
$$\int_S^T p_i(t) \cdot ((1 - \epsilon)f(t, x_i(t), \bar{u}(t)) + \epsilon f(t, x_i(t), u(t)))dt.$$

Now $\dot{x}_i \to \dot{\bar{x}}$ weakly in L^1 and $p_i \to p$ and $x_i \to \bar{x}$ uniformly. Passing to the limit (with the help of the Dominated Convergence Theorem), noting that $\dot{\bar{x}} = f(t, \bar{x}(t), \bar{u}(t))$ and dividing across the resulting inequality by ϵ yields

$$\int_S^T p(t) \cdot (f(t, \bar{x}(t), u(t)) - f(t, \bar{x}(t), \bar{u}(t))) dt \leq 0. \tag{6.20}$$

From (6.17) and the closure properties of the limiting normal cone we deduce that

$$(p(S), -p(T)) \in \lambda \nabla g(\bar{x}(S), \bar{x}(T)) + N_C(\bar{x}(S), \bar{x}(T)). \tag{6.21}$$

It follows from (6.18) that

$$|p(S)| + \lambda = 1. \tag{6.22}$$

All the assertions of the proposition have been verified except that Condition (6.20) is an "integral" form of the Weierstrass Condition and a perturbation term $2\epsilon k_f |p| B$ currently appears in the adjoint inclusion (6.19). Notice however that p and λ have been constructed for a particular $\epsilon > 0$. Take a sequence $\epsilon_j \downarrow 0$. For each j, there are elements p_j and λ_j satisfying (6.18) to (6.22) (when p_j, λ_j, and ϵ_j replace p, λ, and ϵ). A by now familiar convergence analysis yields limits p and λ satisfying (6.18) to (6.22), but with the perturbation absent. We justify replacing the "integral" Weierstrass Condition by a pointwise version of the condition by means of arguments similar to those used at the end of the proof of Proposition 6.4.2.

It remains only to lift the supplementary hypothesis (HS). Suppose (HS) is possibly not satisfied. Replace (P) by

$$\begin{cases} \text{Minimize } g(x(S), x(T)) + z(T) \\ \text{over } x, z \in W^{1,1} \text{ and measurable functions } u \text{ satisfying} \\ \dot{x}(t) = f(t, x(t), u(t)), \quad \dot{z}(t) = |x(t) - \bar{x}(t)|^2 \quad \text{a.e.,} \\ u(t) \in U(t) \quad \text{a.e.,} \\ (x(S), x(T)) \in C, z(S) = 0. \end{cases}$$

Notice that this differs from (P) by addition of a term to the cost function that "smoothly" penalizes deviations of the state trajectory from $\bar{x}$. This new problem satisfies (HS) (in addition to the other hypotheses for Proposition 6.4.3) with reference to the strong local minimizer $(\bar{x}, \bar{z} \equiv 0, \bar{u})$. We may therefore apply the necessary conditions to the new problem, which imply those for the original problem (P). $\square$

We next allow a possibly nonconvex velocity set and a general Lipschitz continuous cost function, provided that a constraint is imposed only on the left endpoint of state trajectories.

Proposition 6.4.4 *Consider the special case of (P) in which the endpoint constraint set C can be expressed*

$$C = C_0 \times R^n$$

for some closed set $C_0 \subset R^n$. Let $(\bar{x}, \bar{u})$ be a strong local minimizer. Then the assertions of Theorem 6.4.1 are valid with $\lambda = 1$ when, in addition to (H1) through (H3) and (S1), we impose merely the hypotheses (S2) and (S3).

Proof. By adjusting $\delta > 0$ we can arrange that $(\bar{x}, \bar{u})$ is minimizing with respect to feasible processes (x, u) for the special case of (P) satisfying $\|x - \bar{x}\|_{L^\infty} \leq \delta$. Define

$$Q := \{x \in W^{1,1} : x(S) \in C_0, \ \dot{x}(t) \in f(t, x(t), U(t)) \text{ a.e } \}.$$

From the Generalized Filippov Selection Theorem (Theorem 2.3.13) we deduce that $\bar{x}$ is a minimizer for the problem

$$(\tilde{P}) \quad \left\{ \begin{array}{c} \text{Minimize } g(x(S), x(T)) \text{ over } x \in Q \\ \text{satisfying } \|x - \bar{x}\|_{L^\infty} \leq \delta. \end{array} \right.$$

We claim that $\bar{x}$ is a minimizer also for

$$(\tilde{P}_r) \quad \left\{ \begin{array}{c} \text{Minimize } g(x(S), x(T)) \text{ over } x \in Q_r \\ \text{satisfying } \|x - \bar{x}\|_{L^\infty} \leq \delta/2, \end{array} \right.$$

where

$$Q_r := \{x \in W^{1,1} : x(S) \in C_0, \ \dot{x}(t) \in \text{co} f(t, x(t), U(t))\}.$$

Indeed, take any $\epsilon > 0$ and $x \in Q_r$ satisfying

$$\|x - \bar{x}\|_{L^\infty} \leq \delta/2.$$

Since the mapping $J : L^\infty([S, T]; R^n) \to R^n$,

$$J(x) = g(x(S), x(T)) \,,$$

is continuous, it follows the Relaxation Theorem (Theorem 2.7.3) that $x' \in Q$ can be found such that $x'(S) = x(S)$,

$$\|x' - x\|_{L^\infty} \leq \delta/2$$

and

$$g(x'(S), x'(T)) < g(x(S), x(T)) + \epsilon.$$

Since x' is a point in Q satisfying $\|x' - \bar{x}\| \leq \delta$, we have

$$g(x'(S), x'(T)) \geq g(\bar{x}(S), \bar{x}(T)).$$

It follows that

$$g(\bar{x}(S), \bar{x}(T)) \leq g(x(S), x(T)) + \epsilon.$$

But $\bar{x}$ satisfies the constraints of $(\tilde{P}_r)$. Since x and ϵ were chosen arbitrarily, we deduce that $\bar{x}$ is a minimizer for $(\tilde{P}_r)$ as claimed.

Making use of the Generalized Filippov Selection Theorem (Theorem 2.3.13) and Carathéodory's Theorem, we deduce that

$$\{\bar{x}, \bar{y} \equiv g(\bar{x}(S), \bar{x}(T)), (\bar{u}_0, \ldots, \bar{u}_n) =$$
$$(\bar{u}, \ldots, \bar{u}), (\lambda_0, \lambda_1, \ldots, \lambda_n) = (1, 0, \ldots, 0)\}$$

is a strong local minimizer for

$$\left\{ \begin{array}{l} \text{Minimize } y(T) \text{ over } x \in W^{1,1}, y \in W^{1,1} \\ \qquad \text{and measurable functions } u_0, \ldots, u_n, \lambda_0, \ldots, \lambda_n \text{ satisfying} \\ \dot{x}(t) = \sum_i \lambda_i(t) f(t, x(t), u_i(t)), \; \dot{y}(t) = 0 \quad \text{a.e.,} \\ (\lambda_0(t), \ldots, \lambda_n(t)) \in \Lambda, u_i(t) \in U(t), i = 0, \ldots, n \quad \text{a.e.,} \\ (x(S), x(T), y(S)) \in \text{ epi } \{g + \Psi_{C_0 \times R^n}\}. \end{array} \right.$$

Here

$$\Lambda := \left\{ (\lambda'_0, \ldots, \lambda'_n) : \lambda'_i \geq 0 \text{ for } i = 0, \ldots, n \text{ and } \sum_i \lambda'_i = 1 \right\}.$$

Ψ_A is the indicator function of the set A. $(\lambda_0, \ldots, \lambda_n)$, $(u_0, \ldots, u_n)$ are regarded as the control variables. Necessary conditions already proved (Proposition 6.4.3) cover this problem since, in particular, the set of "admissible" velocities

$$\left\{ \sum_i \lambda_i f(t, x, u_i) : u_i \in U(t), i = 0, \ldots, n, (\lambda_0, \ldots \lambda_n) \in \Lambda \right\}$$

is convex for all $x \in \bar{x}(t) + \delta B$, a.e. t.

The necessary conditions yield $p \in W^{1,1}$, $d \in R$, and $\alpha \geq 0$ which satisfy

$$(p, d, \alpha) \neq (0, 0, 0),$$
$$-\dot{p}(t) = p(t) f_x(t, \bar{x}(t), \bar{u}(t)) \quad \text{a.e.,} \tag{6.23}$$
$$(p(S), -p(T), d) \in \partial \Psi_{\text{ epi } \{g + \Psi_{C_0 \times R^n}\}} (\bar{x}(S), \bar{x}(T), \bar{y}(S)) \tag{6.24}$$
$$-d = \alpha, \tag{6.25}$$
$$u \to p(t) \cdot f(t, \bar{x}(t), u) \text{ attains its maximum}$$
$$\text{over } U(t) \text{ at } u = \bar{u}(t) \quad \text{a.e.}$$

With the help of the Sum Rule (Theorem 5.4.1), we conclude from (6.24) and (6.25) that

$$(p(S), -p(T)) \in \alpha \partial g(\bar{x}(S), \bar{x}(T)) + N_{C_0}(\bar{x}(S)) \times \{(0)\}. \tag{6.26}$$

The nondegeneracy property of the multipliers ensures that $(p, \alpha) \neq 0$. But in fact $\alpha > 0$. This is because, if $\alpha = 0$, then $p(T) = 0$ by (6.24) and (6.25). It follows from (6.23) that $p \equiv 0$, which is an impossibility. We may scale the multipliers and thereby arrange that $\alpha = 1$. Reviewing our findings, we see that all assertions of the proposition have been proved. $\square$

The next step is to allow for a general endpoint constraint.

Proposition 6.4.5 *The assertions of Theorem 6.4.1 are valid when $(\bar{x}, \bar{u})$ is assumed to be a strong local minimizer and when, in addition to (H1) through (H3) and (S1) of Theorem 6.4.1, Hypotheses (S2) and (S3) of Proposition 6.4.2 are imposed.*

Proof. Reduce $\delta > 0$, if necessary, so that $(\bar{x}, \bar{u})$ is a minimizer with respect to all (x, u)s satisfying $\|x - \bar{x}\|_{L^\infty} \leq \delta$. Consider the set

$$W := \{(x, u, e) : (x, u) \text{ satisfies } \dot{x}(t) = f(t, x(t), u(t)) \quad \text{a.e.,}$$
$$u(t) \in U(t) \quad \text{a.e., } e \in R^n, (x(S), e) \in C \text{ and } \|x - \bar{x}\|_{L^\infty} \leq \delta\}.$$

Define $d_W : W \times W \to R$,

$$d_W((x, u, e), (x', u', e')) = |x(S) - x'(S)| + |e - e'| + \text{meas}\{t : u(t) \neq u'(t)\}.$$

Take a sequence $\epsilon_i \downarrow 0$ such that $\sum_i \epsilon_i < +\infty$ and, for each i, define the function

$$g_i(x, y, x', y') := \max\{g(x, y) - g(\bar{x}(S), \bar{x}(T)) + \epsilon_i^2, |x' - y'|\}.$$

For each i, consider the optimization problem

Minimize $g_i(x(S), e, x(T), e)$ over $(x, u, e) \in W$.

We take note of the following properties of (W, d_W).

(i) d_W defines a metric on W, and (W, d_W) is a complete metric space.

(ii) If $(x_i, u_i, e_i) \to (x, u, e)$ in (W, d_W) then $\|x_i - x\|_{L^\infty} \to 0$.

(iii) The function

$$(x, u, e) \to g_i(x(S), e, x(T), e)$$

is continuous on (W, d_W).

Notice that

$$g_i(\bar{x}(S), \bar{x}(T), \bar{x}(T), \bar{x}(T)) = \epsilon_i^2.$$

Since g_i is nonnegative-valued, it follows that $(\bar{x}, \bar{u}, \bar{x}(T))$ is an "ϵ_i^2-minimizer" for the above problem. According to Ekeland's Variational Principle (Theorem 3.3.1), there exists a sequence $\{(x_i, u_i, e_i)\}$ in W such that for each i,

$$g_i(x_i(S), e_i, x_i(T), e_i) \leq g_i(x(S), e, x(T), e) + \epsilon_i d_W((x, u, e), (x_i, u_i, e_i)) \tag{6.27}$$

for all $(x, u, e) \in W$ and also

$$d_W((x_i, u_i, e_i), (\bar{x}, \bar{u}, \bar{x}(T))) \leq \epsilon_i. \tag{6.28}$$

(6.28) implies that $e_i \to \bar{x}(T)$, $x_i \to \bar{x}$ uniformly, and (since $\sum_i \epsilon_i < +\infty$)

$$J := \text{meas}\{\cup_i J_i\} = 1,$$

for the nested sequence of sets $\{J_i\}$:

$$J_i := \{t : u_j(t) = \bar{u}(t) \text{ for all } j \geq i\}.$$

Define the arc $y_i \equiv e_i$. Notice that $y_i \to \bar{x}(T)$ uniformly. Now introduce the $\mathcal{L} \times \mathcal{B}^m$ measurable function

$$m_i(t, u) := \begin{cases} 1 & \text{if } u \neq u_i(t) \\ 0 & \text{otherwise.} \end{cases}$$

This function has the property

$$\text{meas}\{t : u(t) \neq u_i(t)\} = \int m_i(t, u(t)) dt.$$

We can express the minimizing property (6.27) as follows: for sufficiently large i, $(x_i, y_i, w_i \equiv 0, u_i)$ is a strong local minimizer for the optimal control problem

$$\begin{cases} \text{Minimize } g_i(x(S), y(S), x(T), y(T)) + w(T) \\ \qquad\qquad + \epsilon_i(|x(S) - x_i(S)| + |y(S) - y_i(S)|) \\ \text{over } x, y, w \in W^{1,1} \text{ and measurable functions } u \text{ satisfying} \\ \dot{x}(t) = f(t, x(t), u(t)), \; \dot{y}(t) = 0, \; \dot{w}(t) = \epsilon_i m_i(t, u(t)) \quad \text{a.e.,} \\ u(t) \in U(t) \quad \text{a.e.,} \\ ((x(S), y(S)), w(S)) \in C \times \{0\}. \end{cases}$$

This is a "free right endpoint" problem to which Proposition 6.4.4 is applicable. We are assured then that there exist $p_i \in W^{1,1}$ and $d_i \in R^n$ such that

$$-\dot{p}_i(t) = p_i(t) f_x(t, x_i(t), u_i(t)) \quad \text{a.e.,} \tag{6.29}$$

$$(p_i(S), d_i, -p_i(T), -d_i)$$
$$\in \partial g_i(x_i(S), y_i(S), x_i(T), y_i(T))$$
$$\qquad + \epsilon_i(B \times B) \times \{0, 0\} + N_C(x_i(S), y_i(S)) \times \{0, 0\}, \tag{6.30}$$

$$p_i(t) \cdot f(t, x_i(t), u_i(t)) \geq p_i(t) \cdot f(t, x_i(t), u) - \epsilon_i$$
$$\text{for all } u \in U(t) \text{ a.e.} \tag{6.31}$$

We have denoted by d_i the adjoint arc associated with the y variable, and used the fact that d_i is constant. (Because of the special structure of this problem, the adjoint arc relating to the w variable takes the constant value $-\lambda = -1$). From condition (6.30), we deduce that $\{d_i\}$ and $\{p_i(T)\}$ are bounded sequences. By (6.29), $\{p_i\}$ is uniformly bounded and $\{\dot{p}_i\}$ is uniformly integrably bounded. Following a subsequence extraction then,

$$p_i \to p \text{ uniformly} \quad \text{and} \quad d_i \to d,$$

for some $p \in W^{1,1}$ and $d \in R^n$.

Since meas$\{t : u_i(t) \neq \bar{u}(t)\} \to 0$, we deduce from (6.29) with the help of the Compactness of Trajectories Theorem (Theorem 2.5.3) that

$$-\dot{p}(t) = p(t)f_x(t, \bar{x}(t), \bar{u}(t)) \quad \text{a.e.}$$

For any point t lying in J such that (6.31) is valid for all i (and the set of such points has full measure), we deduce from (6.31) that

$$p(t) \cdot f(t, \bar{x}(t), \bar{u}(t)) \geq p(t) \cdot f(t, \bar{x}(t), u) \quad \text{for all } u \in U(t).$$

We now analyze the Transversality Condition (6.30). The crucial observation to make at this juncture is that, for each i,

$$g_i(x_i(S), y_i(S), x_i(T), y_i(T)) > 0. \tag{6.32}$$

If not, $g_i = 0$ for some i, since the g_is are nonnegative. But then

$$y_i(T) = x_i(T), \quad (x_i(S), x_i(T)) \in C, \quad \|x_i - \bar{x}\|_{L^\infty} \leq \delta,$$

and

$$g(x_i(S), x_i(T)) \leq g(\bar{x}(S), \bar{x}(T)) - \epsilon_i^2,$$

in violation of the optimality of $(\bar{x}, \bar{u})$.

We verify the following estimate for ∂g_i.

$$\partial g_i(x_i(S), y_i(S), x_i(T), y_i(T))$$
$$\subset \{(a, b, e, -e) \in R^n \times R^n \times R^n \times R^n : (a, b) \in \tilde{\lambda}\partial g(x_i(S),$$
$$y_i(S)), \ \tilde{\lambda} \geq 0, \ \text{and} \ \tilde{\lambda} + |e| = 1\}. \tag{6.33}$$

There are two cases to consider.

(a): $x_i(T) = y_i(T)$. In this case (by (6.32))

$$g_i(x, y, x', y') = g(x, y) + \epsilon_i^2 - g(x_i(S), y_i(S))$$

for all (x, y, x', y') near $(x_i(S), y_i(S), x_i(T), y_i(T))$. Consequently (6.33) is true with $e = 0$.

(b): $x_i(T) \neq y_i(T)$. In this case, from the Chain Rule

$$\partial |x - y| \big|_{(x,y)=(x_i(T), y_i(T))} \subset \{(e, -e) : |e| = 1\}.$$

(6.33) now follows from the Max Rule (Theorem 5.5.2). The verification is complete.

From (6.30) and (6.33), we deduce the existence of $\tilde{\lambda}_i \in [0,1]$ such that,

$$(p_i(S), d_i) \in \tilde{\lambda}_i \partial g(x_i(S), y_i(S)) + \epsilon_i(B \times B) + N_C(x_i(S), y_i(S))$$

$$|d_i| + \tilde{\lambda}_i = 1 \quad \text{and} \quad d_i = -p_i(T).$$

By extracting a further subsequence, we can arrange that $\tilde{\lambda}_i \to \lambda$ for some $\lambda \geq 0$. We obtain in the limit as $i \to \infty$ the relationships

$$(p(S), -p(T)) \in \lambda \partial g(\bar{x}(S), \bar{x}(T)) + N_C(\bar{x}(S), \bar{y}(S)) \qquad (6.34)$$

and

$$|p(T)| + \lambda = 1.$$

We have verified the remaining assertions of Theorem 6.4.1, namely, the transversality and the "nontriviality of the multipliers" conditions. □

Proof of the Smooth Maximum Principle (Theorem 6.4.1) Existence of (λ, p) satisfying conditions (i) to (iv) of Theorem 6.4.2 have been confirmed under the supplementary hypotheses that $(\bar{x}, \bar{u})$ is in fact a strong local minimizer and Conditions (S2) and (S3) on the data are satisfied. But we have shown already (Lemma 6.3.1) that these assertions remain valid when the supplementary hypotheses are dropped. The proof is therefore complete. □

6.5   Notes for Chapter 6

The Maximum Principle came to prominence through the book [114], co-authored by Pontryagin, Boltyanskii, Gamkrelidze, and Mischenko, published in Russian in 1961, and in English translation in 1962. It is also referred to as the Pontryagin Maximum Principle because of Pontryagin's role as leader of the research group at the Steklov Institute, Moscow, which achieved this advance. However the first proof is attributed to Boltyanskii [22]. There is a voluminous Russian literature on the original Maximum Principle and extensions, which we make no attempt to summarize here. We refer to Milyutin and Osmolovskii's recent book [102] for contributions of Dubovitskii and Milyutin and their collaborators.

Prominent among researchers in the West who first entered this field were Neustadt, Warga, and Halkin. (See [111] and [147] for monographs covering early contributions of Neustadt and Warga, and [79] for a widely read paper of Halkin. See also [110].) These authors had their distinctive points of view — Neustadt's and Halkin's approach to deriving necessary conditions were close in spirit to that of Dubovitskii–Milyutin and Gamkrelidze, while Warga worked more in a Western tradition of variational analysis associated with L. C. Young and McShane. They all aimed, however, to axiomatize the proof of Pontryagin et al. [114]. The key idea is to show that if a (suitably chosen) convex approximation to some generalized "target set" intersects the interior of a convex approximation to the reachable set at some control u^*, then the reachable set itself intersects the target set. This furnishes a contrapositive proof of the Maximum Principle because the first property is equivalent to "the Maximum Principle conditions are not satisfied at u^*," while the second property implies "u^* is not a minimizer." We regard this as a "dual" approach – the assertion "the Maximum Principle conditions are not satisfied" is essentially a statement about the nonexistence of a hyperplane separating the convex approximations to the target and reachable sets. It is "axiomatic" to the extent that attention focuses on abstract conditions on convex approximations, consistent with these relationships. The elegant mixed Lagrange multiplier rules of Ioffe and Tihomirov [88], yielding necessary conditions in Optimal Control as direct corollaries, were also grounded in this idea.

The challenge of deriving versions of the Maximum Principle covering problems with nonsmooth data was taken up in the 1970s. The techniques used by Clarke to prove his nonsmooth Maximum Principle [32] had a decidedly "primal" flavor, and were a marked departure from the methodologies of Neustadt, Halkin, and Warga. Clarke's proof built on the fact that simple variational arguments can be used to derive necessary conditions of optimality for problems with no right endpoint constraints. He dealt with the troublesome right endpoint constraint by finding a neighboring process that is a minimizer for a perturbed optimization problem with free right endpoint. Necessary conditions for the perturbed, free right endpoint problem are invoked. Necessary conditions are then obtained for the original problem by passage to the limit. This general approach, deriving necessary conditions for a "difficult" optimal control problem via necessary conditions for a simpler perturbed problem and passing to the limit, is a very powerful one and adapts to other contexts; it has been used, for example, to obtain necessary conditions for optimal control problems involving differential inclusion constraints, as we show in Chapter 7, or with impulsive control [141], [128].

On the other hand, Warga extended the earlier dual approach to prove, independently, another kind of nonsmooth Maximum Principle for optimal control problems with data Lipschitz continuous with respect to the state variable x [148], [150]. The role of gradients with respect to x in the adjoint equation and transversality condition is here taken by another kind of "generalized" derivative, namely, a *derivate container*. Because the Optimal Control literature makes reference to this construct, we note here the definition.

Definition *Take a neighborhood $\mathcal{O}$ of a point $\bar{x} \in R^n$ and a Lipschitz continuous mapping $\Lambda : \mathcal{O} \to R^m$. A compact set $\mathcal{L}$ of linear maps from R^n to R^m is said to be a* derivate container *of Λ at $\bar{x}$ if, for every $\epsilon > 0$, there exists a neighborhood $\mathcal{O}_\epsilon$ of $\bar{x}$ and a sequence $\{\Lambda_j : \mathcal{O}_\epsilon \to R^m\}$ of C^1 maps such that*

(i): $\nabla \Lambda_j(x) \in \mathcal{L} + \epsilon B$ for all j and $x \in \mathcal{O}_\epsilon$;

(ii): $\Lambda_j(x) \to \Lambda$ as $j \to \infty$, uniformly over $x \in \mathcal{O}_\epsilon$.

The choice of derivate container can be tailored to a particular application. In some cases, this choice can be exercised to give a different, and more precise, transversality condition than that derived in this chapter. However for technical reasons to do with the fact that the convex hull of a derivate container to a Lipschitz continuous function contains no more information than the generalized Jacobian [149], the adjoint inclusion is essentially the same. The dual approach also yielded a nonsmooth Maximum Principle [83], via a nonsmooth generalization of Ioffe and Tihomirov's earlier mixed multiplier rule.

Recently, Sussmann has revisited the dual approach to proving Maximum Principles. (A representative paper is [132]). Let $(\bar{x}, \bar{u})$ be an optimal process. Scrutiny of the Maximum Principle conditions reveals that the adjoint equation makes sense when we assume that the right side of the differential equation modeling the dynamics is "differentiable" with respect to the state in some sense *merely along the optimal control function.* [132] points to a derivation of the Maximum Principle incorporating what Sussmann refers to as the *Lojasiewicz refinement*, namely, a version of the Maximum Principle in which this weaker differentiability hypothesis replaces the usual ones involving *all* control functions. Another departure from standard versions of the Maximum Principle is to require the data to be differentiable merely *at* the optimal state trajectory instead of near the optimal state trajectory; this is prefigured in earlier multiplier rules of Halkin [80] and Ioffe [85] who merely invoked differentiability of the data at (rather than near) the minimizer under consideration. For extensions to nondifferentiable data, Sussmann uses *semidifferentials*, akin to Warga's derivate containers, but which involve an extra layer of approximation. Primal methods yield a simple alternative proof of the Lojasiewicz refinement for free right endpoint problems (and refinements), though not for general fixed endpoint problems.

According to [132], the price of achieving the Lojasiewicz refinement for fixed endpoint optimal control problems is a hypothesis requiring the solutions of the state equation to be unique for an arbitrary initial condition and control. The Lojasiewicz refinement tells us then that the Maximum Principle remains valid when a Lipschitz continuity hypothesis, resembling standard sufficient conditions for uniqueness of solutions to the state equation (for an arbitrary control and initial state), is replaced by the hypothesis of uniqueness of solutions to the state equation itself. This suggests that the uniqueness condition is in some sense more fundamental to the derivation of necessary conditions in Optimal Control than Lipschitz continuity. On the other hand, examples of optimal control problems can be constructed whose data satisfy the Lipschitz continuity hypotheses for derivation of Clarke's nonsmooth Maximum Principle (these resemble, but are weaker than, standard sufficient conditions for uniqueness), yet for which the uniqueness hypothesis is violated. The question of what hypothesis is the more "fundamental" does not therefore have a simple answer.

In this chapter, essentially three different versions of the Maximum Principle are proved, namely, for necessary conditions for problems with smooth data and free right endpoints, for problems with smooth data and general endpoint constraints, and, finally, for nonsmooth problems with general endpoint constraints. The proof of the first is based on a simple version of the needle variation argument originally devised to prove the Weierstrass Condition and elaborated upon in [114]. (Only one "needle" is required because the right endpoint is free). The treatment of endpoint constraints via proximal analysis of suitable value functions follows [42]. That the nonsmooth Maximum Principle is a simple corollary of the Generalized Euler–Lagrange Condition was shown by Ioffe [86]. The approach is similar to that earlier used by Clarke [38], where the Hamiltonian Inclusion is used as a stepping stone to derive a nonsmooth Maximum Principle. The hypothesis reduction techniques in the proof of the nonsmooth Maximum Principle are those earlier employed by Clarke [38].

Chapter 7

The Extended Euler–Lagrange and Hamilton Conditions

Read Euler, read Euler, he is the master of us all.

–Pierre Simon Laplace

7.1 Introduction

The distinguishing feature of optimal control problems, as compared with traditional variational problems, is the presence of constraints on the velocity variable. We have seen in the previous chapter that necessary conditions in the form of a Maximum Principle can be derived when these constraints are formulated in terms of a differential equation governing the evolution of the state variable, parameterized by control functions:

$$\dot{x}(t) = f(t, x(t), u(t)) \quad \text{a.e.,} \tag{7.1}$$

$$u(t) \in U(t) \quad \text{a.e.} \tag{7.2}$$

Another approach to the derivation of necessary conditions in optimal control is to focus attention on the constraints on the velocity variable implicit in the underlying dynamics of the problem. The choice variables are now taken to be arcs satisfying a differential inclusion

$$\dot{x}(t) \in F(t, x(t)) \quad \text{a.e.}$$

This is a broader framework for studying optimal control problems since constraints described by a parameterized differential equation (7.1) and (7.2) can be reformulated as a differential inclusion constraint by choosing

$$F(t, x) := \{f(t, x, u) : u \in U(t)\}. \tag{7.3}$$

Our goal in this chapter is to derive general necessary conditions for an optimal control problem in which the dynamic constraint takes the form of a differential inclusion:

$$(P) \quad \begin{cases} \text{Minimize } g(x(S), x(T)) \\ \text{over arcs } x \in W^{1,1}([S, T]; R^n) \text{ satisfying} \\ \dot{x}(t) \in F(t, x(t)) \quad \text{a.e.,} \\ (x(S), x(T)) \in C. \end{cases}$$

Here, $[S, T]$ is a given interval, $g : R^n \times R^n \to R$ is a given function, $F : [S, T] \times R^n \rightsquigarrow R^n$ is a given multifunction, and $C \subset R^n \times R^n$ is a given set.

Problem (P) can be regarded as a generalization of traditional variational problems (in one independent variable) to allow for nonsmooth data. This is because (P) can be expressed as

$$(P') \quad \begin{cases} \text{Minimize } g(x(S), x(T)) + \int_S^T L(s, x(s), \dot{x}(s))ds \\ \text{over arcs } x \in W^{1,1}([S, T]; R^n) \text{ satisfying} \\ (x(S), x(T)) \in C, \end{cases}$$

when we choose the cost integrand L to be

$$L(t, x, v) := \Psi_{\mathrm{Gr}F(t,.)}(x, v).$$

(Ψ_G denotes, as usual, the indicator function of the set G.)

This is an equivalent formulation, insofar as the values of the cost of the new problem and of the one it replaces are the same for an arbitrary arc $x \in W^{1,1}$ satisfying

$$\dot{x}(t) \in F(t, x(t)) \quad \text{a.e.}$$

If, on the other hand, an arc x is not feasible for the original problem, i.e.,

$$\dot{x}(t) \notin F(t, x(t))$$

for all t in a subset of positive measure, then the value of the cost for the new problem is infinite and so the arc x is effectively excluded from consideration as a minimizer.

From this perspective, the Maximum Principle, which concerns an inherently nonsmooth problem, was a remarkable achievement. Pontryagin et al. circumvented the need to examine generalized derivatives of the nonsmooth Lagrangian or Hamiltonian, a general theory for which did not exist at the time, by structuring their necessary conditions around the smooth "unmaximized" Hamiltonian function.

By contrast, the necessary conditions now available for optimal control problems with dynamics described by a differential inclusion are closely linked to two decades of extensive research into Nonsmooth Analysis, much of it directed at applications in Optimal Control. Both the statement of these necessary conditions and their proofs confront the nonsmoothness inherent in optimal control head on.

A revealing way of looking at necessary conditions for optimal control problems involving differential inclusions is to regard them as generalizations of conditions from the classical Calculus of Variations. If, in (P'), L were smooth and suitably regular, g smooth, and C expressible in terms of smooth, "nondegenerate" functional inequality constraints, classical variational techniques would supply the following necessary conditions for $\bar{x}$ to be a minimizer: there would exist $p \in W^{1,1}([S, T]; R^n)$ such that

$$(\dot{p}(t), p(t)) = \nabla_{x,v} L(t, \bar{x}(t), \dot{\bar{x}}(t)) \quad \text{a.e. } t \in [S, T], \tag{7.4}$$

$$p(t) \cdot \dot{\bar{x}}(t) - L(t, \bar{x}(t), \dot{\bar{x}}(t))$$
$$\geq p(t) \cdot v - L(t, \bar{x}(t), v) \quad \text{for all } v \in R^n, \text{ a.e. } t \in [S, T] \tag{7.5}$$

and

$$(p(S), -p(T)) \in \nabla g(\bar{x}(S), \bar{x}(T)) + N_C(\bar{x}(S), \bar{x}(T)). \tag{7.6}$$

Of course in the present context L, a characteristic function, is certainly not smooth. Nevertheless, under unrestrictive hypotheses on the data for problem (P), the following analogues of the above conditions can be derived: there exist $\lambda \geq 0$ and $p \in W^{1,1}$, not both zero, such that

$$\dot{p}(t) \in \text{co}\{\xi : (\xi, p(t)) \in N_{\text{Gr } F(t,.)}(\bar{x}(t), \dot{\bar{x}}(t))\} \quad \text{a.e.,} \tag{7.7}$$

$$p(t) \cdot \dot{\bar{x}}(t) \geq p(t) \cdot v \quad \text{for all } v \in F(t, \bar{x}(t)) \quad \text{a.e.,} \tag{7.8}$$

and

$$(p(S), -p(T)) \in \lambda \nabla g(\bar{x}(S), \bar{x}(T)) + N_C(\bar{x}(S), \bar{x}(T)). \tag{7.9}$$

Condition (7.7) is called the Extended Euler–Lagrange Condition because of its close affinity with the classical Euler–Lagrange Condition. This is evident from the fact that, if L is the indicator function of the closed set $\text{Gr } F(t,.)$ and $(x, v) \in \text{Gr } F(t,.)$, we have

$$\partial_{x,v} L(t, x, v) = N_{\text{Gr } F(t,.)}(x, v).$$

Condition (7.7) may be interpreted then as a "partially convexified," non-smooth version of (7.4). (7.9) differs from (7.6) only by the presence of a cost multiplier $\lambda \geq 0$ (to take account of the possibility of certain kinds of degeneracy for variational problems involving endpoint and velocity constraints). Finally (7.8) is simply another way of writing (7.5) when $L(t, ., .)$ is interpreted as the indicator function of $\text{Gr } F(t,.)$.

Necessary conditions along the above lines are the ultimate goal of this chapter (Theorems 7.4.1 and 7.5.1). As a first step towards their derivation, necessary conditions are obtained for a class of variational problems whose cost integrands are finite-valued ("Finite Lagrangian Problems" as we call them); to accomplish this, the finite Lagrangian problems are approximated by optimal control problems with smooth dynamics, to which the smooth Maximum Principle of Chapter 6 may be applied. These preliminary necessary conditions are then extended to cover problem (P), with the help of penalty techniques to take account of the dynamic constraints.

For optimal control problems arising in engineering design, the dynamic constraint usually takes the form of a differential equation parameterized by a control function (7.1) and (7.2). We may think of the necessary conditions of this chapter as "intrinsic" conditions because they depend only on the set of admissible velocities to which the parameterized differential equation gives rise via (7.3). The question therefore needs to be addressed whether the intrinsic necessary conditions have anything to offer over the Maximum Principle which is, after all, specially tailored to these kinds of dynamic constraints. While it is true that the Maximum Principle is often more convenient to apply, the intrinsic conditions have the significant advantage that they cover certain optimal control problems formulated in terms of a differential equation parameterized by control functions, in cases when the control constraint set U is *state* and time dependent, provided of course the multifunction

$$F(t, x) = \{f(t, x, u) : u \in U(t, x)\}$$

satisfies the hypotheses under which the intrinsic conditions are valid. By contrast, the Maximum Principle does not apply to problems with state-dependent control constraint sets.

Even when the control constraint set is not state dependent, the Maximum Principle and the intrinsic conditions are distinct sets of necessary conditions that we can apply to gain information about optimal controls. Examples can be constructed where the Maximum Principle excludes candidates for being a minimizer when the intrinsic conditions fail to do so, and vice versa.

The optimality conditions derived in this chapter are satisfied by arcs that are merely local minimizers for (P). "Local" is understood in the following sense.

Definition 7.1.1 *An arc $\bar{x}$ that satisfies the constraints of (P) is said to be a $W^{1,1}$ local minimizer if there exists $\delta > 0$ such that*

$$g(x(S), x(T)) \geq g(\bar{x}(S), \bar{x}(T))$$

for all arcs x which satisfy the constraints of (P) and also the condition

$$\|x - \bar{x}\|_{W^{1,1}} \leq \delta.$$

This is a variant on the concept of *strong local minimizer*, more frequently encountered in variational analysis, in which the $W^{1,1}$ norm above is replaced by the L^∞ norm.

Recall that the $W^{1,1}$ norm is stronger than the L^∞ norm. It follows that the class of $W^{1,1}$ minimizers is larger than the class of L^∞ minimizers and, therefore, all the optimality conditions derived in this chapter for $W^{1,1}$ local minimizers are also satisfied by strong local minimizers.

7.2 Properties of the Distance Function

In the derivation of necessary conditions to follow, extensive use is made of penalization techniques. The penalty function associated with the velocity constraint

$$\dot{x}(t) \in F(t, x(t)) \quad \text{a.e.}$$

will involve the parameterized distance function

$$\rho_\Gamma(x, v) := \inf\{|v - y| : y \in \Gamma(x)\},$$

in which Γ will be the multifunction $\Gamma(.) = F(t, .)$. We pause to examine properties of $\rho_\Gamma(x, v)$.

Lemma 7.2.1 *Take a multifunction* $\Gamma : R^n \rightsquigarrow R^n$ *and a point* $(\bar{x}, \bar{v}) \in \mathrm{Gr}\,\Gamma$. *Assume that* Γ *has as values nonempty closed sets. Let* $\epsilon > 0$, $r > 0$, $k > 0$, *and* $K > 0$ *be constants such that*

$$\Gamma(x') \cap (\bar{v} + rB) \subset \Gamma(x) + k|x' - x|B \tag{7.10}$$

for all $x', x \in \bar{x} + \epsilon B$ *and*

$$\rho_\Gamma(x, \bar{v}) \leq K|x - \bar{x}| \tag{7.11}$$

for all $x \in \bar{x} + \epsilon B$. *Then:*

(a) *for each* $x \in R^n$, $\rho_\Gamma(x, .)$ *is Lipschitz continuous with Lipschitz constant 1. For each* $v \in \bar{v} + (r/3)B$, $\rho_\Gamma(., v)$ *is Lipschitz continuous on* $\bar{x} + \min\{\epsilon, r/(3K)\}B$ *with Lipschitz constant* k;

(b) *fix elements* $(x, v) \in R^n \times R^n$ *and* $(w, p) \in R^n \times R^n$ *such that* $|x - \bar{x}| < \min\{\epsilon, r/(3K)\}$, $|v - \bar{v}| < r/3$, *and*

$$(w, p) \in \partial \rho_\Gamma(x, v).$$

Then

$$|w| \leq k \quad \text{and} \quad |p| \leq 1.$$

Furthermore

$$v \in \Gamma(x) \text{ implies } (w, p) \in N_{\mathrm{Gr}\,\Gamma}(x, v)$$

and

$$v \notin \Gamma(x) \text{ implies } p = (v - u)/|v - u|$$

for some $u \in \Gamma(x)$ *such that*

$$|v - u| = \min\{|v - y| : y \in \Gamma(x)\}.$$

Proof.

(a): For fixed x, $\rho_\Gamma(x,.)$ is the Euclidean distance function for the set $\Gamma(x)$ and, by the properties of distance functions, is Lipschitz continuous with Lipschitz constant 1. Choose any $x', x \in \bar{x} + \min\{\epsilon, r/(3K)\}B$. Let u' be a closest point to v in $\Gamma(x')$; i.e., $|v - u'| = \rho_\Gamma(x', v)$. Then, by (7.11), we have

$$|u' - \bar{v}| \le |v - \bar{v}| + |v - u'| \le |v - \bar{v}| + |v - \bar{v}| + \rho_\Gamma(x', \bar{v}) \le 2|v - \bar{v}| + K|x' - \bar{x}| \le r.$$

We have shown that $u' \in \bar{v} + rB$. By (7.10), there exists $u \in \Gamma(x)$ such that $|u' - u| \le k|x' - x|$. It follows that

$$\rho_\Gamma(x, v) \le |v - u| \le |v - u'| + |u' - u| \le \rho_\Gamma(x', v) + k|x' - x|.$$

Since the variables x' and x are interchangeable, we conclude that

$$|\rho_\Gamma(x, v) - \rho_\Gamma(x', v)| \le k|x - x'|.$$

This is the desired Lipschitz continuity property.

(b): We have shown that ρ_Γ is Lipschitz continuous on the closed neighborhood $\mathcal{N}$ of $(\bar{x}, \bar{v})$,

$$\mathcal{N} := (\bar{x}, \bar{v}) + (\min\{\epsilon, r/(3K)\}B) \times ((r/3)B)$$

Close scrutiny of the proof of (a) reveals also that

$$(x, v) \in \mathcal{N} \quad \text{implies} \quad |u - \bar{v}| \le r \tag{7.12}$$

for any point $u \in \Gamma(x)$ such that $|v - u| = \min\{|v - u'| : u' \in \Gamma(x)\}$.

Fix (x, v) in the interior of $\mathcal{N}$. Since (as we have shown) ρ_Γ is Lipschitz continuous on a neighborhood of (x, v), it is meaningful to talk of the limiting subdifferential $\partial \rho_\Gamma(x, v)$. Take any $(w, p) \in \partial \rho_\Gamma(x, v)$. We know that there exist $(x_i, v_i) \to (x, v)$, $(w_i, p_i) \to (w, p)$, $\epsilon_i \downarrow 0$, and a sequence of positive numbers M_i such that

$$w_i \cdot (x' - x_i) + p_i \cdot (v' - v_i) \le$$
$$\rho_\Gamma(x', v') - \rho_\Gamma(x_i, v_i) + M_i(|x' - x_i|^2 + |v' - v_i|^2) \tag{7.13}$$

for all $(x', v') \in (x_i, v_i) + \epsilon_i B$.

By part (a), $\rho_\Gamma(., v_i)$ and $\rho_\Gamma(x_i, .)$ are Lipschitz continuous (on appropriate neighborhoods of (x_i, v_i)) with Lipschitz constants k and 1, respectively, for sufficiently large i. We easily deduce from (7.13) that $|w_i| \le k$ and $|p_i| \le 1$. Passing to the limit as $i \to \infty$ gives

$$|w| \le k \quad \text{and} \quad |p| \le 1.$$

Suppose that $v \notin \Gamma(x)$. It follows from (7.10) that $v_i \notin \Gamma(x_i)$ for all i sufficiently large. Let u_i be a closest point to v_i in $\Gamma(x_i)$. In the inequality (7.13), set $x' = x_i$ and $v' = v_i + \delta(u_i - v_i)$. We see that, for $\delta > 0$ sufficiently small,

$$\delta p_i \cdot (u_i - v_i) \leq (1 - \delta)|v_i - u_i| - |v_i - u_i| + M_i \delta^2 |u_i - v_i|^2.$$

Divide across the inequality by δ and pass to the limit as $\delta \downarrow 0$. This gives

$$p_i \cdot (v_i - u_i) \geq |v_i - u_i|.$$

Since $|p_i| \leq 1$ and $v_i - u_i \neq 0$ for i sufficiently large, this relationship implies

$$p_i = (v_i - u_i)/|v_i - u_i|.$$

Now $\{u_i\}$ is a bounded sequence (see (7.12)). By restricting attention to a subsequence then we can arrange that $u_i \to u$ for some $u \in \bar{v} + rB$. Since Γ has as values closed sets, it follows from (7.10) that $u \in \Gamma(x)$. But $v \notin \Gamma(x)$. We conclude that $p_i \to p$ where

$$p := (v - u)/|v - u|. \tag{7.14}$$

Let $\tilde{u}$ be a closest point to v in $\Gamma(x)$. Then, by (7.10), there exist points $\tilde{u}_i \in \Gamma(x_i)$, $i = 1, 2, \ldots$, such that $\tilde{u}_i \to \tilde{u}$. Since u_i is a closest point to v_i in $\Gamma(x_i)$, it follows that $|u_i - v_i| \leq |v_i - \tilde{u}_i|$ for all i. Passing to the limit, we obtain

$$|v - u| \leq |v - \tilde{u}| = \min\{|v - y| : y \in \Gamma(x)\}.$$

Recalling that $u \in \Gamma(x)$, we see that u is a closest point to v in $\Gamma(x)$. The desired representation of p can now be deduced from (7.14).

Next suppose that $v \in \Gamma(x)$. For each i, again take u_i to be a closest point to v_i in $\Gamma(x_i)$ (we allow the possibility that $v_i \in \Gamma(x_i)$ for some values of i). It follows from (7.10) and (7.12) that $u_i \to v$ as $i \to \infty$. Fix i sufficiently large. Choose any (x', u') in $\operatorname{Gr} \Gamma$ close to (x_i, u_i). Now insert $(x', v' = u' + v_i - u_i)$ into (7.13). We arrive at

$$w_i \cdot (x' - x_i) + p_i \cdot (u' - u_i) \leq$$
$$\rho_\Gamma(x', u' + v_i - u_i) - \rho_\Gamma(x_i, v_i) + M_i(|x' - x_i|^2 + |u' - u_i|^2).$$

Notice however that

$$\rho_\Gamma(x', u' + v_i - u_i) \leq |u' + v_i - u_i - u'| = |v_i - u_i| = \rho_\Gamma(x_i, v_i).$$

It follows that

$$w_i \cdot (x' - x_i) + p_i \cdot (u' - u_i) \leq M_i(|x' - x_i|^2 + |u' - u_i|^2)$$

for all (x', u') in $\operatorname{Gr} \Gamma$ sufficiently close to (x_i, u_i). This implies $(w_i, p_i) \in N^P_{\operatorname{Gr} \Gamma}(x_i, u_i)$. However $\operatorname{Gr} \Gamma \cap ((x, v) + \alpha B)$ is a closed set for some $\alpha > 0$.

Recalling that $(x_i, u_i) \overset{\mathrm{Gr}\,\Gamma}{\to} (x, v)$, we deduce from the closure properties of the normal cone that $(w, p) \in N_{\mathrm{Gr}\,\Gamma}(x, v)$. This is what we set out to prove.
□

Lemma 7.2.2 *Take a multifunction* $\Gamma : R^n \rightsquigarrow R^n$ *and a point* $(\bar{x}, \bar{v}) \in \mathrm{Gr}\,\Gamma$. *Assume that* Γ *has as values closed nonempty sets and there exist* $\epsilon > 0$, $\beta > 0$, *and* $k > 0$ *such that*

$$\Gamma(x') \cap (\bar{v} + NB) \subset \Gamma(x) + (k + \beta N)|x' - x|B \qquad (7.15)$$

for all $x', x \in \bar{x} + \epsilon B$, *all* $N \geq 0$. *Then, for any* $v \in R^n$, $\rho_\Gamma(., v)$ *is Lipschitz continuous on* $\bar{x} + \epsilon B$ *with Lipschitz constant* $k(1 + \beta\epsilon) + 2\beta|v - \bar{v}|$.

Proof. Take any $x', x \in \bar{x} + \epsilon B$, and any $v \in R^n$. Let u' be a closest point to v in $\Gamma(x')$. By (7.15), there exist $w' \in \Gamma(x')$ such that $|w' - \bar{v}| \leq k|x' - \bar{x}|$. It follows from the triangle inequality that

$$|v - u'| \leq |v - w'| \leq |v - \bar{v}| + |\bar{v} - w'| \leq$$
$$|v - \bar{v}| + k|x' - \bar{x}| \leq |v - \bar{v}| + k\epsilon$$

and hence
$$|u' - \bar{v}| \leq |u' - v| + |v - \bar{v}| \leq 2|v - \bar{v}| + k\epsilon.$$

A further appeal to (7.15) now gives the existence of $u \in \Gamma(x)$ such that

$$|u - u'| \leq (k + \beta(2|v - \bar{v}| + k\epsilon))|x - x'|.$$

From the triangle inequality,

$$\rho_\Gamma(x, v) \leq |v - u| \leq |v - u'| + |u - u'| \leq$$
$$\rho_\Gamma(x', v) + (k + \beta(2|v - \bar{v}| + k\epsilon))|x' - x|.$$

Since the roles of x' and x can be reversed, we conclude

$$|\rho_\Gamma(x', v) - \rho_\Gamma(x, v)| \leq (k + \beta(2|v - \bar{v}| + k\epsilon))|x' - x|.$$

□

Lemma 7.2.3 *Take a multifunction* $\Gamma : R^n \rightsquigarrow R^n$ *and a point* $(\bar{x}, \bar{v}) \in \mathrm{Gr}\,\Gamma$. *Assume that* Γ *takes as values closed convex sets. Let* $k > 0$ *and* $\epsilon > 0$ *be constants such that*

$$\Gamma(x) \cap (\bar{v} + \epsilon B) \subset \Gamma(x') + k|x - x'|B \quad \text{for all } x, x' \in \bar{x} + \epsilon B. \qquad (7.16)$$

Define
$$R_{\bar{v},\epsilon} := \bar{v} + \epsilon/3B.$$

Take points $(x, v) \in (\bar{x} + \epsilon \min\{1, (12k)^{-1}\}B) \times R_{\bar{v}, \epsilon}$ *and* $(w, q) \in R^n \times R^n$
such that

$$(w, q) \in \partial\rho_\Gamma(x, v) + \{0\} \times (N_{R_{\bar{v}, \epsilon}}(v) + \frac{1}{4}B).$$

Then

$$|w| \leq k \text{ and } v \notin \Gamma(x) \quad \text{implies} \quad |q| \geq \sqrt{5/16}.$$

Proof. (w, q) can be expressed as

$$(w, q) = (a, b) + (0, \alpha(v - \bar{v})/|v - \bar{v}|) + (0, c) \tag{7.17}$$

(the second term on the right term is interpreted as $(0, 0)$ if $v = \bar{v}$) with

$$(a, b) \in \partial\rho_\Gamma(x, v)$$

and $|c| \leq 1/4$. α is a nonnegative number that is zero if $v \notin$ bdy $R_{\bar{v}, \epsilon}$.
We see that $w = a$. Since (7.16) implies

$$\rho_\Gamma(x, \bar{v}) \leq k|x - \bar{x}|, \quad \text{for all } x \in \bar{x} + \epsilon B,$$

we deduce from Lemma 7.2.1 that

$$|w| \leq k.$$

Now suppose that $v \notin \Gamma(x)$. It follows also from Lemma 7.2.1 that

$$b = (v - u)/|v - u|$$

where u is the closest point to v in the convex set $\Gamma(x)$. If

$$v \notin \text{ bdy } R_{\bar{v}, \epsilon}$$

then the parameter α in (7.17) takes value zero and we deduce $q = b + c$.
In this case

$$|q| \geq (|b| - |c|) \geq 3/4 \geq \sqrt{5/16}.$$

So we may assume that v is a boundary point of $R_{\bar{v}, \epsilon}$; i.e.,

$$|v - \bar{v}| = \epsilon/3.$$

Condition (7.16) implies the existence of some $v' \in \Gamma(x)$ such that

$$|\bar{v} - v'| \leq \epsilon/12.$$

The vector $v - u$ is normal to the convex set $\Gamma(x)$ at u. Since $v' \in \Gamma(x)$, we
have

$$(u - v') \cdot (v - u) \geq 0.$$

We deduce that

$$
\begin{aligned}
0 \le (u - v') \cdot (v - u) &= (u - \bar{v}) \cdot (v - u) + (\bar{v} - v') \cdot (v - u) \\
&= (v - \bar{v}) \cdot (v - u) \\
&\quad + (\bar{v} - v') \cdot (v - u) - |v - u|^2 \\
&\le (v - \bar{v}) \cdot (v - u) + \epsilon |v - u|/12 - |v - u|^2 .
\end{aligned}
$$

Since $v \notin \Gamma(x)$ and $u \in \Gamma(x)$, we have $u \ne v$ and so

$$
\frac{(v - u)}{|v - u|} \cdot (v - \bar{v}) \ge |v - u| - \epsilon/12 \ge -\epsilon/12.
$$

However, for any $\alpha' \ge 0$,

$$
\begin{aligned}
|q|^2 &= \left| \frac{(v - u)}{|v - u|} + c + \alpha'(v - \bar{v}) \right|^2 \\
&= \left| \frac{(v - u)}{|v - u|} + c \right|^2 + 2\alpha' \left(\frac{(v - u) \cdot (v - \bar{v})}{|v - u|} + c \cdot (v - \bar{v}) \right) \\
&\quad + \alpha'^2 |v - \bar{v}|^2 \\
&\ge |1 - 1/4|^2 + 2\alpha'(-\epsilon/12 - \epsilon/12) + (\alpha' \epsilon/3)^2 \\
&= (3/4)^2 + (\alpha' \epsilon/3 - 1/2)^2 - 1/4 \\
&\ge 5/16.
\end{aligned}
$$

It follows that, in this case too,

$$
|q| = \left| \frac{(v - u)}{|v - u|} + c + \alpha'(v - \bar{v}) \right| \ge \sqrt{5/16}.
$$

$\square$

7.3 Necessary Conditions for a Finite Lagrangian Problem

In this section, we derive necessary conditions for a smooth variational problem for which the cost integrand (or "Lagrangian") is finite-valued

$$
(FL) \begin{cases} \text{Minimize } J(x) := l(x(S), x(T)) + \int_S^T L(t, x(t), \dot{x}(t)) dt \\ \text{over arcs } x \in W^{1,1}. \end{cases}
$$

These necessary conditions serve primarily as a stepping stone to necessary conditions for problems with dynamic constraints. But necessary conditions for finite Lagrangian problems are of interest also because of the unrestrictive nature of the hypotheses (apart from finiteness of the cost integrand) under which they are valid.

Theorem 7.3.1 *Let $\bar{x}$ be a $W^{1,1}$ local minimizer for (FL). Assume that $J(\bar{x}) < \infty$ and that the following hypotheses are satisfied.*

(H1) l is lower semicontinuous;

(H2) $L(.,x,.)$ is $\mathcal{L} \times \mathcal{B}^n$ measurable for each $x \in R^n$ and $L(t,.,.)$ is lower semicontinuous for a.e. $t \in [S,T]$;

(H3) for every $K > 0$, there exist $\delta > 0$ and $k \in L^1$ such that

$$|L(t,x',v) - L(t,x,v)| \leq k(t)|x' - x|, \quad L(t,\bar{x}(t),v) \geq -k(t)$$

for all $x',x \in \bar{x}(t) + \delta B$ and $v \in \dot{\bar{x}}(t) + KB$.

Then there exists an arc $p \in W^{1,1}$ that satisfies

(i) $\dot{p}(t) \in co\{\eta : (\eta, p(t)) \in \partial L(t, \bar{x}(t), \dot{\bar{x}}(t))\}$ a.e.;

(ii) $(p(S), -p(T)) \in \partial l(\bar{x}(S), \bar{x}(T))$;

(iii) $p(t) \cdot \dot{\bar{x}}(t) - L(t, \bar{x}(t), \dot{\bar{x}}(t)) \geq p(t) \cdot v - L(t, \bar{x}(t), v)$ for all $v \in R^n$ a.e.

Proof. Fix $K > 0$ and let $k(.)$ and δ be the corresponding bound and constant of (H3). We may of course assume that $k(t) \geq 1$ a.e. Let $\eta \in (0, \delta)$ be a constant such that $\bar{x}$ is a minimizer with respect to all competing arcs that satisfy $\|x - \bar{x}\|_{W^{1,1}} \leq \eta$.

The first step of the proof is to find $p \in W^{1,1}$ that satisfies the conditions of the theorem statement, except that (iii) is replaced by a weaker, local version of the condition:

(iii)$'$ $p(t) \cdot \dot{\bar{x}}(t) - L(t, \bar{x}(t), \dot{\bar{x}}(t)) \geq p(t) \cdot v - L(t, \bar{x}(t), v)$
        for all $v \in \dot{\bar{x}}(t) + KB$ a.e.

We can assume without loss of generality that (H3) has been strengthened to:

(H3)$'$ $|L(t,x',v) - L(t,x,v)| \leq k(t)|x' - x|$ and $L(t, \bar{x}(t), v) \geq -k(t)$
        for all $x',x \in R^n$ and $v \in \dot{\bar{x}}(t) + KB$ a.e.

This is because, if (H3)$'$ were not satisfied, we could replace it by

$$L'(t,x,v) := \begin{cases} L(t,x,v) & \text{if } |x - \bar{x}(t)| \leq \delta \\ L(t,\bar{x}(t) + \delta\frac{x-\bar{x}(t)}{|x-\bar{x}(t)|}, v) & \text{otherwise.} \end{cases}$$

The data (L', l) satisfy (H3)$'$ (in addition to (H1) and (H2)). $\bar{x}$ remains a $W^{1,1}$ local minimizer. If some p satisfies (i), (ii), and (iii)$'$ for data (L', l)

at $\bar{x}$, it also satisfies these same conditions for data (L, l) because of their "local" nature. So we may assume that (H3)$'$ is satisfied. Define

$$\begin{aligned}
\tilde{L}(t, w, v) &:= L(t, \bar{x}(t) + w, \dot{\bar{x}}(t) + v) \\
\tilde{l}(x, y) &:= l(\bar{x}(S) + x, \bar{x}(T) + y).
\end{aligned}$$

Choose a sequence of positive numbers $\epsilon_i \to 0$. Define

$$W := \{(\xi, w, v) \in R^n \times L^1 \times L^1 : |v(t)| \le K \text{ a.e.}, \|x_{\xi,v}\|_{W^{1,1}} \le \eta, \; kw \in L^1\},$$

in which $x_{\xi,v}(t) = \xi + \int_S^t v(s)ds$, and

$$\|(\xi, w, v)\|_k := |\xi| + \|kw\|_{L^1} + \|kv\|_{L^1}.$$

For each i, set

$$\tilde{J}_i(\xi, w, v) := \tilde{l}(x_{\xi,v}(S), x_{\xi,v}(T)) + \int_S^T \tilde{L}(t, w(t), v(t))dt + \epsilon_i^{-1} \int_S^T k(t)|x_{\xi,v}(t) - w(t)|^2 dt.$$

Claim *For each i, $(W, \|.\|_k)$ is a complete metric space and $\tilde{J}_i$ is lower semicontinuous on $(W, \|.\|_k)$. There exists a sequence of nonnegative numbers $\alpha_i \to 0$ such that, for each i,*

$$\tilde{J}_i(0, 0, 0) \le \inf_W \tilde{J}_i(\xi, w, v) + \alpha_i^2.$$

We verify the claim. W is a subset of the Banach space

$$\{(\xi, w, v) \in R^n \times L^1 \times L^1 : \|(\xi, w, v)\|_k < \infty\}$$

with norm $\|.\|_k$. We show that it is strongly closed and, for each i, $\tilde{J}_i$ is lower semicontinuous on W. Take an arbitrary sequence $(\xi_j, w_j, v_j) \to (\xi, w, v)$ in $(W, \|.\|_k)$. Write $x_j := x_{\xi_j, v_j}$. Then $x_j \to x_{\xi,v}$ in $W^{1,1}$. Restricting attention to a subsequence, we have $v_j(t) \to v(t)$ a.e. So $|v(t)| \le K$ a.e. and $\|x_{\xi,v}\|_{W^{1,1}} \le \eta$. The limit point (ξ, w, v) then satisfies the conditions confirming membership of W, and so W is strongly closed. This establishes that $(W, \|.\|_k)$ is complete.

Next we show that $\tilde{J}_i$ is lower semicontinuous on W. Again take an arbitrary sequence

$$(\xi_j, w_j, v_j) \to (\xi, w, v) \text{ in } (W, \|.\|_k).$$

Along some subsequence then (we do not relabel), $w_j \to w(t)$ and $v_j(t) \to v(t)$ a.e. By Hypothesis (H3)$'$, the sequence

$$\tilde{L}(t, w_j(t), v_j(t)) + k(t)|w_j(t) - w(t)|, \; j = 1, 2, \dots$$

is bounded below by the integrable function $-k(t) - k(t)|w(t)|$. We may therefore deduce from the lower semicontinuity of $\tilde{L}$ and Fatou's lemma that

$$\liminf_{j \to \infty} \int_S^T \tilde{L}(t, w_j(t), v_j(t))dt$$

$$= \liminf_{j \to \infty} \int_S^T (\tilde{L}(t, w_j(t), v_j(t)) + k(t)|w_j(t) - w(t)|)dt$$

$$\geq \int_S^T \liminf_{j \to \infty} (\tilde{L}(t, w_j(t), v_j(t)) + k(t)|w_j(t) - w(t)|)dt$$

$$\geq \int_S^T \tilde{L}(t, w(t), v(t))dt.$$

Since $x_j(:= x_{\xi_i, v_j}) \to x_{\xi, v}$ uniformly, lower semicontinuity of $\tilde{l}$ gives

$$\liminf_{j \to \infty} \tilde{l}(x_j(S), x_j(T)) \geq \tilde{l}(x_{\xi, v}(S), x_{\xi, v}(T)).$$

It follows that

$$\liminf_{j \to \infty} \tilde{J}_i(\xi_j, w_j, v_j)$$

$$\geq \liminf_{j \to \infty} \tilde{l}(x_j(S), x_j(T)) + \liminf_{j \to \infty} \int_S^T \tilde{L}(t, w_j(t), v_j(t))dt$$

$$+ \liminf_{j \to \infty} \int_S^T \epsilon_i^{-1} k(t)|x_j(t) - w_j(t)|^2 dt$$

$$\geq \tilde{l}(x_{\xi, v}(S), x_{\xi, v}(T)) + \int_S^T \tilde{L}(t, w(t), v(t))dt$$

$$+ \int_S^T \epsilon_i^{-1} k(t)|x_{\xi, v}(t) - w(t)|^2 dt$$

$$= \tilde{J}_i(\xi, w, v).$$

Since this inequality is satisfied for some subsequence of an arbitrary convergent sequence $\{(\xi_j, w_j, v_j)\}$ we conclude that $\tilde{J}_i$ is lower semicontinuous on W.

Define

$$\alpha_i^2 := \tilde{J}_i(0, 0, 0) - \inf_W \tilde{J}_i(\xi, w, v).$$

(Since $(0, 0, 0) \in W$, the right side is nonnegative.)

Choose an arbitrary point $(\xi, w, v) \in W$. Define

$$c := \left(\int_S^T k(t)|w(t) - x_{\xi, v}(t)|^2 dt \right)^{1/2} \quad \text{and} \quad d := \left(\int_S^T k(t)dt \right)^{1/2}.$$

Then

$$\tilde{J}_i(\xi, w, v) \;=\; \tilde{J}_i(\xi, x_{\xi,v}, v) + \int_S^T (\tilde{L}(t, w(t), v(t)) - \tilde{L}(t, x_{\xi,v}(t), v(t)))dt + \epsilon_i^{-1}c^2$$

$$\geq \; \tilde{J}_i(0,0,0) - \int_S^T k(t)|x_{\xi,v}(t) - w(t)|dt + \epsilon_i^{-1}c^2$$

(by the minimizing properties of $(0,0,0)$)
$$\geq \; \tilde{J}_i(0,0,0) - cd + \epsilon_i^{-1}c^2$$

(by the Schwartz inequality)
$$= \; \tilde{J}_i(0,0,0) + \epsilon_i^{-1}(c - \epsilon_i d/2)^2 - \epsilon_i d^2/4$$

$$\geq \; \tilde{J}_i(0,0,0) - \epsilon_i d^2/4.$$

It follows that

$$0 \leq \alpha_i^2 = \tilde{J}_i(0,0,0) - \inf_W \tilde{J}_i(\xi, w, v) \leq \epsilon_i \int_S^T k(t)dt/4.$$

Since $\epsilon_i \to 0$, we conclude that $\alpha_i \to 0$ as $i \to \infty$. The claim is confirmed.

We have shown that, for each i, $(0,0,0)$ is an "α_i^2 minimizer" for $\tilde{J}_i$ over W. By Ekeland's Variational Principle (Theorem 3.3.1), there exists $(\xi_i, w_i, v_i) \in W$ that minimizes

$$J_i(\xi, w, v) := \tilde{J}_i(\xi, w, v) + \alpha_i\|(\xi, w, v) - (\xi_i, w_i, v_i)\|_k$$

over W. Also,

$$\|(\xi_i, w_i, v_i)\|_k \leq \alpha_i.$$

Write $x_i = x_{\xi_i, v_i}$. This last property implies that, for some subsequence, $(w_i, v_i) \to (0,0)$ in $L^1 \times L^1$ and a.e. and $x_i \to 0$ uniformly.

Since, for any i,

$$\tilde{J}_i(0,0,0) + \alpha_i\|(\xi_i, w_i, v_i)\|_k \geq \tilde{J}_i(\xi_i, w_i, v_i),$$

and

$$\tilde{J}_i(0,0,0) = \tilde{l}(0,0) + \int_S^T \tilde{L}(t,0,0)dt,$$

we have

$$\tilde{l}(0,0) + \int_S^T \tilde{L}(t,0,0)dt$$

$$\geq \; \limsup_{i\to\infty} \tilde{J}_i(\xi_i, w_i, v_i)$$

$$\geq \; \liminf_{i\to\infty} \tilde{l}(x_i(S), x_i(T)) + \liminf_{i\to\infty} \int_S^T \tilde{L}(t, w_i(t), v_i(t))dt.$$

But

$$\liminf_{i\to\infty} \tilde{l}(x_i(S), x_i(T)) \geq \tilde{l}(0,0)$$

and

$$\liminf_{i\to\infty} \int_S^T \tilde{L}(t, w_i(t), v_i(t))dt \geq \int_S^T \tilde{L}(t,0,0)dt.$$

It follows from these relationships that

$$\lim_{i\to\infty} \tilde{l}(x_i(S), x_i(T)) = \tilde{l}(0,0)$$

and

$$\liminf_{i\to\infty} \int_S^T (\tilde{L}(t, w_i(t), v_i(t)) - \tilde{L}(t,0,0))dt = 0.$$

But, as before, we can use Fatou's lemma to deduce that

$$\int_S^T (\liminf_{i\to\infty} \tilde{L}(t, w_i(t), v_i(t)) - \tilde{L}(t,0,0))dt$$
$$\leq \liminf_{i\to\infty} \int_S^T (\tilde{L}(t, w_i(t), v_i(t)) - \tilde{L}(t,0,0))dt = 0.$$

Since

$$\liminf_{i\to\infty} \tilde{L}(t, w_i(t), v_i(t)) \geq \tilde{L}(t,0,0) \quad \text{a.e.,}$$

we conclude that

$$\liminf_{i\to\infty} \tilde{L}(t, w_i(t), v_i(t)) = \tilde{L}(t,0,0) \quad \text{a.e.}$$

We pause to sharpen this last relationship.

Claim *We may arrange by subsequence extraction that*

$$\lim_{i\to\infty} \tilde{L}(t, w_i(t), v_i(t)) = \tilde{L}(t,0,0) \quad \text{a.e.}$$

We verify the claim. Since $kw_i \to 0$ (in L^1 and a.e.), and convergence in measure implies a.e. convergence along a subsequence, it suffices to show that $\Delta_i \to 0$ in measure as $i \to \infty$, where

$$\Delta_i(t) := \tilde{L}(t, w_i(t), v_i(t)) - \tilde{L}(t,0,0) + k(t)|w_i(t)|.$$

Note that the functions $t \to \Delta_i(t)$ are bounded below by the integrable, nonpositive function $t \to k_1(t)$

$$k_1(t) := -k(t) - \tilde{L}(t,0,0).$$

This fact is required for application of Fatou's lemma below. Since $kw_i \to 0$ in L^1 and a.e., we have from the preceding analysis that $\liminf_{i \to \infty} \Delta_i(t) \to 0$ a.e. After extracting a subsequence, we have also

$$\lim_{i \to \infty} \int_S^T (\tilde{L}(t, w_i(t), v_i(t)) - \tilde{L}(t, 0, 0))dt = 0.$$

This implies that

$$\int_S^T \Delta_i(t)dt \to 0 \quad \text{as } i \to \infty.$$

Suppose that $\Delta_i \not\to 0$ in measure. Then there exist two positive numbers ϵ and δ and a subsequence of $\{\Delta_i\}$ such that, for each i,

$$\text{meas}(\{t : |\Delta_i(t)| > \epsilon\}) > \delta, \tag{7.18}$$

where "meas" denotes Lebesgue measure on $[S, T]$. Write $A_i^\epsilon = \{t : \Delta_i(t) > \epsilon\}$ and $B_i^r = \{t : \Delta_i(t) < -r\}$. Here r is a positive number whose value is set presently. Note that

$$\liminf_{i \to \infty} \Delta_i(t)\chi_{B_i^r}(t) = 0,$$

where $\chi_{B_i^r}(t)$ equals 1 if $t \in B_i^r$ and 0 otherwise. We have that

$$\liminf_{i \to \infty}(-r \, \text{meas}(B_i^r)) \geq \liminf_{i \to \infty} \int_S^T \Delta_i(t)\chi_{B_i^r}(t)dt$$

$$\geq \int_S^T \liminf_{i \to \infty} \Delta_i(t)\chi_{B_i^r}(t)dt = 0.$$

Hence

$$\limsup_{i \to \infty} \, \text{meas}(B_i^r) = 0.$$

By (7.18) however, $\text{meas}(A_i^\epsilon) > \delta$ for i sufficiently large. Now choose $r > 0$ and an integer N such that

$$r(T - S) + \int_{B_i^r} |k_1(t)|dt < \epsilon\delta/2 \quad \text{for } i \geq N.$$

We have

$$\int_S^T \Delta_i(t)dt = \int_{A_i^\epsilon} \Delta_i(t)dt + \int_{B_i^r} \Delta_i(t)dt + \int_{\{t: -r \leq \Delta_i(t) \leq \epsilon\}} \Delta_i(t)dt$$

$$\geq \epsilon\delta + \int_{B_i^r} k_1(t)dt - r(T - S) > \epsilon\delta/2 \quad \text{for } i \geq N.$$

This contradicts $\int_S^T \Delta_i(t)dt \to 0$ as $i \to \infty$. So $\Delta_i \to 0$ in measure. The claim is confirmed.

We can summarize the above discussion in control theoretic terms. Define

$$\tilde{L}_i(t,x,w,v) := \tilde{L}(t,w,v) + \alpha_i k(t)|w - w_i(t)|$$
$$+ \alpha_i k(t)|v - v_i(t)| + \epsilon_i^{-1} k(t)|x - w|^2,$$
$$\tilde{l}_i(x,y) := \tilde{l}(x,y) + \alpha_i |x - x_i(S)|.$$

The minimizing property of (ξ_i, v_i, w_i) can be expressed as follows: $((x_i, y_i, z_i),$ $(v_i, w_i))$ is a minimizer for the optimal control problem

$$\left\{ \begin{array}{l} \text{Minimize } z(T) + \int_S^T \tilde{L}_i(t,x(t),w(t),v(t))dt \\ \text{over arcs } (x,y,z) \text{ satisfying} \\ \dot{x}(t) = v(t),\ \dot{y}(t) = |v(t)|,\ \dot{z}(t) = 0 \text{ a.e.,} \\ w(t) \in R^n, v(t) \in KB \text{ a.e.,} \\ (x(S),x(T),z(T)) \in \text{epi}\{\tilde{l}_i\},\ y(S) = 0 \text{ a.e.,}\ |x(S)| + y(T) \le \eta. \end{array} \right.$$

Here $x_i(t) := x_{\xi_i,v_i}(t)$, $y_i(t) := \int_S^t |v_i(s)|ds$, and $z_i(t) := \tilde{l}_i(x_i(S),x_i(T))$. We have shown $(v_i, w_i) \to (0,0)$ in L^1 and a.e., $x_i, y_i \to 0$ uniformly, and

$$\tilde{L}(t,w_i(t),v_i(t)) \to \tilde{L}(t,0,0) \text{ a.e.,}$$
$$\tilde{l}(x_i(S),x_i(T)) \to \tilde{l}(0,0).$$

The foregoing optimal control problem is one to which the Smooth Maximum Principle (Theorem 6.4.1), already proved, is applicable. We observe that the differential equation constraint has a right side that is independent of the state variable. Also, since $t \to k(t)|x_i(t) - w_i(t)|$ is an L^1 function, there exist a function $c : [S,T] \times R^n \times R^n \to R^+$ and $\eta' > 0$ such that

(a) the cost integrand $x \to \tilde{L}_i(t,x,v,w)$ is Lipschitz continuous on $x_i(t) + \eta'B$ with Lipschitz constant $c(t,v,w)$ for all $v \in KB$, $w \in R^n$, and a.e. t,

(b) $c(t,v_i(t),w_i(t))$ is integrable,

in accordance with the Lipschitz continuity hypotheses which must be checked for application of Theorem 6.2.1.

Notice that, because the endpoint constraint involving y is inactive at $y \equiv y_i$ and because of the decoupled structure of the cost and dynamics in y and (x,z), the adjoint arc component associated with y must be zero; we therefore drop it from the relationships. The optimality conditions tell us that there exist $p_i \in W^{1,1}$ and $\lambda_i \ge 0$ such that

(A): $-\dot{p}_i(t) = -2\lambda_i \epsilon_i^{-1} k(t)(x_i(t) - w_i(t))$ a.e.;

(B): $(p_i(S), -p_i(T), -\lambda_i) \in N_{\text{epi}\{\tilde{l}_i\}}(x_i(S),x_i(T),\tilde{l}_i(x_i(S),x_i(T)))$;

(C): $(w,v) \mapsto p_i(t){\cdot}v - \lambda_i \tilde{L}_i(t,x_i(t),w,v)$ achieves its maximum at $(w_i(t),v_i(t))$ over all $(w,v) \in R^n \times KB$, a.e.;

(D): $\|p_i\|_{L^\infty} + \lambda_i = 1.$

Condition (C) implies

$$(0, p_i(t)) \in \lambda_i \partial[\tilde{L}(t, w_i(t), v_i(t)) + \Psi_{KB}(v_i(t))]$$
$$+ \lambda_i \alpha_i k(t)(B \times B) - 2\lambda_i \epsilon_i^{-1} k(t)(x_i(t) - w_i(t)).$$

It follows then from Condition (A) that

$$(\dot{p}_i(t), p_i(t)) \quad \in \quad \lambda_i \partial[\tilde{L}(t, w_i(t), v_i(t)) + \Psi_{KB}(v_i(t))]$$
$$+ \quad \lambda_i \alpha_i k(t)(B \times B). \qquad (7.19)$$

Fix $v = v_i(t)$; then $w \to p_i(t) \cdot v_i(t) - \lambda_i \tilde{L}_i(t, x_i(t), w, v_i(t))$ achieves its maximum at $w_i(t)$ over all $w \in R^n$. This implies

$$\dot{p}_i(t) \in \lambda_i \partial_w \tilde{L}(t, w_i(t), v_i(t)) + \lambda_i \alpha_i k(t) B. \qquad (7.20)$$

Fix $w = w_i(t)$. Then $v \to p_i(t) \cdot v - \lambda_i \tilde{L}_i(t, x_i(t), w_i(t), v)$ achieves its maximum at $v_i(t)$ over $v \in KB$. This implies

$$p_i(t) \cdot (v - v_i(t)) \leq \lambda_i \tilde{L}(t, w_i(t), v) - \lambda_i \tilde{L}(t, w_i(t), v_i(t))$$
$$+ \lambda_i \alpha_i k(t)|v - v_i(t)| \text{ for all } v \in KB \quad \text{a.e.} \qquad (7.21)$$

Since $\tilde{L}(t, ., v)$ is Lipschitz continuous with rank $k(t)$ for all $v \in KB$, (7.20) implies $|\dot{p}_i(t)| \leq 2k(t)$. Noting that the p_is are uniformly bounded (see (D)), we can arrange, by limiting attention to a subsequence, that $p_i \to p$ uniformly and $\dot{p}_i \to \dot{p}$ weakly in L^1 for some $p \in W^{1,1}$. We can also ensure that $\lambda_i \to \lambda$ for some $\lambda \geq 0$ such that

$$\|p\|_{L^\infty} + \lambda = 1.$$

(7.21) implies in the limit that

$$p(t) \cdot v \leq \lambda \tilde{L}(t, 0, v) - \lambda \tilde{L}(t, 0, 0) \text{ for } v \in KB \quad \text{a.e.}$$

Observe now that $\lambda > 0$, since otherwise this last relation implies $p(t) \equiv 0$, which contradicts (D). Since $\lambda > 0$, (B) implies, for i sufficiently large,

$$(p_i(S), -p_i(T)) \in \lambda_i \partial \tilde{l}_i(x_i(S), x_i(T))$$

and we conclude that

$$(p(S), -p(T)) \in \lambda \partial \tilde{l}(0, 0).$$

We next verify the Extended Euler–Lagrange Condition (Condition (i) of the theorem statement). By Mazur's Theorem, there exists, for each i, an integer $N_i \geq i$ and a convex combination $\{\lambda_{i1}, \ldots, \lambda_{iN_i}\}$ such that, if we write

$$q_i(t) = \sum_{j=i}^{N_i} \lambda_{ij} \dot{p}_j(t),$$

then

$$q_i \to \dot{p} \quad \text{strongly in } L^1.$$

Appealing to Carathéodory's Theorem, we deduce that, for each i and t, there exists a convex combination $\{\alpha_{i0}(t), \ldots, \alpha_{in}(t)\}$ and integers $0 \leq k_{i0}(t) < \ldots < k_{in}(t)$ such that

$$q_i(t) = \sum_{j=0}^{n} \alpha_{ij}(t) \dot{p}_{i+k_{ij}(t)}(t).$$

A subsequence $\{q_i\}_{i \in S}$ can be chosen (S denotes the index values that are retained) such that

$$q_i(t) \overset{S}{\to} \dot{p}(t) \quad \text{a.e.}$$

Write $A \subset [S, T]$ for the set of full measure:

$$A := \{t : q_i(t) \overset{S}{\to} \dot{p}(t), |\dot{p}_i(t)| \leq 2k(t) \text{ and } v_i(t) \in KB \text{ for all } i$$
$$k(t) < \infty, (w_i(t), v_i(t)) \to (0,0) \text{ and } \tilde{L}(t, w_i(t), v_i(t)) \to \tilde{L}(t, 0, 0)\}.$$

Fix $t \in A$. For any $j \in \{0, \ldots, n\}$, we note that

$$\{\alpha_{ij}(t)\}_{i=1,2,\ldots} \quad \text{and} \quad \{\dot{p}_{i+k_{ij}(t)}(t)\}_{i=1,2,\ldots}$$

are bounded sequences. Consequently we may choose subsequences, with index values the set $S' \subset S$, such that, as $i \overset{S'}{\to} \infty$,

$$\alpha_{ij}(t) \to \alpha_j(t) \quad \text{and} \quad \dot{p}_{i+k_{ij}(t)}(t) \to \tilde{q}_j(t) \quad \text{for } j = 0, \ldots, n$$

for some convex combination $\{\alpha_j(t) : j = 0, \ldots, n\}$ and $\tilde{q}_j(t) \in R^n$. Since $(w_i(t), v_i(t)) \to (0,0)$, $\tilde{L}(t, w_i(t), v_i(t)) + \Psi_{KB}(v_i(t)) \to \tilde{L}(t, 0, 0)$ and $p_i(t) \to p(t)$ as $i \to \infty$, we deduce from (7.19) that

$$(\tilde{q}_j(t), p(t)) \in \lambda \partial \tilde{L}(t, 0, 0) \quad \text{for } j = 0, \ldots, n.$$

It follows that

$$\dot{p}(t) = \sum_{j=0}^{n} \alpha_j q_j(t) \in \text{co}\{\eta : (\eta, p(t)) \in \lambda \partial \tilde{L}(t, 0, 0)\} \quad \text{a.e.}$$

These are the assertions of the theorem, except that they are expressed in terms of a cost multiplier $\lambda > 0$ which is possibly not equal to 1 and that, in the last condition, the inequality holds only for $v \in \dot{\tilde{x}}(t) + KB$.

Take $K_i \to \infty$. Let p_i denote the adjoint arc and $\lambda_i > 0$ the cost multiplier when $K = K_i$. We know that the $\dot{p}_i$s are uniformly integrably bounded and the p_is are uniformly bounded. Now extract subsequences to arrange that p_i converges uniformly to some limit p, $\dot{p}_i$ converges weakly to $\dot{p}$, and $\lambda_i \to \lambda$ for some λ such that $\|p\|_{L^\infty} + \lambda = 1$. Arguing as before, we arrive at our earlier conclusions, but the Weierstrass Condition (Condition (iii) of the theorem statement) is now satisfied. From the Weierstrass Condition however, if $\lambda = 0$, then $p(t) \equiv 0$ which is not possible. So $\lambda > 0$. The final detail of the proof is to scale p and λ so that $\lambda = 1$. The theorem is proved.
□

7.4 The Extended Euler–Lagrange Condition: Nonconvex Velocity Sets

The stage is now set for deriving a general set of necessary conditions for optimal control problems, with dynamic constraint taking the form of a differential inclusion. We revert to the optimal control problem of the introduction:

$$(P) \quad \begin{cases} \text{Minimize } g(x(S), x(T)) \\ \text{over arcs } x \in W^{1,1}([S,T]; R^n) \text{ satisfying} \\ \dot{x}(t) \in F(t, x(t)) \quad \text{a.e.,} \\ (x(S), x(T)) \in C. \end{cases}$$

Here $[S,T]$ is a given interval, $g : R^n \times R^n \to R$ is a given function, $F : [S,T] \times R^n \rightsquigarrow R^n$ is a given multifunction, and $C \subset R^n \times R^n$ is a given closed set.

Theorem 7.4.1 (The Extended Euler–Lagrange Condition: Nonconvex Velocity Sets) *Let $\bar{x}$ be a $W^{1,1}$ local minimizer for (P). Suppose that the following hypotheses are satisfied.*

(G1) g is locally Lipschitz continuous;

(G2) $F(t,x)$ is nonempty for each (t,x), F is $\mathcal{L} \times \mathcal{B}^n$ measurable, $\mathrm{Gr}\, F(t,.)$ is closed for each $t \in [S,T]$, and there exist $k \in L^1$, $\epsilon > 0$, and $\beta \geq 0$, such that, for a.e. $t \in [S,T]$,

$$F(t, x') \cap (\dot{\bar{x}}(t) + NB) \subset F(t,x) + (k(t) + \beta N)|x' - x|B \quad (7.22)$$

for all $N \geq 0$ and $x, x' \in \bar{x}(t) + \epsilon B$.

Then there exist an arc $p \in W^{1,1}([S,T]; R^n)$ and $\lambda \geq 0$ such that

(i) $(\lambda, p) \neq (0,0)$;

(ii) $\dot{p}(t) \in \mathrm{co}\{\eta : (\eta, p(t)) \in N_{\mathrm{Gr}\, F(t,.)}(\bar{x}(t), \dot{\bar{x}}(t))\}$ a.e.;

(iii) $(p(S), -p(T)) \in \lambda \partial g(\bar{x}(S), \bar{x}(T)) + N_C(\bar{x}(S), \bar{x}(T))$;

(iv) $p(t) \cdot \dot{\bar{x}}(t) \geq p(t) \cdot v$ for all $v \in F(t, \bar{x}(t))$ a.e.

Now assume, also, that

$$F(t, x) \text{ does not depend on } t.$$

(In this case we write $F(x)$ in place of $F(t, x)$.) Then, in addition to the above conditions, there exists a constant r such that

(v) $p(t) \cdot \dot{\bar{x}}(t)$ ($= \max_{v \in F(\bar{x}(t))} p(t) \cdot v$) $= r$ a.e.

Remarks

This theorem is the end-product of a sustained effort over many years, to derive necessary conditions for optimal control problems where the dynamic constraint takes the form of a differential inclusion. It is noteworthy both for the unrestrictive nature of the hypotheses under which it is valid and also the precision of the adjoint inclusion (Condition (ii)).

Consider first the hypotheses. Early derivations of necessary conditions were carried out under hypotheses that typically included:

there exists ϵ and $k \in L^1$ such that

$$d_H(F(t, x), F(t, y)) \leq k(t)|x - y| \quad \text{for all } x, y \in \bar{x}(t) + \epsilon B \text{ a.e. } t. \quad (7.23)$$

Here $d_H(A, B)$ is the Hausdorff distance function

$$d_H(A, B) := \max \left\{ \sup_{a \in A} d_B(a), \sup_{b \in B} d_A(b) \right\}.$$

An equivalent statement of the condition,

$$F(t, x') \subset F(t, x) + k(t)|x' - x|B \quad \text{for all } x', x \in \bar{x}(t) + \epsilon B \text{ a.e.},$$

makes explicit the more restrictive nature of Condition (7.23).

In situations where the values of the multifunction F are unbounded sets, Condition (7.23) is usually overly restrictive. This is because, for two unbounded sets that are "close" in an intuitive sense, the Hausdorff distance between them can be very large. Consider, for example, the multifunction $F : R^2 \leadsto R^2$ defined by

$$F(x_1, x_2) := \{(v_1, v_2) : v_2 \leq x_1 v_1\}.$$

Here, values of F are hypographs of linear functions whose slopes depend smoothly on the value of $x = (x_1, x_2)$. A reasonable requirement of a set of necessary conditions for unbounded differential inclusions is that it should allow cases like this. Hypothesis (G2) passes the test. On the other hand, Hypothesis (7.23) is violated because

$$d_H(F(x'), F(x)) = \infty \text{ for } x' \neq x.$$

Consider next the nature of differential inclusion (ii) which governs the adjoint arc p. This is a refined version, in which convexification is carried out with respect to just one variable, of the condition

$$(\dot{p}(t), p(t)) \in \text{co } N_{\text{Gr } F(t,.)}(\bar{x}(t), \dot{\bar{x}}(t)), \tag{7.24}$$

in which convexification is carried out with respect to two variables. The Extended Euler–Lagrange Condition (ii) is to be preferred because it provides more precise information about the adjoint arc. But it is in fact a significant improvement on (7.24) for the following reasons. As we show in Section 7.5, Condition (ii) implies (for convex-valued Fs) an alternative necessary condition, generalizing Hamilton's system of equations. (The same cannot be said of Condition (7.24).) Furthermore, the Extended Euler–Lagrange Condition for (P) has an important role as an analytical tool for the derivation of necessary conditions for optimal control problems of a nonstandard nature (problems involving free end-times, discontinuous state trajectories, etc.). Here it is usual practice to write down the necessary conditions for a suitable auxiliary optimal control problem, which is a special case of (P) and to pass to the limit. Now the Extended Euler–Lagrange Condition (ii), applied to the auxiliary problem, yields an adjoint arc p satisfying

$$|\dot{p}(t)| \leq k(t)|p(t)|,$$

where $k \in L^1$ is as in Hypothesis (G2). This bound can be used to justify the use of weak compactness arguments to obtain an adjoint arc for the nonstandard problem when we pass to the limit. Hypothesis (7.24) yields no such bound in general, curtailing its significance as an analytical tool.

Proof. By reducing the size of ϵ if necessary, we can arrange that $\bar{x}$ is minimizing in relation to arcs x satisfying $\|x - \bar{x}\|_{W^{1,1}} \leq \epsilon$ and that g is Lipschitz continuous on $(\bar{x}(S), \bar{x}(T)) + \epsilon(B \times B)$ with Lipschitz constant k_g. Define S, the set of "admissible" arcs close to $\bar{x}$, to be

$$S := \{x \in W^{1,1} : \dot{x}(t) \in F(t, x(t)) \text{ a.e. }, (x(S), x(T)) \in C, \|x - \bar{x}\|_{W^{1,1}} \leq \epsilon\}$$

and define the functional $\tilde{g} : W^{1,1} \to R$:

$$\tilde{g}(x) := g(x(S), x(T)).$$

The functional $\tilde{g}$ is Lipschitz continuous on S, with Lipschitz constant $2k_g$, with respect to the $W^{1,1}$ norm. We know that $\bar{x}$ is a minimizer for the optimization problem

$$\begin{cases} \text{Minimize } \tilde{g}(x) \\ \text{over } x \in S \subset \{x' \in W^{1,1} : \|x' - \bar{x}\|_{W^{1,1}} \leq \epsilon\}. \end{cases}$$

It follows from the Exact Penalization Theorem (Theorem 3.2.1) that $\bar{x}$ is a minimizer for the problem:

$$\begin{cases} \text{Minimize } \tilde{g}(x) + 2k_g \inf\{||x - y||_{W^{1,1}} : y \in \mathcal{S}\} \\ \text{over arcs } x \text{ that satisfy } ||x - \bar{x}||_{W^{1,1}} \leq \epsilon. \end{cases}$$

We now identify the two possible situations that can arise:

(a): there exist $\epsilon' \in (0, \epsilon)$ and $K > 0$, such that for any $x \in W^{1,1}$ satisfying $||x - \bar{x}||_{W^{1,1}} \leq \epsilon'$, we have

$$\inf\{||x - y||_{W^{1,1}} : y \in \mathcal{S}\} \leq K\left[\int_S^T \rho_{F(t,.)}(x(t), \dot{x}(t))dt + d_C(x(S), x(T))\right];$$

(b): there exists a $W^{1,1}$-convergent sequence of arcs $\bar{x}_i \to \bar{x}$ such that

$$J(\bar{x}_i) < (2i)^{-1} \inf\{||\bar{x}_i - y||_{W^{1,1}} : y \in \mathcal{S}\} \tag{7.25}$$

for $i = 1, 2, \ldots$, where

$$J(x) := \int_S^T \rho_{F(t,.)}(x(t), \dot{x}(t))dt + d_C(x(S), x(T)).$$

Here,

$$\rho_{F(t,.)}(x, v) := \inf_{e \in F(t,x)} |v - e|.$$

We deal first with the more straightforward case (a). Here $\bar{x}$ is a $W^{1,1}$ local minimizer of the minimization problem

$$\begin{cases} \text{Minimize } g(x(S), x(T)) + 2k_g K J(x) \\ \text{over } x \in W^{1,1}. \end{cases}$$

This is an example of the Finite Lagrangian Problem of Section 7.3 with

$$\begin{aligned} L(t, x, v) &:= 2k_g K \rho_{F(t,.)}(x, v) \\ l(x, y) &:= g(x, y) + 2k_g K d_C(x, y). \end{aligned}$$

The hypotheses are satisfied under which Theorem 7.3.1 can be applied. This yields $p \in W^{1,1}$ such that

$$\begin{aligned} \dot{p}(t) &\in \text{co}\{q : (q, p(t)) \in 2k_g K \partial \rho_{F(t,.)}(\bar{x}(t), \dot{\bar{x}}(t))\} \text{ a.e.;} \\ (p(S), -p(T)) &\in \partial g(\bar{x}(S), \bar{x}(T)) + 2k_g \partial d_C(\bar{x}(S), \bar{x}(T)); \\ p(t) \cdot \dot{\bar{x}}(t) &\geq p(t) \cdot v - 2k_g K \rho_{F(t,.)}(\bar{x}(t), v) \text{ for all } v \in R^n \text{ a.e.} \end{aligned}$$

All the claims of the theorem now follow for the above choice of p and for $\lambda = 1$, since

$$\begin{aligned} \partial d_C(\bar{x}(S), \bar{x}(T)) &\subset N_C(\bar{x}(S), \bar{x}(T)) \\ \rho_{F(t,.)}(\bar{x}(t), v) &= 0 \qquad \text{for all } v \in F(t, \bar{x}(t)) \text{ and } t \in [S, T] \end{aligned}$$

and (by Lemma 7.2.1)

$$\partial \rho_{F(t,.)}(\bar{x}(t), \dot{\bar{x}}(t)) \quad \subset \quad N_{\mathrm{Gr}\, F(t,.)}(\bar{x}(t), \dot{\bar{x}}(t)).$$

It remains to address Case (b). Notice first of all that the functional J is continuous on

$$W := \{x \in W^{1,1} : ||x - \bar{x}||_{W^{1,1}} \le \epsilon\}$$

with respect to the strong $W^{1,1}$ topology. Indeed if $x_i \to x$ in W, then $x_i \to x$ uniformly and $\dot{x}_i \to \dot{x}$ strongly in L^1. Consequently, in view of Lemma 7.2.2,

$$
\begin{aligned}
|J(x_i) - J(x)| &\le \int_S^T |\rho_{F(t,.)}(x_i(t), \dot{x}_i(t)) - \rho_{F(t,.)}(x(t), \dot{x}(t))| dt \\
&\quad + |d_C(x_i(S), x_i(T)) - d_C(x(S), x(T))| \\
&\le ((1 + \beta\epsilon)||k||_{L^1} + 2\beta ||\dot{x}_i - \dot{\bar{x}}||_{L^1})||x_i - x||_{L^\infty} \\
&\quad + ||\dot{x}_i - \dot{x}||_{L^1} + 2||x_i - x||_{L^\infty}.
\end{aligned}
$$

The right side has limit zero as $i \to \infty$. Continuity is confirmed.

Define

$$a_i := \inf\{||\bar{x}_i - y||_{W^{1,1}} : y \in \mathcal{S}\}.$$

Since inequality (7.25) is strict and J is nonnegative-valued, $a_i > 0$ for each i. Since $\bar{x}_i \to \bar{x}$ in $W^{1,1}$ and $\bar{x} \in \mathcal{S}$, it follows that $a_i \to 0$ as $i \to \infty$. We see too that $J(\bar{x}_i) \le (2i)^{-1} a_i$, from which we conclude that $\bar{x}_i$ is a "$(2i)^{-1} a_i$-minimizer" of J on W. According to Ekeland's Variational Principle, for each i, there exists $x_i \in W$ such that x_i is a minimizer for

$$
\begin{cases}
\text{Minimize } \int_S^T \rho_{F(t,.)}(x(t), \dot{x}(t)) dt + d_C(x(S), x(T)) \\
\qquad\qquad + i^{-1}(|x(S) - x_i(S)| + \int_S^T |\dot{x}(t) - \dot{x}_i(t)| dt) \\
\text{over arcs } x \text{ that satisfy } ||x - \bar{x}||_{W^{1,1}} \le \epsilon
\end{cases}
$$

and, furthermore,

$$||x_i - \bar{x}_i||_{W^{1,1}} \le a_i/2.$$

Since $a_i := \inf\{||\bar{x}_i - y|| : y \in \mathcal{S}\}$, we have $x_i \notin \mathcal{S}$. It follows that, for each i, either

$$(x_i(S), x_i(T)) \notin C$$

or

$$\dot{x}_i(t) \notin F(t, x_i(t)) \text{ on a set of positive measure.}$$

Notice too that $x_i \to \bar{x}$ strongly in $W^{1,1}$ since $\bar{x}_i \to \bar{x}$ in $W^{1,1}$ and $||x_i - \bar{x}_i||_{W^{1,1}} \to 0$ as $i \to \infty$.

We have here another example of the Finite Lagrangian Problem of Section 7.3, now with

$$
\begin{aligned}
L(t, x, v) &:= \rho_{F(t,.)}(x, v) + i^{-1}|v - \dot{x}_i(t)| \\
l(x, y) &:= d_C(x, y) + i^{-1}|x - x_i(S)|.
\end{aligned}
$$

Note that, for i sufficiently large, $\rho_{F(t,.)}(.,.)$ is Lipschitz continuous on a neighborhood of $(x_i(t), \dot{x}_i(t))$, in view of Lemma 7.2.2.

Now apply Theorem 7.3.1. This yields an adjoint arc p satisfying Conditions (i) to (iii) of Theorem 7.3.1 when x_i replaces $\bar{x}$ and we make the above identifications of l and L. In terms of the scaled multipliers,

$$\lambda_i := (1 + \|p\|_{L^\infty})^{-1} \quad \text{and} \quad p_i := (1 + \|p\|_{L^\infty})^{-1} p,$$

these conditions can be expressed

(A) $\lambda_i + \|p_i\|_{L^\infty} = 1$;

(B) $\dot{p}_i(t) \in \text{co}\{\eta : (\eta, p_i(t)) \in \lambda_i \partial \rho_{F(t,.)}(x_i(t), \dot{x}_i(t)) + \{0\} \times (\lambda_i i^{-1} B)\}$ a.e.;

(C) $(p_i(S), -p_i(T)) \in \lambda_i \partial d_C(x_i(S), x_i(T)) + \{\lambda_i i^{-1} B\} \times \{0\}$;

(D) $p_i(t) \cdot \dot{x}_i(t) - \lambda_i \rho_{F(t,.)}(x_i(t), \dot{x}_i(t)) \geq p_i(t) \cdot v - \lambda_i \rho_{F(t,.)}(x_i(t), v)$
$$- \lambda_i i^{-1} |v - \dot{x}_i(t)|,$$

for all $v \in R^n$ a.e.

Recalling Lemma 7.2.2, we deduce from (B) that

$$|\dot{p}_i(t)| \leq \lambda_i[(1 + \beta\epsilon)k(t) + 2\beta|\dot{x}_i(t) - \dot{\bar{x}}(t)|] \quad \text{a.e.}$$

But $x_i \to \bar{x}$ in $W^{1,1}$ (which implies $\dot{x}_i \to \dot{\bar{x}}$ strongly in L^1) and $\|p_i\|_{L^\infty} \leq 1$. By restricting attention to a subsequence then (we do not relabel), we can arrange that there exists an arc p such that $p_i \to p$ uniformly and $\dot{p}_i \to \dot{p}$ weakly in L^1. By further subsequence extraction, we can arrange also that $\lambda_i \to \lambda'$ for some $\lambda' \geq 0$.

With the help of Mazur's and Carathéodory's Theorems, we deduce from (A) through (D), in the limit, that

$$\|p\|_{L^\infty} + \lambda' = 1,$$
$$\dot{p}(t) \in \text{co}\{\eta : (\eta, p(t)) \in \lambda' \partial \rho_{F(t,.)}(\bar{x}(t), \dot{\bar{x}}(t))\}$$
$$\subset \text{co}\{\eta : (\eta, p(t)) \in N_{\text{Gr } F(t,.)}(\bar{x}(t), \dot{\bar{x}}(t))\} \quad \text{a.e.},$$
$$(p(S), -p(T)) \in \lambda' \partial d_C(\bar{x}(S), \bar{x}(T)) \subset N_C(\bar{x}(S), \bar{x}(T)),$$
$$p(t) \cdot \dot{\bar{x}}(t) \geq p(t) \cdot v \text{ for all } v \in F(t, \bar{x}(t)) \quad \text{a.e. } t.$$

Suppose that $0 \leq \lambda' < 1$. The first condition gives $p \neq 0$. The assertions of the theorem now follow from the other conditions above (with $\lambda = 0$).

It remains to examine the case when $\lambda' = 1$. We show that this cannot arise and therefore need not be considered. Indeed if $\lambda' = 1$ then $p = 0$. Recall however that, for each i, either $(x_i(S), x_i(T)) \notin C$ or $\dot{x}_i(t) \notin F(t, x_i(t))$ on a set of positive measure. We conclude then from (C) and (D) that either

$$|(p_i(S), -p_i(T))| \geq \lambda_i(1 - i^{-1}) \tag{7.26}$$

or

$$|p_i(t)| \geq \lambda_i(1 - i^{-1}) \text{ on a set of positive measure.} \qquad (7.27)$$

If (7.26) is true for infinitely many is, then we obtain in the limit

$$|(p(S), -p(T))| \geq 1.$$

This is impossible since $p = 0$. On the other hand, if (7.27) is valid for infinitely many i, then once again passing to the limit, we deduce $\|p_i\|_{L^\infty} \geq 1$. This contradicts $p = 0$. Since either (7.26) or (7.27) must occur an infinite number of times, we conclude that $\lambda' < 1$.

The assertions of the theorem concerning the case when $F(t, x)$ is independent of t are proved in a broader context in Chapter 8. (See the remarks following Theorem 8.2.1.) □

The following corollary, concerning optimal control problems for which the differential inclusion is independent of x, provides an intermediate step in proving other necessary conditions.

Corollary 7.4.2 *Consider the optimal control problem*

$$\begin{cases} \text{Minimize } l(x(S), x(T)) + \int_S^T L(t, x(t), \dot{x}(t))dt \\ \text{over arcs } x \text{ that satisfy } \dot{x}(t) \in R(t) \text{ a.e.} \end{cases}$$

in which $l : R^n \times R^n \to R$ and $L : [S, T] \times R^n \times R^n \to R$ are given functions and $R : [S, T] \leadsto R^n$ is a given multifunction.

Let $\bar{x}$ be a $W^{1,1}$ local minimizer. Assume

(G1) R has a Borel measurable graph and $R(t)$ is closed for each t;

(G2) $L(.,x,v)$ is Lebesgue measurable for each (x,v), and there exists $\epsilon > 0$ and $k \in L^1$ such that

$$|L(t, x, v) - L(t, x', v')| \leq k(t)|(x, v) - (x', v')|$$

for all $(x, v), (x', v') \in (\bar{x}(t) + \epsilon B) \times (\dot{\bar{x}}(t) + \epsilon B)$;

(G3) l is Lipschitz continuous on a neighborhood of $(\bar{x}(\bar{S}), \bar{x}(\bar{T}))$.

Then there exist an arc p and a constant $\lambda \geq 0$ such that

(i)' $(p, \lambda) \neq (0, 0)$;

(ii)' $\dot{p}(t) \in co\{\eta : (\eta, p(t)) \in \lambda \partial L(t, \bar{x}(t), \dot{\bar{x}}(t)) + \{0\} \times N_{R(t)}(\dot{\bar{x}}(t))\}$ a.e.;

(iii)' $(p(S), -p(T)) \in \lambda \partial l(\bar{x}(S), \bar{x}(T))$.

Proof. Since $\bar{x}$ is a $W^{1,1}$ local minimizer for the above optimization problem, $(\bar{x}, \bar{z}(t) \equiv \int_S^t L(x, \bar{x}(s), \dot{\bar{x}}(s)) ds)$ is a $W^{1,1}$ local minimizer for the optimization problem

$$\begin{cases} \text{Minimize } z(T) + l(x(S), x(T)) \\ \text{over arcs } (x, z) \text{ satisfying} \\ (\dot{x}(t), \dot{z}(t)) \in \tilde{F}(t, x(t)) \quad \text{a.e.,} \\ z(S) = 0, \end{cases}$$

in which

$$\tilde{F}(t, x) := \{(v, \alpha) : v \in R(t) \text{ and } \alpha \geq L(t, x, v)\}.$$

The hypotheses are satisfied under which Theorem 7.4.1 can be applied. We deduce existence of an arc p, $q \in R$ and a constant $\lambda \geq 0$ such that

$$\|p\|_{L^\infty} + |q| + \lambda = 1,$$

$$\dot{p}(t) \in \text{co}\{\eta : (\eta, p(t), -q) \in N_{\text{Gr}\,\tilde{F}(t, \cdot)}(\bar{x}(t), \dot{\bar{x}}(t), L(t, \bar{x}(t), \dot{\bar{x}}(t)))\} \text{ a.e.,}$$

$$(p(S), -p(T)) \in \lambda \partial l(\bar{x}(S), \bar{x}(T)) \quad \text{and} \quad q = \lambda.$$

But

$$\begin{aligned} \text{Gr}\,\tilde{F}(t, \cdot) &= \{(x, v, \alpha) : \alpha \geq L(t, x, v) + \Psi_{R(t)}(v)\} \\ &= \text{epi}\{(x, v) \to L(t, x, v) + \Psi_{R(t)}(v)\} \quad \text{for } t \in [S, T]. \end{aligned}$$

It follows that

$$(\eta, p, -q) \in N_{\text{Gr}\,\tilde{F}(t, \cdot)}(\bar{x}(t), \dot{\bar{x}}(t), \dot{\bar{z}}(t)).$$

This implies that $q \geq 0$ and

$$\begin{aligned} (\eta, p) &\in q\partial(L(t, \bar{x}(t), \dot{\bar{x}}(t)) + \Psi_{R(t)}(\dot{\bar{x}}(t))) \\ &\subset q\partial L(t, \bar{x}(t), \dot{\bar{x}}(t)) + \{0\} \times N_{R(t)}(\dot{\bar{x}}(t)). \end{aligned}$$

To justify the last inclusion we have made use of the Sum Rule (Theorem 5.4.1), taking note of the Lipschitz continuity of $L(t, \cdot, \cdot)$ on a neighborhood of $(\bar{x}(t), \dot{\bar{x}}(t))$. The differential inclusion for p may therefore be replaced by

$$\dot{p}(t) \in \text{co}\{\eta : (\eta, p(t)) \in \lambda \partial L(t, \bar{x}(t), \dot{\bar{x}}(t)) + \{0\} \times N_{R(t)}(\dot{\bar{x}}(t))\}.$$

Since $q = \lambda$ we have $(p, \lambda) \neq 0$. The proposition is proved. $\square$

7.5   The Extended Euler–Lagrange Condition: Convex Velocity Sets

Suppose now that F is convex-valued. To what extent can we relax the hypotheses in other respects so that the assertions of the Extended Euler–Lagrange Condition remain valid? In other areas of nonlinear analysis specialization to the convex case often means that hypotheses of a global nature can be weakened to local ones, so we might expect to be able to replace the global condition (7.22) in Hypothesis (G2) of Theorem 7.4.1 by the local condition:

there exists $\epsilon > 0$ such that, for all $x, x' \in \bar{x}(t) + \epsilon B$,

$$F(t, x') \cap (\dot{\bar{x}}(t) + \epsilon B) \subset F(t, x) + k(t)|x - x'|B \quad \text{a.e.} \tag{7.28}$$

Whether this is an adequate substitute for (7.22) in the convex case remains an open question. However the following theorem tells us that a slight strengthening of (7.28) will serve. In fact the theorem provides two modified versions of (7.28), namely, conditions (a) and (b) below, with the desired consequences.

Theorem 7.5.1 (The Extended Euler–Lagrange Condition: Convex Velocity Sets) *Let $\bar{x}$ be a $W^{1,1}$ local minimizer for (P). Assume that*

$$F(t, x) \text{ is a convex set for each } (t, x) \in [S, T] \times R^n.$$

Assume also that

(G1) g is locally Lipschitz continuous.

(G2)' $F(t, x)$ is nonempty for each (t, x), F is $\mathcal{L} \times \mathcal{B}$ measurable, and Gr $F(t, .)$ is closed for each $t \in [S, T]$.

Assume, furthermore, that either of the following hypotheses is satisfied.

(a) There exists $k \in L^1$ and $\epsilon > 0$ such that for a.e. t

$$F(t, x') \cap (\dot{\bar{x}}(t) + \epsilon k(t)B) \subset$$
$$F(t, x) + k(t)|x - x'|B, \quad \text{for all } x, x' \in \bar{x}(t) + \epsilon B.$$

(b) There exists $k \in L^1$, $\bar{K} > 0$ and $\epsilon > 0$ such that the following two conditions are satisfied for a.e. t.

$$F(t, x') \cap (\dot{\bar{x}}(t) + \epsilon B) \subset F(t, x) + k(t)|x - x'|B,$$
$$\text{for all } x, x' \in \bar{x}(t) + \epsilon B \tag{7.29}$$

and

$$\inf\{|v - \dot{\bar{x}}(t)| : v \in F(t, x)\} \leq \bar{K}|x - \bar{x}(t)|$$
$$\text{for all } x \in \bar{x}(t) + \epsilon B. \tag{7.30}$$

(Notice that Condition (7.30) is superfluous if $k \in L^\infty$, for in this case it is implied by (7.29).)

Then there exist an arc p and a constant $\lambda \geq 0$ such that

(i) $(p, \lambda) \neq (0, 0)$;

(ii) $\dot{p}(t) \in co\{\eta : (\eta, p(t)) \in N_{\mathrm{Gr}\, F(t,.)}(\bar{x}(t), \dot{\bar{x}}(t))\}$ *a.e.;*

(iii) $(p(S), -p(T)) \in \lambda \partial g(\bar{x}(S), \bar{x}(T)) + N_C(\bar{x}(S), \bar{x}(T)).$

Condition (ii) implies

(iv)

$$p(t) \cdot \dot{\bar{x}}(t) \geq p(t) \cdot v \quad \text{for all } v \in F(t, \bar{x}(t)) \text{ a.e.}$$

Now assume, also, that

$$F(t, x) \text{ does not depend on } t.$$

(In this case we write $F(x)$ in place of $F(t, x)$.) Then, in addition to the above conditions, there exists a constant r such that

(v) $p(t) \cdot \dot{\bar{x}}(t) \; (= \max_{v \in F(\bar{x}(t))} p(t) \cdot v) \; = r$ *a.e.*

Proof. We deal first with Condition (iv). Suppose that p satisfies (ii), then for a.e. $t \in [S, T]$ there exists η such that

$$(\eta, p(t)) \in N_{\mathrm{Gr}\, F(t,.)}(\bar{x}(t), \dot{\bar{x}}(t)).$$

But we have assumed F is convex-valued. A straightforward analysis of proximal normals approximating $(\eta, p(t))$ permits us to deduce from this last relationship that

$$p(t) \in N_{F(t, \bar{x}(t))}(\dot{\bar{x}}(t)) \text{ a.e.},$$

where the set on the right is the normal cone to $F(t, \bar{x}(t))$ in the sense of convex analysis. This last relationship can be equivalently expressed

$$p(t) \cdot \dot{\bar{x}}(t) \geq p(t) \cdot v \quad \text{for all } v \in F(t, \bar{x}(t)) \text{ a.e.},$$

which is Condition (iv) of the theorem statement.

We may assume without loss of generality that the function k of Hypothesis (a) (or (b) satisfies $k(t) \geq 1$ a.e. Define

$$R(t) := \dot{\bar{x}}(t) + (\epsilon/3)B.$$

Reduce the size of ϵ if necessary to ensure that $\bar{x}$ is minimizing with respect to arcs x satisfying $\|x - \bar{x}\|_{W^{1,1}} \leq \epsilon$ and that g is Lipschitz continuous on $(\bar{x}(\bar{S}), \bar{x}(\bar{T})) + \epsilon(B \times B)$ with Lipschitz constant k_g. Define $\mathcal{S} \subset W^{1,1}$ and $\tilde{g} : W^{1,1} \to R$:

$$\mathcal{S} := \{x \in W^{1,1} : \dot{x}(t) \in F(x(t)), \dot{x}(t) \in R(t) \text{ a.e.}$$
$$(x(S), x(T)) \in C, \|x - \bar{x}\|_{W^{1,1}} \leq \epsilon\}$$

and
$$\tilde{g}(x) := g(x(S), x(T)).$$

Arguing as in the proof of Theorem 7.4.1 we can show that $\bar{x}$ is a minimizer for the problem

$$\left\{ \begin{array}{l} \text{Minimize } \tilde{g}(x) + 2k_g \inf\{\|x - y\|_{W^{1,1}} : y \in \mathcal{S}\} \\ \text{over arcs } x \text{ satisfying} \\ \dot{x}(t) \in R(t) \text{ a.e. and } \|x - \bar{x}\|_{W^{1,1}} \leq \epsilon. \end{array} \right.$$

There are two possibilities to be considered:

(A): There exist $\epsilon' \in (0, \epsilon)$ and $K' > 0$, such that for any $x \in W^{1,1}$ satisfying
$$\dot{x}(t) \in R(t) \quad \text{a.e.} \quad \text{and} \quad \|x - \bar{x}\|_{W^{1,1}} \leq \epsilon'$$

we have

$$\inf\{\|x - y\|_{W^{1,1}} : y \in \mathcal{S}\} \leq K'[d_C(x(S), x(T)) + \int_S^T \rho_{F(t,.)}(x(t), \dot{x}(t))dt]$$

and

(B): there exists a sequence of arcs $\bar{x}_i$ with $\dot{\bar{x}}_i(t) \in R(t)$ a.e. such that $\bar{x}_i \to \bar{x}$ in $W^{1,1}$ and

$$\int_0^1 \rho_{F(t,.)}(\bar{x}_i(t), \dot{\bar{x}}_i(t))dt + d_C(\bar{x}_i(S), \bar{x}_i(T))$$
$$< (2i)^{-1} \inf\{\|\bar{x}_i - y\|_{W^{1,1}} : y \in \mathcal{S}\}$$

for $i = 1, 2, \ldots$.

Consider first Case (A). Then $\bar{x}$ is a $W^{1,1}$ local minimizer for the problem

$$\left\{ \begin{array}{l} \text{Minimize } g(x(S), x(T)) \\ \qquad + 2k_g K'[d_C(x(S), x(T)) + \int_0^1 \rho_{F(t,.)}(x(t), \dot{x}(t))dt] \\ \text{over arcs } x \text{ satisfying } \dot{x}(t) \in R(t) \text{ a.e.} \end{array} \right.$$

Now apply Lemma 7.2.1. This tells us that $\rho_{F(t,.)}(.,.)$ is Lipschitz continuous with Lipschitz constant $2k(t)$ on $(\bar{x}(t), \dot{\bar{x}}(t)) + (\epsilon/3B) \times (\epsilon/3B)$ if Hypothesis (a) holds and on $(\bar{x}(t), \dot{\bar{x}}(t)) + (\epsilon(3\bar{K})^{-1}B) \times (\epsilon/3B)$ if Hypothesis (b) holds. (Set the parameters of Lemma 7.2.1 to be $r = \epsilon k(t)$ and $K = k(t)$ if (a) holds and $r = \epsilon$ and $K = \bar{K}$ if (b) holds, and remember that $k(t) \geq 1$.) The hypotheses are met under which Corollary 7.4.2 can be applied to this problem. In view of Lemma 7.2.1, part (b), there exist multipliers $p, \lambda \geq 0$ satisfying Conditions (i) to (iii) of the theorem. (Notice that $R(t)$ drops out of the conditions because the constraint $\dot{x}(t) \in R(t)$ is inactive.) This deals with Case (A).

Consider next case (B). Using arguments similar to those employed in the proof of Theorem 7.4.1, we show by means of Ekeland's Variational Principle that there exists $x_i \to \bar{x}$ strongly in $W^{1,1}$ such that, for each i, x_i is a $W^{1,1}$ local minimizer for

$$\begin{cases} \text{Minimize } \int_0^1 \rho_{F(t,.)}(x(t), \dot{x}(t))dt + d_C(x(S), x(T)) \\ \qquad\qquad + i^{-1}(|x(S) - x_i(T)| + \int_0^1 |\dot{x}(t) - \dot{x}_i(t)|dt) \\ \text{over arcs } x \in W^{1,1} \text{ satisfying} \\ \dot{x}(t) \in R(t) \text{ a.e.} \end{cases}$$

Furthermore, $x_i \notin S$. This last property implies that, for each t, either

$$(x_i(s), x_i(T)) \notin C \tag{7.31}$$

or

$$\dot{x}_i(t) \notin F(t, x_i(t)) \text{ for all } ts \text{ in a set of positive measure.} \tag{7.32}$$

Now apply Corollary 7.4.2 to the preceding problem, with reference to the $W^{1,1}$ local minimizer x_i. We conclude that, for each i, there exist multipliers (p_i, λ_i) such that

(A) $\lambda_i + \|p_i\| = 1$;

(B) $\dot{p}_i(t) \in \mathrm{co}\{\eta : (\eta, p(t)) \in \lambda_i \partial \rho_{F(t,.)}(x_i(t), \dot{x}_i(t))$
$\qquad\qquad + \{0\} \times (\lambda_i i^{-1} B) + \{0\} \times N_{R(t)}(\dot{x}_i(t))\}$ a.e.;

(C) $(p_i(S), -p_i(T)) \in \lambda_i \partial d_C(x_i(S), x_i(T)) + \{\lambda_i i^{-1} B\} \times \{0\}$.

We deduce from Lemma 7.2.1 and (B) that

$$|\dot{p}_i(t)| \leq k(t) \text{ a.e.} \quad \text{for i} = 1, 2, \ldots.$$

It follows that the $\dot{p}_i$s are uniformly integrably bounded and the p_is are uniformly bounded. We may therefore arrange, by subsequence extraction, that $p_i \to p$ uniformly and $\dot{p}_i' \to p'$ weakly in L^1 for some $p \in W^{1,1}$. We can ensure by again extracting a subsequence that $\lambda_i \to \lambda'$ for some $\lambda' \geq 0$. The convergence analysis in the proof of Theorem 7.4.1 may now be reproduced to justify passing to the limit in the relationships (A) to (C). We thereby obtain

$$\begin{aligned} \|p\| + \lambda' &= 1, \\ \dot{p}(t) &\in \mathrm{co}\{w : (w, p(t)) \in \lambda' \partial \rho_{F(t,.)}(\bar{x}(t), \dot{\bar{x}}(t))\} \text{ a.e.,} \\ (p(S), -p(T)) &\in \lambda' \partial d_C(\bar{x}(\bar{S}), \bar{x}(\bar{T})). \end{aligned}$$

Suppose first that $0 \leq \lambda' < 1$. In this case $p \neq 0$. Recalling that

$$\partial \rho_{F(t,.)}(\bar{x}(t), \dot{x}(t)) \subset N_{\mathrm{Gr}\, F(t,.)}(\bar{x}(t), \dot{x}(t))$$

and

$$\partial d_C(\bar{x}(S), \bar{x}(T)) \subset N_C(\bar{x}(S), \bar{x}(T))$$

we see that p satisfies Conditions (i) to (iv) of the theorem with $\lambda = 0$.

The remaining case to consider is when $\lambda' = 1$. We complete the proof by showing that this cannot arise. Since $\lambda' = 1$ we conclude $p = 0$. Now in view of Conditions (7.31) and (7.32) (one of which must hold), we can arrange by subsequence extraction that either

(i): $(x_i(S), x_i(T)) \notin C$ for $i = 1, 2, \ldots$

or

(ii): $\dot{x}_i(t) \notin F(t, x_i(t))$ for all ts in a set of positive measure, $i = 1, 2, \ldots$

Suppose first that (i) is true. We deduce from Condition (C) and the properties of the distance function that

$$|(p_i(S), -p_i(T))| \geq \lambda_i(1 - i^{-1}).$$

Passing to the limit, we obtain $|(p(S), -p(T))| \geq 1$. This contradicts $p = 0$ and $\mu = 0$.

Suppose finally (ii) is true. According to Lemma 7.2.3 we know that

$$|p_i(t)| \geq \lambda_i \sqrt{5/16},$$

on a set of positive measure. It follows that

$$\|p\|_\infty \geq \sqrt{5/16}.$$

Once again we have arrived at a contradiction of the fact that $p = 0$.

The assertions of the theorem concerning the case when $F(t, x)$ is independent of t can be deduced from Theorem 8.2.1 of Chapter 8. □

7.6   Dualization of the Extended Euler–Lagrange Condition

Take a function $L : R^n \times R^m \to R \cup \{+\infty\}$ and points $(\bar{x}, \bar{v}) \in \text{dom } L$ and $\bar{p} \in R^n$. Define $H : R^n \times R^m \to R \cup \{-\infty\} \cup \{+\infty\}$ to be the conjugate functional of $L(x, v)$ with respect to v:

$$H(x, p) := \sup_{v \in R^n} \{p \cdot v - L(x, v)\}.$$

Our goal in this section is to verify the following inclusion

$$\{q : (q, \bar{p}) \in \partial L(\bar{x}, \bar{v})\} \subset \text{co}\{q : (-q, \bar{v}) \in \partial H(\bar{x}, \bar{p})\}, \tag{7.33}$$

ate sent

om: Location #:

OCLC Symbol:

T Express Attention:

To Location #	
Hub City:	62
	SAT

under unrestrictive hypotheses on the function L. Notice that, since the set on the right is convex, (7.34) immediately implies

$$\text{co}\{q : (q,\bar{p}) \in \partial L(\bar{x},\bar{v})\} \subset \text{co}\{q : (-q,\bar{v}) \in \partial H(\bar{x},\bar{p})\}. \qquad (7.34)$$

Setting $m = n$ and, for each t, identifying $(\bar{p}, \bar{x}$ and $\bar{v})$ in (7.34) with $(p(t), \bar{x}(t)$ and $\dot{\bar{x}}(t))$, respectively, and identifying L with the indicator of $\text{Gr}\, F(t,.)$, we thereby validate a dual formulation of the Extended Euler–Lagrange Condition of Theorem 7.5.1, namely, the Extended Hamilton Condition:

$$\dot{p}(t) \in \text{co}\{-q : (q,\dot{\bar{x}}(t)) \in H(t,\bar{x}(t),p(t))\}.$$

Here, H denotes the Hamiltonian associated with $F(t,.)$.

The validity of (7.33) is proved under a variety of hypotheses. Our ultimate goal is to prove the relationship under a set of hypotheses that are automatically satisfied in the framework of Theorem 7.5.1.

To prepare the ground, it is necessary to introduce various continuity concepts for the function L, in which we focus attention on the properties of the multifunction $x \rightsquigarrow \text{epi}\, L(x,.)$.

Definition 7.6.1 *Take a function $L : R^n \times R^m \to R \cup \{+\infty\}$. We say that L is* epicontinuous *if for each $x \in R^n$ and each $x_i \to x$ we have*

$$\lim_{i\to\infty} \text{epi}\, L(x_i,.) = \text{epi}\, L(x,.).$$

The defining properties for epicontinuity can be expressed directly in terms of L. The proof of the following characterization is straightforward and is therefore omitted.

Proposition 7.6.2 *Take a function $L : R^n \times R^m \to R \cup \{+\infty\}$. Then L is* epicontinuous *if and only if*

(a) *given any point (x,v) and any sequence $(x_i,v_i) \to (x,v)$, we have*

$$L(x,v) \le \liminf_{i\to\infty} L(x_i,v_i)$$

and

(b) *given any point (x,v) and any sequence $x_i \to x$, there exists a sequence $v_i \to v$ such that*

$$L(x,v) \ge \limsup_{i\to\infty} L(x_i,v_i).$$

Definition 7.6.3 *Take a function $L : R^n \times R^m \to R \cup \{+\infty\}$ and a point $(\bar{x},\bar{v}) \in \text{dom}\, L$. We say that L is* epicontinuous near $(\bar{x},\bar{v})$ *if*

(a) *given any point (x,v) and any sequence $(x_i,v_i) \to (x,v)$, we have*

$$L(x,v) \le \liminf_{i\to\infty} L(x_i,v_i)$$

and

(b)' *there exist neighborhoods* U, V, *and* W *of* $\bar{x}$, $\bar{v}$, *and* $L(\bar{x}, \bar{v})$ *with the property: given any* $(x, v, L(x, v)) \in U \times V \times W$ *and any sequence* $x_i \to x$, *a sequence* $v_i \to v$ *can be found such that*

$$L(x, v) \geq \limsup_{i \to \infty} L(x_i, v_i).$$

Definition 7.6.4 *Given a function* $L : R^n \times R^m \to R \cup \{+\infty\}$ *and a point* $(\bar{x}, \bar{v}) \in \operatorname{dom} L$, *we say that* L *is* locally epi-Lipschitz *near* $(\bar{x}, \bar{v})$ *if there exist neighborhoods* U, V, *and* W *of* $(\bar{x}, \bar{v})$ *and* $L(\bar{x}, \bar{v})$, *respectively, and* $k > 0$ *such that*

$$\operatorname{epi} L(x', .) \cap (V \times W) \subset \operatorname{epi} L(x'', .) + k|x' - x''|B$$

for all $x', x'' \in U$.

The local epi-Lipschitz property defined here is also referred to in the literature as the Aubin continuity property or, in the case that $L(x, .)$ is the indicator function of a set, the pseudo-Lipschitz property.

In the first version of the Dualization Theorem provided here, a Lipschitz continuity hypothesis on the data is imposed via the conjugate functional $H(x, p)$.

Theorem 7.6.5 *Take a function* $L : R^n \times R^m \to R \cup \{+\infty\}$ *and points* $(\bar{x}, \bar{v}) \in \operatorname{dom} L$ *and* $\bar{p} \in R^m$. *Assume that, for some neighborhoods* U *and* P *of* $\bar{x}$ *and* $\bar{p}$, *respectively, the following hypotheses are satisfied.*

(H1): $L(x,.)$ is convex for each $x \in U$;

(H2): L is epicontinuous near $(\bar{x}, \bar{v})$;

(H3): $H(.,p)$ is Lipschitz continuous on U, uniformly with respect to all ps in P.

Then

$$\{q : (q, \bar{p}) \in \partial L(\bar{x}, \bar{v})\} \subset \operatorname{co}\{q : (-q, \bar{v}) \in \partial H(\bar{x}, \bar{p})\}.$$

Proof. In view of (H3) we can arrange, by reducing the size of U and P if required, that H is finite-valued and continuous on $U \times P$. From (H3) it can also be deduced that the set

$$\{q \in R^n : (q, v') \in \partial H(x', p')\}$$

is uniformly bounded as (x', p', v') ranges over $U \times P \times R^m$. In view of (H2), we can arrange by shrinking the neighborhood U if necessary, and choosing suitable neighborhoods V and W of $\bar{v}$ and $L(\bar{x}, \bar{v})$, respectively, that the following assertions are valid: given any $(x, v) \in U \times V$ such that $L(x, v) \in W$ and any sequence $x_i \to x$, there exists a sequence $v_i \to v$ such that

$$\limsup_i L(x_i, v_i) \le L(x, v).$$

These facts are used presently.

Take a point $\bar{q} \in R^n$ that satisfies

$$(\bar{q}, \bar{p}) \in \partial L(\bar{x}, \bar{v}).$$

We must show

$$-\bar{q} \in \mathrm{co}\{q : (q, \bar{v}) \in \partial H(\bar{x}, \bar{p})\}.$$

We note at the outset, however, that it suffices to treat only the case when $(\bar{q}, \bar{p})$ is a proximal normal:

$$(\bar{q}, \bar{p}) \in \partial^P L(\bar{x}, \bar{v}). \tag{7.35}$$

This is because if merely $(\bar{q}, \bar{p}) \in \partial L(\bar{x}, \bar{v})$, there exist sequences $(x_i, v_i) \xrightarrow{L} (\bar{x}, \bar{v})$ and $(q_i, p_i) \to (\bar{q}, \bar{p})$ such that $(q_i, p_i) \in \partial^P L(x_i, v_i)$ for each i. Applying the special case of the theorem gives

$$-q_i \in \mathrm{co}\{q : (q, v_i) \in \partial H(x_i, p_i)\}$$

for each i sufficiently large.

Since H is continuous on $U \times P$, we have $(x_i, p_i) \xrightarrow{H} (\bar{x}, \bar{p})$. We then deduce from the uniform boundedness of the sets $\{q : (q, v_i) \in \partial H(x_i, p_i)\}$ and Carathéodory's Theorem that

$$-\bar{q} \in \mathrm{co}\{q : (q, \bar{v}) \in \partial H(\bar{x}, \bar{p})\}.$$

This confirms that, without loss of generality, we can assume (7.35).

By modifying the neighborhoods U and V if required, we can arrange that V is bounded and, for some constant $\sigma > 0$,

$$M(x, v) \ge 0 \quad \text{for all } (x, v) \in \bar{U} \times \bar{V},$$

where

$$M(x, v) := L(x, v) - L(\bar{x}, \bar{v}) - \bar{q} \cdot (x - \bar{x}) - \bar{p} \cdot (v - \bar{v}) + \sigma |x - \bar{x}|^2 + \sigma |v - \bar{v}|^2. \tag{7.36}$$

Since $M(x, .)$ is lower semicontinuous and strictly convex, for each $x \in \bar{U}$ there exists a unique minimizer v_x over the compact set $\bar{V}$. We can deduce from (H3) that

$$\limsup_{x \to \bar{x}} M(x, v_x) \le 0.$$

By the lower semicontinuity of M, and since $\bar{v}$ is the unique minimizer for $M(\bar{x}, .)$ over $\bar{V}$, we have

$$\lim_{x \to \bar{x}} v_x = \bar{v}. \tag{7.37}$$

For all x sufficiently close to $\bar{x}$ then, v_x is interior to $\bar{V}$ and so, since $M(x, .)$ is convex, v_x is the unique global minimizer. For all such x,

$$0 \le M(x, v_x) = \min_{v \in R^m} M(x, v). \tag{7.38}$$

From (7.36), $\bar{v}$ minimizes

$$v \to L(\bar{x}, v) + \sigma|v - \bar{v}|^2 - \bar{p} \cdot (v - \bar{v}).$$

This implies that $\bar{p} \in \partial_v L(\bar{x}, \bar{v})$, from which we conclude that

$$L(\bar{x}, \bar{v}) = \bar{p} \cdot \bar{v} - H(\bar{x}, \bar{p}). \tag{7.39}$$

Representing $L(x, .)$ at the conjugate function of $H(x, .)$, we obtain

$$\min_{v \in R^m} M(x, v)$$
$$= \min_{v \in R^m} \left\{ \sup_{p \in R^n} [p \cdot v - H(x, p)] - L(\bar{x}, \bar{v}) - \bar{q} \cdot (x - \bar{x}) \right.$$
$$\left. - \bar{p} \cdot (v - \bar{v}) + \sigma|x - \bar{x}|^2 + \sigma|v - \bar{v}|^2 \right\}$$
$$= \min_{v \in R^m} \sup_{p \in R^m} K_x(v, p) + H(\bar{x}, \bar{p}) - \bar{q} \cdot (x - \bar{x}) + \sigma|x - \bar{x}|^2$$

by (7.39). Here

$$K_x(v, p) := (p - \bar{p}) \cdot v + \sigma|v - \bar{v}|^2 - H(x, p).$$

It follows from Proposition 3.4.7 (a version of the Mini-Max Theorem for noncompact domains) that

$$\min_v \sup_p K_x(v, p) = \min_v K_x(v, p_x)$$
$$= (p_x - \bar{p}) \cdot \bar{v} - (4\sigma)^{-1}|p_x - \bar{p}|^2 - H(x, p_x),$$

in which

$$p_x := \bar{p} - 2\sigma(v_x - \bar{v}). \tag{7.40}$$

We also have that

$$\min_v \sup_{p'} K_x(v, p') \ge \min_v K_x(v, p)$$
$$= (p - \bar{p}) \cdot \bar{v} - (4\sigma)^{-1}|p - \bar{p}|^2 - H(x, p)$$

for any p. From (7.38) we deduce that

$$0 \le M(x, v_x) = -(4\sigma)^{-1}|p_x - \bar{p}|^2 - H(x, p_x) + H(\bar{x}, \bar{p})$$
$$- \bar{q} \cdot (x - \bar{x}) + \sigma|x - \bar{x}|^2 + (p_x - \bar{p}) \cdot \bar{v} + H(\bar{x}, \bar{p})$$

and

$$0 = M(\bar{x}, \bar{v}) \geq -(4\sigma)^{-1}|p - \bar{p}|^2 - H(\bar{x}, p) + (p - \bar{p}) \cdot \bar{v} + H(\bar{x}, \bar{p})$$

for all p. It follows from these last two inequalities that

$$H(\bar{x}, p) - H(x, p_x) + (4\sigma)^{-1}(|p - \bar{p}|^2 - |p_x - \bar{p}|^2)$$
$$+ (p_x - \bar{p}) \cdot \bar{v} - \bar{q} \cdot (x - \bar{x}) + \sigma|x - \bar{x}|^2 \geq 0, \quad (7.41)$$

for all x near $\bar{x}$ and all p.

For fixed x close to $\bar{x}$, define the function $\phi_x(z)$, whose argument z is partitioned as $z = (y, p)$, to be

$$\phi_x(z) := H(y, p) - H(x, p_x) + (4\sigma)^{-1}(|p - \bar{p}|^2 - |p_x - \bar{p}|^2)$$
$$\bar{v} \cdot (p_x - p) - \bar{q} \cdot (x - y) + \sigma|x - y|^2.$$

In consequence of the definition of ϕ_x, and also by (7.41),

$$\phi_x(\bar{x}, p) \geq 0 \quad \text{for all } p \text{ and } \phi(x, p_x) = 0. \quad (7.42)$$

By (7.37) and (7.40) we know that $p_x \to \bar{p}$ as $x \to \bar{x}$. We can choose sequences $\epsilon_i \downarrow 0$ and $\delta_i \downarrow 0$ therefore that satisfy $\epsilon_i \delta_i^{-1} \to 0$ as $i \to \infty$ and

$$|x - \bar{x}| \leq \epsilon_i \text{ implies } |p_x - \bar{p}| \leq \delta_i/2.$$

Fix i. For each $x \in \bar{x} + \epsilon_i B$, $x \neq \bar{x}$, apply the Generalized Mean Value Inequality (Theorem 4.5.2) to ϕ_x with

$$z_0 = (x, p_x) \text{ and } Z = \{(\bar{x}, p) : p \in \bar{p} + \delta_i B\}.$$

This supplies $(\zeta'(x), \eta'(x)) \in \partial^P H(y'(x), p'(x))$ for some $(y'(x), p'(x)) \in R^n \times R^m$ such that

$$|y'(x) - x| \leq 2\epsilon_i, |p'(x) - \bar{p}| \leq \epsilon_i + \delta_i$$

and (in view of (7.42))

$$-\epsilon_i|x - \bar{x}| \leq -\epsilon_i|x - \bar{x}| + \inf_{z \in Z} \phi_x(z) - \phi_x(p_x)$$
$$\leq [\zeta'(x) + \bar{q} + 2\sigma(y'(x) - x)] \cdot (\bar{x} - x)$$
$$+ [\eta'(x) - \bar{v} + (2\sigma)^{-1}(p'(x) - \bar{p})] \cdot (p - p_x)$$

for all $p \in \bar{p} + \delta_i B$.

Since $|p_x - \bar{p}| < \delta_i/2$, taking the minimum over p in the final term on the right, we arrive at

$$-\epsilon_i|x - \bar{x}| \leq [\zeta'(x) + \bar{q} + 2\sigma(y'(x) - x)] \cdot (\bar{x} - x)$$
$$- (\delta_i/2)|\eta'(x) - \bar{v} + (2\sigma)^{-1}(p'(x) - \bar{p})|. \quad (7.43)$$

This last inequality is valid for all $x \in \bar{x} + \epsilon_i B$ such that $x \neq \bar{x}$.

Now, $|\zeta'(x)|$ is bounded on $\bar{x} + \epsilon_i B$ by a constant independent of i and of our choice of $y'(x), p'(x)$ (see the remarks at the beginning of the proof). It follows that there exists K, independent of i, such that, for all $x \in \bar{x} + \epsilon_i B$, $x \neq \bar{x}$,

$$-\epsilon_i^2 \leq (K + |\bar{q}| + 4\sigma\epsilon_i)\epsilon_i - (\delta_i/2)|\eta'(x) - \bar{v}| + (\delta_i/2)(2\sigma)^{-1}(\epsilon_i + \delta_i).$$

Since $\epsilon_i \delta_i^{-1} \to 0$ as $i \to \infty$ there exists $\gamma_i \downarrow 0$ such that

$$\sup_{x \in \bar{x} + \epsilon_i B} |\eta'(x) - \bar{v}| < \gamma_i \quad \text{for each } i. \tag{7.44}$$

Inequality (7.43) also tells us that

$$
\begin{aligned}
-\epsilon_i|x - \bar{x}| &\leq [\zeta'(x) + \bar{q} + 2\sigma(y'(x) - x)] \cdot (\bar{x} - x) \\
&\leq (\zeta'(x) + \bar{q}) \cdot (\bar{x} - x) + 4\sigma\epsilon_i|x - \bar{x}|
\end{aligned}
$$

for all $x \in \bar{x} + \epsilon_i B$. This means that

$$0 \leq (\zeta'(x) + \bar{q}) \cdot (\bar{x} - x) + \max_{e \in \epsilon_i(1+4\sigma)B} e \cdot (\bar{x} - x) \tag{7.45}$$

for all $x \in \bar{x} + \epsilon_i B$. (7.44) and (7.45) yield

$$\sup_{\zeta \in \bar{q} + S_i} \zeta \cdot (\bar{x} - x) \geq 0 \tag{7.46}$$

for all $x \in \bar{x} + \epsilon_i B$. Here

$$
\begin{aligned}
S_i := \{\zeta : (\zeta, \bar{v}) \in \partial^P H(x', p') + (\epsilon_i(1 + 4\sigma)B) \times (\gamma_i B) \\
|x' - x| \leq 2\epsilon_i, |p' - \bar{p}| \leq \epsilon_i + \delta_i\}.
\end{aligned}
$$

We conclude from (7.46) that

$$-\bar{q} \in \bar{\text{co}} S_i.$$

But in view of the remarks at the beginning of the proof, the S_is are bounded sets and H is continuous on a neighborhood of $(\bar{x}, \bar{p})$. We deduce from Carathéodory's theorem and the closure properties of ∂H that

$$-\bar{q} \in \text{co}\{\zeta : (\zeta, \bar{v}) : (\zeta, \bar{v}) \in \partial H(\bar{x}, \bar{p})\}.$$

The proof is complete. $\square$

The hypotheses in the above version of the Dualization Theorem include the requirement that the dual function $H(., p)$ is Lipschitz continuous near $\bar{x}$, in some uniform sense. We wish to replace it by the condition that L is locally epi-Lipschitz near $(\bar{x}, \bar{v})$. Unfortunately, the local epi-Lipschitz condition for L concerns the behavior of L on a neighborhood of $(\bar{x}, \bar{v}) \in$ dom L, yet the values of H are affected by the *global* properties of L. We cannot therefore expect, in general, to guarantee regularity properties of H by hypothesizing L is locally epi-Lipschitz.

The local epi-Lipschitz property does however ensure that, for some $\epsilon > 0$, the related function H_ϵ has useful Lipschitz continuity properties, where

$$H_\epsilon(x, p) := \sup\{p \cdot v - L(x, v) : v \in \bar{v} + \epsilon B\}. \tag{7.47}$$

Here ϵ is some parameter. H_ϵ will be recognized as the conjugate function, with respect to the second variable, of the "localization" of L:

$$L_\epsilon(x, v) := \begin{cases} L(x, v) & \text{if } v \in \bar{v} + \epsilon B \\ +\infty & \text{otherwise}. \end{cases}$$

Relevant properties of H_ϵ are recorded in following proposition.

Proposition 7.6.6 *Take a function $L : R^n \times R^m \to R \cup \{+\infty\}$ and a point $(\bar{x}, \bar{v}) \in$ dom L. Assume that*

(i) $L(x, .)$ is convex for each x in a neighborhood of $\bar{x}$;

(ii) L is lower semicontinuous;

(iii) L is locally epi-Lipschitz near $(\bar{x}, \bar{v})$.

Then for each $c > 0$, there exists $\epsilon > 0$ and $\beta > 0$ such that $H_\epsilon(., p)$ is Lipschitz continuous on $\bar{x} + \beta B$, uniformly with respect to all ps in cB.

Proof. Hypothesis (iii) can be expressed: there exist neighborhoods U, V, and W of $\bar{x}$, $\bar{v}$, and $L(\bar{x}, \bar{v})$, respectively, and $k > 0$ such that

$$\text{epi } L(x', .) \cap (V \times W) \subset \text{epi } L(x'', .) + k|x' - x''|B \tag{7.48}$$

for all $x', x'' \in U$. By increasing k if necessary, we can arrange that $k > 1$. Choose α such that $L(\bar{x}, \bar{v}) + \alpha B \subset W$.

Assume that the assertions of the proposition are false. We show that this leads to a contradiction.

Under this assumption, there exists $c > 0$ such that for all $\beta > 0$ and $\epsilon > 0$, the function $H_\epsilon(., p)$ is not Lipschitz continuous on $\bar{x} + \beta B$, uniformly with respect to all ps in cB. We can therefore find a sequence $\beta_i \downarrow 0$ and a positive constant ϵ such that, for all i, $H_\epsilon(., p)$ is not Lipschitz continuous on $\bar{x} + \beta B$ uniformly with respect to $p \in cB$ and

(a) the set $\mathcal{K} := (\bar{x} + \beta_i B) \times (\bar{v} + \epsilon B)$ is included in $U \times V$;

(b) $k\beta_i < \epsilon$;

(c) if (x, v) is in $\mathcal{K}$, then $L(x, v) > L(\bar{x}, \bar{v}) - \alpha$;

(d) $2c\epsilon < \alpha/2$.

(We have invoked Hypothesis (ii) to ensure (c).)

Fix i. Since L is lower semicontinuous on $U \times R^m$, finite at $(\bar{x}, \bar{v})$, and nowhere takes the value $-\infty$, L is bounded below on the compact subset $\mathcal{K}$. By adding a constant to L (this does not affect the assertions of the proposition), we can arrange that $L(x, v) \geq 0$ for all (x, v) in $\mathcal{K}$. Now for each x in $\bar{x} + \beta_i B$, we have by (7.48) that

$$(\bar{v}, L(\bar{x}, \bar{v})) \in \operatorname{epi} L(\bar{x}, \cdot) \cap (V \times W) \subset \operatorname{epi} L(x, \cdot) + k|\bar{x} - x|B.$$

Hence, in particular, there exists v in R^m such that $|v - \bar{v}| \leq k|x - \bar{x}| < k\beta_i < \epsilon$ and $L(x, v) \leq L(\bar{x}, \bar{v}) + k|\bar{x} - x| < +\infty$. This fact, combined with the nonnegativity of L on $\mathcal{K}$, establishes that $H_\epsilon(x, p)$ is finite for all p in R^m and for all x in $\bar{x} + \beta_i B_n$.

By the contradiction hypothesis, we can find points x_i and y_i in $\bar{x} + \beta_i B_n$ (with $x_i \neq y_i$) and p_i in cB such that

$$H_\epsilon(x_i, p_i) - H_\epsilon(y_i, p_i) < -i|x_i - y_i|. \tag{7.49}$$

But a lower semicontinuous function that is locally epi-Lipschitz near a point is locally epi-continuous near the point. Since $\{x_i\}$ converges to $\bar{x}$, it follows that there exists a sequence $\{v_i\}$ converging to $\bar{v}$ such that $\limsup_{i \to \infty} L(x_i, v_i) \leq L(\bar{x}, \bar{v})$. We can assume i sufficently large that v_i is in $\bar{v} + \epsilon B$ and such that $L(x_i, v_i) < L(\bar{x}, \bar{v}) + 0\alpha/2$ (by definition of lim sup).

Now since $v \to L(y_i, v)$ is lower semicontinuous on R^m (by (i)) and since $\bar{v} + \epsilon B$ is a compact set, we have by definition of the real number $H_\epsilon(y_i, p_i)$ that there exists w_i in $\bar{v} + \epsilon B$ such that $H_\epsilon(y_i, p_i) = p_i \cdot w_i - L(y_i, w_i)$. By (7.49), this implies that for all v in $\bar{v} + \epsilon B$,

$$p_i \cdot v - L(x_i, v) - p_i \cdot w_i + L(y_i, w_i) < -i|x_i - y_i|. \tag{7.50}$$

Of course $(w_i, L(y_i, w_i))$ belongs to $\operatorname{epi} L(y_i, .)$. Let us show that it also belongs to $V \times W$. Certainly, w_i belongs to V. Now since v_i is in $\bar{v} + \epsilon B$, using inequality (7.50), we obtain

$$p_i \cdot v_i - L(x_i, v_i) - p_i \cdot w_i + L(y_i, w_i) < -i|x_i - y_i| \leq 0 ,$$

which implies that

$$-L(y_i, w_i) \leq L(x_i, v_i) - p_i \cdot (v_i - \bar{v}) + p_i \cdot (w_i - \bar{v}) \leq L(\bar{x}, \bar{v}) + \frac{\alpha}{2} + 2c\epsilon .$$

Since (y_i, w_i) is in $\mathcal{K}$, we deduce from (c) and (d) that $L(y_i, w_i)$ is in W. We can now use (7.48) (with $x' = y_i$ and $x'' = x_i$) to find $\tilde{v}_i$ in R^m such that

$$\begin{cases} |\tilde{v}_i - w_i| \leq k|x_i - y_i| \\ L(x_i, \tilde{v}_i) - L(y_i, w_i) \leq k|x_i - y_i| . \end{cases} \tag{7.51}$$

There are two cases to consider.

Case A: $|\tilde{v}_i - \bar{v}| \leq \epsilon B$ for an infinite number of index values.

In this case, by extracting a subsequence if necessary, we can arrange that this assertion is valid for all i. Inserting $v = \tilde{v}_i$ into (7.50) gives

$$p_i \cdot \tilde{v}_i - L(x_i, \tilde{v}_i) - p_i \cdot w_i + L(y_i, w_i) < -i|x_i - y_i| \ .$$

Hence by (7.51), we have

$$-(|p_i| + 1)k|x_i - y_i| < -i|x_i - y_i| \ ,$$

which is impossible for large i, because $|p_i| < c$ and $x_i \neq y_i$.

Case B: $|\tilde{v}_i - \bar{v}| > \epsilon$ for all i sufficiently large.

In this case, we deduce from (7.48), in which we set $x' = \bar{x}$ and $x'' = x_i$, that, for i sufficiently large, $u_i \in R^m$ can be found such that

$$\begin{cases} |u_i - \bar{v}| \leq k|x_i - \bar{x}| < k\beta_i < \epsilon, \\ L(x_i, u_i) - L(\bar{x}, \bar{v}) \leq k|x_i - \bar{x}| \ . \end{cases} \tag{7.52}$$

Define v_i' in $\bar{v} + \epsilon B$ according to

$$v_i' = \mu_i u_i + (1 - \mu_i)\tilde{v}_i,$$

where $\mu_i \in [0, 1]$ is chosen such that $|v_i' - \bar{v}| = \epsilon$ (this is possible since $|u_i - \bar{v}| < \epsilon$ and $|\tilde{v}_i - \bar{v}| > \epsilon$). We see that, for i large enough,

$$\begin{aligned} \epsilon = |v_i' - \bar{v}| &= |\mu_i(u_i - \bar{v}) + (1 - \mu_i)(\tilde{v}_i - \bar{v})| \\ &\leq \mu_i|u_i - \bar{v}| + (1 - \mu_i)(|\tilde{v}_i - w_i| + |w_i - \bar{v}|) \\ &\leq \mu_i k|x_i - \bar{x}| + (1 - \mu_i)(k|x_i - y_i| + \epsilon) \ . \end{aligned}$$

It follows that, for i sufficiently large,

$$\mu_i(\epsilon + k|x_i - y_i| - k|x_i - \bar{x}|) \leq k|x_i - y_i|.$$

But $x_i \to \bar{x}$ and $|x_i - y_i| \to 0$. Noting that, for i sufficiently large, $\epsilon - k|x_i - \bar{x}| > 0$ (this follows from (b)), we have, for i sufficiently large, that

$$\mu_i \leq \frac{2k}{\epsilon}|x_i - y_i|. \tag{7.53}$$

Since v_i' is in $\bar{v} + \epsilon B$, we may insert it into (7.50), to get

$$p_i \cdot (\mu_i u_i + (1 - \mu_i)\tilde{v}_i - w_i) - L(x_i, v_i') + L(y_i, w_i) < -i|x_i - y_i|, \tag{7.54}$$

for all i sufficiently large. By convexity, we obtain

$$L(x_i, v_i') \leq \mu_i L(x_i, u_i) + (1 - \mu_i) L(x_i, \tilde{v}_i) \ .$$

Hence, we have, for i sufficiently large,

$$
\begin{aligned}
-L(x_i, v_i') &\geq -\mu_i L(x_i, u_i) - (1 - \mu_i) L(x_i, \tilde{v}_i) \\
&\geq -\mu_i (L(\bar{x}, \bar{v}) + k|x_i - \bar{x}|) - (1 - \mu_i) L(x_i, \tilde{v}_i) \\
&\geq -\mu_i (L(\bar{x}, \bar{v}) + k|x_i - \bar{x}|) - (1 - \mu_i)(L(y_i, w_i) + k|x_i - y_i|) \\
&\geq -\mu_i (L(\bar{x}, \bar{v}) + k|x_i - \bar{x}|) - L(y_i, w_i) - k|x_i - y_i| .
\end{aligned}
$$

To derive these relationships, we have used (7.51) and (7.52), and noted that, since (y_i, w_i) is in $\mathcal{K}$, $\mu_i(L(y_i, w_i) + k|x_i - y_i|) \geq 0$. This inequality combines with (7.54) to give

$$
\begin{aligned}
\mu_i p_i \cdot (u_i - \tilde{v}_i) + p_i \cdot (\tilde{v}_i - w_i) - \mu_i (L(\bar{x}, \bar{v}) + k|x_i - \bar{x}|) \\
- k|x_i - y_i| < -i|x_i - y_i| .
\end{aligned}
$$

We deduce from (7.51) that, for i sufficiently large,

$$
\begin{aligned}
- \mu_i c |u_i - \tilde{v}_i| - ck|x_i - y_i| - \mu_i (L(\bar{x}, \bar{v}) + k|x_i - \bar{x}|) \\
- k|x_i - y_i| < -i|x_i - y_i| .
\end{aligned}
$$

It follows from (7.53),

$$
\begin{aligned}
-\frac{2k}{\epsilon} [c|u_i - \tilde{v}_i| + L(\bar{x}, \bar{v}) + k|x_i - \bar{x}|] \cdot |x_i - y_i| \\
- (c + 1)k|x_i - y_i| < -i|x_i - y_i| .
\end{aligned}
$$

Since $\{u_i\}$ and $\{\tilde{v}_i\}$ are bounded sequences and $\{x_i\}$ converges to $\bar{x}$, this is impossible.

In both Cases (A) and (B), we have arrived at a contradiction. The assertions of the proposition must therefore be true. $\square$

The link that Proposition 7.6.6 provides between epi-Lipschitz conditions on L and Lipschitz continuity properties of H_ϵ motivates the following generalization of Theorem 7.6.5.

Theorem 7.6.7 *Take a function $L : R^n \times R^m \to R \cup \{+\infty\}$ and points $(\bar{x}, \bar{v}) \in \operatorname{dom} L$ and $\bar{p} \in R^m$. Assume that, for some neighborhoods U and P of $\bar{x}$ and $\bar{p}$, respectively, the following hypotheses are satisfied.*

(H1): $L(x,.)$ is convex for each x in a neighborhood of $\bar{x}$;

(H2): L is epicontinuous near $(\bar{x}, \bar{v})$;

(H3): For some $\epsilon > 0$, $H_\epsilon(.\,,p)$ (defined in (7.47)) is Lipschitz continuous on U, uniformly with respect to all ps in P.

Then

$$\{q : (q,\bar{p}) \in \partial L(\bar{x},\bar{v})\} \subset \text{co}\{q : (-q,\bar{v}) \in \partial H(\bar{x},\bar{p})\}.$$

Remark

Under the hypotheses of this theorem, it is not guaranteed that the function H is lower semicontinuous. The limiting subdifferential $\partial H(\bar{z})$ at the point $\bar{z} = (\bar{x},\bar{p}) \in \text{dom}\, H$ was earlier defined only in the case where H is lower semicontinuous. In the present context, $\partial H(\bar{z})$ is defined to be the set of elements ξ having the following properties.

There exist sequences of positive numbers $\{\epsilon_i\}$ and $\{\sigma_i\}$ and convergent sequences $z_i \overset{H}{\to} \bar{z}$ and $\xi_i \to \xi$ such that, for each i,

$$H(z) - H(z_i) \geq \xi_i \cdot (z - z_i) - \sigma_i |z - z_i|^2 \quad \text{for all } z \in z_i + \epsilon_i B.$$

Proof. Take a point $\bar{q}$ such that $(\bar{q},\bar{p}) \in \partial L(\bar{x},\bar{v})$. We must show that

$$-\bar{q} \in \text{co}\{q \in R^n : (q,\bar{v}) \in \partial H(\bar{x},\bar{p})\}.$$

As we have observed,
$$(\bar{q},\bar{p}) \in \partial L_\epsilon(\bar{x},\bar{v}),$$

since L and L_ϵ coincide on a neighborhood of $(\bar{x},\bar{v})$.

But the hypotheses of the preceding theorem are satisfied when L_ϵ replaces L. We deduce that

$$-\bar{q} \in \text{co}\{q : (q,\bar{v}) \in \partial H_\epsilon(\bar{x},\bar{p})\}.$$

It follows from Carathéodory's Theorem and the closure properties of the proximal subdifferential that $\bar{q}$ is a convex combination of $(n+1)$ points of the form

$$q = \lim_i q_i$$

for some sequences $v_i \to \bar{v}$ and $(x_i, p_i) \to (\bar{x},\bar{p})$ such that

$$(-q_i, v_i) \in \partial^P H_\epsilon(x_i, p_i) \quad \text{for all } i. \tag{7.55}$$

Take such a point q. Let $\{v_i\}$, $\{(x_i,p_i)\}$ be sequences associated with q as above.

Claim.
$$H_\epsilon(\bar{x},\bar{p}) = H(\bar{x},\bar{p}) \tag{7.56}$$

and
$$H_\epsilon(x_i, p_i) = H(x_i, p_i) \quad \text{for all } i \text{ sufficiently large.} \tag{7.57}$$

Let us verify these assertions. (7.55) implies that, for each i, $\sigma_i > 0$ can be found such that

$$H_\epsilon(x,p) - H_\epsilon(x_i,p_i)$$
$$\geq -q_i \cdot (x - x_i) + v_i \cdot (p - p_i) - \sigma_i |(x,p) - (x_i,p_i)|^2 \quad (7.58)$$

for all (x,p) in some neighborhood of (x_i, p_i). It follows that, for each i,

$$H_\epsilon(x_i,p) - H_\epsilon(x_i,p_i) \geq v_i \cdot (p - p_i) - \sigma_i |p - p_i|^2,$$

for all p in some neighborhood of p_i.

Since $H_\epsilon(x,.)$ is convex, we deduce that

$$v_i \in \partial H_\epsilon(x_i,.)(p_i),$$

in which ∂H_ϵ denotes the subdifferential of convex analysis. But this implies that

$$H_\epsilon(x_i,p_i) = p_i \cdot v_i - L_\epsilon(x_i,v_i) = p_i \cdot v_i - L(x_i,v_i), \quad (7.59)$$

for i sufficiently large.

For i sufficiently large, the maximum of the concave function $v \to p_i \cdot v - L_\epsilon(x_i,v)$ over $\bar{v} + \epsilon B$ is achieved at the point $v_i \in \text{int}\,\{\bar{v} + \epsilon B\}$. We conclude that v_i achieves the maximum of this function over all $v \in R^n$. It follows that (7.57) is true.

Since H_ϵ is continuous on a neighborhood of $(\bar{x}, \bar{p})$ and $(x_i, v_i) \to (\bar{x}, \bar{v})$, it follows from (7.59) and the lower semicontinuity of L that

$$H_\epsilon(\bar{x},\bar{p}) = \bar{p} \cdot \bar{v} - \lim_i L(x_i,v_i) \leq \bar{p} \cdot \bar{v} - L(\bar{x},\bar{v}) \leq H_\epsilon(\bar{x},\bar{p}).$$

We conclude that $v \to \bar{p} \cdot v - L(\bar{x},v)$ has a local maximum at $v = \bar{v}$. Since $-L(\bar{x},.)$ is concave, this maximum is in fact a *global* maximum. We deduce (7.56). The claim is verified.

Since $H_\epsilon(x,p) \leq H(x,p)$ for all (x,p), we deduce from (7.58) and (7.57) that, for each i sufficiently large, there exists $\sigma_i > 0$ such that

$$H(x,p) - H(x_i,p_i) \geq -q_i \cdot (x - x_i) + v_i \cdot (p - p_i) - \sigma_i |(x,p) - (x_i,p_i)|^2$$

for all (x,p) in some neighborhood of (x_i, p_i). This implies

$$(-q_i, v_i) \in \partial^P H(x_i,p_i)$$

for all i sufficiently large. Since H_ϵ is continuous on a neighborhood of $(\bar{x},\bar{p})$, (7.56) and (7.57) imply

$$\lim_i H(x_i,p_i) = \lim_i H_\epsilon(x_i,p_i) = H_\epsilon(\bar{x},\bar{p}) = H(\bar{x},\bar{p}).$$

We conclude from these relationships that q satisfies

$$(-q, \bar{v}) \in \partial H(\bar{x}, \bar{p}).$$

But $\bar{q}$ is expressible as a convex combination of such qs. It follows that

$$\bar{q} \in \text{co}\{q : (-q, \bar{v}) \in \partial H(\bar{x}, \bar{p})\}.$$

This is what we set out to prove. $\square$

Combining the assertions of Proposition 7.6.6 and Theorem 7.6.7 we arrive at the following version of the Dualization Theorem, whose hypotheses are matched to the framework of Theorem 7.5.1.

Theorem 7.6.8 (Dualization Theorem) *Take a function* $L : R^n \times R^m \to R \cup \{+\infty\}$ *and points* $(\bar{x}, \bar{v}) \in \text{dom}\, L$ *and* $\bar{p} \in R^m$. *Assume that, for some neighborhood* U *of* $\bar{x}$, *the following hypotheses are satisfied.*

(H1): $L(x,.)$ is convex for each $x \in U$;

(H2): L is lower semicontinuous;

(H3): L is locally epi-Lipschitz near $(\bar{x}, \bar{v})$.

Then
$$\{q : (q, \bar{p}) \in \partial L(\bar{x}, \bar{v})\} \subset \text{co}\{q : (-q, \bar{v}) \in \partial H(\bar{x}, \bar{p})\}.$$

Remark

For purposes of dualizing the Extended Euler–Lagrange Condition, only the onesided condition (7.34) is required. We do not use this fact here, but we note Rockafellar [124] has shown that under the hypotheses of Theorem 7.6.8, supplemented by the condition

$$L \text{ is epicontinuous,}$$

the two sets considered do in fact coincide:

$$\text{co}\{q : (q, \bar{p}) \in \partial L(\bar{x}, \bar{v})\} = \text{co}\{q : (-q, \bar{v}) \in \partial H(\bar{x}, \bar{p})\}.$$

7.7   The Extended Hamilton Condition

We return now to the derivation of necessary conditions for problem (P). The analysis of Section 7.6 permits us to augment our earlier necessary conditions with the Extended Hamilton Condition:

$$\dot{p}(t) \in \text{co}\{-q : (q, \dot{\bar{x}}(t)) \in \partial H(t, \bar{x}, p(t))\}.$$

Theorem 7.7.1 (The Extended Hamilton Condition) *Let $\bar{x}$ be a $W^{1,1}$ local minimizer for (P). Assume that*

$$F(t,x) \text{ is a convex set for each } (t,x) \in [S,T] \times R^n.$$

Assume also that

(G1) g is locally Lipschitz continuous;

(G2)' $F(t,x)$ is nonempty for each (t,x), F is $\mathcal{L} \times \mathcal{B}$ measurable, and $\operatorname{Gr} F(t,.)$ is closed for each $t \in [S,T]$.

Assume, furthermore, that either of the following hypotheses is satisfied.

(a) There exists $k \in L^1$ and $\epsilon > 0$ such that for a.e. t

$$F(t,x') \cap (\dot{\bar{x}}(t) + \epsilon k(t)B) \subset F(t,x) + k(t)|x - x'|B$$

for all $x, x' \in \bar{x}(t) + \epsilon B$.

(b) There exists $k \in L^1$, $\bar{K} > 0$, and $\epsilon > 0$ such that the following two conditions are satisfied for a.e. t.

$$F(t,x') \cap (\dot{\bar{x}}(t) + \epsilon B) \subset F(t,x) + k(t)|x - x'|B,$$

for all $x, x' \in \bar{x}(t) + \epsilon B$ and

$$\inf\{|v - \dot{\bar{x}}(t)| : v \in F(t,x)\} \leq \bar{K}|x - \bar{x}(t)|$$

for all $x \in \bar{x}(t) + \epsilon B$.

Then there exist an arc p and a constant $\lambda \geq 0$ such that

(i) $(p, \lambda) \neq 0$;

(ii) $\dot{p}(t) \in co\{\eta : (\eta, p(t)) \in N_{\operatorname{Gr} F(t,.)}(\bar{x}(t), \dot{\bar{x}}(t))\}$ a.e.;

(iii) $(p(S), -p(T)) \in \lambda \partial g(\bar{x}(S), \bar{x}(T)) + N_C(\bar{x}(S), \bar{x}(T))$.

Condition (ii) implies

(iv) $p(t) \cdot \dot{\bar{x}}(t) \geq p(t) \cdot v$ for all $v \in F(t, \bar{x}(t))$ a.e.

and

(v) $\dot{p}(t) \in co\{-q : (q, \dot{\bar{x}}(t)) \in \partial H(t, \bar{x}, p(t))\}$.

Now assume, also, that

$$F(t,x) \text{ does not depend on } t.$$

(In this case we write $F(x)$ in place of $F(t,x)$.) Then, in addition to the above conditions, there exists a constant r such that

(v) $p(t) \cdot \dot{\bar{x}}(t) \ (= \max_{v \in F(\bar{x}(t))} p(t) \cdot v) = r$ *a.e.*

Remark

The interpretation of limiting subdifferential ∂H of the function, $H(t, ., .)$, which under the hypotheses here imposed need not be lower semicontinuous, is attended to in the discussion following Theorem 7.6.7.

A special case of this theorem is the set of necessary conditions incorporating Clarke's Hamilton Inclusion, highlighted in Chapter 1. A frequently cited version invokes, for some $\epsilon > 0$, the hypotheses that

(I): the multifunction $F : [S, T] \times R^n \rightsquigarrow R^n$ has as values closed nonempty convex sets, and F is $\mathcal{L} \times \mathcal{B}^n$ measurable;

(II): there exists $c(.) \in L^1$ such that

$$F(t, x) \subset c(t)B \quad \forall\, x \in \bar{x}(t) + \epsilon B \text{ a.e.};$$

(III): there exist $k(.) \in L$ such that

$$F(t, x) \subset F(t, x') + k(t)|x - x'|B \quad \forall\, x, x' \in \bar{x}(t) + \epsilon B \text{ a.e.};$$

(IV): C is closed and g is Lipschitz continuous on $(\bar{x}(S), \bar{x}(T)) + \epsilon B$,

and asserts the existence of $p \in W^{1,1}([S, T]; R^n)$ and $\lambda \geq 0$, which satisfy Conditions (i) and (iii) of the theorem statement together with the Hamilton inclusion, namely,

$$(-\dot{p}(t), \dot{\bar{x}}(t)) \in \text{co}\, \partial H(t, \bar{x}(t), p(t)) \text{ a.e.}$$

Theorem 7.1.1 refines these necessary conditions. The boundedness hypothesis, Hypothesis (II) is dropped and Hypothesis (III) takes the weaker onesided form (a). On the other hand, the Extended Hamilton Condition (v) is a sharper partially convexified form of the fully convexified Hamilton Inclusion (v). Achieving these refinements greatly adds to the analytic burden of deriving the necessary conditions. A simpler, streamlined proof of the the Hamilton Inclusion, based on infinite dimensional proximal analysis, is given in [53], under the above stronger hypotheses.

Proof. We have already proved that an adjoint arc p exists, satisfying all the conditions of the theorem statement, with the exception of the Extended Hamilton Condition. We show that this condition too is satisfied.

Choose any t at which the conditions (ii) and (iv) are satisfied. Such ts comprise a set of full measure. Define $L : R^n \times R^n \to R^n \cup \{+\infty\}$ to be

$$L(x, v) := \begin{cases} 0 & \text{if } v \in F(t, x) \\ +\infty & \text{if } v \notin F(t, x). \end{cases}$$

L is lower semicontinuous since $F(t,.)$ has a closed graph. $L(x,.)$ is convex for each $x \in R^n$, since F is convex-valued. It can be deduced from Hypotheses (G2)$'$ and either (a) or (b) that L is locally epi-Lipschitz near $(\bar{x}(t), \dot{\bar{x}}(t))$. Since the conjugate function of $L(x,.)$ is the Hamiltonian H associated with F, we conclude from the Dualization Theorem (Theorem 7.6.8) that p satisfies

$$\dot{p}(t) \in \mathrm{co}\{-q : (q, \dot{\bar{x}}(t)) \in \partial H(t, \bar{x}, p(t))\}.$$

This is the desired condition. □

7.8   Notes for Chapter 7

Rockafellar's optimality conditions for fully convex problems of Bolza type revealed how the classical necessary conditions of Euler and Hamilton can be interpreted in terms of subdifferentials (in the sense of convex analysis) of extended-valued Lagrangians and Hamiltonian functions [120], [121]. Despite the limitations of the "fully convex" setting (which, in effect, restricts attention to optimal control problems with linear dynamics and cost integrands jointly convex with respect to state and velocity variables), Rockafellar's conditions provided the template for subsequent necessary conditions of the kind covered in this chapter.

The breakthrough into necessary conditions for broad classes of optimal control problems formulated in terms of a differential inclusion was Clarke's Hamilton Condition of 1976 [31]:

$$(-\dot{p}, \dot{\bar{x}}) \in \mathrm{co}\,\partial H(t, \bar{x}, p). \tag{7.60}$$

A feature of Clarke's derivation, distinguishing it from early necessary conditions for differential inclusion problems, was the natural and intrinsic nature of the hypotheses imposed on the multifunction of the dynamic constraint. This was required to be merely compact, convex-valued, integrably bounded, measurable in time, and Lipschitz continuous in the state variable, with respect to the Hausdorff metric.

Clarke also derived an Euler–Lagrange type condition for problems reducible to free right endpoint problems [34]:

$$(\dot{p}, p) \in \mathrm{co} N_{\mathrm{Gr}\, F(t,.)}(\bar{x}, \dot{\bar{x}}). \tag{7.61}$$

In work reported in [105], Mordukhovich used discrete approximation techniques to derive another form of the Euler–Lagrange Condition, namely,

$$\dot{p} \in \mathrm{co}\{(q, v) : q \cdot v = \max_{v' \in F(t,x)} p \cdot v', (q, p) \in N_{\mathrm{Gr}\, F(t,.)}(\bar{x}, v)\} \tag{7.62}$$

(for bounded convex-valued Fs, Lipschitz continuous in x, and continuous in t), and established its validity for general endpoint conditions. This opened up the possibility of replacing (7.62) by conditions in which $(\dot{p}, p)$ is asserted to lie in a smaller, nonconvex, subset of $\operatorname{co} N_{\operatorname{Gr} F(t,\cdot)}$. Pursuing this line of inquiry for convex-valued Fs, Smirnov [129] and Loewen and Rockafellar [96] established the validity of the Extended Euler–Lagrange Condition

$$\dot{p} \in \operatorname{co}\{q : (q, p) \in N_{\operatorname{Gr} F(t,\cdot)}(\bar{x}, \dot{\bar{x}})\}.$$

This condition coincides with (7.62) when the values of F are strictly convex but, otherwise, is sharper than either (7.60) or (7.62). Loewen and Rockafellar also showed [97] that the adjoint arc p could be chosen additionally to satisfy the Extended Hamilton Condition,

$$\dot{p} \in \operatorname{co}\{q : (-q, \dot{\bar{x}}) \in \partial H(t, \bar{x}, p)\}, \tag{7.63}$$

a more precise version of (7.60), which involves convexification on the right side merely with respect to the first coordinate.

Loewen and Rockafellar highlighted the unsatisfactory nature of traditional Hausdorff metric Lipschitz continuity hypotheses, for applications involving unbounded Fs. These authors introduced, instead, the more appropriate local epi-Lipschitz hypothesis invoked in this Chapter [96].

In subsequent research, Rockafellar [124] showed, using Moreau–Yoshida approximation techniques, that for convex-valued Fs satisfying additional mild hypotheses, the Extended Euler–Lagrange Condition and the Extended Hamilton Condition are in fact equivalent. Ioffe [86], also using Moreau–Yoshida approximation, showed that, under reduced hypotheses matched to the derivation of the Extended Euler–Lagrange Condition, the Extended Euler Condition implies the Extended Hamiltonian Condition. in Section 7.6, we follow the alternative proof from [20] of the "one-sided" dualization theorem, based on an analysis of proximal normals and an application of the Mini-Max Theorem.

Another idea for formulating intrinsic necessary conditions for "convex" differential inclusion problems, closely related to the nonsmooth Maximum Principle, is due to Kaskosz and Lojasiewicz [89]. The authors hypothesize existence of a point-valued selector $f(t, x)$ of $F(t, x)$, which is measurable in t and Lipschitz continuous in x and which satisfies

$$\dot{\bar{x}} = f(t, \bar{x}).$$

(f is a "Lipschitz selector" of F, compatible with the minimizer $\bar{x}$.) The necessary conditions assert the existence of an adjoint arc p satisfying conditions including

$$-\dot{p} \in \operatorname{co} \partial_x p \cdot f(t, x)$$

and

$$p \cdot v = \max_{v \in F(t, \bar{x})} p.v \,.$$

A significant point here is that, for convex-valued Fs, existence of a Lipschitz selector (compatible with $\bar{x}$) is guaranteed under mild hypotheses on

F. So the underlying assumptions, concerning existence of such a selector, are natural and unrestrictive. The Kaskosz–Lojasiewcz type conditions can be derived from the Hamiltonian Inclusion by a simple limiting argument [98].

The foregoing relates to optimal control problems with convex-valued Fs. Of course the Maximum Principle, which predates all these developments, applies to problems with possibly nonconvex velocity sets, in the case that the dynamic constraint can be parameterized as a control-dependent differential equation. Necessary conditions for problems with nonconvex Fs have long been available under hypotheses ("calmness," "controllability," or assumptions about the nature of the endpoint constraints), the implications of which are that minimizers are minimizers also for related problems in which F is replaced by its convex hull. In such cases, necessary conditions can be derived from known necessary conditions for convex F's or by studying local approximations of the mapping of the initial state into the set of admissible state trajectories as in [67]. But the derivation of intrinsic necessary conditions for general classes of problems with nonconvex velocity sets required some new ideas and was a later chapter in the story.

It is relevant at this stage to point out that, for convex Fs, the Generalized Weierstrass Condition

$$p \cdot \dot{\bar{x}} := \max_{v \in F(t, \bar{x})} p \cdot v$$

is merely a consequence of the Extended Euler Condition (7.62). For nonconvex F's however, it is a distinct condition that can have an important role in the elimination of putative minimizers.

Mordukhovich, using discrete approximation techniques, established the validity of the Extended Euler–Lagrange Condition, for nonconvex-valued, bounded Fs, continuous in t and Lipschitz continuous in x [108]. The fact that the Extended Euler–Lagrange Condition remains valid for measurably time-dependent, possibly nonconvex-valued Fs, and can be supplemented by the Weierstrass Condition was shown independently by Ioffe [86] and also by Vinter and Zheng [143] under somewhat stronger hypotheses. An important step was Ioffe and Rockafellar's paper [87] on necessary conditions for nonconvex variational problems with finite Lagrangians. Vinter and Zheng's proof of the Extended Euler–Lagrange and Weierstrass Conditions, based on application of the smooth Maximum Principle to an approximating optimization problem with state free dynamics and passage to the limit, is the one used in this chapter. However techniques of [86] are also exploited, to allow for velocity sets satisfying the mild local epi-Lipschitz hypothesis invoked here.

Constraint reduction and elimination, strongly in evidence in this chapter, are now recognized as a cornerstone of nonsmooth Optimal Control. The idea is already implicit in Rockafellar's original formulation of convex variational problems as generalized problems of Bolza type. In a general

Optimal Control context, it was given prominence by Clarke, who made early and extensive use of exact penalty terms, involving the distance function, to eliminate dynamic and endpoint constraints. In this chapter we use related "approximate" distance function penalty techniques developed by Ioffe [86], to take account of the milder hypotheses on the dynamic constraints invoked here.

The above review has skirted around the following issue. Does a minimizer for an optimal control problem with nonconvex velocity set F satisfy extensions of the Euler–Lagrange Condition or other, related, conditions, in which F is replaced by its convex hull co F? We would expect such necessary conditions to supply additional information in certain cases, because consideration of co F gives rise to a richer class of variations. The question remains open as to whether, for general differential inclusion problems involving nonconvex velocity sets, a "convexified" version of the Extended Euler–Lagrange Condition

$$\dot{p} \in \operatorname{co}\{q : (q,p) \in N_{\operatorname{co} \operatorname{Gr} F(t,.)}(\bar{x},\dot{\bar{x}}). \tag{7.64}$$

is valid. Notice that this would imply the Extended Hamilton Condition (7.63), namely,

$$\dot{p} \in \operatorname{co}\{q : (-q,\dot{\bar{x}}) \in \partial H(t,\bar{x},p)\}$$

(by the dualization results of Section 7.6 and since the Hamiltonians of F and co F coincide), and so, in particular, the Hamiltonian Inclusion (7.60), namely,

$$(-\dot{p},\dot{\bar{x}}) \in \operatorname{co} \partial H(t,\bar{x},p).$$

Apart from cases involving implicit hypotheses ("calmness," etc.), the main special cases when minimizers for (P) are known to satisfy conditions incorporating the Extended Euler–Lagrange Condition for the convexified differential inclusion, or related conditions, are:

(a) (Free right endpoint)

   The Extended Euler–Lagrange Condition (7.64) is valid when $C = C_0 \times R^n$.

(b) (Functional inequality endpoint constraints)

   The Extended Euler–Lagrange Condition (7.64) is valid when $C = C_0 \times C_1$ and C_1 is expressed in terms of a finite number of Lipschitz continuous functional inequality constraints satisfying an appropriate constraint qualification.

(c) (Existence of a Regular Parameterization of co F)

   The Hamiltonian Inclusion (7.60) is valid when co F admits a C^1 parameterization (f,U). This means

$$\operatorname{co} F(t,x) = f(t,x,U),$$

for some set U and point -valued map f satisfying various conditions including the requirement that $f(t, . , u)$ is a C^1 function.

(We have, of course, only highlighted the relevant extra hypotheses.) In Case (a), we use the fact that, for a free endpoint problem, minimizers for (P) are also relaxed minimizers. We can reduce (b) to (a) (for purposes of deriving necessary conditions) by absorbing the functional inequality endpoint constraints into the cost, as indicated in [38], Notes to Chapter 5. Case (c) is discussed below.

Notice that all these cases involve genuine restrictions. Case (b) (and of course Case (a)) exclude fixed endpoints. On the other hand, convex differential inclusions satisfying the usual Lipschitz continuity hypotheses for derivation of necessary conditions possess always Lipschitz parameterizations, but possibly not C^1 parameterizations.

Results of Zhu [156], proved by an analysis of convex approximations of the reachable set in the spirit of Warga's earlier work, come perhaps closest to providing necessary conditions for nonconvex differential inclusion problems, in terms of the convexified differential inclusion. They can be briefly summarized as follows. Take a minimizer for an optimal control problem for which the velocity set F has as values compact, possibly nonconvex sets and is bounded, measurable in t, and Lipschitz continuous in x with respect to the Hausdorff metric. Then a version of the Nonsmooth Maximum Principle of Chapter 6 is satisfied, expressed in terms of any control system

$$\dot{x}(t) \; = \; f(t, x(t), u(t)), \; u(t) \in U,$$

which is a Lipschitz parameterization of the convexified multifunction F; i.e.,

$$\mathrm{co}\, F(t, x) \; = \; \{ f(t, x, u) : u \in U \}.$$

Existence of a Lipschitz parameterization is guaranteed under unrestrictive conditions. If however the optimal control problem admits a C^1 parameterization, then the Maximum Principle, in its smooth manifestation, implies the Hamiltonian Inclusion (7.60). (This, incidently, deals with Case (c) above.) Related results were proved independently by Tuan [137].

Chapter 8

Necessary Conditions for Free End-Time Problems

VLADIMIR: That passed the time.
ESTRAGON: It would have passed in any case.
VLADIMIR: Yes, but not so rapidly.

> – Samuel Beckett, *Waiting for Godot*

8.1 Introduction

Our investigation of the properties of optimal strategies has, up till now, been confined to optimal control problems for which the underlying time interval $[S, T]$ has been fixed. In this chapter, we address a broader class of problems in which the end-times S and T are included among the choice variables:

$$(FT) \quad \begin{cases} \text{Minimize } g(S, x(S), T, x(T)) \\ \text{over intervals } [S, T] \text{ and arcs } x \in W^{1,1}([S, T], R^n) \\ \text{satisfying} \\ \dot{x}(t) \in F(t, x(t)) \quad \text{a.e. } t \in [S, T], \\ (S, x(S), T, x(T)) \in C. \end{cases}$$

Here $g : R \times R^n \times R \times R^n \to R$ is a given function, $F : R \times R^n \rightsquigarrow R^n$ is a given multifunction, and $C \subset R \times R^n \times R \times R^n$ is a given set.

Now, an arc $x \in W^{1,1}([S, T]; R^n)$ that satisfies the constraints of (FT) is customarily denoted $([S, T], x)$, to emphasize the underlying subinterval $[S, T]$.

Important special cases of (FT) are minimum time problems, in which the time duration of a maneuver is the quantity we aim to minimize. Many practical dynamic optimization problems are of this nature. For example, a strategy for the attitude control of an orbiting satellite is bang-bang control, based on elimination of the deviation of the satellite's orientation from its nominal value in minimum time. Optimal control problems associated with evasion or pursuit are usually minimum time problems — escape from, or catch, your adversary as quickly as possible.

Take a minimizer $([\bar{S}, \bar{T}], \bar{x})$ for (FT). Then $\bar{x}$ is a minimizer for the related optimal control problem, in which the end-times are frozen at $\bar{S}$ and $\bar{T}$. So the minimizer certainly satisfies fixed end-time necessary conditions, such as those derived in Chapter 7.

But additional information about the multipliers $p \in W([\bar{S}, \bar{T}]; R^n)$ and $\lambda \geq 0$, featuring in the fixed end-time necessary conditions, is required to take account of the extra degrees of freedom that have been introduced into the optimization problem. The additional information comes in the form of boundary conditions on the Hamiltonian

$$H(t, x, p) := \max_{v \in F(t,x)} p \cdot v \qquad (8.1)$$

(evaluated along $t \to (\bar{x}(t), p(t))$). These boundary conditions are implicit in a generalized transversality condition for free end-time optimal control problems, derived in this chapter:

$$(-\eta, p(\bar{S}), \xi, -p(\bar{T})) \in \lambda \partial g(\bar{S}, \bar{x}(\bar{S}), \bar{T}, \bar{x}(\bar{T})) + N_C(\bar{S}, \bar{x}(\bar{S}), \bar{T}, \bar{x}(\bar{T})) \quad (8.2)$$

in which

$$\eta = H(t, \bar{x}(t), p(t))|_{t=\bar{S}} \quad \text{and} \quad \xi = H(t, \bar{x}(t), p(t))|_{t=\bar{T}}.$$

In one extreme case (fixed end-times), C and g are of the form

$$C = \{(\bar{S}, x_0, \bar{T}, x_1) : (x_0, x_1) \in \tilde{C}\} \quad \text{and} \quad g(S, x_0, T, x_1) = \tilde{g}(x_0, x_1),$$

for some set $\tilde{C} \subset R^n \times R^n$, fixed interval $[\bar{S}, \bar{T}]$, and function $\tilde{g} : R^n \times R^n \to R$ is a given function. Here, (8.2) reduces to the familiar "fixed end-time" transversality condition

$$(p(\bar{S}), -p(\bar{T})) \in \lambda \partial \tilde{g}(\bar{x}(\bar{S}), \bar{x}(\bar{T})) + N_{\tilde{C}}(\bar{x}(\bar{S}), \bar{x}(\bar{T}))$$

and, appropriately, conveys no information about the optimal end-times.

In another extreme case (unconstrained end-times), C and g are expressible as

$$C = \{(\alpha, x_0, \beta, x_1) : (\alpha, \beta) \in R^2, (x_0, x_1) \in \tilde{C}\} \quad \text{and}$$
$$g(S, x_0, T, x_1) = \tilde{g}(x_0, x_1),$$

for some $\tilde{C} \subset R^n \times R^n$ and some $\tilde{g} : R^n \times R^n \to R$. (We assume, for simplicity, that the cost does not depend on the end-times.) In this case the boundary conditions on the Hamiltonian are decoupled from the Generalized Transversality Condition and are simply expressed as

$$H(t, \bar{x}(t), p(t))|_{t=\bar{S}} = 0 \quad \text{and} \quad H(t, \bar{x}(t), p(t))|_{t=\bar{T}} = 0.$$

The interpretation of these boundary conditions on the Hamiltonian is not entirely straightforward however, for the following reason. The set of state trajectories associated with the differential inclusion $\dot{x}(t) \in F(t, x(t))$ is unaffected by arbitrary modifications on a nullset of the "point to multifunction" mapping $t \to F(t, .)$. We can therefore regard the optimal control problem as an optimization problem over state trajectories corresponding to an equivalence class of $F(t,.)$s that differ only on nullset. The boundary conditions on the Hamiltonian are meaningless in this context, because the values of $t \to H(t, \bar{x}(t), p(t))$ at $t = \bar{S}$ and $t = \bar{T}$ are not the same across Fs in an equivalence class.

When the time dependence of $F(t, x)$ is in some sense Lipschitz continuous, the dilemma of how to define boundary conditions on the Hamiltonian is resolved by showing that the equivalence class of functions almost everywhere equal to $t \to H(t, \bar{x}(t), p(t))$ has a continuous representative $r(.)$. We can then express the boundary condition in terms of $r(t)$; i.e., condition (8.2) is satisfied with

$$(\eta, \xi) = (r(\bar{S}), r(\bar{T})).$$

The analysis supplies additional information about the Hamiltonian evaluated along $t \to (\bar{x}(t), p(t))$ (or, more precisely, about $r(.)$): this reduces to the well-known condition that the Hamiltonian is a.e. equal to a constant in the case when F is independent of t.

For measurably time-dependent data, there is no convenient representative of the equivalence class, in terms of which the boundary conditions on the Hamiltonian can conveniently be expressed. In this more general setting, we take a different approach. This involves replacing point evaluation of the Hamiltonian by another operation, namely, calculating the "essential values" of the Hamiltonian. Taking essential values is a generalization of point evaluation which, significantly, is unaffected by modifications on a nullset.

We derive conditions that are satisfied not merely by minimizers for (FT) but by $W^{1,1}$ local minimizers. To clarify exactly what is meant by a "$W^{1,1}$ local minimizer," in circumstances when the underlying time interval $[S, T]$ is a choice variable, the following convention is helpful and is adhered to:

Identify a function $x : [S, T] \to R^n$ with its extension to all of $(-\infty, \infty)$, by constant extrapolation of end values to left and right. For example, given $y \in R^n$ and $t > T$, then $|y - x(t)| := |y - x(T)|$.

In this spirit, given $x \in W^{1,1}([S, T]; R^n)$, $x' \in W^{1,1}([S', T']; R^n)$, we define

$$||x - x'||_{L^\infty} := ||x_e - x'_e||_{L^\infty}, \ ||\dot{x} - \dot{x}'||_{L^1} := ||\dot{x}_e - \dot{x}'_e||_{L^1},$$

etc., where x_e and x'_e are the extensions defined above.

We say that an element $([\bar{S}, \bar{T}], x)$ is a $W^{1,1}$ local minimizer for (FT), if $([\bar{S}, \bar{T}], x)$ satisfies the constraints of (FT) and there exists some $\delta' > 0$ such that

$$g(S, x(S), T, x(T)) \geq g(\bar{S}, \bar{x}(\bar{S}), \bar{T}, \bar{x}(\bar{T}))$$

for all elements $([S, T], x)$ satisfying the constraints of (FT) and also the condition

$$d(([S, T], x), ([\bar{S}, \bar{T}], \bar{x})) \leq \delta'.$$

Here $d(,)$ is the metric

$$d(([S,T],x),([S',T'],x')) := |S - S'| + |T - T'|$$
$$+ |x(S) - x'(S')| + \int_{S \wedge S'}^{T \vee T'} |\dot{x}_e(s) - \dot{x}'_e(s)| ds, \qquad (8.3)$$

in which x_e and x'_e are the extensions of x and x' alluded to above.

Notice that $d(.,.)$ is just the metric induced by the $W^{1,1}([S,T];R^n)$ norm,

$$||x||_{W^{1,1}} = |x(0)| + ||\dot{x}||_{L^1},$$

when it is restricted to pairs of arcs with common domain $[S,T]$. Our terminology is therefore consistent with "fixed end-time" nomenclature.

Observe that, in view of the extension convention, the metric $d(.,.)$ can alternatively be expressed

$$d(([S,T],x),([S',T'],x')) =$$
$$|x(S) - x'(S')| + |S - S'| + |T - T'| + ||\dot{x}_e - \dot{x}'_e||_{L^1(R;R^n)},$$

illustrating the simplification in notation achievable by this means.

8.2   Lipschitz Time Dependence

In this section we derive necessary conditions for the free end-time problem (FT), when Lipschitz continuity hypotheses are invoked regarding time dependence of $F(t,x)$.

The heart of the analysis is a transformation of the independent variable, which generates information about the Hamiltonian evaluated along the minimizing state trajectory and associated adjoint arc. This we now briefly review.

Suppose that $([\bar{S}, \bar{T}], \bar{x})$ is a minimizer for (FT). Fix $\alpha \in (0,1)$. Consider a new *fixed end-time* problem (we call it the "transformed" problem):

$$(C)' \begin{cases} \text{Minimize } g(\tau(\bar{S}), y(\bar{S}), \tau(\bar{T}), y(\bar{T})) \\ \text{over } (\tau, y) \in W^{1,1}([\bar{S}, \bar{T}]; R^{1+n}) \text{ satisfying} \\ (\dot{\tau}, \dot{y}) \in \{(w, wv) : w \in [1 - \alpha, 1 + \alpha], v \in F(\tau, y)\} \text{ a.e.,} \\ (\tau(\bar{S}), y(\bar{S}), \tau(\bar{T}), y(\bar{T})) \in C. \end{cases}$$

The significance of $(C)'$ is that there exists a transformation $\mathcal{G}$ that carries an arbitrary state trajectory (τ, y) satisfying the constraints of $(C)'$ into a state trajectory $([S,T],x)$ satisfying the constraints for (FT). Furthermore this transformation preserves the value of the cost. To be precise, the transformation is

$$\mathcal{G}(\tau, y) = ([\tau(\bar{S}), \tau(\bar{T})], y \circ \psi^{-1}),$$

in which $\psi : [\bar{S}, \bar{T}] \to [\tau(\bar{S}), \tau(\bar{T})]$ is the function

$$\psi(s) = \tau(\bar{S}) + \int_{\bar{S}}^{s} w(\sigma)d\sigma.$$

Here, $w(.)$ is some measurable function such that

$$(\dot{\tau}(s), \dot{y}(s)) \in w(s)(\{1\} \times F(\tau(s), y(s))) \quad \text{a.e. } s \in [\bar{S}, \bar{T}].$$

It follows that, since $(\bar{\tau}(s) \equiv s, \bar{x})$ transforms into $([\bar{S}, \bar{T}], \bar{x})$, $(\bar{\tau}, \bar{x})$ is a minimizer for $(C)'$.

Under appropriate hypotheses on the data for (FT), the data for $(C)'$ satisfy the hypotheses for application of the necessary conditions of Chapter 7. These supply a cost multiplier $\lambda \geq 0$ and adjoint arc components $-r(.)$ and $p(.)$, corresponding to the state components τ and y. Of various conditions satisfied by r and p, particular interest attaches to the Weierstrass and Transversality Conditions:

$$-r(s) + p(s) \cdot \dot{\bar{x}}(s) = \qquad (8.4)$$
$$\max\{w(-r(s) + p(s) \cdot v) : v \in F(t, \bar{x}), w \in [1 - \alpha, 1 + \alpha]\} \text{ a.e.}$$

and

$$(-r(\bar{S}), p(\bar{S}), r(\bar{T}), -p(\bar{T})) \in$$
$$\lambda \partial g(\bar{S}, \bar{x}(\bar{S}), \bar{T}, \bar{x}(\bar{T})) + N_C(\bar{S}, \bar{x}(\bar{S}), \bar{T}, \bar{x}(\bar{T})). \qquad (8.5)$$

Fixing $w = 1$, we recover from (8.4) the Weierstrass Condition for (FT). However, further information is now obtained if we fix $v = \dot{\bar{x}}(t)$. It is

$$H(t, \bar{x}(t), p(t)) = r(t) \quad \text{a.e.}$$

(where $H(t, x, p)$ is the Hamiltonian (8.1)). We see from this relationship and (8.5) that the Transversality Condition provides information about the boundary values of the Hamiltonian evaluated along $(\bar{x}, p)$. The boundary values are interpreted as the endpoints of some absolutely continuous function that coincides with the Hamiltonian almost everywhere.

Of course this analysis is justified only when the data for $(C)'$ satisfy the hypotheses for application of suitable necessary conditions of optimality. Since the independent variable t in (FT) becomes a state variable τ in $(C)'$ and since currently available necessary conditions require the data to be, in some generalized sense "differentiable" with respect to the state variable, the analysis will effectively be limited to problems for which the data are fairly regular regarding their time dependence.

These arguments are reproduced below, but modified to allow for Fs that are unbounded (this necessitates constraining w to lie in some *time-dependent* set

$$w(t) \in [1 - m(t), 1 + m(t)] \quad)$$

and to allow for the fact that $([\bar{S}, \bar{T}], \bar{x})$ is assumed to be merely a $W^{1,1}$ local minimizer (this requires us to introduce more complicated dynamics into $(C)'$).

Theorem 8.2.1 (Free-Time Necessary Conditions: Lipschitz Time Dependence) *Let $([\bar{S}, \bar{T}], \bar{x})$ be a $W^{1,1}$ local minimizer for (FT) such that $\bar{T} > \bar{S}$. Assume that the following hypotheses are satisfied.*

(H1) g is locally Lipschitz continuous and C is a closed set;

(H2) $F(t,x)$ is nonempty for all $(t,x) \in R \times R^n$ and $\operatorname{Gr} F$ is closed;

(H3) there exist $k_F \in L^1$, $\beta \geq 0$, and $\delta > 0$ such that, for a.e. $t \in [\bar{S}, \bar{T}]$,

$$F(t',x') \cap (\dot{\bar{x}}(t) + NB) \subset F(t'',x'') + (k_F(t) + \beta N)|(t',x') - (t'',x'')|B$$

for all $N \geq 0$, $(t',x'), (t'',x'') \in (t, \bar{x}(t)) + \delta B$.

Then there exist absolutely continuous arcs

$$p \in W^{1,1}([\bar{S}, \bar{T}]; R^n) \quad and \quad r \in W^{1,1}([\bar{S}, \bar{T}]; R)$$

and a number $\lambda \geq 0$ such that

(i) $(\lambda, p) \neq (0,0)$;

(ii) $(-\dot{r}(t), \dot{p}(t)) \in \operatorname{co}\{(\zeta, \eta) : ((\zeta, \eta), p(t)) \in N_{\operatorname{Gr} F}((t, \bar{x}(t)), \dot{\bar{x}}(t))\}$
$$a.e. \ t \in [\bar{S}, \bar{T}];$$

(iii) $(-r(\bar{S}), p(\bar{S}), r(\bar{T}), -p(\bar{T})) \in \lambda \partial g(\bar{S}, \bar{x}(\bar{S}), \bar{T}, \bar{x}(\bar{T}))$
$$+ N_C(\bar{S}, \bar{x}(\bar{S}), \bar{T}, \bar{x}(\bar{T}));$$

(iv) $p(t) \cdot \dot{\bar{x}}(t) \geq p(t) \cdot v$ for all $v \in F(t, \bar{x}(t))$ a.e. $t \in [\bar{S}, \bar{T}]$,

(v) $r(t) = H(t, \bar{x}(t), p(t))$ a.e. $t \in [\bar{S}, \bar{T}]$.

Now suppose that

$$F(t,x) \text{ is convex for all } (t,x) \in R \times R^n$$

and (H3) is replaced by the weaker hypothesis:

(H3)' there exist $k_F \in L^1 >$, $\epsilon > 0$, and $\bar{K} \geq 0$ such that, either

(a) $F(t',x') \cap (\dot{\bar{x}}(t) + \epsilon k_F(t)B) \subset F(t'',x'') + k_F(t)|(t',x') - (t'',x'')|B$
for all $(t',x'), (t'',x'') \in (t, \bar{x}(t)) + \epsilon B$ a.e. $t \in [\bar{S}, \bar{T}]$,

or

(b) $\begin{cases} F(t',x') \cap (\dot{\bar{x}}(t) + \epsilon B) \subset F(t'',x'') + k_F(t)|(t',x') - (t'',x'')|B, \\ \inf\{|v - \dot{\bar{x}}(t)| : v \in F(t',x')\} \leq \bar{K}|(t',x') - (t,\bar{x}(t))|, \end{cases}$

for all $(t',x'), (t'',x'') \in (t,\bar{x}(t)) + \epsilon B$ a.e. $t \in [\bar{S},\bar{T}]$.

Then the above assertions remain true and, furthermore, Condition (ii) implies

(ii)' $(\dot{r}(t), -\dot{p}(t)) \in \text{co}\{(\zeta,\eta) : ((\zeta,\eta),\dot{\bar{x}}(t)) \in \partial H(t,\bar{x}(t),p(t))\}$ a.e.

One of the most important aspects of this theorem is its implications for autonomous problems (problems for which $F(t,x)$ is independent of t). In this case, Condition (ii) of the theorem implies that the adjoint arc p satisfies

$$\dot{p}(t) \in \text{co}\{\eta : (\eta,p(t)) \in N_{\text{Gr }F}((\bar{x}(t)),\dot{\bar{x}}(t))\} \quad \text{a.e.}$$

(the Extended Euler Condition). The theorem also tells us (via Conditions (ii), (iii), and (v)) that the Hamiltonian, evaluated along $(\bar{x},p)$, is almost everywhere equal to a constant and supplies supplementary information about the value of this constant.

Proof of Theorem 8.2.1. Assume that, in addition to Hypotheses (H1) and (H2), either (H3) is satisfied or F takes as values convex sets and (H3)' is satisfied.

Since $([\bar{S},\bar{T}],\bar{x})$ is a $W^{1,1}$ local minimizer, there exists $\delta' \in (0,\delta)$ such that $([\bar{S},\bar{T}],\bar{x})$ is a minimizer with respect to all F-trajectories $([S,T],x)$ satisfying the constraints of (FT) and for which

$$d(([S,T],x),([\bar{S},\bar{T}],\bar{x})) < 5\delta'.$$

(See (8.3) for the definition of $d(.,.)$.)

Now choose a twice continuously differentiable function $a : [\bar{S},\bar{T}] \to R^n$ such that

$$a(\bar{S}) = \bar{x}(\bar{S})$$

and

$$d(([\bar{S},\bar{T}],a),([\bar{S},\bar{T}],\bar{x})) \left(= \int_{\bar{S}}^{\bar{T}} |\dot{\bar{x}}(t) - \dot{a}(t)|dt\right) < \delta'.$$

Choose also $\alpha^* \in (0,1/2)$ such that

$$\frac{1 - (1-\alpha^*)^2}{(1-\alpha^*)}\|\dot{a}\|_{L^\infty}|\bar{T} - \bar{S}| \leq \delta'.$$

Fix $\alpha \in (0,\alpha^*)$ and set

$$\rho(s) := \frac{\alpha}{1 + |\dot{a}(s)|}.$$

Consider now the *fixed end-time* optimal control problem:

$$(C) \begin{cases} \text{Minimize } g(\tau(\bar{S}), y(\bar{S}), \tau(\bar{T}), y(\bar{T})) \\ \text{over } (\tau, y, z) \in W^{1,1}([\bar{S}, \bar{T}]; R^{1+n+1}) \text{ satisfying} \\ (\dot{\tau}, \dot{y}, \dot{z}) \in \tilde{F}(s, \tau, y, z) \quad \text{a.e.,} \\ (\tau(\bar{S}), y(\bar{S}), \tau(\bar{T}), y(\bar{T})) \in C, \\ m(\tau(\bar{S}), \tau(\bar{T}), y(\bar{S}), z(\bar{S}), z(\bar{T})) \leq \delta'. \end{cases}$$

Here $\tilde{F} : R \times R \times R^n \times R \rightsquigarrow R \times R^n \times R$ is the multifunction

$$\tilde{F}(s, \tau, y, z) :=$$
$$\{(w, wv, \zeta) : w \in [1 - \rho(s), 1 + \rho(s)], \ v \in F(\tau, y), \ \zeta \geq |wv - \dot{a}(\tau)|\}$$

and

$$m(\tau_0, \tau_1, y_0, z_0, z_1) := |\tau_0 - \bar{S}| + |\tau_1 - \bar{T}| + |y_0 - a(\bar{S})|$$
$$+ \int_{\bar{S}}^{\bar{S} \vee \tau_0} |\dot{a}(\sigma)| d\sigma + \int_{\bar{T} \wedge \tau_1}^{\bar{T}} |\dot{a}(\sigma)| d\sigma + z_1 - z_0.$$

An important feature of these constructions is that $\tilde{F}$ is convex-valued if F is convex-valued.

We claim that

$$(\bar{\tau}(s) \equiv s, \ \bar{x}(.), \ \bar{z}(s) \equiv \int_{\bar{S}}^{s} |\dot{\bar{x}}(\sigma) - \dot{a}(\sigma)| d\sigma)$$

is a minimizer for this problem.

To see this, take any (τ, y, z) satisfying the constraints of (C). In consequence of the Generalized Filippov Selection Theorem (Theorem 2.3.13), there exist measurable functions w and v such that

$$w(s) \in [1 - \rho(s), 1 + \rho(s)] \text{ and } v(s) \in F(\tau(s), y(s)) \text{ a.e.,}$$
$$\dot{\tau}(s) = w(s), \ \dot{y}(s) = w(s)v(s) \text{ a.e.}$$

and

$$|\tau(\bar{S}) - \bar{S}| + |\tau(\bar{T}) - \bar{T}| + |y(\bar{S}) - a(\bar{S})| + \int_{\bar{S}}^{\bar{T}} |w(s)v(s) - \dot{a}(\tau(s))| ds$$
$$+ \int_{\bar{S}}^{\bar{S} \vee \tau(\bar{S})} |\dot{a}(\sigma)| d\sigma + \int_{\bar{T} \wedge \tau(\bar{T})}^{\bar{T}} |\dot{a}(\sigma)| d\sigma \leq \delta'.$$

(We have used the extension convention to evaluate the integrals.) But

$$\int_{\bar{S}}^{\bar{T}} |w(s)v(s) - \dot{a}(\tau(s))| ds = \int_{\bar{S}}^{\bar{T}} w(s)|v(s) - w^{-1}(s)\dot{a}(\tau(s))| ds$$
$$\geq (1 - \alpha) \left[\int_{\bar{S}}^{\bar{T}} |v(s) - w(s)\dot{a}(\tau(s))| ds \right]$$
$$- \left(1 - (1 - \alpha)^2 \right) \|\dot{a}\|_{L^\infty} |\bar{T} - \bar{S}|.$$

It follows that

$$\int_{\bar{S}}^{\bar{T}} |v(s) - w(s)\dot{a}(\tau(s))|ds \leq \frac{1}{(1-\alpha)}\delta + \frac{1-(1-\alpha)^2)}{(1-\alpha)}||\dot{a}||_{L^\infty}|\bar{T} - \bar{S}|.$$

Then

$$|\tau(\bar{S}) - \bar{S}| + |\tau(\bar{T}) - \bar{T}| + |y(\bar{S}) - a(\tau(\bar{S}))|$$

$$+ \int_{\bar{S}}^{\bar{T}} |v(s)/w(s) - \dot{a}(\tau(s))|w(s)ds$$

$$+ \int_{\bar{S}}^{\bar{S}\vee\tau(\bar{S})} |\dot{a}(\sigma)|d\sigma + \int_{\bar{T}\wedge\tau(\bar{T})}^{\bar{T}} |\dot{a}(\sigma)|d\sigma$$

$$\leq (1 + \frac{1}{(1-\alpha)})\delta' + \frac{1-(1-\alpha)^2}{(1-\alpha)}||\dot{a}||_{L^\infty}|\bar{T} - \bar{S}| < 4\delta'.$$

Consider now the transformation $\psi : [\bar{S}, \bar{T}] \to [S, T]$,

$$S = \tau(\bar{S}), \quad T = \tau(\bar{T}), \quad \psi(s) = \tau(\bar{S}) = \int_{\bar{S}}^{s} w(\sigma)d\sigma.$$

ψ is strictly increasing and invertible. ψ and ψ^{-1} are absolutely continuous functions with essentially bounded derivatives. It can be deduced that $x(t) := y \circ \psi^{-1}(t)$ is absolutely continuous and satisfies

$$\dot{x}(t) \in F(t, x(t)) \quad \text{a.e. } t \in [S, T],$$
$$(\tau(\bar{S}), y(\bar{S}), \bar{T}, y(\bar{T})) = (S, x(S), T, x(T))$$

and

$$|S - \bar{S}| + |T - \bar{T}| + |x(S) - a(\bar{S})| + \int_{S\wedge\bar{S}}^{T\vee\bar{T}} |\dot{x}(t) - \dot{a}(t)|dt < 4\delta'.$$

In terms of the metric $d(.,.)$,

$$d(([S, T], x)([\bar{S}, \bar{T}], a)) < 4\delta'.$$

With the help of the triangle inequality, we deduce

$$d(([S, T], x)([\bar{S}, \bar{T}], \bar{x})) \leq$$
$$d(([S, T], x)([\bar{S}, \bar{T}], a)) + d(([S, T], a)([\bar{S}, \bar{T}], \bar{x})) < 5\delta'.$$

By the minimizing property of $([\bar{S}, \bar{T}], \bar{x})$ then,

$$g(\tau(\bar{S}), y(\bar{S}), \tau(\bar{T}), y(\bar{T})) = g(S, x(S), T, x(T)) \geq g(\bar{S}, \bar{x}(\bar{S}), \bar{T}, \bar{x}(\bar{T})).$$

But $(\bar{\tau}, \bar{x}, \bar{z})$ satisfies the constraints of (C) and has cost $g((\bar{S}, \bar{x}(\bar{S}), \bar{T}, \bar{x}(\bar{T})))$. It follows that $(\bar{\tau}, \bar{x}, \bar{z})$ is a minimizer for (C), as claimed.

As a prelude to applying the necessary conditions of Chapter 7 to (C), we check that the data for this problem satisfy the relevant hypotheses.

If F satisfies (H2) and (H3), then $\tilde{F}$ is an $\mathcal{L} \times \mathcal{B}^{1+n+1}$ measurable multifunction that takes as values closed nonempty sets and $\tilde{F}(s,.)$ has a closed graph for each s. Furthermore there exist $\beta_1 > 0$, $k_1 \in L^1$, and $\epsilon_1 > 0$ such that, for a.e. $s \in [\bar{S}, \bar{T}]$,

$$\tilde{F}(s, \tau', y', z') \cap ((\dot{\bar{\tau}}(s), \dot{\bar{x}}(s), \dot{\bar{z}}(s)) + NB)$$
$$\subset \tilde{F}(s, \tau'', y'', z'') + (k_1(s) + \beta_1 N)|(\tau', y', z') - (\tau'', y'', z'')|B$$

for all (τ', y', z'), $(\tau'', y'', z'') \in (s, \bar{x}(s), \bar{z}(s)) + \epsilon_1 B$.

(It is precisely to ensure this estimate that we have chosen $\rho(t)$ to be small when $|\dot{\bar{x}}(t)|$ is large.)

On the other hand, if $\tilde{F}$ satisfies (H3)$'$ and F is convex-valued then, as we have observed, $\tilde{F}$ is convex-valued. It can be shown that, provided the parameter α is chosen sufficiently small, there exist $k_2 \in L^1$, $\epsilon_2 > 0$, and K_2 such that, for a.e. $s \in [\bar{S}, \bar{T}]$,

$$\tilde{F}(s, \tau', y', z') \cap ((\dot{\bar{\tau}}(s), \dot{\bar{x}}(s), \dot{\bar{z}}(s)) + \epsilon_2 B)$$
$$\subset \tilde{F}(s, \tau'', y'', z'') + k_2(s)|(\tau', y', z') - (\tau'', y'', z'')|B,$$

$$\inf\{|d - (1, \dot{\bar{x}}(s), \dot{\bar{z}}(s))| \; : \; d \in \tilde{F}(\tau', y', z')\} \leq$$
$$K_2|(\tau', y', z') - (s, \bar{x}(s), \bar{z}(s))|$$

for all (τ', y', z'), $(\tau'', y'', z'') \in (s, \bar{x}(s), \bar{z}(s)) + \epsilon_2 B$.

We have justified application of Theorems 7.4.1 and 7.7.1 to (C), with reference to the minimizer $(\bar{\tau}, \bar{x}, \bar{z})$. We deduce that there exist $\lambda \geq 0$, and adjoint arc components $-r$, p, and γ such that

$$(\lambda, (r, p, \gamma)) \neq (0, (0, 0, 0)),$$

and, for a.e. $s \in [\bar{S}, \bar{T}]$,

$$(-\dot{r}(s), \dot{p}(s), \dot{\gamma}(s)) \; \in \; \mathrm{co}\{(\eta_1, \eta_2, \eta_s) : ((\eta_1, \eta_2, \eta_s), (-r(s), p(s), \gamma(s)))$$
$$\subset \; N_{\mathrm{Gr}\tilde{F}(s,\cdot,\cdot,\cdot)}((s, \bar{x}(s), \bar{z}(s)), (1, \dot{\bar{x}}(s), \dot{\bar{z}}(s)))\} \quad (8.6)$$

and

$$w[-r(s) + p(s) \cdot v + \gamma(s)|wv - \dot{a}(s)|] \leq$$
$$- r(s) + p(s) \cdot \dot{\bar{x}}(s) + \gamma(s)|\dot{\bar{x}}(s) - \dot{a}(s)| \quad (8.7)$$

for all $v \in F(s, \bar{x}(s))$ and $w \in [1 - \rho(s), 1 + \rho(s)]$.

The Tranversality Condition yields the information that $\gamma(\bar{T}) = 0$ and

$$(-r(\bar{S}), p(\bar{S}), +r(\bar{T}), -p(\bar{T}))$$
$$\in N_C(\bar{S}, \bar{x}(\bar{S}), \bar{T}, \bar{x}(\bar{T})) + \lambda \partial g(\bar{S}, \bar{x}(\bar{S}), \bar{T}, \bar{x}(\bar{T})).$$

(Note that the constraint

$$m(\tau(\bar{S}), \tau(\bar{T}), y(\bar{S}), z(\bar{S}), z(\bar{T})) \leq \delta'$$

is inactive for arcs with endpoints near those of $(\bar{\tau}, \bar{x}, \bar{z})$, and can therefore be ignored so far as the analysis of the Transversality Condition is concerned.)

A careful analysis of the Euler–Lagrange Condition (8.6), based on examining proximal normals and applying the Sum Rule (Theorem 5.4.1), reveals that $\dot{\gamma} \equiv 0$ and

$$(-\dot{r}(s), \dot{p}(s)) \in$$
$$\mathrm{co}\,\{(\eta_1, \eta_2) : ((\eta_1, \eta_2), p(t)) \in N_{\mathrm{Gr}F}((s, \bar{x}(s)), \dot{\bar{x}}(s)) + |\gamma(s)|B\} \quad \text{a.e.}$$

From the foregoing,

$$\gamma \equiv 0.$$

We deduce from Condition (8.7) that, for a.e. $s \in [\bar{S}, \bar{T}]$,

$$p(s) \cdot \dot{\bar{x}}(s) = \max_{v \in F(s, \bar{x}(s))} p(s) \cdot v \quad (\,= H(s, \bar{x}(s), p(s))\,)$$

and

$$r(s) = H(s, \bar{x}(s), p(s)).$$

This last condition implies that $r \equiv 0$ if $p \equiv 0$. Since $\gamma \equiv 0$ and $(\lambda, r, p, \gamma) \neq (0,0,0,0)$, it follows that $(\lambda, p) \neq (0,0)$. All the relevant relationships have been confirmed. The proof is complete. $\square$

8.3 Essential Values

The operation of taking the "essential value" of a given real-valued function on the real line is a generalization of point evaluation of a continuous function. Its most important property is that the essential values of a function are unaltered if the function is adjusted on a set of Lebesgue measure zero. This is exploited to interpret necessary conditions for free end-time optimal control problems with measurably time-dependent data, specifically to make sense of boundary conditions on the Hamiltonian.

Definition 8.3.1 *Take an open interval $I \subset R$, an essentially bounded function $f : I \to R$, and a point $t \in I$. The essential value of f at t is the set*

$$\underset{\tau \to t}{\mathrm{ess}}\, f(\tau) := [a^-, a^+],$$

where

$$a^- := \lim_{\delta \downarrow 0} \mathrm{ess\,inf}_{t-\delta \leq \tau \leq t+\delta} f(\tau)$$

and

$$a^+ := \lim_{\delta \downarrow 0} \mathrm{ess\,sup}_{t-\delta \leq \tau \leq t+\delta} f(\tau).$$

The operation of taking essential values generates a multifunction

$$t \to \operatorname*{ess}_{\tau \to t} f(\tau)$$

taking as values closed, possibly unbounded, intervals. Salient properties of essential values are summarized in the following proposition.

Proposition 8.3.2 *Take an open interval $I \subset R$ and a set $A \subset R^n$.*

(i) If an essentially bounded function $f : I \to R$ has left and right limits $f(t^-)$ and $f(t^+)$ at a point $t \in I$, then

$$\operatorname*{ess}_{\tau \to t} f(\tau) = [\alpha^-, \alpha^+],$$

where

$$\alpha^- := \min\{f(t^-), f(t^+)\} \quad \text{and} \quad \alpha^+ := \max\{f(t^-), f(t^+)\}.$$

It follows that, if f is continuous at t, then

$$\operatorname*{ess}_{\tau \to t} f(\tau) = \{f(t)\}.$$

(ii) If $f : I \to R$ and $g : I \to R$ are two essentially bounded functions such that

$$f(t) \geq g(t) \quad a.e.,$$

then, for each $t \in I$,

$$\operatorname*{ess}_{\tau \to t} f(\tau) \geq \operatorname*{ess}_{\tau \to t} g(\tau).$$

It follows that, if f and g coincide almost everywhere, then

$$\operatorname*{ess}_{\tau \to t} f(\tau) = \operatorname*{ess}_{\tau \to t} g(\tau).$$

(iii) For any essentially bounded, measurable function $f : I \to R$, $\xi \in R$, $t \in I$, and $\sigma_i \downarrow 0$ such that

$$\lim_{i \to 0} \sigma_i^{-1} \int_t^{t+\sigma_i} f(\sigma) d\sigma = \xi,$$

we have

$$\xi \in \operatorname*{ess}_{\sigma \to t} f(\sigma).$$

(iv) Take a function $d : I \times A \to R$ such that $d(.\,, x)$ is essentially bounded for each x and $d(\tau, \cdot)$ is continuous on A, uniformly with respect to $\tau \in I$. Then for any convergent sequences $x_i \xrightarrow{A} x$, $t_i \xrightarrow{I} t$, and $\xi_i \to \xi$ such that

$$\xi_i \in \operatorname*{ess}_{\tau \to t_i} d(\tau, x_i) \quad \text{for all } i,$$

we have that

$$\xi \in \operatorname*{ess}_{\tau \to t} d(\tau, x).$$

Proof. Properties (i) through (iii) of the essential value are more or less immediate consequences of the definition.

Consider (iv). Fix $\Delta > 0$ such that $(t - \Delta, t + \Delta) \subset I$. Then, by uniform continuity, there exists $\epsilon_i \downarrow 0$ such that

$$\xi_i \geq \text{ess inf}_{\,t_i - \Delta/2 \leq \tau \leq t_i + \Delta/2}\, d(\tau, x_i) \geq \text{ess inf}_{\,t - \Delta \leq \tau \leq t + \Delta}\, d(\tau, x) - \epsilon_i$$

for all i sufficiently large. It follows that

$$\xi = \lim_i \xi_i \geq \text{ess inf}_{\,t - \Delta \leq \tau \leq t + \Delta}\, d(\tau, x).$$

Likewise we show that

$$\xi \leq \text{ess sup}_{\,t - \Delta \leq \tau \leq t + \Delta}\, d(\tau, x).$$

Since these relationships are true for all $\Delta > 0$ we conclude that

$$\xi \in \text{ess}_{\tau \to t}\, d(\tau, x).$$

$\square$

8.4   Measurable Time Dependence

In this section, we derive necessary conditions for (FT), under hypotheses that require the differential inclusion to have right side $F(t, x)$ which is merely measurable regarding its time dependence.

The motivation for treating the measurable time-dependence case is partly to unify the theory of necessary conditions for fixed and free end-time optimal control problems. A framework that requires the dynamic constraint to be merely measurable with respect to time is widely adopted for fixed end-time problems. Why should extra regularity be required for free end-time problems?

But there are also practical reasons for developing a theory of free end-time problems, which allows the "dynamic constraint" to be discontinuous with respect to time. Optimal control problems arising in resource economics, for example, typically require us to find harvesting/investment strategies to minimize a cost that involves an integral cost of the form

$$-\int_0^T c(t, x(t), u(t))dt.$$

Here, c is a given function representing the rate of return on harvesting effort and investment. It is natural in this context to consider problems in which T is a choice variable; the company can choose the harvesting period. Special cases are of interest, in which c is discontinuous with respect to time, to take account, for example, of abrupt changes in interest rates (reflecting, perhaps, penalties incurred for late completion). When the integral cost term is absorbed into the dynamics by means of "state augmentation" the resulting dynamics will be discontinuous with respect to time. Furthermore, since we can expect an optimal strategy to terminate at the instant when there is an abrupt unfavorable change in the rate of return, it is not satisfactory to develop a theory in which it is assumed that the optimal end-time occurs at a point of continuity of the data; rather we should allow for the possibility that discontinuities and optimal end-times interact.

The first set of free end-time necessary conditions is centered on the Extended Euler–Lagrange Condition. In these conditions, $H(t, x, p)$, as usual, denotes the Hamiltonian:

$$H(t,x,p) = \sup_{v \in F(t,x)} p \cdot v.$$

Theorem 8.4.1 (Free End-Time Euler–Lagrange Condition: Measurable Time Dependence) *Let $([\bar{S}, \bar{T}], \bar{x})$ be a $W^{1,1}$ local minimizer for (FT) such that $\bar{T} - \bar{S} > 0$. Assume that, for some $\delta > 0$, the following hypotheses are satisfied.*

(H1) g is locally Lipschitz and C is a closed set.

(H2) $F(t, x)$ is a nonempty closed set for all $(t, x) \in R \times R^n$, $\mathrm{Gr}\, F(t, .)$ is closed for each $t \in R$, and F is $\mathcal{L} \times \mathcal{B}^n$ measurable.

(H3) There exist $\beta \geq 0$ and $k_F \in W^{1,1}([\bar{S}, \bar{T}]; R)$ such that, for a.e. $t \in [\bar{S}, \bar{T}]$,

$$F(t, x') \cap (\dot{\bar{x}}(t) + NB) \subset F(t, x'') + (k_F(t) + \beta N)|x' - x''|B$$

for all $N \geq 0$, $x', x'' \in \bar{x}(t) + \delta B$.

(H4) There exist $c_0 \geq 0$, $k_0 \geq 0$, and $\delta_0 > 0$ such that, for a.e. $t \in [\bar{S} - \delta_0, \bar{S} + \delta_0]$,

$$\begin{cases} F(t, x) \in c_0 B & \text{for all } x \in \bar{x}(\bar{S}) + \delta B \\ F(t, x') \subset F(t, x'') + k_0|x' - x''|B & \text{for all } x', x'' \in \bar{x}(\bar{S}) + \delta B. \end{cases}$$

(H5) There exist $c_1 \geq 0$, $k_1 \geq 0$, and $\delta_1 > 0$ such that, for a.e. $t \in [\bar{T} - \delta_1, \bar{T} + \delta_1]$,

$$\begin{cases} F(t, x) \in c_1 B & \text{for all } x \in \bar{x}(\bar{T}) + \delta B \\ F(t, x') \subset F(t, x'') + k_1|x' - x''|B & \text{for all } x', x'' \in \bar{x}(\bar{T}) + \delta B. \end{cases}$$

Then there exist $p \in W^{1,1}([\bar{S}, \bar{T}]; R^n)$ and real numbers $\lambda \geq 0$, ξ, and η such that *(i)* $(\lambda, p) \neq (0,0)$;

(ii) $\dot{p}(t) \in \text{co}\{\zeta : (\zeta, p(t)) \in N_{\text{Gr} F(t,\cdot)}(\bar{x}(t), \dot{\bar{x}}(t))\}$ *a.e.* $t \in [\bar{S}, \bar{T}]$;

(iii) $(-\xi, p(\bar{S}), \eta, -p(\bar{T}))$

$$\in \lambda \partial g(\bar{S}, \bar{x}(\bar{S}), \bar{T}, \bar{x}(\bar{T})) + N_C(\bar{S}, \bar{x}(\bar{S}), \bar{T}, \bar{x}(\bar{T}));$$

(iv) $p(t) \cdot \dot{\bar{x}}(t) \geq p(t) \cdot v$ *for all* $v \in F(t, \bar{x}(t))$ *a.e.* $t \in [\bar{S}, \bar{T}]$;

(v) $\xi \in \text{ess}_{t \to \bar{S}} H(t, \bar{x}(\bar{S}), p(\bar{S}))$;

(vi) $\eta \in \text{ess}_{t \to \bar{T}} H(t, \bar{x}(\bar{T}), p(\bar{T}))$.

If the initial time is fixed (i.e., $C = \{\bar{S}\} \times \tilde{C}\}$ for some $\bar{S} \in R$ and some set $\tilde{C} \subset R^{n+1+n}$), then the above assertions (except (v)) remain true when (H4) is dropped from the hypotheses. If the final time is fixed then the assertions (except (vi)) remain true when (H5) is dropped from the hypotheses.

It is clear from the concluding assertions that this theorem captures the fixed end-time necessary condition Theorem 7.4.1 as a special case. (When both end-times are fixed, we can drop both Hypotheses (H4) and (H5) and (iii) reduces to the customary Transversality Condition for $p(\bar{S})$ and $p(\bar{T})$.)

We now assume that the velocity set $F(t, x)$ is convex. As for fixed end-time problems, it is possible to derive the necessary conditions when the Lipschitz continuity hypothesis (H3) is replaced by a weaker local hypothesis. A version of the Dualization Theorem (Theorem 7.6.8) permits us also to augment the Extended Euler–Lagrange Condition with the Extended Hamilton Condition.

Theorem 8.4.2 Free End-Time Hamilton Condition: Measurable Time Dependence) *Let $([\bar{S}, \bar{T}], \bar{x})$ be a $W^{1,1}$ local minimizer for (FT), such that $\bar{T} > \bar{S}$. Assume that*

$$F(t, x) \text{ is convex for each } (t, x).$$

Assume also that, for some $\delta > 0$, the following hypotheses are satisfied.

(H1) g is locally Lipschitz continuous and C is a closed set.

(H2) $F(t, x)$ is a nonempty closed set for all $(t, x) \in R \times R^n$, $\text{Gr} F(t, .)$ is closed for each $t \in R$, and F is $\mathcal{L} \times \mathcal{B}^n$ measurable,

(H3) There exist $K \geq 0$, $\epsilon > 0$, and $k_F \in W^{1,1}([\bar{S}, \bar{T}]; R)$ such that either

(a) $F(t, x) \cap (\dot{\bar{x}}(t) + \epsilon k_F(t) B) \subset F(t, x') + k_F(t)|x - x'|B$,

$$\text{for all } x, x' \in \bar{x}(t) + \delta B \text{ a.e. } t \in [\bar{S}, \bar{T}]$$

or

(b) $\begin{cases} F(t,x) \cap (\dot{\bar{x}}(t) + \epsilon B) \subset F(t,x') + k_F(t)|x-x'|B, \\ \inf\{|v - \dot{\bar{x}}(t)| : v \in F(t,x)\} \leq K|x - \bar{x}(t)| \end{cases}$

$$\text{for all } x, x' \in \bar{x}(t) + \delta B \text{ a.e. } t \in [\bar{S}, \bar{T}].$$

(H4) *There exist* $c_0 \geq 0$, $k_0 \geq 0$, *and* $\delta_0 > 0$ *such that, for a.e.* $t \in$ $[\bar{S} - \delta_0, \bar{S} + \delta_0]$,

$$\begin{cases} F(t,x) \in c_0 B \quad \text{for all } x \in \bar{x}(\bar{S}) + \delta B \\ F(t,x') \subset F(t,x'') + k_0|x' - x''|B \quad \text{for all } x', x'' \in \bar{x}(\bar{S}) + \delta B. \end{cases}$$

(H5) *There exist* $c_1 \geq 0$, $k_1 \geq 0$, *and* $\delta_1 > 0$ *such that, for a.e.* $t \in$ $[\bar{T} - \delta_1, \bar{T} + \delta_1]$,

$$\begin{cases} F(t,x) \in c_1 B \quad \text{for all } x \in \bar{x}(\bar{T}) + \delta B \\ F(t,x') \subset F(t,x'') + k_1|x' - x''|B \quad \text{for all } x', x'' \in \bar{x}(\bar{T}) + \delta B. \end{cases}$$

Then there exist an arc $p \in W^{1,1}([\bar{S}, \bar{T}]; R^n)$ *and real numbers* $\lambda \geq 0$, ξ, *and* η *such that*

(i) $(\lambda, p) \neq (0, 0)$;

(ii) $\dot{p}(t) \in \mathrm{co}\{\zeta : (\zeta, p(t)) \in N_{\mathrm{Gr}F(t,\cdot)}(\bar{x}(t), \dot{\bar{x}}(t))\}$ a.e. $t \in [\bar{S}, \bar{T}]$;

(iii) $(-\xi, p(\bar{S}), \eta, -p(\bar{T})) \in$

$$\lambda \partial g(\bar{S}, \bar{x}(\bar{S}), \bar{T}, \bar{x}(\bar{T})) + N_C(\bar{S}, \bar{x}(\bar{S}), \bar{T}, \bar{x}(\bar{T}));$$

(iv) $\xi \in \mathrm{ess}_{t \to \bar{S}} H(t, \bar{x}(\bar{S}), p(\bar{S}))$;

(v) $\eta \in \mathrm{ess}_{t \to \bar{T}} H(t, \bar{x}(\bar{T}), p(\bar{T}))$.

Under the stated hypotheses, (ii) *implies that* p *also satisfies the conditions*

(I) $\dot{p}(t) \in \mathrm{co}\{-\zeta : (\zeta, \dot{\bar{x}}(t)) \in \partial H(t, \bar{x}(t), p(t))\}$ a.e. $t \in [\bar{S}, \bar{T}]$

and

(II) $p(t) \cdot \dot{\bar{x}}(t) \geq p(t) \cdot v$ *for all* $v \in F(t, \bar{x}(t))$ a.e. $t \in [\bar{S}, \bar{T}]$.

If the initial time is fixed (i.e., $C = \{\bar{S}\} \times \tilde{C}$ *for some* $\bar{S} \in R$ *and some set* $\tilde{C} \subset R^{n+1+n}$), *then the above assertions (except* (iv)) *remain true when* (H4) *is dropped from the hypotheses. If the final time is fixed, then the assertions (except* (v)) *remain true when* (H5) *is dropped from the hypotheses.*

8.5   Proof of Theorem 8.4.1

The proof breaks down into several steps.

Step 1 (A special free right end-time problem):   Take a point $\bar{S} \in R$, a number $\epsilon \geq 0$, a set $\tilde{C} \in R^n$, and functions $g_1 : R^n \to R$, $g_2 : R \times R^n \to R$, and $g_3 : R \to R$. Let $([\bar{S}, \bar{T}], \bar{x})$ be a $W^{1,1}$ local minimizer for

$$\begin{cases} \text{Minimize } g_1(x(\bar{S})) + g_2(T, x(T)) + \epsilon g_3(T) \\ \text{over arcs } ([\bar{S}, T], x) \text{ satisfying} \\ \dot{x}(t) \in F(t, x(t)) \text{ a.e.,} \\ x(\bar{S}) \in \tilde{C}. \end{cases}$$

Assume that the data for this optimization problem, regarded as a special case of (FT), satisfy Hypotheses (H1) to (H3) and (H5). Assume further that g_1 is Lipschitz continuous, g_2 is twice continuously differentiable, and g_3 is Lipschitz continuous.

We show that there exist $p \in W^{1,1}([\bar{S}, \bar{T}]; R^n)$ and $\lambda \geq 0$ such that

(A) $(\lambda, p) \neq (0, 0)$,

(B) $\dot{p}(t) \in \text{co}\{\zeta : (\zeta, p(t)) \in N_{\text{Gr} F(t, \cdot)}(\bar{x}(t), \dot{\bar{x}}(t))\}$ a.e.,

(C) $p(\bar{S}) \in \lambda \partial g_1(\bar{x}(\bar{S})) + N_{\tilde{C}}(\bar{x}(\bar{S}))$, $-p(\bar{T}) = \lambda \nabla_x g_2(\bar{T}, \bar{x}(\bar{T}))$,

(D) $p(t) \cdot \dot{\bar{x}}(t) \geq p(t) \cdot v$ for all $v \in F(t, \bar{x}(t))$ a.e., and

(E) $\lambda \nabla_t g_2(\bar{T}, \bar{x}(\bar{T})) \subset \text{ess}_{t \to \bar{T}} H(t, \bar{x}(\bar{T}), p(\bar{T})) + \lambda \epsilon k_3 B$,

in which k_3 is a Lipschitz constant for g_3.

Conditions (A) through (D) follow directly from the fixed end-time necessary conditions (Theorem 7.4.1). It remains to verify Condition (E).

For $\sigma > 0$ sufficiently small $([\bar{S}, \bar{T} - \sigma], \bar{x}|_{[\bar{S}, \bar{T} - \sigma]})$ must have cost not less than $([\bar{S}, \bar{T}], \bar{x})$ since $([\bar{S}, \bar{T}], \bar{x})$ is a $W^{1,1}$ local minimizer. It follows that

$$g_2(\bar{T} - \sigma, \bar{x}(\bar{T} - \sigma)) + \epsilon g_3(\bar{T} - \sigma) \geq g_2(\bar{T}, \bar{x}(\bar{T})) + \epsilon g_3(\bar{T}).$$

Since g_2 is C^2 near $(\bar{T}, \bar{x}(\bar{T}))$ there exists $r_1 > 0$ such that for all σ sufficiently small

$$\begin{aligned} 0 &\leq g_2(\bar{T} - \sigma, \bar{x}(\bar{T} - \sigma)) - g_2(\bar{T}, \bar{x}(\bar{T})) + \epsilon k_3 \sigma \\ &= g_2(\bar{T} - \sigma, \bar{x}(\bar{T})) - \int_{\bar{T} - \sigma}^{\bar{T}} \dot{\bar{x}}(t) \, dt - g_2(\bar{T}, \bar{x}(\bar{T})) + \epsilon k_3 \sigma \\ &\leq -\nabla_t g_2(\bar{T}, \bar{x}(\bar{T})) \sigma - \int_{\bar{T} - \sigma}^{\bar{T}} \nabla_x g_2(\bar{T}, \bar{x}(\bar{T})) \cdot \dot{\bar{x}}(t) \, dt + r_1 \sigma^2 + \epsilon k_3 \sigma. \end{aligned}$$

We know however that $-p(\bar{T}) = \lambda \nabla_x g_2(\bar{T}, \bar{x}(\bar{T}))$, so

$$\limsup_{\sigma \downarrow 0} \sigma^{-1} \int_{\bar{T} - \sigma}^{\bar{T}} p(\bar{T}) \cdot \dot{\bar{x}}(t) \, dt \geq \lambda \nabla_t g_2(\bar{T}, \bar{x}(\bar{T})) - \lambda \epsilon k_3. \tag{8.8}$$

By (H5), we have, for a.e. $t \in [\bar{T} - \sigma, \bar{T}]$,

$$\dot{x}(t) \in F(t, \bar{x}(t)) \subset F(t, \bar{x}(\bar{T})) + k_1 |\bar{x}(t) - \bar{x}(\bar{T})| B \subset F(t, \bar{x}(\bar{T})) + k_1 c_1 \sigma B.$$

It follows that, for a.e. $t \in [\bar{T} - \sigma, \bar{T}]$,

$$\begin{aligned} p(\bar{T}) \cdot \dot{x}(t) \ \leq \ & \max\{p(\bar{T}) \cdot v : v \in F(t, \bar{x}(\bar{T}))\} \\ & + k_1 c_1 \sigma |p(\bar{T})| \quad \text{a.e. } t \in [\bar{T} - \sigma, \bar{T}]. \end{aligned}$$

We conclude from (8.8) that

$$\lim_{\sigma \downarrow 0} \operatorname*{ess\,sup}_{\bar{T} - \sigma \leq t \leq \bar{T} + \sigma} \{H(t, \bar{x}(\bar{T}), p(\bar{T}))\} \geq \lambda \nabla_t g_2(\bar{T}, \bar{x}(\bar{T})) - \lambda \epsilon k_3. \tag{8.9}$$

Select a measurable function $\xi : [\bar{T}, \bar{T} + \delta_1] \to R^n$ such that $\xi(t) \in F(t, \bar{x}(\bar{T}))$ a.e. and

$$p(\bar{T}) \cdot \xi(t) = \max_{v \in F(t, \bar{x}(\bar{T}))} p(\bar{T}) \cdot v, \quad \text{a.e. } t \in [\bar{T}, \bar{T} + \delta_1].$$

By Filippov's Existence Theorem (Theorem 2.3.13), there exists $\sigma' > 0$ and an F-trajectory y on $[\bar{T}, \bar{T} + \sigma']$ such that $y(\bar{T}) = \bar{x}(\bar{T})$ and

$$\int_{\bar{T}}^{\bar{T}+\sigma} |\dot{y}(t) - \xi(t)| dt \leq (1/2) e^{k_1 \sigma} k_1 c_1 \sigma^2$$

for all $\sigma \in [0, \sigma']$.

We construct an F-trajectory $([\bar{S}, \bar{T} + \sigma'], x)$ by concatenating $([\bar{S}, \bar{T}], \bar{x})$ and $([\bar{T}, \bar{T} + \sigma'], y)$. Since this F-trajectory satisfies the constraints of the optimal control problem, its cost cannot exceed that of $([\bar{S}, \bar{T}], \bar{x})$ (for σ sufficiently small). It follows that

$$0 \leq g_2(\bar{T} + \sigma, y(\bar{T} + \sigma)) - g_2(\bar{T}, \bar{x}(\bar{T})) + \epsilon k_3 \sigma$$

for all $\sigma \in [0, \sigma']$. But then, since g_2 is twice continuously differentiable near $(\bar{T}, \bar{x}(\bar{T}))$ there exists $r_1 \geq 0$ such that

$$\begin{aligned} 0 \ \leq \ & \nabla_t g_2(\bar{T}, \bar{x}(\bar{T})) \sigma + \int_{\bar{T}}^{\bar{T}+\sigma} \nabla_x g_2(\bar{T}, \bar{x}(\bar{T})) \cdot \dot{y}(t) dt + r_1 \sigma^2 + \epsilon k_3 \sigma \\ \leq \ & \nabla_t g_2(\bar{T}, \bar{x}(\bar{T})) \sigma + \int_{\bar{T}}^{\bar{T}+\sigma} \nabla_x g_2(\bar{T}, \bar{x}(\bar{T})) \cdot \xi(t) \, dt \\ & + (r_1 + (1/2) e^{k_1 \sigma} k_1 c_1) \sigma^2 + \epsilon k_3 \sigma. \end{aligned}$$

Since

$$\lambda \nabla_x g_2(\bar{T}, \bar{x}(\bar{T})) \cdot \xi(t) = -H(t, \bar{x}(\bar{T}), p(\bar{T})) \quad \text{a.e. } t \in [\bar{T}, \bar{T} + \delta_1],$$

these relationships imply that

$$\lim_{\sigma \downarrow 0} \operatorname*{ess\,inf}_{\bar{T} - \sigma \leq t \leq \bar{T} + \sigma} H(t, \bar{x}(\bar{T}), p(\bar{T})) \leq \lambda \nabla_t g_2(\bar{T}, \bar{x}(\bar{T})) + \lambda \epsilon k_3.$$

The above inequality and (8.9) imply (E). Step 1 is complete.

Step 2: Take $\bar{S} \in R$, $\tilde{C} \subset R^{1+n}$, and $\tilde{g} : R^{1+n+n} \to R$. Let $([\bar{S}, \bar{T}], \bar{x})$ be a minimizer for

$$\begin{cases} \text{Minimize } \tilde{g}(T, x(\bar{S}), x(T)) \\ \text{over arcs } ([\bar{S}, T], x) \text{ satisfying} \\ \dot{x}(t) \in F(t, x(t)) \text{ a.e.,} \\ (T, x(\bar{S})) \in \tilde{C}. \end{cases}$$

It is assumed that the data for this optimization problem, regarded as a special case of (FT), satisfy hypotheses (H1) to (H3) and (H5).

We find p, λ, and η satisfying Conditions (i) to (iv) and (vi) (for any $\xi \in R$).

Take a sequence $K_i \uparrow \infty$. Define

$$J_i(T, \tau, x, y) := \tilde{g}(\tau(\bar{S}), x(\bar{S}), y(\bar{S})) + K_i[|x(T) - y(T)|^2 + |\tau(T) - T|^2].$$

Let $\delta' \in (0, \delta \wedge \delta_1)$ be such that $([\bar{S}, \bar{T}], \bar{x})$ is a minimizer with respect to arcs $([\bar{S}, T], x)$ satisfying

$$d(([\bar{S}, T], x), ([\bar{S}, \bar{T}], \bar{x})) \leq \delta'.$$

Consider the following sequence of optimization problems, $i = 1, 2, \ldots$.

$$\begin{cases} \text{Minimize } J_i(T, \tau, x, y) \\ \text{over arcs } ([\bar{S}, T], (\tau, x, y)) \text{ satisfying} \\ (\dot{\tau}(t), \dot{x}(t), \dot{y}(t)) \in \{0\} \times F(t, x(t)) \times \{0\} \text{ a.e.,} \\ (\tau(\bar{S}), x(\bar{S})) \in C, \\ d(([\bar{S}, T], x), ([\bar{S}, \bar{T}], \bar{x})) \leq \delta'. \end{cases}$$

Define W to be the set of all arcs $([\bar{S}, T], (\tau, x, y))$ satisfying the constraints of the above optimization problem.

With respect to the metric $d([\bar{S}, T], (\tau, x, y)), ([\bar{S}, T'], (\tau', x', y'))$, W is complete and the cost function $J_i(T, \tau, x, y)$ is continuous.

(Notice that $d(., .)$ denotes the metric (8.3) both for the set of n-vector arcs comprising the domain of the preceding optimization problem and for the $(1 + n + 1)$-vector arcs in W.)

Take any $([\bar{S}, T], (\tau, x, y))$ in W. τ and y are constant functions; we write their values also τ and y. After a reduction in the size of δ', if required, we have, by local optimality,

$$\tilde{g}(\tau, x(\bar{S}), x(\tau)) \geq \tilde{g}(\bar{T}, \bar{x}(\bar{S}), \bar{x}(\bar{T})).$$

It follows that

$$\begin{aligned} J_i(([\bar{S}, T], (\tau, x, y))) &\geq \tilde{g}(\tau, x(\bar{S}), x(\tau)) - k_g|y - x(\tau)| + \\ &\qquad K_i(|x(T) - y|^2 + |\tau - T|^2) \\ &\geq \tilde{g}(\tau, x(\bar{S}), x(\tau)) - k_g|y - x(T)| \\ &\qquad - k_g|x(T) - x(\tau)| + K_i(|x(T) - y|^2 + |\tau - T|^2) \\ &\geq \tilde{g}(\bar{T}, x(\bar{S}), \bar{x}(\bar{T})) - k_g(1 + c_1^2)/(4K_i). \end{aligned}$$

Here, k_g is a Lipschitz constant for $\tilde{g}$; c_1 is the constant of Hypothesis (H5). For each i, define

$$\epsilon_i := k_g^{1/2}(1 + c_1^2)^{1/2}/2K_i^{1/2}.$$

Clearly, $\epsilon_i \downarrow 0$.

Since $([\bar{S}, T], (\tau, x, y))$ is an element in W and $([\bar{S}, \bar{T}], (\bar{\tau} \equiv \bar{T}, \bar{x}, \bar{y} \equiv \bar{x}(\bar{T}))$ has cost $\tilde{g}(\bar{T}, \bar{x}(\bar{S}), \bar{x}(\bar{T}))$, it follows from the preceding inequality that, for each i,

$$J_i((\bar{T}, (\bar{\tau}, \bar{x}, \bar{y})) \leq \inf_{([\bar{S}, T]),(\tau, x, y) \in W} J_i(T, \tau, x, y) + \epsilon_i^2.$$

We have shown, for each i, $([\bar{S}, \bar{T}], \bar{\tau}, \bar{x}, \bar{y})$ is an ϵ_i^2 minimizer. Now apply Ekeland's Principle. This asserts that, for each i, there exists

$$([\bar{S}, T_i], (\tau_i, x_i, y_i)) \in W$$

which minimizes

$$J_i(T, \tau, x, y) + \epsilon_i d(([\bar{S}, T], (\tau, x, y)), ([\bar{S}, T_i], (\tau_i, x_i, y_i)))$$

over $([\bar{S}, T], (\tau, x, y)) \in W$ and

$$d(([\bar{S}, T_i], (\tau_i, x_i, y_i)), ([\bar{S}, \bar{T}]), (\bar{\tau}, \bar{x}, \bar{y})) \leq \epsilon_i.$$

Since $\epsilon_i \downarrow 0$, we can arrange, by subsequence extraction, that

$$T_i \to \bar{T}, \ \tau_i \to \bar{T}, \ y_i \to \bar{x}(\bar{T}),$$

$$x_i \to \bar{x} \text{ uniformly,}$$

and

$$\dot{x}_i \to \dot{\bar{x}} \quad \text{a.e. and also strongly in } L^1.$$

(Our "extension" convention is used to interpret these notions of convergence.)

The above relationships can be interpreted as follows. For i sufficiently large, $([\bar{S}, T_i], (\tau_i, x_i, y_i))$ is a $W^{1,1}$ local minimizer for

$$\left\{ \begin{array}{l} \text{Minimize } \tilde{g}(\tau(\bar{S}), x(\bar{S}), y(\bar{S})) + K_i|x(T) - y(T)|^2 + K_i|\tau(T) - T|^2 \\ \qquad + \epsilon_i(|\tau(\bar{S}) - \tau_i| + |x(\bar{S}) - x_i(\bar{S})| + |y(\bar{S}) - y_i|) + \epsilon_i(|T - T_i| \\ \qquad + \int_{\bar{S}}^T |\dot{x}(t) - \dot{x}_i(t)|dt + \int_T^{T \vee T_i} |\dot{x}_i|dt) \\ \text{over arcs } ([\bar{S}, T], (\tau, x, y)) \text{ satisfying} \\ (\dot{\tau}(t), \dot{x}(t), \dot{y}(t)) \in \{0\} \times F(t, x(t)) \times \{0\} \text{ a.e.,} \\ (\tau(\bar{S}), x(\bar{S})) \in \tilde{C}. \end{array} \right.$$

We have arrived at an optimization problem to which the necessary conditions derived in Step 1 are applicable, following elimination of the integral cost term by state augmentation. For each i, they yield $\sigma_i \in R$, $p_i \in W^{1,1}([\bar{S}, T_i]; R^n)$, $r_i \in R^n$, and $\lambda_i \geq 0$ such that

(A) $\lambda_i + |\sigma_i| + ||p_i||_{L^\infty} + |r_i| = 1;$

(B) $\dot{p}_i(t) \in \mathrm{co}\{\zeta : (\zeta, p_i(t)) \in N_{\mathrm{GrF}(t,\cdot)}(x_i(t), \dot{x}_i(t))\}$
$$+\{0\} \times (\lambda_i \epsilon_i B), \quad \text{a.e. } t \in [\bar{S}, T_i];$$

(C) $p_i(t) \cdot \dot{x}_i(t) \geq p_i(t) \cdot v - \lambda_i \epsilon_i |v - \dot{x}_i(t)|$
$$\text{for all } v \in F(t, x_i(t)), \quad \text{a.e. } t \in [\bar{S}, T_i];$$

(D) $(\sigma_i, p_i(\bar{S}), r_i) \in \lambda_i \partial \tilde{g}(\tau_i, x_i(\bar{S}), y_i) + N_{\tilde{C}}(\tau_i, x_i(\bar{S})) \times \{0\} + \lambda_i \epsilon_i B;$

(E) $-2\lambda_i K_i(\tau_i - T_i) \in \mathrm{ess}_{t \to T_i} H_i(t, x_i(T_i), p_i(T_i)) + \lambda_i \epsilon_i (1 + c_1) B;$

(F) $-(\sigma_i, p_i(T_i), r_i) = \lambda_i(2K_i(\tau_i - T_i), 2K_i(x_i(T_i) - y_i), -2K_i(x_i(T_i) - y_i)).$

Here
$$H_i(t, x, p) = \sup\{p \cdot v - \epsilon_i |v - \dot{x}_i(t)| : v \in F(t, x)\}.$$

Eliminating $2\lambda_i K_i(x_i(T_i) - y_i)$ and $2\lambda_i K_i(\tau_i - T_i)$ from these relationships we obtain

$$(\sigma_i, p_i(\bar{S}), -p_i(T_i)) \in \lambda_i \partial \tilde{g}(\tau_i, x_i(\bar{S}), y_i) + N_{\tilde{C}}(\tau_i, x_i(\bar{S})) \times \{0\} + \lambda_i \epsilon_i B$$
$$\sigma_i \in \underset{t \to T_i}{\mathrm{ess}} \; H_i(t, x_i(T_i), p_i(T_i)) + \lambda_i \epsilon_i (1 + c_1) B.$$

We deduce from (A) and (B) that the p_is are uniformly bounded and their derivatives are integrably bounded. So there exists $p \in W^{1,1}$ such that, along a subsequence,

$$p_i \to p \text{ uniformly} \quad \text{and} \quad \dot{p}_i \to \dot{p} \text{ weakly in } L^1.$$

A further subsequence extraction ensures that

$$\lambda_i \to \lambda, \; \sigma_i \to \eta,$$

for some $\lambda \geq 0$ and η.

We obtain the stated necessary conditions for the special case under consideration, with multipliers p, λ, and η, by passing to the limit as $i \to \infty$ in the preceding relationships. The convergence analysis follows the pattern set in previous proofs, but we now also make use of the uniform closure properties of the set of essential values (see Part (iv) of Proposition 8.3.2).

Step 3 (Necessary conditions for a general fixed left end-time problem): Take $\bar{S} \in R$, $\tilde{C} \subset R^{1+n+n}$, and $\tilde{g} : R \times R^n \times R^n \to R$. Let $([\bar{S}, \bar{T}], \bar{x})$ be a $W^{1,1}$minimizer for

$$\begin{cases} \text{Minimize } \tilde{g}(T, x(\bar{S}), x(T)) \\ \text{over arcs } ([\bar{S}, T], x) \text{ satisfying} \\ \dot{x}(t) \in F(t, x(t)) \quad \text{a.e.,} \\ (T, x(\bar{S}), x(T)) \in \tilde{C}. \end{cases}$$

Regarding this as a special case of (FT), we assume that Hypotheses (H1) to (H3) and (H5) are satisfied.

We find p, λ, and η such that Conditions (i) to (iv) and (vi) are satisfied (for any $\xi \in R$).

Take $\epsilon_i \downarrow 0$. Let $\delta' \in (0, \delta \wedge \delta_1)$ be such that $([\bar{S}, \bar{T}], \bar{x})$ is a minimizer with respect to all arcs $([\bar{S}, T], x)$ satisfying

$$d(([\bar{S}, T], x), ([\bar{S}, \bar{T}], \bar{x})) \le \delta'.$$

For each i, consider the optimization problem

$$\begin{cases} \text{Minimize } l_i(T, x(\bar{S}), y(\bar{S}), x(T), y(T)) \\ \text{over arcs } ([\bar{S}, T], (x, y)) \text{ satisfying} \\ (\dot{x}(t), \dot{y}(t)) \in F(t, x(t)) \times \{0\} \quad \text{a.e.,} \\ (T, x(\bar{S}), y(\bar{S})) \in C, \\ d(([\bar{S}, T], (x, y)), ([\bar{S}, \bar{T}], (\bar{x}, \bar{y}))) \le \delta' \end{cases}$$

in which

$$l_i(T, x, y, x', y') := \max\{\tilde{g}(T, x, y) - \tilde{g}(\bar{T}, \bar{x}(\bar{S}), \bar{x}(\bar{T})) + \epsilon_i^2, \; |y' - x'|\}.$$

The function l_i is nonnegative and

$$l_i(\bar{T}, \bar{x}(\bar{S}), \bar{x}(\bar{T}), \bar{x}(\bar{T}), \bar{x}(\bar{T})) = \epsilon_i^2.$$

This implies that $([\bar{S}, \bar{T}], \bar{x}, \bar{y} \equiv \bar{x}(\bar{T}))$ is an ϵ_i^2 minimizer.

Denote by W the space of elements $([\bar{S}, T], (x, y))$ satisfying the constraints of this optimization problem, equipped with the metric $d(., .)$.

W is complete and the functional $l : W \to R$ is continuous. For each i, we deduce from Ekeland's Theorem that there exists an element $([\bar{S}, T_i], (x_i, y_i))$ in W, which is a $W^{1,1}$ local minimizer for

$$\begin{cases} \text{Minimize } l_i(T, x(\bar{S}), y(\bar{S}), x(T), y(T)) + \epsilon_i[|T - T_i| + |x(\bar{S}) - x_i(\bar{S})| \\ \quad + \int_{\bar{S}}^T |\dot{x}(t) - \dot{x}_i(t)| dt + \int_T^{T_i \vee T} |\dot{x}_i(t)| dt + |y(T) - y_i|] \\ \text{over arcs } ([\bar{S}, T], (x, y)) \text{ satisfying} \\ (\dot{x}(t), \dot{y}(t)) \in F(t, x(t)) \times \{0\} \text{ a.e.,} \\ (T, x(\bar{S}), y(\bar{S})) \in \tilde{C} \end{cases}$$

and

$$d(([\bar{S}, T_i], (x_i, y_i)), ([\bar{S}, \bar{T}], (\bar{x}, \bar{y}))) \le \epsilon_i.$$

Now apply the necessary conditions of Step 2, following elimination of integral cost terms by state augmentation. These yield $p_i \in W^{1,1}([\bar{S}, \bar{T}])$, $r_i \in R^n$, $\lambda_i \ge 0$, and $\xi_i \in R$ such that

(A) $\lambda_i + \|p_i\|_{L^\infty} + |r_i| = 1$;

(B) $\dot{p}_i(t) \in \text{co}\{\zeta : (\zeta, p_i(t)) \in \{0\} \times \lambda_i \epsilon_i B + N_{\text{Gr}F(t, \cdot)}(x_i(t), \dot{x}_i(t))\}$ a.e.,

(C) $p_i(t) \cdot \dot{x}_i(t) \geq p_i(t) \cdot v - \lambda_i \epsilon_i |v - \dot{x}_i(t)|$ for all $v \in F(t, x_i(t))$ a.e.;

(D) $(\xi_i, p_i(\bar{S}), r_i, -p_i(T_i), -r_i) \in \lambda_i \partial l_i(T_i, x_i(\bar{S}), y_i, x_i(T_i), y_i) +$
    $N_{\tilde{C}}(T_i, x_i(\bar{S}), y_i) \times \{(0,0)\} + \lambda_i \epsilon_i (B \times B \times B) \times \{(0,0)\} +$
    $\lambda_i \epsilon_i c_i B \times \{(0,0,0,0)\};$

(E) $\xi_i \in \operatorname{ess}_{t \to T_i} H(t, x(T_i), p_i(T_i)) + 2\epsilon_i c_1 B.$

Note that, for i sufficiently large,

$$l_i(T_i, x_i(\bar{S}), y_i, x_i(T_i), y_i) > 0, \qquad (8.10)$$

since, otherwise, $y_i = x_i(T_i)$ and $\tilde{g}(T_i, x_i(\bar{S}), x_i(T_i)) < \tilde{g}(\bar{T}, \bar{x}(\bar{S}), \bar{x}(\bar{T}))$, in contradiction of the optimality of $([\bar{S}, \bar{T}], (\bar{x}, \bar{y}))$.

Observe also that there exists $\lambda_i \in [0, 1]$ such that

$$\partial l_i(T_i, x_i(\bar{S}), y_i, x_i(T_i), y_i) \subset \lambda \partial \tilde{g}(T_i, x_i(\bar{S}), y_i) \times \{(0,0)\} +$$
$$(1 - \lambda)\{(0,0,0)\} \times \{(e, -e) : e \in R^n, |e| = 1\}. \qquad (8.11)$$

In the case when

$$\tilde{g}(T_i, x_i(\bar{S}), y_i) - \tilde{g}(\bar{T}, \bar{x}(\bar{S}), \bar{x}(\bar{T})) + \epsilon_i^2 > |x_i(T_i) - y_i|,$$

this relationship is clearly true with $\lambda = 1$. On the other hand, in the case when

$$|x_i(T_i) - y_i| \geq \tilde{g}(T_i, x_i(\bar{S}), y_i) - \tilde{g}(\bar{T}, \bar{x}(\bar{S}), \bar{x}(\bar{T})) + \epsilon_i^2,$$

Condition (8.10) implies

$$|x_i(T_i) - y_i| > 0$$

and we can then deduce (8.11) from the Max Rule (Theorem 5.5.2) and Chain Rule (Theorem 5.5.1).

From (D) and (A) we now conclude that

$$(\xi_i, p_i(\bar{S}), -p_i(T_i)) \in \lambda_i \partial \tilde{g}(T_i, x_i(\bar{S}), y_i) +$$
$$\lambda_i \epsilon_i (B \times B \times B) + \lambda_i \epsilon_i c_1 B \times \{(0,0)\} + N_{\tilde{C}}(T_i, x_i(\bar{S}), y_i)$$

and

$$\lambda_i + \|p_i\|_{L^\infty} + |p_i(T_i)| = 1.$$

In view of Conditions (A) and (B), the p_is are uniformly bounded and the $\dot{p}_i$s are uniformly integrably bounded. The ξ_is and λ_is, too, are bounded. We can therefore arrange by subsequence extraction that

$$\xi_i \to \xi, \quad \lambda_i \to \lambda$$

and

$$p_i \to p \quad \text{uniformly}$$

as $i \to \infty$, for some $\xi \in R^n$, $\lambda \geq 0$, and $p \in W^{1,1}$.

Our previous analysis justifies passing to the limit in (A) through (E). The resulting relationships confirm the assertions of Theorem 8.4.1. Notice, in particular, that (A) yields

$$\lambda + ||p||_{L^\infty} + |p(\bar{T})| = 1$$

for the limiting multipliers, a condition which ensures that

$$\lambda + ||p||_{L^\infty} \neq 0.$$

Step 4 (A fixed right end-time problem): Fix $\bar{T} \in R$. Take $\tilde{C} \subset R^{1+n+n}$, $\tilde{g} : R \times R^n \times R^n \to R$. Let $([\bar{S}, \bar{T}], \bar{x})$ be a $W^{1,1}$ local minimizer for the optimization problem

$$\begin{cases} \text{Minimize } \tilde{g}(S, x(S), x(\bar{T})) \\ \text{over arcs } ([S, \bar{T}], x) \text{ satisfying} \\ \dot{x}(t) \in F(t, x(t)) \quad \text{a.e.,} \\ (S, x(S), x(\bar{T})) \in \tilde{C}. \end{cases}$$

Regarding this as a special case of (FT), we assume that Hypotheses (H1) to (H4) are satisfied. We find $p \in W^{1,1}([\bar{S}, \bar{T}], R^n)$, $\lambda \geq 0$, and $\eta \in R$ such that Conditions (i) to (v) of Theorem 8.4.1 are satisfied (for any $\eta \in R$).

Define

$$\bar{S}' := -\bar{T}, \ \bar{T}' := -\bar{S}, \ \bar{x}'(s) := \bar{x}(-s)$$
$$F'(s, x) := -F(-s, x), \ g'(s, x, y) := \tilde{g}(-s, x, y)$$
$$C' := \{(s, x, y) : (-s, x, y) \in \tilde{C}\}.$$

By considering a change of independent variable $s = -t$ we readily deduce that $([\bar{S}', \bar{T}'], \bar{x}')$ is a $W^{1,1}$ local minimizer for the following minimization problem in which the left end-time is fixed at $\bar{S}'$:

$$(P_1) \begin{cases} \text{Minimize } g'(T', x'(T'), x'(\bar{S}')) \\ \text{over arcs } ([\bar{S}', T'], x') \text{ satisfying} \\ \dot{x}'(s) \in F'(s, x'(s)) \quad \text{a.e.,} \\ (T', x'(T'), x'(\bar{S}')) \in C'. \end{cases}$$

This is a special case of (FT) to which the necessary conditions derived in Step 3 are applicable. We thereby obtain p', λ', and η' satisfying Conditions (i) to (iv) and (vi) (for arbitrary $\xi' \in R$), in relation to the data for problem (P_1). It is a simple matter to deduce that

$$p(t) := -p'(-t), \quad \lambda := \lambda' \quad \text{and} \quad \eta := \eta'$$

satisfy Conditions (i) to (v) in relation to the data for the original problem.

Step 5 (Conclusion of the proof): Let $([\bar{S}, \bar{T}], \bar{x})$ be a $W^{1,1}$ local minimizer for (FT) (with no restrictions on the nature of the data). Assume Hypotheses (H1) to (H5) are satisfied.

Step 4 has provided general necessary conditions for problems with a free left end-time. We now use these to repeat Steps 1 to 3 (modified to allow for a free left end-time), to obtain necessary conditions for the free right end-time problem and then for general end-time constraints. The arguments are almost the same as before. The only difference is that we have to carry forward the extra information concerning the free left end-time now available to us. In this way existence of a set of multipliers satisfying Conditions (i) to (vi) of the theorem statement is confirmed, merely under Hypotheses (H1) to H5).

Specifically, the optimization problems considered in Steps 1 through 3 are replaced by

$$\begin{cases} \text{Minimize } g_1(S, x(S)) + g_2(T, x(T)) + \epsilon g_3(T) \\ \text{over arcs } ([S, T], x) \text{ satisfying} \\ \dot{x}(t) \in F(t, x(t)) \quad \text{a.e.,} \\ (S, x(S)) \in C_1 \end{cases}$$

(g_1 and g_3 are Lipschitz continuous functions and g_2 is a function that is twice continuously differentiable on a neighborhood of the point $\bar{T}$);

$$\begin{cases} \text{Minimize } g(S, x(S), T, x(T)) \\ \text{over arcs } ([S, T], x) \text{ satisfying} \\ \dot{x}(t) \in F(t, x(t)) \quad \text{a.e.,} \\ (S, x(S), T) \in C_2; \end{cases}$$

and

$$\begin{cases} \text{Minimize } g(S, x(S), T, x(T)) \\ \text{over arcs } ([S, T], x) \text{ satisfying} \\ \dot{x}(t) \in F(t, x(t)) \quad \text{a.e.,} \\ (S, x(S), T, x(T)) \in C_3. \end{cases}$$

The last optimization problem is of course (FT) (with no restrictions on the nature of the data).

It remains only to comment on the concluding assertions of the theorem. A review of our earlier arguments reveals that Hypothesis (H4) is not required if the left end-time is fixed. In this case, (iii) is valid for any ξ (provided we drop (vi), since, without (H4) to ensure that the Hamiltonian is essentially bounded near the left end-time, $\text{ess}_{t \to \bar{S}} H(t, \bar{x}(\bar{S}), p(\bar{S}))$ may not be defined). Likewise (H5) is not required if the right end-time is fixed. The proof is complete. $\square$

8.6   Proof of Theorem 8.4.2

Take $l : R \times R^n \times R \times R^n \to R$ and $L : R \times R^n \times R^n \to R$. Define

$$\tilde{F}(t,x) := \begin{cases} \dot{\bar{x}}(t) + \epsilon B & \text{for } \bar{S} + \delta_0 \le t \le \bar{T} - \delta_1 \\ F(t,x) & \text{otherwise.} \end{cases}$$

$$\tilde{L}(t,x,v) := \begin{cases} L(t,x,v) & \text{for } \bar{S} + \delta_0 \le t \le \bar{T} - \delta_1 \\ 0 & \text{otherwise.} \end{cases}$$

Consider now the following optimal control problem, in which the dynamic constraint is "state free," except near $\bar{S}$ and $\bar{T}$.

$$(D) \begin{cases} \text{Minimize } l(S, x(S), T, x(T)) + \int_S^T \tilde{L}(t, x(t), \dot{x}(t)) dt \\ \text{over arcs } ([S,T], x) \text{ satisfying} \\ \dot{x}(t) \in \tilde{F}(t, x(t)) \quad \text{a.e. } t \in [S, T]. \end{cases}$$

Proposition 8.6.1 *Let $([\bar{S}, \bar{T}], \bar{x})$ be a minimizer for (D) such that $\bar{T} > \bar{S}$. Assume that F satisfies Hypotheses (H2), (H4), and (H5) of Theorem 8.4.2, for some $\delta > 0$. Assume also*

(G1) $L(.,x,v)$ *is Lebesgue measurable for each (x,v) and there exist $\epsilon' > 0$ and $k \in L^1$ such that, for a.e. $t \in [\bar{S} + \delta_0, \bar{T} - \delta_1]$,*

$$|L(t,x,v) - L(t,x',v')| \le k(t)|(x,v) - (x',v')|$$

for all $(x,v), (x',v') \in (\bar{x}(t) + \delta B) \times (\dot{\bar{x}}(t) + \epsilon' B)$.

(G2) l *is locally Lipschitz continuous.*

   Then there exist an arc $p \in W^{1,1}([\bar{S}, \bar{T}]; R^n)$ and numbers ξ and η such that

(i) $\dot{p}(t) \in co\{\zeta : (\zeta, p(t)) \in \partial \tilde{L}(t, \bar{x}(t), \dot{\bar{x}}(t))$
$$+ N_{\mathrm{Gr}\, \tilde{F}(t,\cdot)}(\bar{x}(t), \dot{\bar{x}}(t))\} \quad a.e. \ t \in [\bar{S}, \bar{T}];$$

(ii) $(-\xi, p(\bar{S}), \eta, -p(\bar{T})) \in \partial l(\bar{S}, \bar{x}(\bar{S}), \bar{T}, \bar{x}(\bar{T}));$

(iii) $\xi \in \mathrm{ess}_{t \to \bar{S}}\, H(t, \bar{x}(\bar{S}), p(\bar{S}));$

(iv) $\eta \in \mathrm{ess}_{t \to \bar{T}}\, H(t, \bar{x}(\bar{T}), p(\bar{T})).$

In Conditions (iii) and (iv) above, H denotes the Hamiltonian corresponding to $\dot{x} \in F(t,x)$:
$$H(t,x,p) := \max_{v \in F(t,x)} p \cdot v.$$

Proof. To prove this proposition, eliminate the integral cost term by state augmentation. Apply Theorem 8.4.1 and express the resulting conditions directly in terms of the data for problem $(FT)'$. The analysis is similar to that used in the fixed time case (see the proof of Proposition 7.4.2). $\square$

Let $\delta' \in (0, \min\{\delta, \delta_1, \delta_2\})$ be such that $([\bar{S}, \bar{T}], \bar{x})$ is a minimizer for (FT) over the class of arcs satisfying the constraints of (FT) and also

$$d(([S, T], x), ([\bar{S}, \bar{T}], \bar{x})) \leq \delta'.$$

Define

$$\mathcal{S} := \{([S, T], x) : \dot{x}(t) \in F(t, x(t)) \cap \tilde{F}(t, x(t)) \text{ a.e.,}$$
$$(S, x(S), T, x(T)) \in C, \, d(([S, T], x), ([\bar{S}, \bar{T}], \bar{x})) \leq \delta'\}$$

and

$$\tilde{g}(([S, T], x)) := g(S, x(S), T, x(T)).$$

$([\bar{S}, \bar{T}], \bar{x})$ minimizes $\tilde{g}$ over $\mathcal{S}$. The function $\tilde{g}$ satisfies

$$\tilde{g}(([S, T], x)) - \tilde{g}(([S', T'], x')) \leq k_g d(([S, T], x), ([S', T'], x')),$$

in which k_g is a Lipschitz constant for g. According to the Exact Penalization Principle then, $([\bar{S}, \bar{T}], \bar{x})$ is also a minimizer for the problem

$$\left\{ \begin{array}{l} \text{Minimize } \tilde{g}([S, T], x) + k_g \inf\{d(([S, T], x), ([S', T'], y)) : ([S', T'], y) \in \mathcal{S}\} \\ \quad \text{over arcs } ([S, T], x) \text{ satisfying} \\ \dot{x}(t) \in \tilde{F}(t, x(t)) \quad \text{a.e. } t \in [S, T], \\ d(([S, T], x), ([\bar{S}, \bar{T}], \bar{x})) \leq \delta'. \end{array} \right.$$

Define

$$\rho_F(t, x, v) := \inf\{|v - w| : w \in F(t, x)\}.$$

There are two possibilities to consider.

(a): There exist $\epsilon' \in (0, \delta')$ and $K' > 0$ such that, for any $\tilde{F}$-trajectory $([S, T], x)$ satisfying $d(([S, T], x), ([\bar{S}, \bar{T}], \bar{x})) \leq \epsilon'$, we have

$$\inf\{d(([S, T], x), ([S', T'], y)) : ([S', T'], y) \in \mathcal{S}\}$$
$$\leq K' \left(d_C(S, x(S), T, x(T)) + \int_S^T \rho_F(t, x(t), \dot{x}(t)) dt \right).$$

(b): There exists a sequence of $\tilde{F}$-trajectories $\{([\bar{S}_i, \bar{T}_i], \bar{x}_i)\}$ such that

$$d_C(\bar{S}_i, \bar{x}_i(\bar{S}_i), \bar{T}_i, \bar{x}_i(\bar{T}_i)) + \int_{\bar{S}_i}^{\bar{T}_i} \rho_F(t, \bar{x}_i(t), \dot{\bar{x}}_i(t)) dt$$
$$< (2i)^{-1} \inf\{d(([\bar{S}_i, \bar{T}_i], \bar{x}_i), ([S', T'], y)) : ([S', T'], y) \in \mathcal{S}\}$$

for each i, and

$$d(([\bar{S}_i, \bar{T}_i], \bar{x}_i), ([\bar{S}, \bar{T}], \bar{x})) \to 0 \quad \text{as } i \to \infty.$$

Consider (a). In this case, $([\bar{S}, \bar{T}], \bar{x})$ is a $W^{1,1}$ local minimizer of the optimal control problem

$$\begin{cases} \text{Minimize } g(S, x(S), T, x(T)) + k_g K'[d_C(S, x(S), T, x(T)) \\ \qquad\qquad\qquad\qquad\qquad\qquad\qquad + \int_S^T \rho_F(t, x(t), \dot{x}(t)) dt] \\ \text{over arcs } ([S, T], x) \text{ satisfying} \\ \dot{x}(t) \in \tilde{F}(t, x(t)) \quad \text{a.e. } t \in [S, T]. \end{cases}$$

The hypotheses are satisfied under which Proposition 8.6.1 can be applied to this problem, when we set $L(t, x, v) = \rho_F(t, x, v)$ and

$$l(S, x_0, T, x_1) = k_g K' d_C(S, x_0, T, x_1).$$

Notice, for t near $\bar{S}$ or $\bar{T}$,

$$\begin{aligned} \tilde{H}(t, x, v) &= \sup\{p \cdot v - 2k_g K' \rho_F(t, x, v) : v \in \tilde{F}(t, x)\} \\ &= \sup\{p \cdot v - 2k_g K' \rho_F(t, x, v) : v \in F(t, x)\} \\ &= \sup\{p \cdot v : v \in F(t, x)\} \\ &= H(t, x, p). \end{aligned}$$

Notice also that

$$N_{\mathrm{Gr}\, \tilde{F}(t,\cdot)}(\bar{x}(t), \dot{\bar{x}}(t)) = \begin{cases} \{(0,0)\} & \text{a.e. } t \in [\bar{S} + \delta_0, \bar{T} - \delta_1] \\ N_{\mathrm{Gr}\, F(t,\cdot)} & \text{a.e. } t \notin [\bar{S} + \delta_0, \bar{T} - \delta_1]. \end{cases}$$

We deduce from Lemma 7.2.1 that

$$\partial \tilde{L}(t, \bar{x}(t), \dot{\bar{x}}(t)) \subset \begin{cases} N_{\mathrm{Gr}\, F(t,\cdot)}(\bar{x}(t), \dot{\bar{x}}(t) & \text{a.e. } t \in [\bar{S} + \delta_0, \bar{T} - \delta_1] \\ \{(0,0)\} & \text{a.e. } t \in [\bar{S} + \delta_0, \bar{T} - \delta_1]. \end{cases}$$

It follows that the adjoint arc p of Proposition 8.6.1 satisfies

$$\dot{p}(t) \in \mathrm{co}\,\{\eta : (\eta, p(t)) \in N_{\mathrm{Gr}\, \tilde{F}(t,\cdot)}\} \quad \text{a.e. } t \in [\bar{S}, \bar{T}].$$

The other assertions of the theorem follow also from Proposition 8.6.1.

Consider next Case (b). In this case we show, by means of Ekeland's Theorem, that there exists a sequence $\{([S_i, T_i], x_i)\}$ in $\mathcal{S}$ such that

$$d(([S_i, T_i], x_i), ([\bar{S}, \bar{T}], \bar{x})) \to 0$$

and, for each i, $([S_i, T_i], x_i)$ is a local minimizer for

$$\begin{cases} \text{Minimize } \int_S^T \rho_F(t, x(t), \dot{x}(t)) dt + d_C(S, x(S), T, x(T)) + i^{-1} e_i(S, T, x) \\ \text{over arcs } ([S, T], x) \text{ satisfying} \\ \dot{x}(t) \in \tilde{F}(t, x(t)) \text{ a.e. } t \in [S, T]. \end{cases}$$

Here,

$$e_i(S,T,x) := |S - S_i| + |T - T_i| + \int_{S_i}^{S \vee S_i} |\dot{x}_i(s)| ds + \int_{T \wedge T_i}^{T_i} |\dot{x}_i(s)| ds$$

$$+ \int_S^T |\dot{x}(s) - \dot{x}_i(s)| ds.$$

Furthermore,

$$(S_i, T_i, x_i) \notin \mathcal{S}. \tag{8.12}$$

For each i we apply Proposition 8.6.1 to the preceding problem. We deduce existence of p_i, λ_i, ξ_i and η_i satisfying Conditions (i) to (v) when $([S_i, T_i], x_i)$ replaces $([\bar{S}, \bar{T}], \bar{x})$. Extracting subsequences and passing to the limit we recover p, λ, ξ, and η satisfying Conditions (i) to (v) of Theorem 8.4.2. Condition (8.12), which implies that either

$$(S_i, x_i(S_i), T_i, x_i(T_i)) \notin C$$

or

$$\text{meas}\{t \in [\bar{S} + \delta_0, \bar{T} - \delta_1] : \dot{x}_i(t) \notin F(t, x_i(t))\} > 0.$$

is used, as in the proof of Theorem 7.5.1, to ensure nondegeneracy of the multipliers.

Conditions (I) and (II) of the theorem statement can be deduced from Condition (ii), with the help of the Dualization Theorem. (See Theorem 7.6.8.) The hypotheses can be relaxed if either the left or right endpoints are fixed, as indicated at the end of the proof of Theorem 8.4.1. The proof is complete. □

8.7 A Free End-Time Maximum Principle

We have derived necessary conditions for free end-time optimal control problems in which the dynamic constraint is expressed as a differential inclusion. A similar analysis provides necessary conditions for free end-time problems, in the form of a Maximum Principle, when the dynamics are modeled instead by a control-dependent differential equation.

$$(FT)' \begin{cases} \text{Minimize } g(S, x(S), T, x(T)) \\ \text{over intervals } [S, T], \text{ arcs } x \in W^{1,1}([S, T]; R^n) \\ \qquad \text{and measurable functions } u : [S, T] \to R^m \text{ satisfying} \\ \dot{x}(t) = f(t, x(t), u(t)) \quad \text{a.e. } t \in [S, T], \\ u(t) \in U(t) \quad \text{a.e. } t \in [S, T], \text{ and} \\ (S, x(S), T, x(T)) \in C, \end{cases}$$

for which the data comprise functions $g : R^{1+n+1+n} \to R$ and $f : R \times R^n \times R^m \to R^n$, a nonempty multifunction $U : R \rightsquigarrow R^m$, and a closed set $C \subset R^{1+n+1+n}$.

Earlier terminology is modified to emphasize that the endpoints of the underlying time interval $[S,T]$ are now choice variables. A *process* is taken to be a triple $([S,T],x,u)$ in which $[S,T]$ is an interval, x is an element in $W^{1,1}([S,T];R^n)$ (the *state trajectory*), and u (the *control function*) is a measurable R^m-valued function on $[S,T]$ satisfying, for a.e. $t \in [S,T]$,

$$\begin{aligned} \dot{x}(t) &= f(t,x(t),u(t)) \\ u(t) &\in U(t). \end{aligned}$$

A process $([\bar{S},\bar{T}],\bar{x},\bar{u})$, that satisfies the constraints of $(FT)'$ is said to be a $W^{1,1}$ local minimizer if there exists $\delta' > 0$ such that

$$g(S,x(S),T,x(T)) \geq g(\bar{S},\bar{x}(\bar{S}),\bar{T},\bar{x}(\bar{T}))$$

for every process $([S,T],x,u)$ that satisfies the constraints of $(FT)'$ and also

$$d(([S,T],x),([\bar{S},\bar{T}],\bar{x})) \leq \delta'.$$

Here, $d(.,.)$ is the metric (8.3).

Let $\mathcal{H}$ denote the Unmaximized Hamiltonian function for $(FT)'$:

$$\mathcal{H}(t,x,p,u) := p \cdot f(t,x,u).$$

Theorem 8.7.1 (Free End-Time Maximum Principle: Lipschitz Time Dependence) *Let $([\bar{S},\bar{T}],\bar{x},\bar{u})$ be a $W^{1,1}$ local minimizer for (FT)'. Assume that*

(H1) g is locally Lipschitz continuous;

(H2) for fixed x, $f(.,x,.)$ is $\mathcal{L} \times \mathcal{B}^m$ measurable. There exist $\delta > 0$ and a function $k : [\bar{S},\bar{T}] \times R^m \to R$ such that $t \to k(t,\bar{u}(t))$ is integrable and, for a.e. $t \in [\bar{S},\bar{T}]$,

$$|f(t'',x'',u) - f(t',x',u)| \leq k(t,\bar{u}(t))\,|(t'',x'') - (t',x')|$$

for all $(t'',x''),(t',x') \in (t,\bar{x}) + \delta B$, a.e. $t \in [\bar{S},\bar{T}]$;

(H3) $U(t) = U$ for all $t \in R$, for some Borel set $U \subset R^m$.

Then there exist $p \in W^{1,1}([\bar{S},\bar{T}];R^n)$, $r \in W^{1,1}([\bar{S},\bar{T}];R)$, and $\lambda \geq 0$ such that

(i) $(p,\lambda) \neq (0,0)$;

(ii) $(\dot{r}(t),-\dot{p}(t)) \in \text{co } \partial_{t,x}\mathcal{H}(t,\bar{x}(t),p(t),\bar{u}(t))$ a.e. $t \in [\bar{S},\bar{T}]$;

(iii) $(-r(\bar{S}),p(\bar{S}),+r(\bar{T}),-p(T))$
* $\in \lambda\partial g(\bar{S},\bar{x}(\bar{S}),\bar{T},\bar{x}(\bar{T})) + N_C(\bar{S},\bar{x}(\bar{S}),\bar{T},\bar{x}(\bar{T}));$*

(iv) $\mathcal{H}(t,\bar{x}(t),p(t),\bar{u}(t)) = \max_{u \in U} \mathcal{H}(t,\bar{x}(t),p(t),u)$ a.e. $t \in [\bar{S},\bar{T}]$;

(v) $\mathcal{H}(t, \bar{x}(t), p(t), \bar{u}(t)) = r(t)$ a.e. $t \in [\bar{S}, \bar{T}]$.

Proof. Choose δ' such that $([\bar{S}, \bar{T}], \bar{x}, \bar{u})$ is a minimizer for $(FT)'$, with respect to processes $([S, T], x, u)$ satisfying the constraints of $(FT)'$ and also

$$d(([S, T], x), ([\bar{S}, \bar{T}], \bar{x})) < 2\delta'.$$

Take a C^2 function $a : [\bar{S}, \bar{T}] \to R^n$ such that

$$a(\bar{S}) = \bar{x}(\bar{S})$$

and

$$d(([\bar{S}, \bar{T}], \bar{x}), ([\bar{S}, \bar{T}], a)) \quad \left(= \int_{\bar{S}}^{\bar{T}} |\dot{\bar{x}}(t) - \dot{a}(t)| dt \right) \quad < \delta'.$$

Consider the fixed end-time optimal control problem:

$$(R) \quad \begin{cases} \text{Minimize } g(\tau(\bar{S}), y(\bar{S}), \tau(\bar{T}), y(\bar{T})) \\ \text{over processes } ([S, T], (\tau, y, z), (v, \alpha)) \text{ satisfying} \\ (\dot{\tau}(s), \dot{y}(s), \dot{z}(s)) = \tilde{f}(\tau(s), y(s), v(s), \alpha(s)) \quad \text{a.e. } s \in [\bar{S}, \bar{T}], \\ v(s) \in U \text{ a.e.,} \\ \alpha(s) \in [0.5, 1.5] \text{ a.e.,} \\ (\tau(\bar{S}), y(\bar{S}), \tau(\bar{T}), y(\bar{T})) \in C, \\ m(\tau(\bar{S}), \tau(\bar{T}), y(\bar{S}), z(\bar{S}), z(\bar{T})) \leq \delta', \end{cases}$$

where

$$\tilde{f}(\tau, y, v, \alpha) := (\alpha, \alpha f(\tau, y, v), \alpha |f(\tau, y, v) - \dot{a}(\tau)|)$$

and

$$m(\tau_0, \tau_1, y_0, z_0, z_1) := |\tau_0 - \bar{S}| + |\tau_1 - \bar{T}| + |y_0 - a(\bar{S})|$$
$$+ \int_{\bar{S}}^{\bar{S} \vee \tau_0} |\dot{a}(\sigma)| d\sigma + \int_{\bar{T} \wedge \tau_1}^{\bar{T}} |\dot{a}(\sigma)| d\sigma + z_1 - z_0.$$

Here (τ, y, z) and (α, v) are partitions of the state and control vectors, respectively. Then

$$((\bar{\tau}(s) \equiv s, \bar{x}, \bar{z}(s) = \int_S^s |\dot{\bar{x}}(\sigma) - \dot{a}(\sigma)| d\sigma), (\bar{u}, \bar{\alpha}(s) \equiv 1))$$

is a minimizer for (R) To see this, take any process $([S, T], (\tau, y, z), (v, \alpha))$ that is feasible for problem (R) and consider the transformation $\psi : [\bar{S}, \bar{T}] \to [S, T]$:

$$S = \tau(\bar{S}), \quad T = \tau(\bar{T}), \quad \psi(s) := \tau(\bar{S}) + \int_{\bar{S}}^s \alpha(s) ds.$$

We find that ψ is a strictly increasing, Lipschitz continous function, with Lipschitz continuous inverse. It can be deduced that $x(.) : [S,T] \to R^n$ and $u(.) : [S,T] \to R^m$ defined by

$$x(.) := y \circ \psi^{-1}(.) \quad \text{and} \quad u(.) := v \circ \psi^{-1}(.)$$

are absolutely continuous and measurable functions, respectively, which satisfy

$$(S, x(S), T, x(T)) = (\tau(\bar{S}), x(\bar{S}), \tau(\bar{T}), x(\bar{T})),$$

$$\dot{x}(t) = f(t, x(t), u(t)) \quad \text{a.e. } t \in [S,T],$$

$$u(t) \in U \quad \text{a.e. } t \in [S,T]$$

and

$$d(([S,T], x), ([\bar{S}, \bar{T}], a)) \leq \delta'.$$

By the triangle inequality,

$$d(([S,T], x), ([\bar{S}, \bar{T}], \bar{x})) \leq 2\delta'.$$

It follows from the foregoing relationships that $([S,T], x, u)$ is a feasible process for the original optimal control problem. In view of the minimizing properties of $([\bar{S}, \bar{T}], \bar{x}, \bar{u})$, we have

$$g(\tau(\bar{S}), y(\bar{S}), \tau(\bar{T}), y(\bar{T})) = g(S, x(S), T, x(T))$$
$$\geq g(\bar{S}, \bar{x}(\bar{S}), \bar{T}, \bar{x}(\bar{T})) = g(\bar{\tau}(\bar{S}), \bar{y}(\bar{S}), \tau(\bar{T}), \bar{y}(\bar{T})).$$

It follows that the process $((\bar{\tau}, \bar{x}, \bar{z}), (\bar{u}, \bar{\alpha}))$ is a minimizer for (R), as claimed.

The hypotheses are satisfied for application of the necessary conditions of Theorem 6.2.1 to (R), with reference to the minimizer $((\bar{\tau}, \bar{x}, \bar{z}), (\bar{u}, \bar{\alpha}))$. We deduce that there exists $p \in W^{1,1}([\bar{S}, \bar{T}]; R^n)$, $r \in W^{1,1}([\bar{S}, \bar{T}]; R)$, and $\lambda \geq 0$, not all zero, such that

$$(\dot{r}(s), -\dot{p}(s)) = \text{co } \partial_{t,x} \mathcal{H}(s, \bar{x}(s), p(s), \bar{u}(s)) \quad \text{a.e.},$$

$$(-r(\bar{S}), p(\bar{S}), r(\bar{T}), -p(\bar{T})) \in \lambda \partial g(\bar{S}, \bar{x}(\bar{S}), \bar{T}, \bar{x}(\bar{T})) + N_C(\bar{S}, \bar{x}(\bar{S}), \bar{T}, \bar{x}(\bar{T})),$$

and

$$p(s) \cdot f(s, \bar{x}(s), \bar{u}(s)) - r(s) \geq \alpha(p(s) \cdot f(s, \bar{x}(s), u) - r(s))$$

$$\text{for all } u \in U \text{ and } \alpha \in [0.5, 1.5] \quad \text{a.e.}$$

The last relationship implies

$$\mathcal{H}(s, \bar{x}(s), p(s), \bar{u}(s)) = \max_{u \in U} \mathcal{H}(s, \bar{x}(s), p(s), u) \quad \text{a.e.},$$

$$\mathcal{H}(s, \bar{x}(s), p(s), \bar{u}(s)) = r(s) \quad \text{a.e.}$$

(Notice that, since $m < \delta'$, the last-listed constraint in (R) has no effect on the transversality condition and the adjoint variable associated with the z variable is zero.)

All the assertions of the theorem have been confirmed. $\square$

Theorem 8.7.2 (The Free End-Time Maximum Principle: Measurable Time Dependence) *Let* $([\bar{S}, \bar{T}], \bar{x}, \bar{u})$ *be a* $W^{1,1}$ *local minimizer for (FT)$'$ such that* $\bar{T} - \bar{S} > 0$. *Assume that there exist* $\delta > 0$ *such that*

(H1) g is locally Lipschitz continuous and C is a closed set;

(H2) $\mathrm{Gr}\, U(.)$ is a Borel set;

(H3) for each $x \in R^n$, $f(., x, .)$ is $\mathcal{L} \times \mathcal{B}^m$ measurable. There exists a function $k(.,.) : R \times R^m \to R$ such that $t \to k(t, \bar{u}(t))$ is integrable and

$$|f(t, x', u') - f(t', x'', u')| \le k(t, u)|x' - x''|$$

for all $(t', x'), (t'', x'') \in (t, \bar{x}(t)) + \delta B$, a.e. $t \in [\bar{S}, \bar{T}]$;

(H4) there exist $c_0 \ge 0$, $\delta_0 > 0$, and $k_0 \ge 0$ such that, for a.e. $t \in [\bar{S} - \delta_0, \bar{S} + \delta_0]$,

$$|f(t, x', u)| \le c_0 \quad and \quad |f(t, x', u') - f(t', x'', u')| \le k_0 |x' - x''|$$

for all $x', x'' \in \bar{x}(\bar{S}) + \delta B$ and $u \in U(t)$;

(H5) there exist $c_1 \ge 0$, $\delta_1 > 0$, and $k_1 \ge 0$ such that, for a.e. $t \in [\bar{T} - \delta_1, \bar{T} + \delta_1]$,

$$|f(t, x', u)| \le c_1 \quad and \quad |f(t, x', u) - f(t', x'', u)| \le k_1 |x' - x''|$$

for all $x', x'' \in \bar{x}(\bar{S}) + \delta B$ and $u \in U(t)$.

Then there exist $p \in W^{1,1}([\bar{S}, \bar{T}]; R^n)$ and real numbers $\lambda \ge 0$, ξ, and η such that

(i) $(\lambda, p) \ne (0, 0)$;

(ii) $-\dot{p}(t) \in \mathrm{co}\partial_x \mathcal{H}(t, \bar{x}(t), p(t), \bar{u}(t))$ a.e. $t \in [\bar{S}, \bar{T}]$;

(iii) $(-\xi, p(\bar{S}), \eta, -p(\bar{T})) \in \lambda \partial g(\bar{S}, \bar{x}(\bar{S}), \bar{T}, \bar{x}(\bar{T})) + N_C(\bar{S}, \bar{x}(\bar{S}), \bar{T}, \bar{x}(\bar{T}))$;

(iv) $\mathcal{H}(t, \bar{x}(t), p(t), \bar{u}(t)) = \max_{u \in U} \mathcal{H}(t, \bar{x}(t), p(t), u)$ a.e. $t \in [\bar{S}, \bar{T}]$;

(v) $\xi \in \mathrm{ess}_{t \to \bar{S}}\, H(t, \bar{x}(\bar{S}), p(\bar{S}))$;

(vi) $\eta \in \mathrm{ess}_{t \to \bar{T}}\, H(t, \bar{x}(\bar{T}), p(\bar{T}))$.

If the initial time is fixed (i.e., $C = \{\bar{S}\} \times \tilde{C}$ for some $\bar{S} \in R$ and some $\tilde{C} \subset R^{n+1+n}$), then the above assertions (except (v)) remain true when (H4) is dropped from the hypotheses. If the final time is fixed then the assertions (except (vi)) remain true when (H5) is dropped from the hypotheses.

Proof. This theorem is a straightforward consequence of Theorem 8.2.1. In the proof of Theorem 6.2.1, we saw how the Maximum Principle for optimal control problems with nonsmooth data can be deduced from the Extended Euler–Lagrange Condition, when the end-times are fixed. The steps involved were:

(i): construct a nested sequence of discrete approximations $\{U_j(t) : \bar{S} \leq t \leq \bar{T}\}_{j=1}^{\infty}$ to the control constraint sets $U(t)$, $\bar{S} \leq t \leq \bar{T}$, and

(ii): apply the Extended Euler–Lagrange Condition (and accompanying conditions) to each of a sequence of optimal control problems, in which the dynamic constraint is the differential inclusion

$$\dot{x} \in F_j(t,x) := f(t,x,U_j(t))$$

and recover the Maximum Principle in the limit, as $j \to \infty$.

Recall that the convergence analysis was carried out with the help of an interim hypothesis.

There exist $c(.) \in L^1$ and $k(.) \in L^1$ such that

$$|f(t,\bar{x}(t),u)| \leq c(t) \quad \text{and} \quad |f(t,x',u) - f(t,x'',u)| \leq k(t)|x' - x''|$$

for all $x', x'' \in \bar{x}(t) + \delta B$ and $u \in U(t)$, a.e. $t \in [\bar{S},\bar{T}]$

which was, eventually, removed.

This pattern of arguments can be followed in the broader context of free end-time problems. The difference of course is that, now, we apply necessary conditions centered on the free end-time Extended Euler–Lagrange Condition (namely Theorem 8.4.1) in Step (ii). The Transversality Condition, which takes account of the free end-times, carries across to the Maximum Principle, under the interim hypothesis. The procedure earlier devised to eliminate the interim hypothesis can be used, for free end-time problems, only to eliminate it on the subinterval $[\bar{S} + \delta_0, \bar{T} - \delta_1]$; the interim hypothesis (requiring $c(.)$ and $k(.)$ to be $L\infty$ functions) must be retained "near" the end-times as asserted in the theorem statement. □

8.8   Notes for Chapter 8

In this chapter we have derived two kinds of necessary conditions for free end-time optimal control problems or, to be more precise, problems for which the end-times are included among the choice variables.

Free end-time necessary conditions of the first kind give information about the nature of the Hamiltonian on the interior and at the endpoints of the optimal time interval under consideration, in particular, "Constancy of the Hamiltonian" for autonomous dynamics. They are restricted to problems with dynamic constraints that are Lipschitz continuous with respect to time. The approach followed here, to introduce a change of independent variable that replaces the free end-time problem by a fixed end-time problem and to apply the fixed end-time necessary conditions to the transformed problem, is substantially that employed by Clarke [38]. The underlying idea is implicit in earlier literature, however. (See, e.g., [60]).

The fact that necessary conditions of the first kind (including Constancy of the Hamiltonian) are valid for arcs that are merely $W^{1,1}$ local minimizers, a matter of some past speculation, is a new result.

Free time necessary conditions of the second kind give information about the Hamiltonian merely at the optimal end-times, but cover problems for which the time dependence of the dynamic constraint is no longer assumed Lipschitz continuous. They are typically derived by reworking the proofs of the related fixed end-time necessary conditions to allow explicitly for variations of end-times as well as state trajectories.

In the case of continuous time dependence, free time necessary conditions including boundary conditions on the Hamiltonian are evident in the early Russian literature. They are also to be found in Berkovitz' book [19].

Free time necessary conditions for problems with data which is measurably time-dependent were first derived by Clarke and Vinter [49] [50], who introduced the concept of "essential value" of an almost everywhere defined function, to make sense of the boundary condition on the Hamiltonian in this case. Extensions to problems with state constraints were carried out by Clarke, Loewen, and Vinter [55] and by Rowland and Vinter [127], who employed a more refined analysis to allow for active state constraints at optimal end-times.

Chapter 9

The Maximum Principle for State Constrained Problems

Between a rock and a hard place.

– Anonymous

9.1 Introduction

In this chapter, we return to the framework of Chapter 6, in which the dynamic constraint takes the form of a differential equation parameterized by control functions. Our goal is to extend the earlier derived necessary conditions of optimality, in the form of a Maximum Principle, to allow for pathwise constraints on the state trajectories.

The optimal control problem of interest is now

$$(P) \begin{cases} \text{Minimize } g(x(S), x(T)) \\ \text{over } x \in W^{1,1}([S,T]; R^n) \text{ and measurable functions } u \text{ satisfying} \\ \dot{x}(t) = f(t, x(t), u(t)) \quad \text{a.e.,} \\ u(t) \in U(t) \quad \text{a.e.,} \\ h(t, x(t)) \leq 0 \text{ for all } t \in [S,T], \\ (x(S), x(T)) \in C, \end{cases}$$

the data for which comprise: an interval $[S,T]$, functions $g : R^n \times R^n \to R$, $f : [S,T] \times R^n \times R^m \to R^n$, and $h : [S,T] \times R^n \to R$, a multifunction $U : [S,T] \rightsquigarrow R^m$, and a closed set $C \subset R^n \times R^n$.

The new ingredient in problem (P) is the pathwise state constraint:

$$h(t, x(t)) \leq 0 \quad \text{for all } t \in [S,T].$$

We have chosen to formulate it as a scalar functional inequality constraint partly because this is a kind of state constraint frequently encountered in engineering applications and partly because it is a convenient starting point for deriving necessary conditions for other types of state constraints (multiple state constraints, implicit state constraints, etc.) of interest.

As before, we refer to a measurable function $u : [S,T] \to R^m$ that satisfies $u(t) \in U(t)$ a.e., as a *control function*. The set of control functions is written $\mathcal{U}$. A *process* (x,u) comprises a control function u and an arc $x \in W^{1,1}([S,T]; R^n)$ that is a solution to the differential equation $\dot{x}(t) = f(t, x(t), u(t))$ a.e. A *state trajectory* x is the first component of some process (x, u). A process (x, u) is said to be *feasible* if the state trajectory x satisfies the endpoint constraint $(x(S), x(T)) \in C$ and the state constraint $h(t, x(t)) \leq 0$, for all $t \in [S, T]$.

Consistent with earlier terminology, a process $(\bar{x}, \bar{u})$ is said to be a $W^{1,1}$ *local minimizer* if there exists $\delta > 0$ such that the process $(\bar{x}, \bar{u})$ minimizes $g(x(S), x(T))$ over all feasible processes (x, u) satisfying

$$||x - \bar{x}||_{W^{1,1}} \leq \delta.$$

If the $W^{1,1}$ norm is replaced by the L^∞ norm, $(\bar{x}, \bar{u})$ is called a *strong local minimizer*.

What effect does the state constraint have on necessary conditions of optimality? We might expect that it can be accommodated by a Lagrange multiplier term

$$\int_S^T h(s, x(s)) m(s) ds$$

added to the cost. That is to say, if $(\bar{x}, \bar{u})$ is a minimizer for (P), then for some appropriately chosen, nonnegative valued function m satisfying the *Complementary Slackness Condition*:

$$m(t) = 0 \quad \text{for } t \in \{s : h(s, \bar{x}(s)) < 0\}, \tag{9.1}$$

$(\bar{x}, \bar{u})$ satisfies first-order necessary conditions of optimality also for the state constraint-free problem

$$\begin{cases} \text{Minimize} g(x(S), x(T)) + \int_S^T h(s, x(s)) m(s) ds \\ \text{over } x \in W^{1,1} \text{ and measurable functions } u \text{ satisfying} \\ \dot{x} = f \quad \text{a.e., } u(t) \in U(t) \quad \text{a.e. and,} (x(S), x(T)) \in C. \end{cases}$$

Reducing this problem to one with no integral cost term by state augmentation and applying the state constraint-free Maximum Principle we arrive at the following set of conditions: there exist $q \in W^{1,1}$ and $\lambda \geq 0$ such that

$$(q, \lambda, m) \neq (0, 0, 0); \tag{9.2}$$

$$-\dot{q}(t) = q(t) f_x(t, \bar{x}(t), \bar{u}(t)) - \lambda h_x(t, x(t)) m(t) \quad \text{a.e.;} \tag{9.3}$$

$$(q(S), -q(T)) \in N_C(\bar{x}(S), \bar{x}(T)) + \lambda \nabla g(\bar{x}(S), \bar{x}(T)); \tag{9.4}$$

$$q(t) \cdot f(t, \bar{x}(t), \bar{u}(t)) = \max_{u \in \Omega} q(t) \cdot f(t, \bar{x}(t), u) \quad \text{a.e.} \tag{9.5}$$

These relationships capture the essential character of necessary conditions for problems with state constraints, although one modification is required: in order to derive broadly applicable necessary conditions we must allow the multiplier q (about whose precise nature we have up to this point been rather coy) to be the "derivative of a function of bounded variation." That is to say, the conditions assert the existence of a nondecreasing function of bounded variation $\nu : [S, T] \to R$ such that the preceding relationships apply with (9.1) through (9.3) replaced by

$$\nu \text{ is constant on any subinterval of } \{t : h(t, x(t)) < 0\}, \tag{9.6}$$

$$(q, \lambda, \nu) \neq (0, 0, 0), \tag{9.7}$$

and, for all $t \in (S, T]$,

$$-q(t) = -q(S) + \int_S^t q(s) f_x(s, \bar{x}(s), \bar{u}(s)) ds - \int_{[S,t]} h_x(s, \bar{x}(s)) d\nu(s). \tag{9.8}$$

(We have absorbed λ into the state constraint multiplier: $d\nu = \lambda m(t) dt$.) This is not surprising since, if h is a continuous function, problem (P) can be set up as an optimization problem over pairs of elements (x, u) satisfying (among other conditions) the constraint

$$G(x) \in P^-,$$

where $G : W^{1,1} \to C([S, T])$ is the function

$$G(x)(t) = h(t, x(t))$$

and P^- is the cone of nonpositive-valued continuous functions on $[S, T]$. One expects for such problems a multiplier rule to apply, involving a multiplier ξ in the negative polar cone of P^-, regarded as a subset of the topological dual space $C([S, T])$, which satisfies

$$< \xi, G(\bar{x}) > = 0.$$

Such a multiplier is represented by a nondecreasing function ν of bounded variation according to

$$< \xi, y > = \int_S^T y(s) d\nu(s) \quad \text{for all } y \in C([S, T]).$$

The anticipated relationships then are (9.4) to (9.8).

Of course the adjoint arc q, which satisfies the integral equation (9.8), is a function of bounded variation itself. It is customary to aim for necessary conditions that differ from those just outlined in one small respect. A further modification to the necessary conditions, which is really just a matter of redefinition, is motivated by the desire to supply relationships involving an absolutely continuous adjoint arc. The new adjoint arc, p, is obtained simply by subtracting the "troublesome" bit from q:

$$p(t) := \begin{cases} q(t) - \int_{[S,t)} h_x(s, \bar{x}(s)) \nu(ds) & t \in [S, T) \\ q(T) - \int_{[S,T]} h_x(s, \bar{x}(s)) \nu(ds) & t = T. \end{cases}$$

We summarize the relationship between p and q briefly as

$$q = p + \int h_x d\nu. \tag{9.9}$$

The conditions now become: there exist $p \in W^{1,1}$, $\lambda \geq 0$, and a nondecreasing function of bounded variation ν such that

$(p, \lambda, \nu) \neq 0$;

$-\dot{p}(t) = q(t) f_x(t, \bar{x}(t), \bar{u}(t))$ a.e.;

$(p(S), -q(T)) \in N_C(\bar{x}(S), \bar{x}(T)) + \lambda \nabla g(\bar{x}(S), \bar{x}(T))$;

$q(t) \cdot f(t, \bar{x}(t), \bar{u}(t)) = \max_{u \in \Omega} q(t) \cdot f(t, \bar{x}(t), u)$ a.e.;

ν is constant on any subinterval of $\{t : h(t, \bar{x}(t)) < 0\}$.

In these conditions, the function q is determined from p and ν according to (9.9).

The final touch is to express the conditions in terms of the regular Borel measure μ on $[S, T]$ associated with the function of bounded variation ν: μ is the unique regular Borel measure such that $\mu(I) = \int_I d\nu(t)$ for all closed subintervals $I \subset [S, T]$. The change then is to replace $\int h_x d\nu(s)$ by $\int h_x \mu(ds)$.

Our aims in this chapter are twofold. The first is to provide necessary conditions for problem (P) under general hypotheses, which allow the function f in the dynamical constraint to be nonsmooth with respect to the x variable (nonsmooth dynamics). The second is to provide a self-contained proof of a special case of the Maximum Principle for state constrained problems in which the dynamics are smooth. As in our earlier treatment of state constraint-free problems, the smooth Maximum Principle applied to various intermediate problems has a pivotal role in the derivation of all the necessary conditions for state constrained problems covered in this book, including a general "nonsmooth" version of the Maximum Principle itself.

In what follows, we replace $h_x(s, \bar{x}(s))$ by a selector γ of some suitably defined subdifferential of $x \to h(t, x)$, to allow for a nonsmooth state constraint functional h. For convenience, we refer to $(p, \lambda, \mu, \gamma)$ as "multipliers" for (P). (Strictly speaking, the multipliers are (q, λ, μ) — those associated with the differential inclusion constraint $\dot{x} \in F(t, x)$, the cost, and the state constraint, respectively.)

9.2   Convergence of Measures

As earlier discussed, the presence of state constraints requires us to consider "multipliers" that are elements in the topological dual $C^*([S, T]; R^k)$ of the space of continuous functions $C([S, T]; R^k)$ with supremum norm. (Here

$[S, T]$ is a given interval.) The norm on $C^*([S, T], R^k)$, written $\|\mu\|_{\text{T.V.}}$, is the induced norm. The set of elements in $C^*([S, T]; R)$ taking nonnegative values on nonnegative-valued functions in $C([S, T]; R)$ is denoted $C^\oplus(S, T)$.

As is well known, elements $\mu \in C^*([S, T]; R^k)$ can be identified with the set of finite regular vector-valued measures on the Borel subsets of $[S, T]$. We loosely refer then to elements $\mu \in C^*([S, T]; R^k)$ as "measures." Notice that, for $\mu \in C^\oplus(S, T)$, $\|\mu\|_{\text{T.V.}}$, as defined above, coincides with the total variation of μ, $\int_{[S,T]} \mu(ds)$, as the notation would suggest.

The support of a measure $\mu \in C^*([S, T]; R^k)$, written $\text{supp}\{\mu\}$, is the smallest closed subset $A \subset [S, T]$ with the property that for all relatively open subsets $B \subset [S, T] \setminus A$ we have $\mu(B) = 0$.

Given $\mu \in C^\oplus(S, T)$, a μ-continuity set is a Borel subset $B \subset [S, T]$ for which $\mu(\text{bdy } B) = 0$. Take $\mu \in C^\oplus(S, T)$. Then there is a countable set $S \subset (S, T)$, such that all sets of the form $[a, b]$, $[a, b)$, $(a, b]$ with $a, b \in ([S, T] \setminus S)$ are μ-continuity sets.

Given a weak* convergent sequence $\mu_i \to \mu$ in $C([S, T]; R^k)$, there exists a countable subset $S \subset (S, T)$ such that

$$\int_{[S,t)} \mu_i(ds) \to \int_{[S,t)} \mu(ds)$$

for all $t \in ([S, T] \setminus S)$.

Take a weak* convergent sequence $\mu_i \to \mu$ in $C^\oplus(S, T)$. Then

$$\int_B \mu(dt) \leq \liminf_{i \to \infty} \int_B \mu_i(dt)$$

for any relatively open subset $B \subset [S, T]$. Also,

$$\int_B h(t)\mu(dt) = \lim_{i \to \infty} \int_B h(t)\mu_i(dt)$$

for any $h \in C([S, T]; R^k)$ and any μ-continuity set B.

Take closed subsets A and A_i, $i = 1, 2, \ldots$ of $[S, T] \times R^n$. We denote by $A(.) : [S, T] \rightsquigarrow R^n$ the multifunction

$$A(t) := \{a : (t, a) \in A\}.$$

The multifunctions $A_i(.)$ are likewise defined. The following proposition has an important role in justifying limit-taking in "measure" relationships of the kind

$$\eta_i(dt) = \gamma_i(t)\mu_i(dt), \quad i = 1, 2, \ldots,$$

in which the sequence of Borel measurable functions $\{\gamma_i(.)\}$ satisfies

$$\gamma_i(t) \in A_i(t) \quad \mu_i \text{ a.e.}$$

Conditions are given under which we can conclude

$$\eta_0(dt) \; = \; \gamma_0(t)\mu_0(dt),$$

where η_0 and μ_0 are weak* limits of $\{\eta_i\}$ and $\{\mu_i\}$, respectively, and γ_0 is a Borel measurable function satisfying

$$\gamma_0(t) \in A(t) \quad \mu_0 \text{ a.e.} \tag{9.10}$$

Proposition 9.2.1 *Take a weak* convergent sequence $\{\mu_i\}$ in $C^{\oplus}(S,T)$, a sequence of Borel measurable functions $\{\gamma_i : [S,T] \to R^n\}$, and a sequence of closed sets $\{A_i\}$ in $[S,T] \times R^n$. Take also a closed set A in $[S,T] \times R^n$, and a measure $\mu \in C^{\oplus}(S,T)$.*

Assume that $A(t)$ is convex for each $t \in \text{dom } A(.)$ and that the sets A and $A_1, A_2, \ldots$ are uniformly bounded. Assume further that

$$\limsup_{i\to\infty} A_i \subset A,$$

$$\gamma_i(t) \in A_i(t) \quad \mu_i \text{ a.e. for } i = 1, 2, \ldots$$

and

$$\mu_i \to \mu_0 \quad weakly*.$$

Define $\eta_i \in C^*([S,T]; R^k)$

$$\eta_i(dt) \; := \; \gamma_i(t)\mu_i(dt).$$

Then, along a subsequence,

$$\eta_i \to \eta_0 \quad weakly*,$$

for some $\eta_0 \in C^*([S,T]; R^k)$ such that

$$\eta_0(dt) \; = \; \gamma_0(t)\mu_0(t),$$

in which γ_0 is a Borel measurable function that satisfies

$$\gamma_0(t) \in A(t) \quad \mu_0 \text{ a.e.}$$

Proof. Since the sets A_i, $i = 1, 2, \ldots$, are uniformly bounded there exists some constant K such that

$$|\gamma_i(t)| \leq K \quad \mu_i \text{ a.e.}$$

But $\{\mu_i\}$ is a weak* convergent sequence. It follows that $\{\|\eta_i\|_{\text{T.V.}}\}$ is a bounded sequence. Along a subsequence then, $\eta_i \to \eta_0$ (weakly*) for some $\eta_0 \in C^*([S,T]; R^k)$. Given any $\phi \in C([S,T]; R^n)$

$$\begin{aligned}
\left| \int_{[S,T]} \phi(t)\eta_0(dt) \right| &= \lim_i \left| \int_{[S,T]} \phi(t)\gamma_i(t)\mu_i(dt) \right| \\
&\leq K \lim_i \int_{[S,T]} |\phi(t)|\mu_i(dt) \\
&= K \int_{[S,T]} |\phi(t)|\mu_0(dt),
\end{aligned}$$

by weak* convergence. This inequality, which holds for every continuous ϕ, implies that η_0 is absolutely continuous with respect to μ_0. By the Radon–Nicodym theorem then, there exists an R^n-valued, Borel measurable, and a μ_0-integrable function $\gamma_0 : [S, T] \to R^n$ such that

$$\int_E \eta_0(dt) = \int_E \gamma_0(t)\mu_0(dt)$$

for all Borel subsets $E \subset [S, T]$.

For each i we express the n-vector valued measure η_i in terms of its components $\eta_i = (\eta_{i1}, \ldots, \eta_{in})$. Let $\eta_{ij} = \eta_{ij}^+ - \eta_{ij}^-, j = 1, \ldots, n$, be the Jordan decomposition of η_{ij}. For each j, $\{\eta_{ij}^+\}_{i=1}^\infty$ and $\{\eta_{ij}^-\}_{i=1}^\infty$ are bounded in total variation, since the η_is are bounded in total variation. By limiting attention to a further subsequence then, we can arrange that $\lim_i \eta_{ij}^+ = \eta_j^+$, $\lim_i \eta_{ij}^- = \eta_j^-$, for some $\eta_j^+, \eta_j^- \in C^\oplus(S, T)$ and $j = 1, \ldots, n$. (Here limits are interpreted as weak* limits.) Since, for each $\phi = (\phi_1, \ldots, \phi_n) \in C([S, T]; R^n)$,

$$\int \phi(t) \cdot \eta_0(dt) = \lim_i \int \phi(t) \cdot \eta_i(dt)$$
$$= \lim_i \int \sum_j \phi_j(t) \, [\eta_{ij}^+ - \eta_{ij}^-](dt)$$
$$= \int \sum_j \phi_j(t) \, [\eta_j^+ - \eta_j^-](dt),$$

we see that $\eta_0 = ((\eta_1^+ - \eta_1^-), \ldots, (\eta_n^+ - \eta_n^-))$. Let $\mathcal{C}$ denote the class of Borel sets that are continuity sets of $\eta_1^+, \ldots, \eta_n^+, \eta_1^-, \ldots, \eta_n^-$, and μ_0. Then for any $E \subset \mathcal{C}$ and $h \in C([S, T]; R^n)$,

$$\int_E \eta_i(dt) \to \int_E \eta_0(dt) \quad \text{and} \quad \int_E h(t)\mu_i(dt) \to \int_E h(t)\mu_0(dt). \quad (9.11)$$

Fix a positive integer j. Define the set $A^j \subset [S, T] \times R^n$ to be

$$A^j := (A + j^{-1}B) \cap ([S, T] \times R^n).$$

Then, since the sets A_i, $i = 1, 2, \ldots$, are uniformly bounded and

$$\limsup_{i \to \infty} A_i \subset A,$$

we have $A_i \subset A^j$, for all i sufficiently large.

Take any relatively open set $E \subset [S, T] \setminus \operatorname{dom} A^j(.)$. Then, since $\eta_i(dt) = \gamma_i\mu_i(dt)$ and $\operatorname{supp}\{\eta_i\} \subset \operatorname{dom} A^j(.)$, for i sufficiently large, we have

$$0 \geq \liminf_{i \to \infty} \int_E \eta_{ij}^+(dt) \geq \int_E \eta_j^+(dt) \geq 0.$$

It follows that $\eta_j^+(E) = 0, j = 1, \ldots, n$. Likewise, we show that $\eta_j^-(E) = 0, j = 1, \ldots, n$.

We conclude that $\eta_0(E) = 0$ for all open sets $E \subset [S, T] \setminus \text{dom } A^j(.)$. This means that

$$\text{supp }\{\eta_0\} \subset \text{dom } A^j(.).$$

Fix $q \in R^n$. Then, for $r > 0$ sufficiently large, the function

$$s_q(t) := \begin{cases} \max\{q \cdot d : d \in A^j(t)\} & \text{if } A^j(t) \neq \emptyset \\ r & \text{otherwise} \end{cases}$$

is upper semicontinuous and bounded on $[S, T]$. By a well-known property of upper semicontinuous functions, there exists a sequence of continuous functions $\{c_q^k : [S, T] \to R\}_{k=1}^{\infty}$ such that

$$s_q(t) \leq c_q^k(t) \quad \text{for all } t \in [S, T], \ k = 1, 2, \ldots$$

and

$$s_q(t) = \lim_{k \to \infty} c_q^k(t) \quad \text{for all } t \in [S, T]. \tag{9.12}$$

Choose any $E \subset C$. For each i sufficiently large,

$$\begin{aligned} q \cdot \int_E \eta_i(dt) &= \int_E q \cdot \gamma_i(t)\mu_i(dt) \\ &= \int_{E \cap \text{dom } A^j(\cdot)} q \cdot \gamma_i(t)\mu_i(dt) \\ &\leq \int_E c_q^k(t)\mu_i(dt). \end{aligned}$$

By (9.11) and since c_q^k is continuous, we obtain in the limit as $i \to \infty$

$$\int_E q \cdot \gamma_0(t)\mu_0(dt) \leq \int_E c_q^k(t)\mu_0(dt).$$

Since C generates the Borel sets, we readily deduce that this inequality is valid for all Borel sets E. It follows that,

$$q \cdot \gamma_0(t) \leq c_q^k(t) \quad \mu_0 \text{ a.e.}$$

From (9.12) then,

$$q \cdot \gamma_0(t) \leq s_q(t) \ (= \ \max\{q \cdot d : d \in A^j(t)\}) \quad \mu_0 \text{ a.e.}$$

By σ-additivity, this last relationship holds for all q belonging to some countable dense subset of R^n. Since A^j is a bounded set, the mapping $q \to c_q^k(t)$ is continuous for each $t \in \text{dom } A^j$. We conclude that it is true for all $q \in R^m$. But then, since $A^j(t)$ is closed and convex for each $t \in \text{dom } A^j$, we have

$$\gamma_0(t) \in A^j(t) \quad \mu_0 \text{ a.e.}$$

Finally we observe that, since A is closed set,

$$\gamma_0(t) \in \cap_j A^j(t) = A(t) \quad \mu_0 \text{ a.e.}$$

We have shown that (along some subsequence) $\eta_i \to \eta_0$ (weakly*) for some η_0 which can be expressed $\eta_0(dt) = \gamma_0(t)\mu_0(dt)$, where γ_0 is a Borel measurable function satisfying

$$\gamma_0(t) \in A(t) \quad \mu_0 \text{ a.e.}$$

This is what we set out to prove. $\square$

9.3 The Maximum Principle for Problems with State Constraints

Denote by $\mathcal{H}$ the Unmaximized Hamiltonian:

$$\mathcal{H}(t,x,p,u) := p \cdot f(t,x,u).$$

Define the following partial subdifferential $\partial_x^> h(t,x)$.

$$\partial_x^> h(t,x) := \text{co}\{\xi : \text{ there exists } (t_i,x_i) \overset{h}{\to} (t,x) \text{ such that}$$
$$h(t_i,x_i) > 0 \text{ for all } i \text{ and } \nabla_x h(t_i,x_i) \to \xi\}. \quad (9.13)$$

Theorem 9.3.1 (The Maximum Principle for State Constrained Problems) *Let $(\bar{x},\bar{u})$ be a $W^{1,1}$ local minimizer for (P). Assume that, for some $\delta > 0$, the following hypotheses are satisfied.*

(H1) $f(.,x,.)$ is $\mathcal{L} \times \mathcal{B}^m$ measurable, for fixed x. There exists a Borel measurable function $k(.,.) : [S,T] \times R^m \to R$ such that $t \to k(t,\bar{u}(t))$ is integrable and

$$|f(t,x,u) - f(t,x',u)| \leq k(t,u)|x - x'|$$

for all $x, x' \in \bar{x}(t) + \delta B, u \in U(t)$, a.e.;

(H2) $\text{Gr } U$ is a Borel set;

(H3) g is Lipschitz continuous on $(\bar{x}(S), \bar{x}(T)) + \delta B$;

(H4) h is upper semicontinuous and there exists $K > 0$ such that

$$|h(t,x) - h(t,x')| \leq K|x - x'|$$

for all $x, x' \in \bar{x}(t) + \delta B, t \in [S,T]$.

Then there exist $p \in W^{1,1}([S,T]; R^n)$, $\lambda \geq 0$, $\mu \in C^{\oplus}(S,T)$, and a Borel measurable function $\gamma : [S,T] \to R^n$ satisfying

$$\gamma(t) \in \partial_x^> h(t, \bar{x}(t)) \quad \mu \text{ a.e.,}$$

such that

(i) $(p, \mu, \lambda) \neq (0,0,0)$;

(ii) $-\dot{p}(t) = \text{co } \partial_x \mathcal{H}(t, \bar{x}(t), q(t), \bar{u}(t))$ *a.e.;*

(iii) $(p(S), -q(T)) \in \lambda \partial g(\bar{x}(S), \bar{x}(T)) + N_C(\bar{x}(S), \bar{x}(T))$;

(iv) $\mathcal{H}(t, \bar{x}(t), q(t), \bar{u}(t)) = \max_{u \in U(t)} \mathcal{H}(t, \bar{x}(t), q(t), u)$;

(v) $\text{supp}\{\mu\} \subset I(\bar{x})$.

Now assume, also, that

$$f(t, x, u), \ h(t, x), \ and \ U(t) \ are \ independent \ of \ t.$$

Then, in addition to the above conditions, there exists a constant r such that

(vi) $\mathcal{H}(t, \bar{x}(t), q(t), \bar{u}(t)) = r \quad$ *a.e.*

Here $q := p + \int \gamma(s)\mu(ds)$; i.e.,

$$q(t) := \begin{cases} p(t) + \int_{[S,t)} \gamma(s)\mu(ds) & \text{for } t \in [S,T) \\ p(T) + \int_{[S,T]} \gamma(s)\mu(ds) & \text{for } t = T \end{cases}$$

and

$$I(\bar{x}) := \{t : h(t, \bar{x}(t)) = 0\}.$$

Remarks

(a): We have added Condition (v) to the list of conditions, for emphasis. It is in fact implied by the condition

$$\gamma(t) \in \partial_x^> h(t, \bar{x}(t)) \quad \mu \text{ a.e.,}$$

since, if $t \notin I(\bar{x})$, $\partial_x^> h(t, \bar{x}(t)) = \emptyset$.

(b): The necessary conditions of Theorem 9.3.1 are of interest only when the state constraint is nondegenerate, in the sense that $0 \notin \partial_x^> h(t, \bar{x}(t))$ for all t such that $h(t, \bar{x}(t)) = 0$. This is because the necessary conditions are automatically satisfied along any arc $\bar{x}$ such that

$$0 \in \partial_x^> h(t', \bar{x}(t')) \quad \text{and} \quad h(t', \bar{x}(t')) = 0,$$

for some time t', with the choice of multipliers $\mu = \delta_{\{t'\}}$ (the unit measure concentrated on $\{t'\}$), $p(t) \equiv 0$, and $\lambda = 0$. Unless we are able, a priori to exclude the existence of such arcs, the necessary conditions convey no useful information about a minimizer.

(c): We draw attention to the fact that a state constraint function $h(t, x)$ is required merely to be upper semicontinuous in t. This is a useful feature of the formulation of the state constraint problem adopted here, since it allows us to consider problems in which the state constraint is imposed merely on some closed subset $I \subset [S, T]$:

$$\tilde{h}(t, x(t)) \leq 0 \quad \text{for all } t \in I.$$

If $\tilde{h}(t, x)$ is a bounded continuous function such that $\tilde{h}(t, .)$ is Lipschitz continuous (uniformly with respect to t in $[S, T]$), then the constraint can be equivalently expressed as

$$h(t, x(t)) \leq 0 \quad \text{for all } t \in [S, T],$$

where

$$h(t, x) = \begin{cases} \tilde{h}(t, x) & \text{if } t \in I \\ r & \text{otherwise.} \end{cases}$$

Here r is some fixed, strict lower bound on the values of $\tilde{h}$.

(d): Multiple state constraints

$$h_i(t, x(t)) \leq 0, \quad \text{for } i = 1, \ldots, k, \tag{9.14}$$

are easily accommodated in the above necessary conditions, by choosing the scalar state constraint functional h to be

$$h(t, x) := \max\{h_1(t, x), \ldots, h_k(t, x)\}.$$

Using the Max Rule (Theorem 5.5.2) to estimate $\partial_x^> h$ in this case, we deduce from Theorem 9.3.1 the following corollary.

(Multiple State Constraints) *Let $(\bar{x}, \bar{u})$ be a minimizer for a variant on (P) in which the single state constraint $h(t, x(t)) \leq 0$ is replaced by the collection of constraints (9.14).*

Assume that Hypotheses (H1) through (H3) are satisfied and that (H4) is also satisfied when h_i replaces h for $i = 1, \ldots, k$. Then there exist $p \in W^{1,1}([S, T]; R^n)$, $\lambda \geq 0$, $\mu_i \in C^\oplus(S, T), i = 1, \ldots, k$ and Borel measurable functions $\gamma_i : [S, T] \to R^n$ such that

$$\gamma_i(t) \in \partial_x^> h_i(t, \bar{x}(t)) \quad \mu_i \text{ a.e.,}$$

$$(p, \mu_i, \ldots, \mu_k, \lambda) \neq 0,$$

$$\text{supp}\{\mu_i\} \subset \{t : h_i(t, \bar{x}(t)) = 0\}$$

and Conditions (ii) to (iv) (and also (vi) for time-independent data) of Theorem 9.3.1 are satisfied with q defined to be

$$q := p + \int \sum_{i=1}^{n} \gamma_i(s)\mu_i(ds).$$

(e): Consider next another variation on problem (P). It is to replace the state constraint by the implicit constraint

$$x(t) \in X(t), \tag{9.15}$$

in which $X : [S, T] \rightsquigarrow R^n$ is a given multifunction.

Theorem 9.3.1 can be adapted to handle this kind of state constraint also, when X satisfies a suitable "constraint qualification."

(CQ) X is a lower semicontinuous multifunction; i.e.,

$$X(t) \subset \liminf_{s \to t} X(s) \quad \text{for all } t \in [S, T]$$

and $\text{co}\bar{N}_{X(t)}(\bar{x}(t))$ is pointed for each $t \in [S, T]$.

Here,

$$\bar{N}_{X(t)}(\bar{x}(t)) := \limsup\{N_{X(s)}(y) : (s, y) \xrightarrow{\text{Gr } X} (t, \bar{x}(t))\}.$$

The term "pointed" is interpreted in the sense:

a convex cone $K \subset R^k$ is said to be pointed when for any nonzero elements $d_1, d_2 \in K$,

$$d_1 + d_2 \neq 0.$$

The statement of Theorem 9.3.1 can be modified to cover optimal control problems with an implicit state constraint (9.15), when (CQ) is satisfied, as follows. In (P) take the state constraint function h to be

$$h(t, x) = d_{X(t)}(x), \tag{9.16}$$

where, as usual, $d_{X(t)}$ denotes the Euclidean distance function. Fix $t \in [S, T]$. It can be deduced from the definition of the subdifferential (9.13), with the help of Lemma 4.8.3, that every $\xi \in \partial_x^> h(t, \bar{x}(t))$ is expressible as

$$\xi = \sum_{i=0}^{n} \lambda_i \xi_i$$

for some vectors $\xi_i \in \bar{N}_X(\bar{x}(t)) \cap \{\eta \in R^n : |\eta| = 1\}, i = 0, \ldots, n$ and some $\{\lambda_i\}_{i=0}^n \in \Lambda$, where

$$\Lambda := \{\lambda_0', \ldots, \lambda_n' : \lambda_i' \geq 0 \text{ for all } i \text{ and } \sum_i \lambda_i' = 1\}.$$

In view of the pointedness assumption, we conclude that

$$\partial_x^> h(t, \bar{x}(t)) \subset \bar{N}_X(\bar{x}(t)) \cap \{\xi : \xi \neq 0\} \tag{9.17}$$

for all $t \in [S, T]$.

Now let $(p, \mu, \lambda, \gamma)$ be the "multipliers" of Theorem 9.3.1 for the choice of h (9.16). Define $\eta \in C^*([S, T]; R^n)$ according to $d\eta(t) = \gamma(t)\mu(dt)$. Since (see (9.17)) $\gamma(t) \neq 0$ μ a.e., we have $\eta \neq 0$ if $\mu \neq 0$. But $(p, \mu, \lambda) \neq 0$. It follows that

$$(p, \eta, \lambda) \neq (0, 0, 0). \tag{9.18}$$

Relationship (9.17) also tells us that $\gamma(t) \in \partial_x^> h(t, \bar{x}(t))$ μ a.e., whence

$$\int_{[S,T]} \xi(t) \cdot \eta(dt) \leq 0, \tag{9.19}$$

for all $\xi \in C([S, T]; R^n)$ such that $\xi(t) \in (\bar{N}_{X(t)}(\bar{x}(t)))^*$ η a.e. Here $(\bar{N}_{X(t)}(\bar{x}(t)))^*$ is the negative polar of $\bar{N}_{X(t)}(\bar{x}(t))$:

$$(\bar{N}_{X(t)}(\bar{x}(t)))^* := \{\xi : \xi \cdot \gamma \leq 0 \text{ for all } \gamma \in \bar{N}_{X(t)}(\bar{x}(t))\}.$$

Restating the assertions of Theorem 9.3.1 in terms of (p, η, λ) in place of γ and (p, μ, λ) and noting (9.18) and (9.19) we arrive at another version of the Maximum Principle for state constrained problems.

(The Maximum Principle for Implicit State Constraints) *Let $(\bar{x}, \bar{u})$ be a minimizer for a variant on (P) in which the state constraint $h(t, x(t)) \leq 0$ is replaced by*

$$x(t) \in X(t) \quad \text{for all } t \in [S, T].$$

Assume that hypotheses (H1) to (H3) are satisfied and that the "state constraint" multifunction $X : [S, T] \rightsquigarrow R^n$ satisfies Hypothesis (CQ). Then there exist $p \in W^{1,1}([S, T]; R^n)$, $\lambda \geq 0$, and $\eta \in C^*([S, T]; R^n)$ such that

$$\int_{[S,T]} \xi(t) \cdot \eta(dt) \leq 0$$

for all $\xi \in C([S, T]; R^n)$ satisfying $\xi(t) \in (\bar{N}_{X(t)}(\bar{x}(t)))^*$ η a.e.,

$$(p, \eta, \lambda) \neq 0,$$

$$\text{supp}\{\eta\} \subset \{t : (t, \bar{x}(t)) \in \text{bdy Gr } X(t)\},$$

and Conditions (ii) to (iv) of Theorem 9.3.1 are satisfied, where now

$$q = p + \int \eta(dt).$$

If

$$f(t, x, u), \ X(t), \ and \ U(t) \ are \ independent \ of \ t,$$

then, in addition to the above conditions, Condition (vi) is satisfied for some constant r.

What is the nature of the constraint qualification (CQ)? It is, essentially, the requirement that the "tangent cone" to the state constraint set be not too small. To be precise, in the case that $X(t)$ is independent of t (write it X),

(CQ)′ X is closed and int $\bar{T}_D(\bar{x}(t))$ is nonempty for each $t \in [S, T]$

is a sufficient condition for (CQ). Here $\bar{T}_D$ denotes the Clarke tangent cone. "Sufficiency" is easily deduced from the fact that $\bar{T}_D$ is the negative polar of the limiting normal cone N_D. (See Theorem 4.10.7.)

9.4 Derivation of the Maximum Principle for State Constrained Problems from the Euler–Lagrange Condition

In this section, we first show that, without loss of generality, it suffices to prove a special case of the Maximum Principle (for state constrained problems), in which additional hypotheses are in force. We then show that, in this special case, the Maximum Principle is essentially a corollary of the Euler–Lagrange Condition of Chapter 10.

We omit the proof of the constancy of the Hamiltonian condition (vi) for autonomous problems, because it follows so closely the proof already provided in the state constraint-free case (see the proof of Theorem 8.7.1). Briefly, condition (vi) is proved by applying the Maximum Principle (without this extra condition) to a reformulation of the problem in which "time" has the role of a state component. The only significant difference is that we apply the Maximum Principle for problems with state constraints to the transformed problem.

Lemma 9.4.1 (Hypothesis Reduction) *Suppose that the assertions of Theorem 9.3.1 are valid, when Hypotheses (H1) to (H4) are supplemented by these additional hypotheses.*

(R1): $(\bar{x}, \bar{u})$ is a strong local minimizer (not merely a $W^{1,1}$ local minimizer).

(R2): For a.e. $t \in [S,T]$, $U(t)$ is a finite set and

$$f(t, \bar{x}(t), u) \neq f(t, \bar{x}(t), v) \quad for\ u, v \in U(t),\ u \neq v.$$

(R3): There exist $k_f \in L^1$ and $c_f \in L^1$ such that

$$|f(t, x, u) - f(t, x', u)| \leq k_f(t)|x - x'|$$

and

$$|f(t, x, u)| \leq c_f(t)$$

for all $x, x' \in \bar{x}(t) + \delta B$, $u \in U(t)$, a.e. $t \in [S,T]$.

Then the assertions of Theorem 9.3.1 remain valid in the absence of these extra hypotheses.

Proof. Assume that the assertions of the theorem are valid under the extra hypotheses. Let us suppose that $(\bar{x}, \bar{u})$ is a strong local minimizer, and that (in addition to (H1) through (H4)) (R1) and (R3) are in force but that Hypothesis (R2) is possibly violated. Set

$$L := \exp\left(\int_S^T k(t, \bar{u}(t))dt\right) + K,$$

in which $k(t, \cdot)$ and K are the Lipschitz constants of Hypotheses (H1) and (H4), respectively. The significance of L is that, for any $p \in W^{1,1}$, $\mu \in C^{\oplus}(s, T)$, $\lambda \geq 0$, and measurable function $\gamma : [S,T] \to R^n$ satisfying $\gamma(t) \in \partial_x^> h(t, \bar{x}(t))$ μ a.e.,

$$|p(S)| + \|\mu\|_{\mathrm{T.V.}} + \lambda = 1$$

and

$$\dot{p}(t) \in \mathrm{co}\, \partial \mathcal{H}(t, \bar{x}t, \bar{u}(t), p(t) + \int_{[S,t)} \gamma(s)\mu(ds)) \quad \text{a.e.},$$

we have

$$|p(t)| \leq L \quad \text{for all } t \in [S,T].$$

Let $\{S_j\}$ be an increasing family of finite subsets of LB such that $LB \subset S_j + j^{-1}B$ for all j. Fix j.

For each $s \in S_j$ we may select a measurable function $w_s(\cdot)$ such that $w_s(t) \in U(t)$ a.e. and

$$H(t, \bar{x}(t), s) \leq s \cdot f(t, \bar{x}(t), w_s(t)) + j^{-1} \quad \text{a.e.}$$

Here

$$H(t, x, s) := \sup\{s \cdot f(t, x, u) : u \in U(t)\}.$$

Mimicking the reordering procedure employed in the proof of Lemma 6.3.1, we can construct from the functions $\bar{u} \cup \{w_s : s \in S_j\}$ a finite collection of measurable selectors $\{v_0, \dots, v_N\}$ of U with the following properties:

$$w_1, w_2 \in U_j \text{ and } w_1 \neq w_2 \text{ implies}$$
$$f(t, \bar{x}(t), w_1) \neq f(t, \bar{x}(t), w_2) \quad \text{a.e.}$$

We can also arrange that, for each $s \in S_j$, there exists a selector $\tilde{w}_s(t)$ for U_j such that

$$H(t, \bar{x}(t), s) \leq s \cdot f(t, \bar{x}(t), \tilde{w}_s(t)) + j^{-1} \quad \text{a.e.} \tag{9.20}$$

Here, U_j is the measurable multifunction

$$U_j(t) := \{v_0(t), \ldots, v_N(t)\}.$$

Of course,

$$\bar{u}(t) \in U_j \quad \text{a.e.}$$

Now consider a new problem in which the multifunction U is replaced by U_j. For the new problem, $(\bar{x}, \bar{u})$ is a strong local minimizer because $(\bar{x}, \bar{u})$ remains "admissible" and the new control constraint set $U_j(t)$ is a subset of $U(t)$ a.e. Now however Hypothesis (R2) is satisfied.

The special case of Theorem 9.3.1, whose validity is postulated under hypotheses satisfied by the data for the new problem, yields multipliers p, μ, and λ with the properties listed in Theorem 9.3.1, except that the "maximization of the Hamiltonian" condition here takes the form

$$q(t) \cdot \dot{\bar{x}}(t) = H_j(t, \bar{x}(t), q(t)) \quad \text{a.e.,} \tag{9.21}$$

where

$$H_j(t, x, s) := \sup\{s \cdot f(t, x, u) : u \in U_j(t)\}$$

and $q = p + \int \gamma \mu(ds)$.

We know $(p, \mu, \lambda) \neq (0, 0, 0)$. It may be deduced from the adjoint inclusion (ii) of Theorem 9.3.1 and Gronwall's Lemma, however, that $(p(S), \mu, \lambda) \neq (0, 0, 0)$. We may arrange then, by scaling the multipliers, that

$$|p(S)| + \|\mu\|_{\text{T.V.}} + \lambda = 1. \tag{9.22}$$

For each t at which Condition (iv) of Theorem 9.3.1 is valid choose $s \in S_j$ such that

$$q(t) \in \{s\} + j^{-1}B.$$

Since $c_f(t)$ (see supplementary hypothesis (R3)) is a Lipschitz constant for $H(t, \bar{x}(t), .)$ and $H_j(t, \bar{x}(t), .)$ and in view of (9.4.1), we have

$$\begin{aligned}
H_j(t, \bar{x}(t), q(t)) &\geq H_j(t, \bar{x}(t), s) - j^{-1}c_f(t) \\
&\geq s \cdot f(t, \bar{x}(t), \tilde{w}_s(t)) - j^{-1}c_f(t) \\
&\geq H(t, \bar{x}(t), s) - j^{-1} - j^{-1}c_f(t) \\
&\geq H(t, \bar{x}(t), q(t)) - j^{-1} - 2j^{-1}c_f(t).
\end{aligned}$$

We conclude from (9.21) then that

$$q(t) \cdot \dot{\bar{x}}(t) - H(t, \bar{x}(t), q(t)) \geq -j^{-1}(1 + 2c_f(t)) \quad \text{a.e.} \tag{9.23}$$

Since the multipliers satisfy the normalization condition (9.22), the ps are uniformly bounded (as j ranges over the positive integers), the $\dot{p}$s are uniformly integrably bounded, the μs are uniformly bounded in total variation, and the λs are bounded. A routine convergence analysis, now incorporating the results of Proposition 9.2.1, provides, in the limit, multipliers p, μ, γ, and λ in accordance with the assertions of the theorem. In particular (9.23) permits us to conclude that

$$q(t) \cdot \dot{\bar{x}}(t) \geq \sup\{q(t) \cdot f(t, \bar{x}(t), u) : u \in U(t)\} \quad \text{a.e.}$$

in which

$$q = p + \int \gamma\mu(ds).$$

This is the maximization of the Hamiltonian condition. We have verified the necessary conditions for a strong minimizer under Hypotheses (H1) to (H4), (R1), and (R3).

Next suppose that $(\bar{x}, \bar{u})$ is a strong local minimizer, Hypotheses (H1) to (H4) and (R1) are satisfied, but Hypothesis (R3) is possibly violated. For each j define

$$U_j(t) := \{u \in U(t) : k(t, u) \leq k(t, \bar{u}(t)) + j, f(t, \bar{x}(t), u) \leq |\dot{\bar{x}}(t)| + j\}$$

and consider the problem obtained by replacing $U(t)$ in problem (P) by $U_j(t)$. For each j the data for this new problem satisfy (iii) and so the special case of the necessary conditions just proved applies. We are assured then of the existence of multipliers p, μ, and λ and a Borel measurable function γ (they all depend on j) with the properties listed in the theorem, except that the "maximization of the Hamiltonian" condition now takes the form

$$q(t) \cdot \bar{x}(t) \geq q(t) \cdot f(t, \bar{x}(t), u) \quad \text{for all } u \in U_j(t), \text{ a.e. } t \in [S, T]$$

$(q = p + \int \gamma\mu(dt))$. A further exercise in convergence analysis, in which we make use of the fact that, for any $t \in [S, T]$,

$$u(t) \in U(t) \text{ implies } u(t) \in U_j(t) \text{ for } j \text{ sufficiently large,}$$

justifies all the assertions of Theorem 9.3.1, under Hypotheses (H1) to (H4) and also (R1) (the requirement that $(\bar{x}, \bar{u})$ is a strong local minimizer).

To complete the proof, assume that $(\bar{x}, \bar{u})$ is merely a $W^{1,1}$ local minimizer. In this case, $(\bar{x}, \bar{y} \equiv 0, \bar{u})$ is a strong local minimizer for

$$
\begin{cases}
\text{Minimize } g(x(S), x(T)) \\
\text{over } x \in W^{1,1}, y \in W^{1,1} \text{ and measurable functions } u \text{ satisfying} \\
\dot{x}(t) = f(t, x(t), u(t)), \; \dot{y}(t) = |f(t, x(t), u(t)) - \dot{\bar{x}}(t)|, \\
u(t) \in U(t) \quad \text{a.e.,} \\
h(t, x(t)) \leq 0 \quad \text{for all } t \in [S, T], \\
(x(S), x(T), y(S), y(T)) \in C \times \{0\} \times R^n.
\end{cases}
$$

The most recently proved special case of the necessary conditions may be applied to this problem, with reference to the strong local minimizer $(\bar{x}, \bar{u})$. The adjoint arc component r associated with the y variable is constant, because the right sides of the differential equations governing the state are independent of y. But then $r = 0$ since the right endpoint of y is unconstrained. All reference to the extra variable y then drops out of the conditions, which reduce to the assertions of Theorem 9.3.1 relating to $(\bar{x}, \bar{u})$, now regarded as a $W^{1,1}$ local minimizer for the original problem (P). $\square$

It is now a straightforward task to deduce the Maximum Principle from the Extended Euler–Lagrange Condition of Chapter 10 (Theorem 10.3.1).

According to Lemma 9.4.1, we may assume, without loss of generality, that Hypotheses (R2) and (R3) are in force (in addition to (H1) through (H4)). Define the multifunction $F : [S, T] \times R^n \rightsquigarrow R^n$ to be

$$F(t, x) := \{ f(t, x, u) : u \in U(t) \}.$$

We note that, since $U(t)$ is here assumed to be a finite nonempty set, F takes as values compact nonempty sets. $F(\cdot, x)$ is measurable for each $x \in R^n$ and

$$F(t, x) \subset F(t, x') + k_f(t)|x - x'| \quad \text{for all } x, x' \in \bar{x}(t) + \delta B \quad \text{a.e.}$$

The state trajectory $\bar{x}$ is a $W^{1,1}$ local minimizer for

$$\left\{ \begin{array}{l} \text{Minimize } g(x(S), x(T)) \\ \text{over } x \in W^{1,1}([S, T]; R^n) \text{ satisfying} \\ \dot{x}(t) \in F(t, x(t)) \quad \text{a.e. } t \in [S, T], \\ h(t, x(t)) \le 0 \quad \text{for all } t \in [S, T], \\ (x(S), x(T)) \in C. \end{array} \right.$$

The hypotheses are satisfied under which Theorem 10.3.1 applies. This yields a Borel measurable function $\gamma : [S, T] \to R^n$, $p \in W^{1,1}$, $\mu \in C^{\oplus}([S, T]; R^n)$, and $\lambda \ge 0$ such that, if we write $q := p + \int \gamma \mu(ds)$, all the conditions of Theorem 9.3.1 are satisfied, except that in place of the adjoint inclusion (ii) we have

$$\dot{p}(t) \in \text{co}\{\xi : (\xi, q(t)) \in N_{\text{Gr}F(t, \cdot)}(\bar{x}(t), \dot{\bar{x}}(t))\} \quad \text{a.e. } t \in [S, T]. \quad (9.24)$$

But in view of Hypotheses (R2) and (R3), for each t there exists $\alpha > 0$ (α will depend on t) such that

$$\text{Gr}F(t, .) \cap ((\bar{x}(t), \dot{\bar{x}}(t)) + \alpha B) = \text{Gr}f(t, ., \bar{u}(t)) \cap ((\bar{x}(t), \dot{\bar{x}}(t)) + \alpha B).$$

It follows that

$$N_{\text{Gr}F(t, \cdot)}(\bar{x}(t), \dot{\bar{x}}(t)) = N_{\text{Gr}f(t, ., \bar{u}(t))}(\bar{x}(t), \dot{\bar{x}}(t)) \quad \text{for all } t.$$

But for a function $d : R^n \to R^n$ and a point $y \in R^n$ such that d is Lipschitz continuous on a neighborhood of y we have

$$(q', q) \in N_{\mathrm{Gr}d(\cdot)}(y, d(y)) \quad \text{implies} \quad -q' \in \partial[q \cdot d(x)]|_{x=y}.$$

(See Proposition 5.4.2.) We conclude now from (9.24) that

$$
\begin{aligned}
-\dot{p}(t) &\in -\mathrm{co}\{-\xi : \xi \in \partial_x \mathcal{H}(t, \bar{x}(t), q(t), \bar{u}(t)) \\
&= \mathrm{co}\partial_x \mathcal{H}(t, \bar{x}(t), q(t), \bar{u}(t)) \quad \text{a.e.}
\end{aligned}
$$

This is the required adjoint inclusion. Proof of the theorem is complete. $\square$

9.5 A Smooth Maximum Principle for State Constrained Problems

In this section, we provide a self-contained proof of a special case of the Maximum Principle, Theorem 9.3.1, applicable to problems with "smooth" dynamics. The problem of interest remains:

$$(P) \begin{cases} \text{Minimize } g(x(S), x(T)) \\ \text{over } x \in W^{1,1}([S,T]; R^n) \text{ and measurable functions } u \text{ satisfying} \\ \dot{x}(t) = f(t, x(t), u(t)) \quad \text{a.e.,} \\ u(t) \in U(t) \quad \text{a.e.,} \\ h(t, x(t)) \le 0 \quad \text{for all } t \in [S, T], \\ (x(S), x(T)) \in C, \end{cases}$$

with data an interval $[S, T]$, functions $g : R^n \times R^n \to R$, $f : [S,T] \times R^n \times R^m \to R^n$, and $h : [S,T] \times R^n \to R$, a multifunction $U : [S,T] \rightsquigarrow R^m$, and a closed set $C \subset R^n \times R^n$.

Denote by $\mathcal{H}$ the Unmaximized Hamiltonian:

$$\mathcal{H}(t, x, p, u) := p \cdot f(t, x, u).$$

Theorem 9.5.1 (A Smooth Maximum Principle for State Constrained Problems) *Let $(\bar{x}, \bar{u})$ be a $W^{1,1}$ local minimizer for (P). Assume that, for some $\delta > 0$, the Hypotheses (H1) to (H4) of Theorem 9.3.1 are satisfied, namely:*

(H1): $f(., x, .)$ is $\mathcal{L} \times \mathcal{B}$ measurable for fixed x. There exists a Borel measurable function $k(.,.) : [S,T] \times R^m \to R$ such that $k(t, \bar{u}(t))$ is integrable and

$$|f(t, x, u) - f(t, x', u)| \le k(t, u)|x - x'|$$

for all $x, x' \in \bar{x}(t) + \delta B$, $u \in U(t)$, a.e.;

(H2): $\mathrm{Gr}\, U$ is $\mathcal{L} \times \mathcal{B}$ measurable;

(H3): g is Lipschitz continuous on $(\bar{x}(S), \bar{x}(T)) + \delta B$;

(H4): h is upper semicontinuous and there exists $K > 0$ such that

$$|h(t, x) - h(t, x')| \leq K|x - x'| \quad \text{for all } x, x' \in \bar{x}(t) + \delta B, \ t \in [S, T].$$

Assume furthermore that

(S1): $f(t, ., u)$ is continuously differentiable on $\bar{x}(t) + \delta B$ for all $u \in U(t)$, a.e.

Then there exist $p \in W^{1,1}([S, T]; R^n)$, $\lambda \geq 0$, $\mu \in C^{\oplus}(S, T)$, and a Borel measurable function $\gamma : [S, T] \to R^n$ satisfying

$$\gamma(t) \in \partial_x^{>} h(t, \bar{x}(t)) \quad \mu \ a.e.,$$

such that

(i) $(p, \mu, \lambda) \neq 0$,

(ii) $-\dot{p}(t) = \mathcal{H}_x(t, \bar{x}(t), q(t), \bar{u}(t))$ a.e.,

(iii) $(p(S), -q(T)) \in \lambda \partial g(\bar{x}(S), \bar{x}(T)) + N_C(\bar{x}(S), \bar{x}(T))$,

(iv) $\mathcal{H}(t, \bar{x}(t), q(t), \bar{u}(t)) = \max_{u \in U(t)} \mathcal{H}(t, \bar{x}(t), q(t), u)$ a.e.,

(v) $\text{supp}\{\mu\} \subset I(\bar{x})$,

where

$$q(t) := \begin{cases} p(t) + \int_{[S,t)} \gamma(s)\mu(ds) & \text{if } t \in [S, T) \\ p(T) + \int_{[S,T]} \gamma(s)\mu(ds) & \text{if } t = T, \end{cases}$$

and

$$I(\bar{x}) := \{t \in [S, T] : h(t, \bar{x})(t) = 0\}.$$

The partial subdifferential $\partial_x^{>} h$, referred to in this proposition, was defined earlier. (See (9.13).)

Proof. The proof is built up in stages, each summarized as a proposition, in which we validate the assertions of the Maximum Principle for state constrained problems under progressively weaker hypotheses.

The first case treated is that when the velocity set is compact and convex, the cost function is smooth, and no endpoint constraints are applied, $(\bar{x}, \bar{u})$ is required to be a strong local minimizer (not merely a $W^{1,1}$ local minimizer) and the necessary conditions are expressed in terms of a coarser subdifferential for $h(t, .)$ than the subdifferential $\partial_x^{>} h$ employed in Theorem 9.3.1.

Proposition 9.5.2 *Let $(\bar{x}, \bar{u})$ be a strong local minimizer for (P). Assume that, in addition to Hypotheses (H1) through (H4) and (S1) of Theorem 9.5.1, the following hypotheses are satisfied.*

(S2): There exist $k_f \in L^1$ and $c_f \in L^1$ such that

$$|f(t, x, u) - f(t, x', u)| \leq k_f(t)|x - x'|$$

and

$$|f(t, x, u)| \leq c_f(t)$$

for all $x, x' \in \bar{x}(t) + \delta B$, $u \in U(t)$ a.e.;

(S3): $f(t, x, U(t))$ is a compact convex set for all $x \in \bar{x}(t) + \delta B$ and $t \in [S, T]$;

(S4): g is C^1 on a neighborhood of $(\bar{x}(S), \bar{x}(T))$,

(S5): $C = R^n \times R^n$.

Then there exist $p \in W^{1,1}$, $\lambda \geq 0$, $\mu \in C^{\oplus}(S, T)$, and a Borel measurable function γ such that

$$\gamma(t) \in \bar{\partial}_x h(t, \bar{x}(t)) \quad \mu \text{ a.e.}$$

and Conditions (i) through (v) of Theorem 9.5.1 are satisfied.

The new partial subdifferential employed in these conditions is

$$\bar{\partial}_x h(t, x) := \mathrm{co}\{\lim \eta_i : \eta_i = \nabla_x h(t_i, x_i) \,\forall i \text{ and } (t_i, x_i) \xrightarrow{h} (t, y)\}.$$

(Notice $\bar{\partial}_x h$ is a possibly larger set than $\partial_x^> h$, because the qualifier $h(t_i, x_i) > 0$ is omitted from the definition.)

Proof. By reducing $\delta > 0$, if necessary, we arrange that $(\bar{x}, \bar{u})$ is a minimizer with respect to all (x, u)s satisfying

$$\|x - \bar{x}\|_{L^\infty} \leq \delta,$$

and g is C^1 on $(\bar{x}(S), \bar{x}(T)) + \sqrt{2}\delta B$.

Step 1 (Linearization): We show that, for any $u \in \mathcal{U}$ and $y \in W^{1,1}$ such that

$$\dot{y}(t) = f_x(t, \bar{x}(t), \bar{u}(t))y(t) + (f(t, \bar{x}(t), u(t)) - f(t, \bar{x}(t), \bar{u}(t))) \quad \text{a.e.}$$

we have

$$\max\{ \nabla g(\bar{x}(S), \bar{x}(T)) \cdot (y(S), y(T)),$$
$$\max\{\xi \cdot y(t) : \xi \in \bar{\partial}_x h(t, \bar{x}(t)), t \in I(\bar{x})\} \} \geq 0. \quad (9.25)$$

(The second max term on the left side is interpreted as $-\infty$ if $I(\bar{x}) = \emptyset$.)

Fix $\xi \in R^n$ and $u \in \mathcal{U}$. Take a sequence $\epsilon_i \downarrow 0$ and, for each i, consider the differential equation

$$\begin{cases} \dot{x}(t) &= f(t, x(t), \bar{u}(t)) + \epsilon_i \Delta f(t, x(t), u(t)) \\ x(0) &= \bar{x}(0) + \epsilon_i \xi, \end{cases}$$

in which

$$\Delta f(t, x, u) := f(t, x, u) - f(t, x, \bar{u}(t)).$$

By the Generalized Filippov Existence Theorem (Theorem 2.4.3), for all i sufficiently large there exists a solution $x_i \in W^{1,1}([S,T]; R^n)$ satisfying $\|x_i - \bar{x}\|_{L^\infty} < \delta$. Also, if we define

$$y_i(t) := \epsilon_i^{-1}(x_i(t) - x(t)),$$

the y_is are uniformly bounded and the $\dot{y}_i$s are uniformly integrably bounded. For a subsequence then,

$$y_i(t) \to y(t) \quad \text{uniformly,}$$

for some $y \in W^{1,1}$. Subtracting the differential equation for $\bar{x}$ from that for x_i, dividing across by ϵ_i and integrating gives

$$\epsilon_i^{-1}(x_i - \bar{x})(t) = \xi +$$
$$\int_S^t \epsilon_i^{-1}(f(s, x_i(s), \bar{u}(s)) - f(s, \bar{x}(s), \bar{u}(s)))ds + \int_S^t \Delta f(s, x_i(s), u(s))ds.$$

With the aid of the Dominated Convergence Theorem and making use of Hypothesis (S2), we obtain in the limit

$$y(t) = \xi + \int_S^t f_x(s, \bar{x}(s), \bar{u}(s))y(s)ds + \int_S^t \Delta f(s, \bar{x}(s), u(s))ds.$$

The arc y may be interpreted as the solution to the differential equation

$$\begin{align} \dot{y}(t) &= f_x(t, \bar{x}(t), \bar{u}(t))y(t) + \Delta f(t, \bar{x}(t), u(t)) \qquad (9.26) \\ y(S) &= \xi. \end{align}$$

Under the convexity hypothesis (S3), and in view of the Generalized Filippov Selection Theorem (Theorem 2.3.13), for each i there exists a control function $\tilde{u}_i$ such that

$$\dot{x}_i(t) = f(t, x_i(t), \tilde{u}_i(t)) \quad \text{a.e.}$$

Since $(\bar{x}, \bar{u})$ is a strong local minimizer and $(x_i = \bar{x} + \epsilon y_i, \tilde{u}_i)$ is a feasible process satisfying $\|x_i - \bar{x}\|_\infty \le \delta$, we must have (for sufficiently large i)

$$\epsilon_i^{-1} \max\{g((\bar{x}(S), \bar{x}(T)) + \epsilon_i(y_i(S), y_i(T))) - g(\bar{x}(S), \bar{x}(T))),$$
$$\max_{t \in [S,T]} h(t, \bar{x}(t) + \epsilon_i y_i(t))\} \ge 0. \qquad (9.27)$$

This is because, otherwise, we could reduce the cost of $(\bar{x}, \bar{u})$ without violating the state constraint.

Consider now index values i for which the state constraint is binding or violated:

$$\max_{t \in [S,T]} h(t, \bar{x}(t) + \epsilon_i y_i(t)) \geq 0. \tag{9.28}$$

There are two cases to which we need attend:

Case (i): There are at most a finite number of index values such that (9.28) is true.

From (9.27), for all i sufficiently large,

$$\epsilon_i^{-1}(g((\bar{x}(S), \bar{x}(T)) + \epsilon_i(y_i(S), y_i(T))) - g(\bar{x}(S), \bar{x}(T))) \geq 0.$$

Since g is differentiable at $(\bar{x}(S), \bar{x}(T))$ we may pass to the limit and obtain

$$\nabla g(\bar{x}(S), \bar{x}(T)) \cdot (y(S), y(T)) \geq 0.$$

In this case (9.25) has been confirmed.

Case (ii): There are an infinite number of index values such that (9.28) is satisfied.

By restricting attention to a subsequence we can find $t_i \in [S, T]$ such that

$$h(t_i, \bar{x}(t_i) + \epsilon_i y_i(t_i)) \geq 0$$

for all i. Extract a subsequence such that $t_i \to \bar{t}$ for some $\bar{t} \in [S, T]$. By upper semicontinuity, and since $h(\bar{t}, \bar{x}(\bar{t})) \leq 0$,

$$
\begin{aligned}
0 \geq h(\bar{t}, \bar{x}(\bar{t})) &\geq \limsup_{i \to \infty} h(t_i, \bar{x}(t_i) + \epsilon_i y_i(t_i)) \\
&\geq \liminf_{i \to \infty} h(t_i, \bar{x}(t_i) + \epsilon_i y_i(t_i)) \geq 0.
\end{aligned}
$$

It follows that

$$h(t_i, \bar{x}(t_i) + \epsilon_i y_i(t_i)) \to h(\bar{t}, \bar{x}(\bar{t})) \ (= 0). \tag{9.29}$$

Let $L = \|y\|_{L^\infty}$. By the Mean Value Theorem (Theorem 4.5.3), for each i there exists $\xi_i \in \mathrm{co} \partial_x h(t_i, z_i)$ for some $z_i \in \bar{x}(t_i) + \epsilon_i LB$ such that

$$
\begin{aligned}
\xi_i \cdot y_i(t_i) &= \epsilon_i^{-1}(h(t_i, \bar{x}(t_i) + \epsilon_i y_i(t_i)) - h(t_i, \bar{x}(t_i))) \\
&\geq \epsilon_i^{-1}(h(t_i, \bar{x}(t_i) + \epsilon_i y_i(t_i)) - 0) \\
&\geq 0.
\end{aligned}
$$

Since, however, elements in co $\partial_x h$ can be expressed as convex combinations of limits of neighboring gradients of $h(t_i, .)$, we can find z_i' such that $|z - z_i'| \leq \epsilon_i$ and

$$\nabla_x h(t_i, z_i') \cdot y_i(t_i) \geq -\epsilon_i.$$

But $h(t_i, z_i) \to h(\bar{t}, \bar{x}(\bar{t}))$ by (9.29) and the uniform Lipschitz continuity of $h(t, .)$. Following a further subsequence extraction then, we have $\nabla_x h(t_i, z_i) \to \xi$ (for some $\xi \in R^n$). Taking account of the continuity of y, we deduce from the definition of $\bar{\partial}_x h$ that $\xi \in \bar{\partial}_x h(\bar{t}, \bar{x}(\bar{t}))$ and $\xi \cdot y(\bar{t}) \geq 0$. In this case too then, (9.25) is confirmed.

Step 2 (Dualization): We show that there exist $\lambda \geq 0$, $\mu \in C^{\oplus}(S, T)$, and a Borel measurable function γ such that

$$\lambda + \|\mu\|_{\text{T.V.}} = 1,$$
$$\text{supp}\{\mu\} \subset I(\bar{x}) \text{ and } \gamma(t) \in \bar{\partial}_x h(t, \bar{x}(t)) \quad \mu \text{ a.e.},$$
$$\lambda \nabla g(\bar{x}(S), \bar{x}(T)) \cdot (y(S), y(T))$$
$$+ \int_{[S,T]} y(t) \cdot \gamma(t) \mu(dt) \geq 0, \qquad (9.30)$$

for all $u \in U$ and all $y \in W^{1,1}$ such that

$$\dot{y}(t) = f_x(t, \bar{x}(t), \bar{u}(t)) y(t) + \Delta f(t, \bar{x}(t), u(t)).$$

Define sets

$$E := \{y \in W^{1,1} : \dot{y}(t) = f_x(t, \bar{x}(t), \bar{u}(t)) y(t) + \Delta f(t, \bar{x}(t), u(t))$$
$$\text{for some } u \in \mathcal{U}\}$$
$$F := \{(\lambda, \eta) \in R \times C^* : \lambda \geq 0, \eta(dt) = \gamma(t) \mu(dt)$$
$$\text{for some } \mu \in C^{\oplus}(S, T) \text{ and some Borel measurable}$$
$$\gamma \text{ such that } \gamma(t) \in \bar{\partial}_x h(t, \bar{x}(t)) \quad \mu \text{ a.e.}$$
$$\text{supp}\{\mu\} \subset I(\bar{x}) \text{ and } \lambda + \|\mu\| = 1\}$$

and also the function $G : C \times (R \times C^*) \to R$,

$$G(y, (\lambda, \eta)) := \lambda \nabla g(\bar{x}(S), \bar{x}(T)) \cdot (y(S), y(T)) + \int_{[S,T]} y(t) \cdot \eta(dt).$$

Fix $y \in E$ and write

$$a := \nabla g(\bar{x}(S), \bar{x}(T)) \cdot (y(S), y(T))$$
$$b := \max\{\xi \cdot y(t) : \xi \in \bar{\partial}_x h(t, \bar{x}(t)), t \in I(\bar{x})\}.$$

According to the linearization analysis, $a \geq 0$ or $b \geq 0$. If $a \geq 0$ set $\bar{\lambda} = 1$, $\bar{\eta} = 0$. If $b \geq 0$ set $\bar{\lambda} = 0$. Let $(\bar{\xi}, \bar{t})$ achieve the maximum in b.

(Such a maximizer exists since the set $\{(\xi, t) : \xi \in \bar{\partial}_x h(t, \bar{x}(t)), t \in I(\bar{x})\}$ is compact.) Set $\bar{\gamma} = \bar{\xi}$, $\bar{\mu} = \delta_{\{\bar{t}\}}$, and $\eta(dt) = \bar{\gamma}(t)\bar{\mu}(dt)$. In either case $(\bar{\lambda}, \bar{\eta}) \in F$ and $G(y, (\bar{\lambda}, \bar{\eta})) \geq 0$. It follows that

$$\inf_{y \in E} \sup_{(\lambda, \eta) \in F} G(y, (\lambda, \eta)) \geq 0.$$

However $G(., (\lambda, \eta))$ is a strongly continuous linear functional on $W^{1,1}$, and $G(y, (.,.))$ is a weak* continuous linear functional on $R \times C^*$. E is convex, by Hypothesis (S3). It can also be shown, with the help of Proposition 9.2.1, that F is convex and weak* compact. By the One-Sided Mini-Max Theorem (Theorem 3.4.5) then, there exist $\lambda \geq 0$, a Borel measurable γ, and $\mu \in C^{\oplus}(S, T)$ with support in $I(\bar{x})$, such that $\gamma(t) \in \bar{\partial}_x h(t, \bar{x}(t))$ μ a.e., $\lambda + \|\mu\|_{T.V} = 1$, and

$$\inf_{y \in E} G(y, (\lambda, \gamma\mu(dt))) \geq 0.$$

Otherwise expressed

$$\lambda \nabla g(\bar{x}(S), \bar{x}(T)) \cdot (y(S), y(T)) + \int_{[S,T]} y(t) \cdot \gamma(t)\mu(dt) \geq 0$$

for all $y \in E$. This is what we set out to prove.

Step 3 (Conclusion): Let $p \in W^{1,1}$ be the solution of the differential equation

$$-\dot{p}(t) = (p(t) + \int_{[S,t)} \gamma(s)\mu(ds))f_x(t, \bar{x}(t), \bar{u}(t)) \quad \text{a.e.}$$

$$-p(T) = \lambda \nabla_{x_1} g(\bar{x}(S), \bar{x}(T)) + \int_{[S,T]} \gamma(s)\mu(ds).$$

Here γ, μ, and λ are the multipliers supplied by the above dualization analysis. (We have written $\nabla g := (\nabla_{x_0} g, \nabla_{x_1} g)$.)

Define

$$q(t) := \begin{cases} p(t) + \int_{[S,t)} \gamma(s)\mu(ds) & \text{if } t \in [S, T) \\ p(T) + \int_{[S,T]} \gamma(s)\mu(ds) & \text{if } t = T. \end{cases}$$

Take any y, u, and ξ satisfying (9.26). Add the term

$$\int_S^T q(t) \cdot (\dot{y}(t) - f_x(t, \bar{x}(t), \bar{u}(t))y(t) - \Delta f(t, \bar{x}(t), u(t)))dt$$

(which is zero) to the left side of inequality (9.30). This gives

$$\lambda \nabla g(\bar{x}(S), \bar{x}(T)) \cdot (y(S), y(T)) + \int_{[S,T]} y(t) \cdot \gamma(t)\mu(dt)$$

$$+ \int_S^T q(t) \cdot (\dot{y}(t) - f_x(t, \bar{x}(t), \bar{u}(t))y(t) - \Delta f(t, \bar{x}(t), \bar{u}(t)))dt \geq 0.$$

An application of the integration by parts formula yields

$$\int_S^T q(t) \cdot \dot{y}(t)dt =$$

$$-\int_S^T \dot{p}(t) \cdot y(t)dt - \int_{[S,T]} y(t) \cdot \gamma(t)\mu(dt) + q(T) \cdot y(T) - p(S) \cdot \xi.$$

This identity and the preceding inequality combine to give

$$-\int_S^T (\dot{p}(t) + q(t)f_x(t, \bar{x}(t), \bar{u}(t)))y(t)dt$$

$$-\int_S^T q(t) \cdot \Delta f(t, \bar{x}(t), u(t))dt + (\lambda\nabla_{x_0}g(\bar{x}(S), \bar{x}(T)) - p(S)) \cdot \xi \geq 0.$$

Taking note of the relationships satisfied by p, we arrive at

$$-\int_S^T q(t) \cdot (f(t, \bar{x}(t), u(t)) - f(t, \bar{x}(t), \bar{u}(t)))dt$$

$$+(\lambda\nabla_{x_0}g(\bar{x}(S), \bar{x}(T)) - p(S)) \cdot \xi \geq 0.$$

This inequality must hold for all $u \in \mathcal{U}$ and all $\xi \in R^n$. It can be shown that, in consequence,

$$\mathcal{H}(t, \bar{x}(t), q(t), \bar{u}(t)) = \max_{u \in U(t)} \mathcal{H}(t, \bar{x}(t), q(t), u) \quad \text{a.e.}$$

and

$$p(S) = \lambda\nabla_{x_0}g(\bar{x}(S), \bar{x}(T)).$$

All assertions of the proposition have been confirmed. □

In the next stage, endpoint constraints are introduced; i.e., we justify dropping (S5) from the hypotheses invoked in Proposition 9.5.2.

Proposition 9.5.3 *The assertions of Proposition 9.5.2 are valid when, in addition to (H1) through (H4) and (S1), we impose merely Hypotheses(S2) through (S4).*

Proof. Fix $\epsilon \in (0, 1]$. Adjust $\delta > 0$ (the constant in the hypotheses) if necessary so that $(\bar{x}, \bar{u})$ is minimizing with respect to all feasible processes (x, u) satisfying $\|x - \bar{x}\|_{L^\infty} \leq \delta$. Now embed (P) (augmented by the constraint $\|x - \bar{x}\|_{L^\infty} \leq \delta$) in a family of problems $\{P(a) : a \in R^n \times R^n\}$.

$$P(a) \begin{cases} \text{Minimize } g(x(S), x(T)) \\ \text{over } x \in W^{1,1} \text{ and measurable functions } u \text{ satisfying} \\ \dot{x}(t) = (1 - \epsilon)f(t, x(t), \bar{u}(t)) + \epsilon f(t, x(t), u(t)) \quad \text{a.e.,} \\ u(t) \in U(t) \quad \text{a.e.,} \\ h(t, x(t)) \leq 0 \quad \text{for all } t \in [S, T], \\ (x(S), x(T)) \in C + \{a\}, \\ \|x - \bar{x}\|_{L^\infty} \leq \delta. \end{cases}$$

Since $f(t, x, U(t))$ is convex,

$$(1 - \epsilon)f(t, x(t), \bar{u}(t)) + \epsilon f(t, x(t), U(t)) \subset f(t, x(t), U(t)) \quad \text{a.e.}$$

We deduce that $(\bar{x}, \bar{u})$ is a minimizer for $P(0)$. We temporarily impose the hypothesis:

(U): if (x, u) is a minimizer for $P(0)$ then $x = \bar{x}$.

Denote by $V(a)$ the infimum cost of $P(a)$. (Set $V(a) = +\infty$ if there exist no (x, u)s satisfying the constraints of $P(a)$.)

Relevant properties of V, straightforward consequences of the closure properties of the set of F-trajectories of convex differential inclusions (see Section 2.5), are as follows.

(a): If $V(a) < +\infty$ then $P(a)$ has a minimizer.

(b): V is a lower semicontinuous function on $R^n \times R^n$.

(c): If $a_i \to 0$ and $V(a_i) \to V(0)$ and if (x_i, u_i) is a minimizer for $P(a_i)$ for each i, then $x_i \to \bar{x}$ uniformly and $\dot{x}_i \to \dot{\bar{x}}$ weakly in L^1 as $i \to \infty$.

Since V is lower semicontinuous and $V(0) < +\infty$, there exists a sequence $a_i \to 0$ such that $V(a_i) \to V(0)$ as $i \to \infty$ and V has a proximal subgradient ξ_i at a_i for each i. This means that, for each i, there exists $\alpha_i > 0$ and $M_i > 0$ such that

$$V(a) - V(a_i) \geq \xi_i \cdot (a - a_i) - M_i |a - a_i|^2 \tag{9.31}$$

for all $a \in \{a_i\} + \alpha_i B$.

In view of the above listed properties of V, $P(a_i)$ has a minimizer (x_i, u_i) for each i and $x_i \to \bar{x}$ uniformly. By eliminating initial terms in the sequence we may arrange then that $\|x_i - \bar{x}\|_{L^\infty} < \delta/2$ for all i.

Take any (x, u) that satisfies $\|x - x_i\|_{L^\infty} < \delta/2$ and

$$\dot{x}(t) = (1 - \epsilon)f(t, x(t), \bar{u}(t)) + \epsilon f(t, x(t), u(t)) \quad \text{a.e.}$$

Choose an arbitrary point $c \in C$. Notice that

$$(x(S), x(T)) \in C + ((x(S), x(T)) - c).$$

This means that (x, u) is feasible process for $P((x(S), x(T)) - c)$. The cost of (x, u) cannot be smaller than the infimum cost $V((x(S), x(T)) - c)$. Otherwise expressed,

$$g(x(S), x(T)) \geq V((x(S), x(T)) - c). \tag{9.32}$$

For each i define

$$c_i := (x_i(S), x_i(T)) - a_i.$$

Since (x_i, u_i) solves $P(a_i)$ $(= P((x_i(S), x_i(T)) - c_i))$ we have

$$g(x_i(S), x_i(T)) = V((x_i(S), x_i(T)) - c_i). \qquad (9.33)$$

Now define the function

$$
\begin{aligned}
J_i((x, u), c) \quad := \quad & g(x(S), x(T)) - \xi_i \cdot ((x(S), x(T)) - c) \\
& + M_i(|(x(S), x(T)) - (x_i(S), x_i(T))|^2 + |c - c_i|^2).
\end{aligned}
$$

Identifying a with $(x(S), x(T)) - c$ and a_i with $(x_i(S), x_i(T)) - c_i$, we deduce from (9.31), together with (9.32) and (9.33), that

$$J_i((x, u), c) \geq J_i((x_i, u_i), c_i) \qquad (9.34)$$

for all $c \in C$ and all (x, u)s satisfying

$$
\begin{cases}
\dot{x}(t) = (1 - \epsilon)f(t, x(t), \bar{u}(t)) + \epsilon f(t, x(t), u(t)) \quad \text{a.e.,} \\
u(t) \in U(t) \quad \text{a.e.,} \\
h(t, x(t)) \leq 0 \quad \text{for all } t \in [S, T], \\
\|x - \bar{x}\|_{L^\infty} \leq \delta/2.
\end{cases}
$$

Set $(x, u) = (x_i, u_i)$ in (9.34). The inequality implies

$$-\xi_i \cdot (c - c_i) \leq M_i|c - c_i|^2 \quad \text{for all } c \in C.$$

We conclude that

$$-\xi_i \in N_C^P((x_i(S), x_i(T)) - a_i).$$

Next set $c = c_i$. We see that (x_i, u_i) is a strong local minimizer for

$$
\begin{cases}
\text{Minimize } g(x(S), x(T)) + M|(x(S), x(T)) - (x_i(S), x_i(T))|^2 \\
\qquad\qquad - \xi_i \cdot (x(S), x(T)) \\
\text{over } x \in W^{1,1} \text{ and measurable functions } u \text{ satisfying} \\
\dot{x}(t) = (1 - \epsilon)f(t, x(t), \bar{u}(t)) + \epsilon f(t, x(t), u(t)) \quad \text{a.e.,} \\
u(t) \in U(t) \quad \text{a.e.,} \\
h(t, x(t)) \leq 0 \quad \text{for all } t \in [S, T].
\end{cases}
$$

This is an "endpoint constraint-free" optimal control problem to which the special case of the Maximum Principle, summarized as Proposition 9.5.2 is applicable. We deduce existence of $p_i \in W^{1,1}$, $\mu_i \in C^\oplus(S, T)$, a Borel measurable function γ_i satisfying

$$\gamma_i(t) \in \bar{\partial}_x h(t, x_i(t)) \quad \mu_i \text{ a.e.,}$$

such that

$$\lambda_i + \|\mu_i\|_{T.V} = 1,$$

$$-\dot{p}_i(t) = q_i(t)((1-\epsilon)f_x(t, x_i(t), \bar{u}(t))$$
$$+\epsilon f_x(t, x_i(t), u_i(t))), \tag{9.35}$$
$$q_i(t) \cdot \dot{x}_i(t) \geq q_i(t) \cdot ((1-\epsilon)f(t, x_i(t), \bar{u}(t))$$
$$+\epsilon f(t, x_i(t), u)) \text{ for all } u \in U(t), \tag{9.36}$$
$$(p_i(S), -q_i(T)) \in \lambda_i \nabla g(x_i(S), x_i(T))$$
$$+N_C((x_i(S), x_i(T)) - a_i), \tag{9.37}$$
$$\text{supp}\{\mu\} \subset I(x_i).$$

In these relationships

$$q_i(t) := \begin{cases} p_i(t) + \int_{[S,t)} \gamma_i(s)\mu_i(ds) & t \in [S,T) \\ p_i(T) + \int_{[S,T]} \gamma_i(s)\mu_i(ds) & t = T. \end{cases}$$

Now scale p, λ, and μ so that

$$|p_i(S)| + \lambda_i + \|\mu_i\|_{T.V} = 1.$$

It follows from the above listed properties of V that $x_i \to \bar{x}$ uniformly. Since $\{p_i(S)\}$ is a bounded sequence, we deduce from (9.35), that the p_is are uniformly bounded and $\dot{p}_i$s are uniformly integrably bounded. Along a subsequence then, $p_i \to p$ uniformly for some $p \in W^{1,1}$. Since $\{\lambda_i\}$ and $\{\|\mu_i\|_{T.V.}\}$ are bounded sequences we may arrange by yet another subsequence extraction that $\mu_i \to \mu$ (weakly*), $\lambda_i \to \lambda$, and $\gamma_i(s)\mu_i(ds) \to \gamma(s)\mu(ds)$ (weakly*), for some $\mu \in C^\oplus(S,T)$, $\lambda \geq 0$, and some Borel measurable function $\gamma : [S,T] \to R^n$ satisfying

$$\text{supp}\{\mu\} \subset I(\bar{x}) \quad \text{and} \quad \gamma(t) \in \bar{\partial}_x h(t, \bar{x}(t)) \quad \mu \text{ a.e.}$$

To arrive at these properties we have used Proposition 9.2.1, identifying A_i with the set $B(x_i)$ and A with the set $B(\bar{x})$, where

$$B(x) := \text{Gr}\{t \to \bar{\partial}_x h(t, x(t))\} \cap (\{t : h(t, x(t)) \geq 0\} \times R^n).$$

Write $q = p + \int \gamma\mu(ds)$. We deduce from (9.35) with the help of the Compactness of Trajectories Theorem that p satisfies

$$-\dot{p}(t) \in q(t)f_x(t, \bar{x}(t), \bar{u}(t)) + 2\epsilon q(t)c_f(t)B. \tag{9.38}$$

From (9.36) we get that, for an arbitrary $u \in \mathcal{U}$,

$$\int_S^T q_i(t) \cdot \dot{x}_i(t)dt \geq \int_S^T q_i(t) \cdot ((1-\epsilon)f(t, x_i(t), \bar{u}(t)) + \epsilon f(t, x_i(t), u(t)))dt.$$

Examine now limit-taking in the relationship. The right side converges to the same expression with $\bar{x}$ replacing x_i and q replacing q_i, by the Dominated Convergence Theorem. The left side can be expanded as

$$\int_S^T p_i(t) \cdot \dot{x}_i(t)dt + \int_S^T \int_{[S,t)} \gamma_i(s)\mu_i(s) \cdot \dot{x}_i(t)dt.$$

Concerning the first, we note that $\int p_i \cdot \dot{x}_i dt \to \int p \cdot \dot{x} dt$ since $p_i \to p$ uniformly and $\dot{x}_i \to \dot{x}$ weakly. The second term requires a little attention. Parts integration gives

$$\int_S^T \int_{[S,t]} (\gamma_i(s)\mu_i(ds)) \cdot \dot{x}_i(t) dt =$$
$$-\int_{[S,T]} x_i(t) \cdot \gamma_i \mu_i(dt) + x_i(T) \cdot \int_{[S,T]} \gamma_i \mu_i(ds).$$

Since $\gamma_i \mu_i(dt) \to \gamma\mu(dt)$ weakly* and $x_i \to \bar{x}$ (uniformly), the expression on the right converges to

$$-\int_{[S,T]} x(t) \cdot \gamma(t)\mu(dt) + x(T) \cdot \int_{[S,T]} \gamma(s)\mu(ds)$$
$$= \int_S^T \int_{[S,t]} \gamma(s)\mu(ds) \cdot \dot{\bar{x}}(t) dt.$$

(Another parts integration has been carried out.) It follows that

$$\int_S^T q(t) \cdot \dot{\bar{x}}(t) dt \geq \int_S^T q(t) \cdot ((1 - \epsilon)f(t, \bar{x}(t), \bar{u}(t)) + \epsilon f(t, \bar{x}(t), u(t))) dt.$$

Canceling terms and dividing by ϵ ($\bar{x}$ satisfies $\dot{\bar{x}}(t) = f(t, \bar{x}(t), \bar{u}(t))$ remember), we obtain

$$\int_S^T q(t) \cdot (f(t, \bar{x}(t), u(t)) - f(t, \bar{x}(t), \bar{u}(t))) dt \leq 0.$$

This inequality is valid for any $u \in \mathcal{U}$. From (9.37),

$$(p(S), -q(T)) \in \lambda \nabla g(\bar{x}(S), \bar{x}(T)) + N_C(\bar{x}(S), \bar{x}(T)),$$

since $z \to N_C(z)$ has a closed graph.

Now recall that the function γ and the multipliers (p, λ, μ) have been constructed for a particular $\epsilon > 0$. Take a sequence $\epsilon_j \downarrow 0$. For each j there are sequences $\{p_j\}$, $\{\lambda_j\}$, $\{\mu_j\}$, and $\{\gamma_j\}$ of elements satisfying the above relationships (when ϵ_j replaces ϵ in (9.38)). A by now familiar convergence analysis yields limits p, λ, μ, and γ that fulfill the assertions of the proposition, except that the "maximization of the Hamiltonian" condition (9.5) is here expressed in "integral form." Replacing it by a pointwise version of the condition is justified along the lines of the earlier, state constraint-free, analysis.

The final task is to lift the temporary hypothesis (U). Suppose (U) is possibly not satisfied. Replace (P) by $(\tilde{P})$.

$$(\tilde{P}) \begin{cases} \text{Minimize } g(x(S), x(T)) + z(T) \\ \text{over } x, z \in W^{1,1} \text{ and measurable functions } u \text{ satisfying} \\ \dot{x}(t) = f(t, x(t), u(t)), \ \dot{z}(t) = |x(t) - \bar{x}(t)|^2 \quad \text{a.e.,} \\ u(t) \in U(t) \quad \text{a.e.,} \\ h(t, x(t)) \le 0 \quad \text{for all } t \in [S, T], \\ (x(S), x(T)) \in C, \ z(S) = 0. \end{cases}$$

Notice that $(\tilde{P})$ "smoothly" penalizes deviations of the state trajectory from $\bar{x}$. This new problem satisfies (U) (in addition to the other hypotheses invoked in Step 2), with reference to the strong local minimizer $(\bar{x}, \bar{z} \equiv 0, \bar{u})$. The necessary conditions for $(\tilde{P})$ imply those for (P). $\square$

Necessary conditions of optimality are now derived for a mini-max optimal control problem.

$$(R) \begin{cases} \text{Minimize } \tilde{g}(x(S), x(T), \max_{t \in [S,T]} h(t, x(t))) \\ \text{over } x \in W^{1,1} \text{ and measurable functions } u \text{ satisfying} \\ \dot{x}(t) = f(t, x(t), u(t)) \quad \text{a.e.,} \\ u(t) \in U(t) \quad \text{a.e.,} \\ (x(S), x(T)) \in C_0 \times R^n, \end{cases}$$

in which (besides the previously specified data), $C_0 \subset R^n$ is a given closed set and $\tilde{g}: R^n \times R^n \times R \to R$ is a given function.

Proposition 9.5.4 *Let $(\bar{x}, \bar{u})$ be a strong local minimizer for (R). Assume that (in addition to Hypotheses (H1), (H2), (H4), and (S1) of Theorem 9.5.1) the data for problem (R) satisfy Hypotheses (S2), (S3)', and (G) below.*

(S2): There exist $k_f \in L^1$ and $c_f \in L^1$ such that

$$|f(t, x, u) - f(t, x', u)| \le k_f(t)|x - x'|$$

and

$$|f(t, x, u)| \le c_f(t)$$

for all $x, x' \in \bar{x}(t) + \delta B$, $u \in U(t)$ a.e. $t \in [S, T]$,

(S3)': $f(t, x, U(t))$ is compact for all $x \in \bar{x}(t) + \delta B$ a.e.,

(G): $\tilde{g}$ is Lipschitz continuous on a neighborhood of

$$(\bar{x}(S), \bar{x}(T), \max_{t \in [S,T]} h(t, \bar{x}(t)))$$

and $\tilde{g}$ is monotone in the z variable, in the sense that

$$z' \geq z \quad \text{implies} \quad \tilde{g}(t, x, z') \geq \tilde{g}(t, x, z),$$

for each $(t, x) \in [S, T] \times R^n$.

Then there exist $p \in W^{1,1}$, $\mu \in C^{\oplus}(S, T)$, *and a Borel measurable function* γ *such that*

$$\text{supp}\{\mu\} \subset \{t : h(t, \bar{x}(t)) = \max_{s \in [S,T]} h(s, \bar{x}(s))\},$$

$$\gamma(t) \in \bar{\partial}_x h(t, \bar{x}(t)) \quad \mu \text{ a.e.,}$$
$$-\dot{p}(t) = q(t) f_x(t, \bar{x}(t), \bar{u}(t)) \quad \text{a.e.,}$$

$$(p(S), -q(T), \int_{[S,T]} \mu(ds)) \in$$
$$\partial \tilde{g}(\bar{x}(S), \bar{x}(T), \max_{t \in [S,T]} h(t, \bar{x}(t))) + N_{C_0}(\bar{x}(S)) \times \{(0, 0)\},$$

$$q(t) \cdot f(t, \bar{x}(t), \bar{u}(t)) = \max_{u \in U(t)} q(t) \cdot f(t, \bar{x}(t), u) \quad \text{a.e.}$$

Here

$$q(t) = \begin{cases} p(t) + \int_{[S,t)} \gamma(s) \mu(ds) & \text{if } t \in [S, T) \\ p(T) + \int_{[S,T]} \gamma(s) \mu(ds) & \text{if } t = T. \end{cases}$$

Proof. By adjusting $\delta > 0$ we can arrange that $(\bar{x}, \bar{u})$ is minimizing with respect to processes (x, u) satisfying the constraints of (R) and also $\|x - \bar{x}\|_{L^\infty} \leq \delta$. Define

$$Q := \{x \in W^{1,1} : x(S) \in C, \dot{x}(t) \in f(t, x(t), U(t))\}.$$

By the Generalized Filippov Selection Theorem (Theorem 2.3.13), $\bar{x}$ is a minimizer for the problem

$$\begin{cases} \text{Minimize } \tilde{g}(x(S), x(T), \max_{t \in [S,T]} h(t, x(t))) \\ \text{over arcs } x \in Q \text{ satisfying } \|x - \bar{x}\|_{L^\infty} < \delta. \end{cases}$$

But, in view of the Relaxation Theorem, any arc x in the set

$$Q_r := \{x \in W^{1,1} : x(S) \in C, \dot{x}(t) \in \text{co} f(t, x(t), U(t))\}$$

that satisfies $\|x - \bar{x}\|_{L^\infty} < \delta$ can be approximated by an arc y in Q satisfying $\|y - \bar{x}\|_{L^\infty} < \delta$, arbitrarily closely with respect to the supremum norm. Since the mapping

$$x \to \tilde{g}(x(S), x(T), \max_{t \in [S,T]} h(t, x(t)))$$

is continuous on a neighborhood of $\bar{x}$ (with respect to the supremum norm topology), $\bar{x}$ is a minimizer also for the optimization problem

$$\begin{cases} \text{Minimize } \tilde{g}(x(S), x(T), \max_{t \in [S,T]} h(t, x(t))) \\ \text{over } x \in Q_r \text{ and } \|x - \bar{x}\|_{L^\infty} < \delta. \end{cases}$$

Write $\bar{z} := \max_{t \in [S,T]} h(t, \bar{x}(t))$. By the Generalized Filippov Selection Theorem (Theorem 2.3.13) and Carathéodory's Theorem, and in view of the monotonicity property of $\tilde{g}$,

$$\{\bar{x}, \bar{y} \equiv \tilde{g}(\bar{x}(S), \bar{x}(T), \bar{z}), \bar{z}, (\bar{u}_0, \dots, \bar{u}_n) \equiv (\bar{u}, \dots, \bar{u}),$$
$$(\lambda_0, \lambda_1, \dots, \lambda_n) \equiv (1, 0, \dots, 0)\}$$

is a strong local minimizer for the optimization problem

$$\begin{cases}
\text{Minimize } y(T) \\
\text{over } x \in W^{1,1}, \ y \in W^{1,1}, \ z \in W^{1,1}, \text{ and measurable} \\
\text{functions } u_0, \dots, u_n, \ \lambda_0, \dots, \lambda_n \text{ satisfying} \\
\dot{x}(t) = \sum_i \lambda_i(t) f(t, x(t), u_i(t)), \ \dot{y}(t) = 0, \ \dot{z}(t) = 0 \quad \text{a.e.,} \\
(\lambda_0(t), \dots, \lambda_n(t)) \in \Lambda, \ u_i(t) \in U(t), \ i = 0, \dots, n \quad \text{a.e.,} \\
h(t, x(t)) - z(t) \leq 0 \quad \text{for all } t \in [S, T], \\
(x(S), x(T), z(S), y(S)) \in \text{epi}\{\tilde{g} + \Psi_{C_0 \times R^n \times R}\}.
\end{cases}$$

Here

$$\Lambda := \{\lambda'_0, \dots, \lambda'_n : \lambda'_i \geq 0 \text{ for } i = 0, \dots, n \text{ and } \sum_i \lambda'_i = 1\}$$

and Ψ_A is the indicator function of the set A. $(\lambda_0, \dots, \lambda_n)$ and $(u_0, \dots, u_n)$ are regarded as control variables. This a problem to which the special case of the Maximum Principle, summarized as Proposition 9.5.3, is applicable. It follows that there exist $p \in W^{1,1}$, $d \in R$, $r \in R$, $\lambda \geq 0$, $\mu \in C^{\oplus}(S, T)$, and a measurable function γ such that

$$(p, d, r, \mu, \lambda) \neq 0,$$
$$\gamma(t) \in \bar{\partial}_x h(t, \bar{x}(t)) \quad \mu \text{ a.e. and } \text{supp}\{\mu\} \subset \{t : h(t, \bar{x}(t)) = \bar{z}\},$$
$$-\dot{p}(t) = q(t) f_x(t, \bar{x}(t), \bar{u}(t)) \quad \text{a.e.,} \tag{9.39}$$
$$(p(S), -q(T), r, d) \in \partial \Psi_{\text{epi}\{\tilde{g} + \Psi_{C_0 \times R^n \times R}\}}(\bar{x}(S), \bar{x}(T), \bar{z}, \bar{y}), \tag{9.40}$$
$$-d = \lambda, \quad -[r - \int_{[S,T]} \mu(dt)] = 0, \tag{9.41}$$
$$q(t) \cdot f(t, \bar{x}(t), \bar{u}(t)) = \max_{u \in U(t)} q(t) \cdot f(t, \bar{x}(t), u) \quad \text{a.e.}$$

However (9.40) and (9.41) imply

$$(p(S), -q(T), \int_{[S,T]} \mu(ds)) \in \lambda \partial \tilde{g}(\bar{x}(S), \bar{x}(T), \bar{z}) + N_{C_0}(\bar{x}(S)) \times \{(0, 0)\}. \tag{9.42}$$

From the above relationships and the nontriviality property of the multipliers, we deduce that $(p, \mu, \lambda) \neq 0$. But in fact $\lambda > 0$. This is because, if $\lambda = 0$, then by (9.42) $\mu = 0$ and by (9.39) and (9.40) $p \equiv 0$, an impossibility. By scaling the multipliers we can arrange that $\lambda = 1$. Reviewing our

findings, we see that all the assertions of the proposition have been proved.
□

The next stage is to generalize the version of the Maximum Principle in Proposition 9.5.2 to allow for a general endpoint constraint. We also sharpen the necessary conditions, replacing the subdifferential $\bar{\partial}_x h$ of Proposition 9.5.2 by the more refined subdifferential $\partial_x^> h$ of Theorem 9.5.1.

Proposition 9.5.5 *The assertions of Theorem 9.5.1 are valid when $(\bar{x}, \bar{u})$ is assumed to be a strong local minimizer and when, in addition to (H1) through (H4) and (S1) of Theorem 9.5.1, the Hypotheses (S2) and (S3)' of Proposition 9.5.4, namely,*

(S2): there exist $k_f \in L^1$ and $c_f \in L^1$ such that

$$|f(t, x, u) - f(t, x', u)| \ \leq \ k_f(t)|x - x'|$$

and

$$|f(t, x, u)| \ \leq \ c_f(t)$$

for all $x, x' \in \bar{x}(t) + \delta B, \ u \in U(t)$ a.e.; $t \in [S, T]$;

(S3)': $f(t, x, U(t))$ is compact for all $x \in \bar{x}(t) + \delta B$ a.e.

are imposed.

Proof. Reduce $\delta > 0$ if necessary so that $(\bar{x}, \bar{u})$ is a minimizer with respect to all (x, u)s satisfying $\|x - \bar{x}\|_{L^\infty} \leq \delta$. Consider the set

$$W := \{(x, u, e) : (x, u) \text{ satisfies } \dot{x}(t) = f(t, x(t), u(t)),$$
$$u(t) \in U(t) \text{ a.e.}, \ e \in R^n, \ (x(S), e) \in C, \text{ and } \|x - \bar{x}\|_{L^\infty} \leq \delta\}$$

and define $d_W : W \times W \to R$,

$$d_W((x, u, e), (x', u', e'))$$
$$= \ |x(S) - x'(S)| + |e - e'| + \text{meas}\{t : u(t) \neq u'(t)\}.$$

Choose $\epsilon_i \downarrow 0$ and, for each i, define the function

$$\tilde{g}_i(x, y, x', y', z) := \ \max\{g(x, y) - g(\bar{x}(S), \bar{x}(T)) + \epsilon_i^2, z, |x' - y'|\}.$$

Now consider the optimization problem

$$\text{minimize } \{\tilde{g}_i(x(S), e, x(T), e, \max_{t \in [S, T]} h(t, x(t))) : (x, u, e) \in W\}.$$

We omit the straightforward verification of the following properties of W and d_W.

(i) d_W defines a metric on W, and (W, d_W) is a complete metric space;

(ii) if $(x_i, u_i, e_i) \to (x, u, e)$ in (W, d_W) then $\|x_i - x\|_{L^\infty} \to 0$;

(iii) the function

$$(x, u, e) \to \tilde{g}_i(x(S), e, x(T), e, \max_{t \in [S,T]} h(t, x(t)))$$

is continuous on (W, d_W).

Notice that

$$\tilde{g}_i(\bar{x}(S), \bar{x}(T), \bar{x}(T), \bar{x}(T), \max_{t \in [S,T]} h(t, \bar{x}(t))) = \epsilon_i^2.$$

Since $\tilde{g}_i$ is nonnegative-valued it follows that $(\bar{x}, \bar{u}, \bar{x}(T))$ is an "ϵ_i^2-minimizer" for the above minimization problem.

According to Ekeland's Theorem, there exists a sequence $\{(x_i, u_i, e_i)\}$ in W such that for each i,

$$\tilde{g}_i(x_i(S), e_i, x_i(T), e_i, \max_{t \in [S,T]} h(t, x_i(t)))$$

$$\leq \tilde{g}_i(x(S), e, x(T), e, \max_{t \in [S,T]} h(t, x(t)))$$

$$+\epsilon_i d_W((x, u, e), (x_i, u_i, e_i)) \tag{9.43}$$

for all $(x, u, e) \in W$ and also

$$d_W((x_i, u_i, e_i), (\bar{x}, \bar{u}, \bar{x}(T))) \leq \epsilon_i. \tag{9.44}$$

We can arrange by subsequence extraction that $\sum_i \epsilon_i < \infty$. Then (9.44) implies that $e_i \to \bar{x}(T)$, $x_i \to \bar{x}$, uniformly and

$$\text{meas}\{J_i\} \to 1$$

for the increasing sequence of sets $\{J_i\}$:

$$J_i := \{t : u_j(t) = \bar{u}(t) \text{ for all } j \geq i\}.$$

Define the arc $y_i \equiv e_i$. Then $y_i \to \bar{x}(T)$ uniformly. With the aid of the $\mathcal{L} \times \mathcal{B}$ measurable function

$$m_i(t, u) := \begin{cases} 1 & \text{if } u \neq u_i(t) \\ 0 & \text{otherwise,} \end{cases}$$

which has the property $\text{meas}\{t : u(t) \neq u_i(t)\} = \int m_i(t, u(t))dt$, we can express the minimization property (9.43) as follows: $(x_i, y_i, w_i \equiv 0, u_i)$ is a strong local minimizer for the optimal control problem

$$\begin{cases} \text{Minimize } \tilde{g}_i(x(S), y(S), x(T), y(T), \max_{t \in [S,T]} h(t, x(t))) \\ \quad +\epsilon_i[|x(S) - x_i(S)| + |y(S) - y_i(S)| + w(T)] \\ \text{over } x, y, w \in W^{1,1} \text{ and measurable functions } u \text{ satisfying} \\ \dot{x}(t) = f(t, x(t), u(t)), \ \dot{y}(t) = 0, \ \dot{w}(t) = m_i(t, u(t)) \quad \text{a.e.,} \\ u(t) \in U(t) \quad \text{a.e.,} \\ (x(S), y(S), w(S)) \in C \times \{0\}. \end{cases}$$

This is an example of the optimal control problem to which the special case of the Maximum Principle, summarized as Proposition 9.5.4 applies. We deduce the existence of $p_i \in W^{1,1}$, $d_i \in R^n$, $r_i \in R$, $\mu_i \in C^{\oplus}(S,T)$, and a measurable function γ_i satisfying

$$\gamma_i(t) \in \bar{\partial}_x h(t, x_i(t)) \quad \mu \text{ a.e.,}$$
$$\text{supp}\{\mu_i\} \subset \{t : h(t, x_i(t)) = \max_{s \in [S,T]}\{h(s, x_i(s))\}\},$$
$$-\dot{p}_i(t) = q_i(t) f_x(t, x_i(t), u_i(t)), \tag{9.45}$$
$$(p_i(S), d_i, -q_i(T), -d_i, \int_{[S,T]} \mu_i(dt)),$$
$$\in \partial \tilde{g}_i(x_i(S), y_i(S), x_i(T), y_i(T), \max\{h(t, x_i(t))\}),$$
$$+ \epsilon_i(B \times B) \times \{(0,0,0)\}$$
$$+ N_C(x_i(S), y_i(S)) \times \{(0,0,0)\}, \tag{9.46}$$
$$q_i(t) \cdot f(t, x_i(t), u_i(t)) \geq$$
$$\max_{u \in U(t)} q_i(t) \cdot f(t, x_i(t), u) - \epsilon_i \quad \text{a.e.} \tag{9.47}$$

In the above relationships $q_i := p_i + \int \gamma_i \mu_i(ds)$. (To derive these conditions, we have identified p, d, and r as the adjoint variables associated with the x, y, and w variables, respectively. We have also noted that d is a constant and $r = -1$.)

From Condition (9.46) we deduce that $\{\|\mu_i\|_{T.V}\}$, $\{d_i\}$, and $\{p_i(T)\}$ are bounded sequences. By (9.45) $\{p_i\}$ is uniformly bounded and $\{\dot{p}_i\}$ is uniformly integrably bounded. Invoking Proposition 9.4.1, we deduce that, following a subsequence extraction,

$$p_i \to p \text{ uniformly,} \quad d_i \to d$$

and

$$\mu_i \to \mu, \quad \gamma_i \mu_i(dt) \to \gamma\mu(dt) \text{ weakly}^*,$$

for some $p \in W^{1,1}$, $d \in R^n$, $\mu \in C^{\oplus}(S,T)$, and some Borel measurable function γ, as $i \to \infty$. Furthermore

$$\text{supp}\{\mu\} \subset \{t : h(t, \bar{x}(t)) = \max_{s \in [S,T]} h(s, \bar{x}(s))\}$$

and

$$\gamma(t) \in \bar{\partial}_x h(t, \bar{x}(t)) \quad \mu \text{ a.e.}$$

Write $q = p + \int \gamma\mu(ds)$.

For each i let $A_i \subset [S,T]$ be the measurable subset of points t at which (9.45) and (9.47) are satisfied. Define

$$B := \{t \in [S,T] : \int_{[S,t)} \gamma_i \mu_i(ds) \to \int_{[S,t)} \gamma\mu(ds)\}.$$

Now set

$$J := (\cup_i J_i) \cap (\cap_i A_i) \cap B.$$

Then the subset J has full measure.

Since meas$\{t : u_i(t) \neq \bar{u}(t)\} \to 0$ we deduce from (9.45), with the help of the Compactness of Trajectories Theorem (Theorem 2.5.3), that

$$-\dot{p}(t) = q(t) f_x(t, \bar{x}(t), \bar{u}(t)) \quad \text{a.e.}$$

For any $t \in J$ and $u \in U(t)$ we deduce from (9.47) that

$$q(t) \cdot f(t, \bar{x}(t), \bar{u}(t)) \geq q(t) \cdot f(t, \bar{x}(t), u).$$

Let us now analyze the Transversality Condition (9.46). An important implication is that

$$\tilde{g}_i(x_i(S), y_i(S), x_i(T), y_i(T), \max_{s \in [S,T]} h(s, x_i(s))) > 0 \qquad (9.48)$$

for all i sufficiently large. This is because, if $\tilde{g}_i = 0$ for some sufficiently large i,

$$y_i(T) = x_i(T), \ (x_i(S), x_i(T)) \in C, \ \max_{s \in [S,T]} h(s, x_i(s)) \leq 0, \ \|x_i - \bar{x}\|_{L^\infty} \leq \delta$$

and

$$g(x_i(S), x_i(T)) \leq g(\bar{x}(S), \bar{x}(T)) - \epsilon_i^2,$$

in violation of the optimality of $(\bar{x}, \bar{u})$.

Define

$$z_i = \max_{s \in [S,T]} h(s, x_i(s)).$$

We verify the following estimate for $\partial \tilde{g}_i$:

$$\partial \tilde{g}_i(x_i(S), y_i(S), x_i(T), y_i(T), z_i) \subset$$
$$\{(a, b, e, -e, c) \in R^n \times R^n \times R^n \times R^n \times R : \exists \tilde{\lambda} \geq 0$$
$$\text{such that } \tilde{\lambda} + |e| = 1 \text{ and}$$
$$(a, b, c) \in \tilde{\lambda} \partial \max\{g(x, y) - g(x_i(S), y_i(S))$$
$$+ \epsilon_i^2, z\}|_{(x_i(S), y_i(S), z_i)}\}. \qquad (9.49)$$

There are two cases to consider.

Case (a): $x_i(T) = y_i(T)$. In this case (by (9.48))

$$\tilde{g}_i(x, y, x', y', z) = \max\{g(x, y) - g(x_i(S), y_i(S)) + \epsilon_i^2, z\}$$

near $(x_i(S), y_i(S), x_i(T), y_i(T), z_i)$. Consequently (9.49) is true with $e = 0$.

Case (b): $x_i(T) \neq y_i(T)$. In this case, from the Chain Rule

$$\partial |x - y| \mid_{(x_i(T), y_i(T))} \in \{(e, -e) : |e| = 1\}.$$

(9.49) then follows from the Max Rule (Theorem 5.5.2) for limiting subdifferentials.

Noting also that, again by the Max Rule,

$$\partial \max\{g(x, y) - g(x_i(S), y_i(S)) + \epsilon_i^2, z\} \mid_{(x_i(S), y_i(S), z_i)}$$
$$\subset \{(\alpha \partial g(x_i(S), y_i(S)), (1 - \alpha)) : \alpha \in [0, 1]\}, \qquad (9.50)$$

we deduce from (9.46) and (9.49) that there exist $\tilde{\lambda}_i \geq 0$ and $\alpha_i \in [0, 1]$ such that

$$\begin{aligned}
d_i &= -q_i(T) \\
\tilde{\lambda} + |q_i(T)| &= 1 \qquad\qquad (9.51) \\
(p_i(T), -q_i(T)) &\in \alpha_i \tilde{\lambda}_i \partial g(x_i(S), y_i(S)) \\
&\qquad\qquad + N_C(x_i(S), y_i(S)), \quad (9.52) \\
\|\mu_i\|_{\text{T.V.}} &= (1 - \alpha_i)\tilde{\lambda}_i.
\end{aligned}$$

Observe also that

$$\mu_i = 0 \quad \text{if} \ \ z_i \leq 0,$$

by (9.46), since $z_i \leq 0$ implies

$$\tilde{g}_i(x, y, x', y', z) := \max\{g(x, y) - g(\bar{x}(S), \bar{x}(T)) + \epsilon_i^2, |x' - y'|\}$$

for (x, y, x', y', z) near $(x_i(S), y_i(S), x_i(T), y_i(T), z_i)$.

Now choose $\lambda_i = \alpha_i \tilde{\lambda}_i$. It follows from (9.51) and (9.52) that

$$\lambda_i + \|\mu_i\|_{T.V} + |q_i(T)| = 1. \qquad (9.53)$$

Along a subsequence $\lambda_i \to \lambda$, for some $\lambda \geq 0$.

In the limit as $i \to \infty$ we obtain from (9.52) and (9.53) the conditions

$$(p(S), -q(T)) \in \lambda \partial g(\bar{x}(S), \bar{x}(T)) + N_C(\bar{x}(S), \bar{x}(T))$$

and

$$\lambda + \|\mu\|_{\text{T.V.}} + |q(T)| = 1.$$

Surveying these relationships we see that the proposition is proved, except that $\bar{\partial}_x h$ replaces $\partial_x^> h$. To attend to this remaining detail we must examine two possible outcomes of the sequence construction we have undertaken.

(A): $\max_{t \in [S,T]} h(t, x_i(t)) > 0$ for at most a finite number of is. For this possibility $\mu_i = 0$ for all i sufficiently large, by (9.46) and (9.48). Then the preceding convergence analysis gives $\mu = 0$ and $\gamma \in \partial_x^> h(t, \bar{x}(t))$ μ a.e., trivially.

(B): $\max_{t\in[S,T]} h(t, x_i(t)) > 0$ for an infinite number of is. Now we can arrange, by a further subsequence extraction, that

$$\max_{t\in[S,T]} h(t, x_i(t)) > 0 \quad \text{for all } i. \tag{9.54}$$

We have

$$\text{supp}\{\mu_i\} \subset \{t : h(t, x_i(t)) = \max_{s\in[S,T]} h(s, x_i(s))\}$$

and, since the inequality (9.54) is strict, the condition $\gamma_i \in \bar{\partial}_x h(t, x_i(t))$ can be replaced by the more precise relationship

$$\gamma_i(t) \in \partial_x^> h(t, x_i(t)) \quad \mu_i \text{ a.e.}$$

The fact that we can arrange, in the limit, that

$$\gamma(t) \in \partial_x^> h(t, \bar{x}(t))$$

now follows from Proposition 9.2.1, when we identify the sets A_i and A of this proposition with the sets $B(x_i)$ and $B(\bar{x})$, respectively, where

$$B(x) := \{(t, \xi) : \xi \in \partial_x^> h(t, x(t))\}.$$

$\square$

Proof of Theorem 9.5.1. The assertions of Theorem 9.5.1 have been confirmed under the extra hypotheses that $(\bar{x}, \bar{u})$ is in fact a strong local minimizer and Conditions (S2) and (S3) on the data are satisfied. But it was established in Lemma 9.4.1 that these extra hypotheses may be imposed without loss of generality. The proof is complete. $\square$

9.6 Notes for Chapter 9

Additional material on convergence properties of Borel measures may be found in [21]. Proposition 9.2.1, whose role is to ensure the preservation of necessary conditions involving measure multipliers under limit-taking, is due to Pappas and Vinter [140].

Necessary conditions for state constrained optimal control problems with optimal state trajectories lying completely in the boundary of the state constraint region were derived in [114]. An early version of the Maximum Principle for optimal control problems with state constraints, which made allowance for a finite number of boundary and interior arcs was obtained by Gamkrelidze [75], under strong regularity hypotheses on the optimal control. A general Maximum Principle for state constrained problems involving measure multipliers was proved by Dubovitskii and Milyutin [60], making no a priori assumptions about the structure of optimal controls. There is a substantial and continuing Russian literature on necessary conditions for such problems and also for more complicated problems involving mixed constraints of the type

$$h(t, x(t), u(t)) = 0.$$

See [102] and [2]. In the West, measure multiplier necessary conditions for problems with unilateral state constraints were also derived by Neustadt [110] and Warga (whose early contributions are reworked in [147]).

Nonsmooth Maximum Principles for state constrained problems were proved by Warga [150], Vinter and Pappas [140], Clarke [38], and Ioffe [83]. The Maximum Principle for problems involving a functional inequality state constraint, proved in this chapter, is a refinement of Clarke's necessary conditions. The significant difference is that, here, the more precise subdifferential $\partial_x^> h$ of the inequality constraint function h (introduced by Clarke to derive necessary conditions for boundary arcs) appears in place of the subdifferential $\bar{\partial}_x h$, for problems with general endpoint constraints. It is this feature that permits us to derive necessary conditions for problems with implicit state constraints, as a simple corollary.

Chapter 10

Necessary Conditions for Differential Inclusion Problems with State Constraints

It seems to me that Berlin would not be at all suitable for me while M. Euler is there.

– Joseph Louis Lagrange (giving a reason for declining a post at the Berlin Academy)

10.1 Introduction

In this chapter, we continue our investigation of necessary conditions for optimal control problems with pathwise state constraints. Now, however, the class of optimal control problems considered is one in which the dynamic constraint is formulated as a differential inclusion.

$$(P) \begin{cases} \text{Minimize } g(x(S), x(T)) \\ \text{over absolutely continuous arcs } x : [S, T] \to R^n \text{ satisfying} \\ \dot{x}(t) \in F(t, x(t)) \quad \text{a.e.,} \\ (x(S), x(T)) \in C, \\ h(t, x(t)) \le 0 \quad \text{for all } t \in [S, T]. \end{cases}$$

Here $g : R^n \times R^n \to R$ and $h : [S, T] \times R^n \to R$ are given functions, $F : [S, T] \times R^n \rightsquigarrow R^n$ is a given multifunction, and $C \subset R^n \times R^n$ is a given set.

We derive optimality conditions for (P), which are state constrained analogues of the necessary conditions of Chapter 7. We also provide optimality conditions for the related free end-time problem, which generalize (to a state constrained context) the necessary conditions of Chapter 8. In all cases, the techniques we have used to derive necessary conditions can be refined to allow for the state constraints. A key role in the analysis of this chapter is played by the Smooth Maximum Principle for state constrained problems.

The fact that state constraint-free necessary conditions can be adapted to allow for state constraints, by the introduction of measure Lagrange multipliers, has been known since the late 1950s. More recently, degenerate aspects of these traditional necessary conditions have come to light: they give no useful information about minimizers in certain cases of interest, notably when the initial state is fixed and lies in the boundary of the state constrained region. The chapter concludes with a discussion of the degeneracy phenomenon and techniques for overcoming it.

As in the Maximum Principle of Chapter 9, the state constraint contributes a term

$$p + \int \gamma(t)\mu(dt)$$

to the necessary conditions considered here, in which μ is a regular nonnegative measure (on the Borel subsets of $[S, T]$) and γ is a Borel measurable function satisfying

$$\gamma(t) \in \partial_x^> h(t, \bar{x}(t)) \quad \mu \text{ a.e.} \tag{10.1}$$

Here $\bar{x}$ is the $W^{1,1}$ local minimizer under consideration and $\partial_x^> h(t, x)$ is the subdifferential:

$$\partial_x^> h(t, x) := \text{co}\{\xi : \exists\, (t_i, x_i) \xrightarrow{h} (t, x) \text{ such that} \tag{10.2}$$
$$h(t_i, x_i) > 0 \text{ for all } i \text{ and } \nabla_x h(t_i, x_i) \to \xi\}.$$

Recall that, because of the presence of the qualifier $h(t_i, x_i) > 0$ in the definition of above subgradient, the inclusion (10.1) implies the "complementary slackness" condition,

$$\text{supp}\, \mu \subset \{t : h(t, \bar{x}(t)) = 0\}.$$

In what follows, "feasible" is to be interpreted as "satisfies the constraints of the optimal control problem at hand."

10.2   A Finite Lagrangian Problem

As a first step towards optimality conditions for (P), we derive necessary conditions for a variational problem in which a pathwise constraint is imposed on the state trajectories but not on the velocities.

$$(F) \begin{cases} \text{Minimize } J(x) := l(x(S), x(T)) + \int_S^T L(t, x(t), \dot{x}(t)) dt \\ \text{over arcs } x \text{ satisfying} \\ h(t, x(t)) \leq 0 \quad \text{for all } t \in [S, T]. \end{cases}$$

Here $[S, T]$ is a given interval and $l : R^n \times R^n \to R \cup \{+\infty\}$, $L : [S, T] \times R^n \times R^n \to R$, and $h : [S, T] \times R^n \to R$ are given functions.

Notice that l is allowed to be extended-valued, so (F) implicitly incorporates the "endpoint constraint,"

$$(x(S), x(T)) \in \{(\alpha, \beta) : l(\alpha, \beta) < +\infty\}.$$

Theorem 10.2.1 Let $\bar{x}$ be a $W^{1,1}$ local minimizer for (F) such that $J(\bar{x}) < +\infty$. Assume that the following hypotheses are satisfied.

(H1) l is lower semicontinuous.

(H2) $L(.,x,.)$ is $\mathcal{L} \times \mathcal{B}$ measurable for each x and $L(t,.,.)$ is lower semicontinuous for a.e. $t \in [S,T]$.

(H3) For every $N > 0$ there exist $\delta > 0$ and $k \in L^1$ such that, for a.e. $t \in [S,T]$,

$$|L(t,x',v) - L(t,x,v)| \leq k(t)|x' - x|, \quad L(t,\bar{x}(t),v) \geq -k(t)$$

for all $x',x \in \bar{x}(t) + \delta B$ and $v \in \dot{\bar{x}}(t) + NB$.

(H4) There exist $\epsilon > 0$ and a constant k_h such that h is upper semicontinuous on $\{(t,x) : t \in [S,T], x \in \bar{x}(t) + \epsilon B\}$ and

$$|h(t,x) - h(t,x')| \leq k_h|x - x'|$$

for all $x,x' \in \bar{x}(t) + \epsilon B$ and $t \in [S,T]$.

Then there exist an arc $p \in W^{1,1}([S,T];R^n)$, $\lambda \geq 0$, a measure $\mu \in C^{\oplus}(S,T)$, and a μ-integrable function $\gamma : [S,T] \to R^n$ such that

(i) $\lambda + \|p\|_{L^\infty} + \|\mu\|_{T.V.} = 1$;

(ii) $\dot{p}(t) \in co\{\eta : (\eta, p(t) + \int_{[S,t)} \gamma(s)\mu(ds), -\lambda)$

$$\in N_{\text{epi}\{L(t,\cdot,\cdot)\}}(\bar{x}(t), \dot{\bar{x}}(t), L(t,\bar{x}(t),\dot{\bar{x}}(t)))\} \quad a.e.;$$

(iii) $(p(S), -[p(T) + \int_{[S,T]} \gamma(s)\mu(ds)], -\lambda)$

$$\in N_{\text{epi}\{l\}}(\bar{x}(S), \bar{x}(T), l(\bar{x}(S), \bar{x}(T)));$$

(iv) $(p(t) + \int_{[S,t)} \gamma(s)\mu(ds)) \cdot \dot{\bar{x}}(t) - \lambda L(t,\bar{x}(t),\dot{\bar{x}}(t)) \geq$

$$\max_{v \in R^n}\{(p(t) + \int_{[S,t)} \gamma(s)\mu(ds)) \cdot v - \lambda L(t,\bar{x}(t),v)\} \quad a.e.;$$

(v) $\gamma(t) \in \partial_x^> h(t,\bar{x}(t)) \quad \mu$ a.e.

Here $\partial_x^> h(t,x)$ is the subdifferential (10.2).

Remarks

(a): If the state constraint $h(t,x(t)) \leq 0$ is dropped from (P) then, under Hypotheses (H1) to (H3), we may deduce from Theorem 10.2.1 the existence of $p \in W^{1,1}$ such that

(ii)' $\dot{p}(t) \in co\{\eta : (\eta, p(t)) \in \partial L(t,\bar{x}(t),\dot{\bar{x}}(t))\}$ a.e.

(∂L denotes the limiting subdifferential of $L(t,.,.)$);

(iii)' $(p(S), -p(T)) \in \partial l(\bar{x}(S), \bar{x}(T))$;

(iv)' $(p(t)) \cdot \dot{\bar{x}}(t) - L(t, \bar{x}(t), \dot{\bar{x}}(t))$

$$\geq \max_{v \in R^n} \{ p(t) \cdot v - L(t, \bar{x}(t), v) \} \text{ a.e.}$$

We thereby recover Theorem 7.3.1 of Chapter 7 as a special case. To see this, set $h \equiv -1$ (so that the state constraint is automatically satisfied). Condition (v) implies $\mu \equiv 0$. We must have $\lambda > 0$ (for if $\lambda = 0$, then $\|p\|_{L^\infty}$ by (i), yet (iv) implies $p \equiv 0$, a contradiction). Scaling the multipliers so that $\lambda = 1$ and noting the definition of the limiting subgradient, we arrive at (ii)' to (iv)'.

(b): The need to formulate Conditions (ii) and (iii) in terms of limiting normal cones to epigraph sets arises because $L(t, ., ., .)$ and l are possibly not Lipschitz continuous. If we further hypothesize that l is Lipschitz continuous near $(\bar{x}(S), \bar{x}(T))$ and $L(t, ., ., .)$ is Lipschitz continuous near $(\bar{x}(t), \dot{\bar{x}}(t))$ for a.e. t, the assertions of Theorem 10.2.1 are true with Conditions (ii) and (iii) strengthened to

(ii)'' $\dot{p}(t) \in \text{co}\{\eta : (\eta, p(t) + \int_{[S,t)} \gamma(s)\mu(ds)) \in \lambda\partial L(t, \bar{x}(t), \dot{\bar{x}}(t))\}$ a.e.;

(iii)'' $(p(S), -[p(T) + \int_{[S,T]} \gamma(s)\mu(ds)]) \in \lambda l(\bar{x}(S), \bar{x}(T))$.

This is a consequence of the fact that, if $f : R^k \to R \cup \{+\infty\}$ is finite and Lipschitz continuous on a neighborhood of $x \in R^k$, then

$$N_{\text{epi}\{f\}}(x, f(x)) = \{(\alpha\xi, -\alpha) : \alpha \geq 0, \xi \in \partial f(x)\}.$$

Proof. Fix $N > 0$ and let $k(.)$ and δ be the corresponding bounds and constant of (H3). Define

$$\begin{aligned}
\tilde{L}(t, w, v) &:= L(t, \bar{x}(t) + w, \dot{\bar{x}}(t) + v), \\
\tilde{l}(x, y) &:= l(\bar{x}(S) + x, \bar{x}(T) + y), \\
\tilde{h}(t, x) &:= h(t, \bar{x}(t) + x).
\end{aligned}$$

Choose a positive sequence $\epsilon_i \downarrow 0$. Define the subset of $R^n \times L^1 \times L^1$:

$$W := \{(\xi, v, w) : |v(t)| \leq K, \tilde{h}(t, x_{\xi,v}(t)) \leq 0$$
$$\text{for all } t \in [S, T], \|x_{\xi,v}\|_{W^{1,1}} \leq \epsilon\},$$

where $x_{\xi,v}(t) := \xi + \int_S^t v(s)ds$.
For each i, define the norm $\| \cdot \|_k$ on W to be

$$\|(\xi, v, w)\|_k := |\xi| + \|kv\|_{L^1} + \|kw\|_{L^1}$$

and define the functional $\tilde{J}_i$ on W

$$\begin{aligned}
\tilde{J}_i(\xi, v, w) &:= \tilde{l}(x_{\xi,v}(S), x_{\xi,v}(T)) \\
&+ \int_S^T \tilde{L}(t, w(t), v(t))dt + \epsilon_i^{-1} \int_S^T k(t)|x_{\xi,v}(t) - w(t)|^2 dt.
\end{aligned}$$

It can be shown that $(W, \|\cdot\|_k)$ is a complete metric space and, for each i, the functional $\tilde{J}_i$ is lower semicontinuous on $(W, \|\cdot\|_k)$. Furthermore, there exists a positive sequence $\alpha_i \downarrow 0$ such that for each i,

$$\tilde{J}_i(0,0,0) \leq \inf_{(\xi,v,w)\in W} \tilde{J}_i(\xi, v, w) + \alpha_i^2.$$

Since, for each i, $(0,0,0)$ is an "α_i^2 minimizer" for $\tilde{J}_i$ over W, Ekeland's Variational Principle tells us that there exists an element $(\xi_i, v_i, w_i) \in W$ which minimizes

$$\tilde{J}_i(\xi, v, w) + \alpha_i \|(\xi, v, w) - (\xi_i, v_i, w_i)\|_k$$

over W and

$$\|(\xi_i, v_i, w_i)\|_k \leq \alpha_i.$$

Write $x_i = x_{\xi_i, v_i}$. We can arrange by subsequence extraction that $(v_i, w_i) \to 0$ in L^1 and a.e., $x_i \to 0$ in $W^{1,1}$,

$$\lim_{i \to \infty} \tilde{l}(x_i(S), x_i(T)) = \tilde{l}(0,0)$$

and

$$\lim_{i \to \infty} \tilde{L}(t, w_i(t), v_i(t)) = \tilde{L}(t, 0, 0) \quad \text{a.e.}$$

The arguments justifying these assertions are the same as those previously used in Section 7.3. Define

$$\tilde{L}_i(t, x, v, w) := \tilde{L}(t, v, w) + \epsilon_i^{-1} k(t)|x - w|^2 +$$
$$\alpha_i k(t)|v - v_i(t)| + \alpha_i k(t)|w - w_i(t)|,$$
$$\tilde{l}_i(x, y) := \tilde{l}(x, y) + \alpha_i|x - x_i(S)|.$$

The minimizing property of (ξ_i, v_i, w_i) can be expressed as follows:
$((x_i, y_i, z_i), (v_i, w_i))$ is a strong local minimizer for the optimal control problem

$$\begin{cases} \text{Minimize } z(T) + \int_S^T \tilde{L}_i(t, x(t), w(t), v(t))dt \\ \text{over processes } ((x, y, z), (v, w)) \text{ satisfying} \\ \dot{x}(t) = v(t), \ \dot{y} = |v(t)|, \ |\dot{z}(t) = 0 \quad \text{a.e.,} \\ v(t) \in NB, \ w(t) \in R^n, \\ \tilde{h}(t, x(t)) \leq 0 \quad \text{for all } t \in [S, T], \\ (x(S), x(T), z(T)) \in \text{epi}\{\tilde{l}_i\}, \ y(S) = 0, \ |x(S)| + y(T) \leq \epsilon. \end{cases}$$

Here $x_i(t) := x_{\xi_i, v_i}(t)$, $y_i(t) := \int_S^t |v_i(s)|ds$, and $z_i(t) \equiv \tilde{l}_i(x_i(S), x_i(T))$.

This is an optimal control problem to which the Smooth Maximum Principle (Theorem 9.5.1) is applicable.

Notice that, because the right endpoint constraint on y is inactive at $y = y_i(t)$ and because of the decoupled structure of the cost and dynamics in y and (x, z), the adjoint arc component associated with y must be zero; we therefore drop it from the relationships.

The optimality conditions tell us that there exist an arc $p_i \in W^{1,1}$, a constant $\lambda_i \geq 0$, a measure $\mu_i \in C^{\oplus}(S, T)$, and a μ_i-integrable function γ_i such that

(A) $\|p_i\|_{L^\infty} + \lambda_i + \|\mu_i\|_{T.V.} = 1$;

(B) $-\dot{p}_i(t) = -2\lambda_i \epsilon_i^{-1} k(t)(x_i(t) - w_i(t))$ a.e.;

(C) $(p_i(S), -[p_i(T) + \int_{[S,T]} \gamma_i(s)\mu_i(ds)], -\lambda_i)$

       $\in N_{\text{epi}\{\tilde{l}_i\}}(x_i(S), x_i(T), \tilde{l}_i(x_i(S), x_i(T)))$;

(D) $(v, w) \mapsto (p_i(t) + \int_{[S,t)} \gamma_i(s)\mu_i(ds)) \cdot v - \lambda_i \tilde{L}_i(t, x_i(t), w, v)$ achieves its maximum at $(v_i(t), w_i(t))$ over all $(v, w) \in NB \times R^n$ a.e.,

(E) $\gamma_i(t) \in \partial_x^{>} \tilde{h}(t, x_i(t))$ μ_i a.e.

(B) and (D) imply that

$$(\dot{p}_i(t), p_i(t) + \int_{[S,t)} \gamma_i(s)\mu_i(ds))$$

$$\in \lambda_i \partial[\tilde{L}(t, w_i(t), v_i(t)) + \Psi_{NB}(v_i(t))] + \lambda_i \alpha_i k(t)(B \times B).$$

Fix $v = v_i(t)$. From (B) and (D),

$$\dot{p}_i(t) \in \lambda_i \partial_w \tilde{L}(t, w_i(t), v_i(t)) + \lambda_i \alpha_i k(t) B. \tag{10.3}$$

Fix $w = w_i(t)$. (D) and (E) imply that

$$(p_i(t) + \int_{[S,t)} \gamma_i(s)\mu_i(ds)) \cdot v_i(t) - \lambda_i \tilde{L}(t, w_i(t), v_i(t)) \tag{10.4}$$

$$\geq (p_i(t) + \int_{[S,t)} \gamma_i(s)\mu_i(ds)) \cdot v - \lambda_i \tilde{L}(t, w_i(t), v) - \lambda_i \alpha_i k(t)|v - v_i(t)|.$$

for all $v \in NB$.

Since $\tilde{L}(t, ., v)$ is Lipschitz continuous with rank $k(t)$ for all $v \in NB$, (10.3) implies that $|\dot{p}_i(t)| \leq 2k(t)$ for sufficiently large i. In view of (A), we can arrange, by extracting a suitable subsequence, that

$$p_i \to p \text{ uniformly} \quad \text{and} \quad \dot{p}_i \to \dot{p} \text{ weakly in } L^1$$

for some arc p. After extracting another subsequence, we have

$$\lambda_i \to \lambda, \quad \mu_i \to \mu \text{ weakly}^*, \quad \gamma_i d\mu_i \to \gamma d\mu \text{ weakly}^*$$

for some constant $\lambda \geq 0$, some measure $\mu \in C^\oplus(S,T)$, and some μ-integrable function γ. Furthermore,

$$g(t) \in \partial_x^\gtrless \tilde{h}(t,0) \quad \mu \text{ a.e.}$$

by (E) and Proposition 9.2.1. This is (v). We note that from (10.3) and (10.4) that

$$\dot{p}(t) \in \lambda \partial_w \tilde{L}(t,0,0). \tag{10.5}$$

$$-\lambda \tilde{L}(t,0,0) \geq (p(t) + \int_{[S,t]} \gamma(s)\mu(ds))v - \lambda \tilde{L}(t,0,v) \quad \text{for all } v \in NB.$$

Assertion (i) of the theorem is valid because $\|p_i\|_{L^\infty} + \|\mu_i\|_{T.V.} + \lambda_i = 1$.

We now confirm Assertions (ii) and (iii) of the theorem. The following conditions may be obtained from Mazur's and Carathéodory's Theorems.

For each i, there exist integers $0 \leq k_{i0} \leq \cdots \leq k_{in}$ and a "convex combination" $\{\alpha_{i0}, \ldots, \alpha_{kn}\}$ such that $\sum_{j=0}^n \alpha_{ij} \dot{p}_{i+k_{ij}}(t)$, $i = 1, 2, \ldots$, converges strongly in L^1 to $\dot{p}(t)$.

We can arrange by subsequence extraction that convergence is a.e. Fix t outside a suitable measure zero set. For each j, α_{ij}, $i = 1, 2, \ldots$ and $\dot{p}_{i+k_{ij}}(t)$, $i = 1, 2, \ldots$ are bounded. A further subsequence extraction ensures that (for each j) they have limits α_j and $\tilde{p}_j(t)$. Since

$$v_i(t) \to 0, \quad w_i(t) \to 0, \quad \tilde{L}(t, w_i(t), v_i(t)) + \Psi_{NB}(v_i(t)) \to \tilde{L}(t,0,0) \quad \text{a.e.} ,$$

$$p_i(t) \to p(t) \text{ uniformly,}$$

and

$$\tilde{l}_i(x_i(S), x_i(T)) \to \tilde{l}(0,0),$$

we have

$$\begin{cases} (\tilde{p}_j(t), p(t) + \int_{[S,t]} \gamma(s)\mu(ds)) \in \lambda \partial \tilde{L}(t,0,0) & \text{for } j = 0, \ldots, n, \\ (p(S), -[p(T) + \int_{[S,T]} \gamma(s)\mu(ds)]) \in \lambda \partial \tilde{l}(0,0), \end{cases}$$

if $\lambda > 0$, or

$$\begin{cases} (\tilde{p}_j(t), p(t) + \int_{[S,t]} \gamma(s)\mu(ds)) \in \partial^\infty \tilde{L}(t,0,0) & \text{for } j = 0, \ldots, n, \\ (p(S), -[p(T) + \int_{[S,T]} \gamma(s)\mu(ds)]) \in \partial^\infty \tilde{l}(0,0), \end{cases}$$

if $\lambda = 0$.

In either case we deduce

$$\begin{cases} (\tilde{p}_j(t), p(t) + \int_{[S,t]} \gamma(s)\mu(ds), -\lambda) \in N_{\text{epi}\{\tilde{L}(t,\cdot,\cdot)\}}(0, 0, \tilde{L}(t,0,0)) \\ \qquad\qquad\qquad\qquad\qquad\qquad\qquad \text{for } j = 0, \ldots, n, \\ (p(S), -[p(T) + \int_{[S,T]} \gamma(s)\mu(ds)], -\lambda) \in N_{\text{epi}\{\tilde{l}\}}(0, 0, \tilde{l}(0,0)). \end{cases}$$

But the α_js define a convex combination. It follows that, a.e.,

$$\dot{p}(t) = \sum_{j=0}^{n} \alpha_j \tilde{p}_j(t) \in$$

$$\text{co}\{\eta : (\eta, p(t) + \int_{[S,t)} \gamma(s)\mu(ds), -\lambda) \in N_{\text{epi}\{\tilde{L}(t,\cdot,\cdot)\}}(0, 0, \tilde{L}(t, 0, 0))\}.$$

These are precisely the assertions of the theorem, except that, in (iv), the inequality holds only for $v \in \dot{\bar{x}}(t) + NB$.

Take $N_i \to \infty$. Let $(p_i, \lambda_i, \mu_i, \gamma_i)$ denote the multipliers when $N = N_i$. We deduce from (10.5) that the $\dot{p}_i$s are uniformly integrably bounded. We can then show as before that along a subsequence $p_i \to p$ uniformly and $\dot{p}_i \to \dot{p}$ weakly in L^1 to some arc p. Arguing as before, we arrive at a strengthened form of Condition (iv) of the theorem statement in which R^n replaces NB, for some $\lambda \geq 0$, some measure $\mu \in C^{\oplus}(S, T)$, and some μ-integrable function γ. This is Condition (iv). The proof is complete. $\square$

10.3   The Extended Euler–Lagrange Condition for State Constrained Problems: Nonconvex Velocity Sets

We now build on the preceding analysis of finite Lagrangian problems, to derive necessary conditions for the optimal control problem with pathwise state constraints, introduced in Section 10.1.

$$(P) \begin{cases} \text{Minimize } g(x(S), x(T)) \\ \text{over absolutely continuous arcs } x : [S, T] \to R^n \text{satisfying} \\ \dot{x}(t) \in F(t, x(t)) \quad \text{a.e.,} \\ (x(S), x(T)) \in C, \\ h(t, x(t)) \leq 0 \quad \text{for all } t \in [S, T], \end{cases}$$

in which $[S, T]$ is a fixed interval, $g : R^n \times R^n \to R$ and $h : [S, T] \times R^n \to R$ are given functions, $F : [S, T] \times R^n \leadsto R^n$ is a given multifunction, and $C \subset R^n \times R^n$ is a given set.

The following theorem is an extension of Theorem 7.4.1 to allow for state constraints. It asserts (under appropriate hypotheses) the validity of the Extended Euler–Lagrange, Weierstrass, and Transversality Conditions for problems in which F is possibly nonconvex valued.

Theorem 10.3.1 (Necessary Conditions for State Constrained Problems: Nonconvex Velocity Sets) *Let $\bar{x}$ be a $W^{1,1}$ local minimizer for (P). Suppose that, for some $\epsilon > 0$, the following hypotheses are satisfied*

(G1) g is Lipschitz continuous on a neighborhood of $(\bar{x}(S), \bar{x}(T))$;

(G2) $F(t,x)$ *is nonempty, for each* (t,x). *F is* $\mathcal{L} \times \mathcal{B}$ *measurable,* Gr $F(t,.)$ *is closed for each* $t \in [S,T]$, *and there exist* $k \in L^1$ *and* $\beta \geq 0$ *such that, for a.e.* $t \in [S,T]$,

$$F(t,x') \cap (\dot{\bar{x}}(t) + NB) \subset F(t,x) + (k(t) + \beta N)|x' - x|\bar{B}$$

for all $N \geq 0$ *and* $x, x' \in \bar{x}(t) + \epsilon\bar{B}$.

(G3) h *is upper semicontinuous on* $\{(t,x) : t \in [S,T], x \in \bar{x}(t) + \epsilon B\}$ *and there is a constant* k_h *such that*

$$|h(t,x) - h(t,x')| \leq k_h|x - x'|$$

for all t *in* $[S,T]$ *and all* $x, x' \in \bar{x}(t) + \epsilon B$.

Then there exist an arc $p \in W^{1,1}([S,T]; R^n)$, *a nonnegative number* λ, *a measure* $\mu \in C^{\oplus}(S,T)$, *and a* μ-*integrable function* γ, *such that*

(i) $\lambda + \|p\|_{L^\infty} + \|\mu\|_{T.V.} = 1$;

(ii) $\dot{p}(t) \in co\{\eta : (\eta, p(t) + \int_{[S,t)} \gamma(s)\mu(ds)) \in N_{\text{Gr } F(t,.)}(\bar{x}(t), \dot{\bar{x}}(t))\}$ *a.e.;*

(iii) $(p(S), -[p(T)+\int_{[S,T]} \gamma(s)\mu(ds)]) \in \lambda\partial g(\bar{x}(S), \bar{x}(T))+N_C(\bar{x}(S), \bar{x}(T))$;

(iv) $(p(t) + \int_{[S,t)} \gamma(s)\mu(ds)) \cdot \dot{\bar{x}}(t) \geq (p(t) + \int_{[S,t)} \gamma(s)\mu(ds)) \cdot v$
   for all $v \in F(t,\bar{x}(t))$ *a.e.;*

(v) $\gamma(t) \in \partial_x^{>} h(t,\bar{x}(t))$ μ *a.e.*

Finally, assume that

$$F(t,x) \text{ does not depend on } t.$$

Then all the above assertions remain valid when Conditions (i) through (v) are supplemented by the condition

(vi) $p(t) + \int_{[S,t)} \gamma(s)\mu(ds) \cdot \dot{\bar{x}}(t)$

$$(= \max_{v \in F(\bar{x}(t))} p(t) + \int_{[S,t)} \gamma(s)\mu(ds) \cdot v) = r \quad a.e.,$$

for some constant r.

Remarks

(a): The one-sided Lipschitz continuity hypothesis (G2) is a refinement of the Lipschitz continuity hypothesis

$$F(t,x') \subset F(t,x) + k(t)|x' - x|B$$

typically invoked in the earlier necessary conditions literature. The advantages of the less restrictive hypothesis (G2) in situations when the multifunction F is unbounded were discussed in Chapter 7.

(b): In (P), the pathwise state constraint has been formulated as a single functional inequality constraint:

$$h(t, x(t)) \leq 0 \quad \text{for all } t \in [S, T]$$

which is operative on the entire time interval $[S, T]$. This formulation subsumes problems involving multiple state constraints, problems for which the state constraint is imposed merely on a specified closed subset of $[S, t]$ and problems in which the state constraint takes the implicit form:

$$x(t) \in X(t) \tag{10.6}$$

for some multifunction $X : [S, T] \rightsquigarrow R^n$. Techniques for reformulating pathwise state constraints such as these as a single functional inequality constraint on $[S, T]$ are discussed at length after the statement of the Nonsmooth Maximum Principle (Theorem 9.3.1.). These techniques were described for optimal control problems in which the dynamic constraint takes the form of a family of differential equations, but are applicable also in the context of problems involving differential inclusions.

Here, we comment merely on the implications of Theorem 10.3.1 for problems involving the implicit state constraint (10.6). Suppose that, in (P),

$$x(t) \in X(t) \quad \text{replaces} \quad h(t, x(t)) \leq 0.$$

Then a version of Theorem 10.3.1 can be proved, when the data satisfy (G1), (G2), and (in place of (G3)) the following hypothesis is imposed on $X(.)$.

(G3)′ $X(.)$ is a lower semicontinuous multifunction and $\mathrm{co}\bar{N}_{X(t)}(\bar{x}(t))$ is pointed for each $t \in [S, T]$,

in which

$$\bar{N}_{X(t)}(\bar{x}(t)) := \limsup\{N_{X(s)}(y) : (s, y) \overset{\mathrm{Gr}\,X}{\to} (t, \bar{x}(t))\}.$$

Applying Theorem 10.3.1 and choosing as state constraint functional

$$h(t, x) = d_{X(t)}(t),$$

we deduce necessary conditions for problems involving the implicit state constraint (10.6), which assert the existence of $p \in W^{1,1}$, a nonnegative number λ, and an element $\eta \in C^*([S, T]; R^n)$ satisfying

$$\int_{[S,T]} \xi(t) \cdot \eta(dt) \leq 0$$

for all

$$\xi \in \{\, \xi' \in C([S,T]; R^n) \,:\, \xi'(t) \in (\bar{N}_{X(t)}(\bar{x}(t)))^* \quad \eta \text{ a.e.}\,\},$$

such that Conditions (i) to (iv) of Theorem 10.3.1 are satisfied, when $p + \int \gamma \mu(dt)$ is replaced by

$$p + \int \eta(dt).$$

Proof of Theorem 10.3.1. By reducing the size of ϵ if necessary we can arrange that $\bar{x}$ is minimizing with respect to feasible arcs x satisfying $\|x - \bar{x}\|_{W^{1,1}} \leq \epsilon$ and that g is Lipschitz continuous on

$$(\bar{x}(S), \bar{x}(T)) + \epsilon(B \times B)$$

with Lipschitz constant k_g. Define $\mathcal{S}$, the set of "feasible" arcs close to $\bar{x}$, to be

$$\mathcal{S} := \{x \in W^{1,1} : \dot{x}(t) \in F(t, x(t)) \text{ a.e.,}$$
$$h(t, x(t)) \leq 0, (x(S), x(T)) \in C, \|x - \bar{x}\|_{W^{1,1}} \leq \epsilon\}$$

and define the functional $\tilde{g} : W^{1,1} \to R$:

$$\tilde{g}(x) := g(x(S), x(T)).$$

Evidently, $\tilde{g}$ is Lipschitz continuous on $\mathcal{S}$, with Lipschitz constant $2k_g$, with respect to the $W^{1,1}$ norm. We know that $\bar{x}$ is a minimizer for

$$\begin{cases} \text{Minimize } \tilde{g}(x) \\ \text{over } x \in \mathcal{S} \subset \{x \in W^{1,1} : h(t, x(t)) \leq 0, \|x - \bar{x}\|_{W^{1,1}} \leq \epsilon\}. \end{cases}$$

It follows from the Exact Penalization Theorem (Theorem 3.2.1) that $\bar{x}$ is a minimizer for the problem:

$$\begin{cases} \text{Minimize } \tilde{g}(x) + 2k_g \inf\{\|x - y\|_{W^{1,1}} : y \in \mathcal{S}\} \\ \text{over arcs } x \in W^{1,1}([S,T]; R^n) \text{ satisfying} \\ h(t, x(t)) \leq 0 \quad \text{for all } t \in [S, T], \\ \|x - \bar{x}\|_{W^{1,1}} \leq \epsilon. \end{cases}$$

There are two possible situations to be considered:

(a): there exist $\epsilon' \in (0, \epsilon)$ and $K > 0$, such that for any $x \in W^{1,1}$ satisfying $h(t, x(t)) \leq 0$ for all $t \in [S, T]$ and $\|x - \bar{x}\|_{W^{1,1}} \leq \epsilon'$ we have

$$\inf\{\|x - y\|_{W^{1,1}} : y \in \mathcal{S}\} \leq K[\int_S^T \rho_F(t, x(t), \dot{x}(t))dt + d_C(x(S), x(T))];$$

(b): there exists a sequence of arcs $\bar{x}_i$ such that $h(t, \bar{x}_i(t)) \leq 0$ for all $t \in [S, T]$ and $i = 1, 2, \ldots,$ $\bar{x}_i \to \bar{x}$ in $W^{1,1}$ as $i \to \infty$ and

$$J(\bar{x}_i) < (2i)^{-1} \inf\{\|\bar{x}_i - y\|_{W^{1,1}} : y \in \mathcal{S}\} \qquad (10.7)$$

for $i = 1, 2, \ldots,$ where

$$J(x) := \int_S^T \rho_F(t, x(t), \dot{x}(t))dt + d_C(x(S), x(T)).$$

We deal first with the more straightforward Case (a). Here, $\bar{x}$ is a $W^{1,1}$ local minimizer of $g(x(S), x(T)) + 2k_g K J(x)$ over $x \in W^{1,1}$ satisfying $h(t, x(t)) \leq 0$. This is the finite Lagrangian problem (F) discussed in Section 10.2 with

$$L(t, x, v) := 2k_g K \rho_F(t, x, v)$$
$$l(x, y) := g(x, y) + 2k_g K d_C(x, y).$$

The hypotheses are satisfied for existence of an arc p, a nonnegative number λ, a measure $\mu \in C^\oplus(S, T)$, and a μ-integrable function γ satisfying (i) to (v) of Theorem 10.2.1. Conditions (i) and (v) of Theorem 10.2.1 are clearly satisfied. (iv) is obtained by noticing $\rho_F(t, \bar{x}(t), v) = 0$ if $v \in F(t, \bar{x}(t))$. That (ii) and (iii) are also satisfied may be deduced from Comments (b) following the statement of Theorem 10.3.1 and Lemma 7.2.2.

It remains to address Case (b). Notice first of all that the functional J is continuous on

$$W := \{x \in W^{1,1} : \|x - \bar{x}\|_{W^{1,1}} \leq \epsilon, h(t, x(t)) \leq 0\}$$

with respect to the strong $W^{1,1}$ topology. Indeed if $x_i \to x$ in W, then $x_i \to x$ uniformly and $\dot{x}_i \to \dot{x}$ strongly in L^1. Consequently, in view of Lemma 7.2.2,

$$
\begin{aligned}
|J(x_i) - J(x)| &\leq \int_S^T |\rho_F(t, x_i, \dot{x}_i) - \rho_F(t, x, \dot{x})|dt \\
&\quad + |d_C(x_i(S), x_i(T)) - d_C(x(S), x(T))| \\
&\leq ((1 + \beta)\|k\|_{L^1} + 2\beta\|\dot{x} - \dot{\bar{x}}\|_{L^1})\|x_i - x\|_{L^\infty} \\
&\quad + \|\dot{x}_i - \dot{x}\|_{L^1} + 2\|x_i - x\|_{L^\infty}.
\end{aligned}
$$

The right side has limit zero as $i \to \infty$. Continuity is confirmed.

Define
$$a_i := \inf\{\|\bar{x}_i - y\|_{W^{1,1}} : y \in \mathcal{S}\}.$$

Since the inequality (10.7) is strict and J is nonnegative valued, $a_i > 0$ for all i. Since $\bar{x}_i \to \bar{x}$ in $W^{1,1}$ and $\bar{x} \in \mathcal{S}$, it follows that $a_i \to 0$ as $i \to \infty$. We see too that $J(\bar{x}_i) \leq (2i)^{-1}a_i$, from which we conclude that $\bar{x}_i$

is a "$(2i)^{-1}a_i$-minimizer" of J on W. According to Ekeland's Theorem, for each i there exists $x_i \in W$ such that x_i is a minimizer for

$$\begin{cases} \text{Minimize } \int_S^T \rho_F(x(t), \dot{x}(t))dt + d_C(x(S), x(T)) \\ \qquad\qquad + i^{-1}(|x(S) - x_i(S)| + \int_S^T |\dot{x}(t) - \dot{x}_i(t)|dt) \\ \text{over arcs } x \in W^{1,1} \text{ satisfying} \\ h(t, x(t)) \leq 0 \quad \text{for all } t \in [S, T] \end{cases}$$

and

$$\|x_i - \bar{x}_i\|_{W^{1,1}} \leq a_i/2.$$

Recalling that

$$a_i := \inf\{\|\bar{x}_i - y\| : y \in S\},$$

we conclude from (10.7) that $x_i \notin S$. It follows that, for each i, either $(x_i(S), x_i(T)) \notin C$ or $\dot{x}_i(t) \notin F(t, x_i(t))$ on a set of positive measure. Notice too that $x_i \to \bar{x}$ strongly in $W^{1,1}$ since $\bar{x}_i \to \bar{x}$ in $W^{1,1}$ and $\|x_i - \bar{x}_i\|_{W^{1,1}} \to 0$ as $i \to \infty$.

We have here another example of the finite Lagrangian problem (F) of Section 10.2, in which

$$L(t, x, v) := \rho_F(x, v) + i^{-1}|v - \dot{x}_i(t)|,$$
$$l(x, y) := d_C(x, y) + i^{-1}|x - x_i(S)|.$$

For i sufficiently large, $\rho_F(t, \cdot, \cdot)$ is Lipschitz continuous on a neighborhood of $(x_i(t), \dot{x}_i(t))$. (See Lemmas 7.2.1 and 7.2.2.) By Theorem 10.2.1 and subsequent comments we can find $(p_i, \lambda_i, \mu_i, \gamma_i)$ such that

(A) $\lambda_i + \|p_i\|_{L^\infty} + \|\mu_i\|_{T.V.} = 1$;

(B) $\dot{p}_i(t) \in \text{co}\{\eta : (\eta, p_i(t) + \int_{[S,t)} \gamma_i(s)\mu_i(ds)) \in \lambda_i \partial \rho_F(t, , x_i(t), \dot{x}_i(t))$

$$+ \{0\} \times (\lambda_i i^{-1} B)\} \quad \text{a.e.};$$

(C) $(p_i(S), -[p_i(T) + \int_{[S,T]} \gamma_i(s)\mu_i(ds)]) \in \lambda_i \partial d_C(x_i(S), x_i(T))$

$$+ \lambda_i i^{-1} B \times \{0\};$$

(D) $(p_i(t) + \int_{[S,t)} \gamma_i(s)\mu_i(ds)) \cdot \dot{x}_i(t) - \lambda_i \rho_F(t, x_i(t), \dot{x}_i(t)) \geq$

$$(p_i(t) + \int_{[S,t)} \gamma_i(s)\mu_i(ds)) \cdot v - \lambda_i \rho_F(t, x_i(t), v) - \lambda_i i^{-1}|v - \dot{x}_i(t)|$$

for all $v \in R^n$ a.e.;

(E) $\gamma_i(t) \in \partial_x^> h(t, x_i(t))$ μ_i a.e.

We deduce from (B), with the help of Lemma 7.2.2, that

$$|\dot{p}_i(t)| \leq \lambda_i[(1 + \beta\epsilon)k(t) + 2\beta|\dot{x}_i(t) - \dot{\bar{x}}(t)|].$$

But $x_i \to \bar{x}$ in $W^{1,1}$ (which implies $\dot{x}_i \to \dot{\bar{x}}$ strongly in L^1) and $\|p_i\|_{L^\infty} \le 1$. It follows that there exists an arc p such that $p_i \to p$ uniformly and $\dot{p}_i \to \dot{p}$ weakly in L^1. By further subsequence extraction we can arrange that

$$\lambda_i \to \lambda', \quad \mu_i \to \mu \text{ weakly}^* \quad \text{and} \quad \gamma_i d\mu_i \to \gamma d\mu \text{ weakly}^*,$$

for some $\lambda' \ge 0$, some regular nonnegative measure μ, and some μ-integrable function γ satisfying

$$\gamma(t) \in \partial_x^> h(t, \bar{x}(t)) \quad \mu \text{ a.e.}$$

If $0 \le \lambda' < 1$, then $\|p\|_{L^\infty} + \|\mu\|_{T.V.} = 1 - \lambda' > 0$ and

$$\dot{p}(t) \in \text{co}\{\eta : (\eta, p(t) + \int_{[S,t)} \gamma(s)\mu(ds)) \in \lambda' \partial \rho_F(t, \bar{x}(t), \dot{\bar{x}}(t))\} \quad \text{a.e.,}$$

$$(p(S), -[p(T) + \int_{[S,T]} \gamma(s)\mu(ds)]) \in \lambda' \partial d_C(\bar{x}(S), \bar{x}(T)),$$

$$(p(t) + \int_{[S,t)} \gamma(s)\mu(ds)) \cdot \dot{\bar{x}}(t) \ge (p(t) + \int_{[S,t)} \gamma(s)\mu(ds)) \cdot v$$

for all $v \in F(t, \bar{x}(t))$ a.e.

By scaling the multipliers we can arrange that $\|p\|_{L^\infty} + \|\mu\|_{T.V.} = 1$. Since

$$\lambda' \partial \rho_F(t, \bar{x}(t), \dot{\bar{x}}(t)) \subset N_{\text{Gr } F(t, .)}(\bar{x}(t), \dot{\bar{x}}(t))$$

and

$$\lambda' d_C(\bar{x}(S), \bar{x}(T)) \subset N_C(\bar{x}(S), \bar{x}(T)),$$

we arrive at Conditions (i) to (iv) of the theorem, with $\lambda' = 0$.

The remaining alternative to consider is $\lambda' = 1$. We now show that this cannot arise and thereby complete the proof of the theorem. Indeed if $\lambda' = 1$, then $\|p\|_{L^\infty} + \|\mu\|_{T.V.} = 0$. Recall however that, for each i, either $(x_i(S), x_i(T)) \notin C$ or $\dot{x}_i(t) \notin F(t, x_i(t))$ on a set of positive measure. We conclude then from (C) and (D) that either

$$|(p_i(S), -[p_i(T) + \int_{[S,T]} \gamma_i(s)\mu_i(ds)])| \ge \lambda_i(1 - i^{-1}) \tag{10.8}$$

or

$$|p_i(t) + \int_{[S,t)} \gamma_i(s)\mu_i(ds)| \ge \lambda_i(1 - i^{-1}) \text{ on a set of positive measure.} \tag{10.9}$$

If (10.8) is true for infinitely many i, then

$$|(p(S), -[p(T) + \int_{[S,T]} \gamma(s)\mu(ds)])| \ge 1.$$

This is impossible since $p = 0$ and $\mu = 0$.

If, on the other hand, (10.9) is valid for infinitely many i, then

$$|p_i(t) + \int_{[S,t)} \gamma_i(s)\mu_i(ds)| \leq |p_i(t)| + \int_S^t |\gamma_i(s)|\mu_i(ds) \leq |p_i(t)| + k_h \|\mu_i\| \quad \text{a.e.}$$

(k_h is a Lipschitz constant for $h(t,.)$) and hence

$$\|p_i\|_{L^\infty} = \max_{t \in [S,T]} |p_i(t)| \geq \lambda_i(1 - i^{-1}) - k_h \|\mu_i\|_{T.V.}.$$

Since $\|\mu_i\|_{T.V.} \to 0$ we obtain in the limit $\|p\|_{L^\infty} \geq 1$. But this is not possible since $\|p\|_{L^\infty} = 0$. Since either (10.8) or (10.9) must occur an infinite number of times, we have arrived at the desired contradiction.

Finally, the additional information supplied by the theorem in the case when $F(t,x)$ is independent of t is implied by Theorem 10.5.1 below. $\square$

10.4   Necessary Conditions for State Constrained Problems: Convex Velocity Sets

Now suppose that F is convex valued. For state constraint-free problems, we have already seen that the Extended Euler–Lagrange and Weierstrass Conditions are valid under local hypotheses and, on the other hand, these conditions can be supplemented by the Extended Hamilton Condition. We now show that the same is true in the presence of state constraints.

Theorem 10.4.1 (Necessary Conditions for State Constrained Problems: Convex Velocity Sets) *Let $\bar{x}$ be a $W^{1,1}$ local minimizer for (P). Assume that*

$$F(t,x) \text{ is a convex set for each } (t,x) \in [S,T] \times R^n.$$

Assume further that, for some $\epsilon > 0$, Hypotheses (G1) and (G3) of Theorem 10.3.1 are satisfied, in addition to the following weaker version of (G2).

(G2)′ There exists $k \in L^1$ and $\bar{K} > 0$ such that either

(a) $F(t,x') \cap (\dot{\bar{x}}(t) + \epsilon k(t)B) \subset F(t,x) + k(t)|x - x'|B$
    for all $x, x' \in \bar{x}(t) + \epsilon B$ a.e.,

or

(b) $F(t,x') \cap (\dot{\bar{x}}(t) + \epsilon B) \subset F(t,x) + k(t)|x - x'|B$,
    $\inf\{|v - \dot{\bar{x}}(t)| : v \in F(t,x)\} \leq \bar{K}|x - \bar{x}(t)|$
    for all $x, x' \in \bar{x}(t) + \epsilon B$ a.e.

Then there exist an arc $p \in W^{1,1}$, a constant $\lambda \geq 0$, a measure $\mu \in C^{\oplus}(S,T)$, and a μ-integrable function γ such that

(i) $\lambda + \|p\|_{L^{\infty}} + \|\mu\|_{T.V.} = 1$;

(ii) $\dot{p}(t) \in co\{\eta : (\eta, p(t) + \int_{[S,t)} \gamma(s)\mu(ds)) \in N_{Gr\ F(t,.)}(\bar{x}(t), \dot{\bar{x}}(t))\}$ *a.e.*;

(iii) $(p(S), -[p(T) + \int_{[S,T]} \gamma(s)\mu(ds)]) \in \lambda \partial g(\bar{x}(S), \bar{x}(T)) + N_C(\bar{x}(S), \bar{x}(T))$;

(iv) $\gamma(t) \in \partial_x^{>} h(t, \bar{x}(t))$ μ *a.e.*

Under the stated hypotheses, (ii) implies that (p, μ, γ) also satisfy the conditions

(I) $\dot{p}(t) \in co\{-\xi : (\xi, \dot{\bar{x}}(t)) \in \partial H(t, \bar{x}(t), p(t) + \int_{[S,t)} \gamma(s)\mu(ds))\}$ *a.e.*

and

(II) $(p(t) + \int_{[S,t)} \gamma(s)\mu(ds)) \cdot \dot{\bar{x}}(t) \geq (p(t) + \int_{[S,t)} \gamma(s)\mu(ds)) \cdot v$

$$\text{for all } v \in F(t, \bar{x}(t))   \text{a.e.}$$

Finally, assume that

$$F(t, x) \text{ does not depend on } t.$$

Then all the above assertions remain valid when Conditions (i) to (v) are supplemented by the condition

(vi) $(p(t) + \int_{[S,t)} \gamma(s)\mu(ds)) \cdot \dot{\bar{x}}(t)$

$$(= \max_{v \in F(\bar{x}(t))} (p(t) + \int_{[S,t)} \gamma(s)\mu(ds)) \cdot v) = r   \text{a.e.},$$

for some constant r.

(Here,

$$H(t, x, p) = \max\{p \cdot v : v \in F(t, x)\}$$

and ∂H is the limiting subdifferential of $H(t, ., .)$.)

The theorem is proved with the help of the following proposition, a simple consequence of Theorem 10.3.1, which provides necessary conditions for a class of optimal control problems in which the dynamic constraint does not depend on the state variable.

$$(\tilde{P}) \begin{cases} \text{Minimize } l(x(S), x(T)) + \int_S^T L(t, x(t), \dot{x}(t))dt \\ \text{over arcs } x \text{ satisfying} \\ \dot{x}(t) \in R(t)   \text{a.e.}, \\ h(t, x(t)) \leq 0   \text{for all } t \in [S, T]. \end{cases}$$

Here $[S, T]$ is a given subinterval, $l : R^n \times R^n \to R$, $L : [S,T] \times R^n \times R^n \to R$, and $h : [S, T] \times R^n \to R$ are given functions, and $R : [S, T] \rightsquigarrow R^n$ is a given multifunction.

Proposition 10.4.2 *Let $\bar{x}$ be a $W^{1,1}$ local minimizer for $(\tilde{P})$. Assume that, for some $\epsilon > 0$,*

(G1) R has a Borel measurable graph and $R(t)$ is closed for each t;

(G2) $L(.,x,v)$ is Lebesgue measurable for each (x,v), and there exists $k \in L^1$ such that

$$|L(t,x,v) - L(t,x',v')| \leq k(t)|(x,v) - (x',v')|$$

for all $(x,v),(x',v') \in (\bar{x}(t) + \epsilon B) \times (\dot{\bar{x}}(t) + \epsilon B)$;

(G3) l is Lipschitz continuous on a neighborhood of $(\bar{x}(\bar{S}), \bar{x}(\bar{T}))$;

(G4) h is upper semicontinuous on a relative neighborhood of Gr $\bar{x}$ in $[S,T] \times R^n$ and there exists $k_h \geq 0$ such that

$$|h(t,x) - h(t,x')| \leq k_h|x - x'|$$

for all $x,x' \in \bar{x}(t) + \epsilon B$ and $t \in [S,T]$.

Then there exist an arc $p \in W^{1,1}$, a constant $\lambda \geq 0$, a measure $\mu \in C^{\oplus}(S,T)$, and a μ-integrable function γ such that

(i) $\lambda + \|p\|_{L^\infty} + \|\mu\|_{T.V.} = 1$;

(ii) $\dot{p}(t) \in co\{\eta : (\eta, p(t) + \int_{[S,t)} \gamma(s)\mu(ds)) \in \lambda\partial L(t,\bar{x}(t),\dot{\bar{x}}(t)) +$
$$\{0\} \times N_{R(t)}(\dot{\bar{x}}(t))\} \quad a.e.;$$

(iii) $(p(S), -[p(T) + \int_{[S,T]} \gamma(s)\mu(ds)]) \in \lambda\partial l(\bar{x}(S),\bar{x}(T))$;

(iv) $\gamma(t) \in \partial_x^{\geq} h(t,\bar{x}(t)) \quad \mu$ a.e.

Proof of Proposition 10.4.2. Since $\bar{x}$ is a $W^{1,1}$ local minimizer for $(\tilde{P})$, $(\bar{x}, \bar{z}(t) \equiv \int_S^t L(x,\bar{x}(s),\dot{\bar{x}}(s))ds)$ is a $W^{1,1}$ local minimizer for

$$\begin{cases} \text{Minimize } z(T) + l(x(S), x(T)) \\ \text{over arcs } (x,z) \text{ satisfying} \\ (\dot{x}(t), \dot{z}(t)) \in \tilde{F}(t, x(t)) \quad a.e., \\ z(S) = 0, \\ h(t, x(t)) \leq 0 \quad \text{for all } t \in [S,T], \end{cases}$$

in which

$$\tilde{F}(t,x) := \{(v,\alpha) : v \in R(t) \text{ and } \alpha \geq L(t,x,v)\}.$$

The hypotheses are satisfied under which Theorem 10.3.1 may be applied. We deduce the existence of an arc $p \in W^{1,1}$, $q \in R$, $\lambda \geq 0$, a measure $\mu \in C^{\oplus}(S,T)$, and a μ-integrable function γ such that

$$\lambda + |q| + \|p\|_{L^\infty} + \|\mu\|_{T.V.} = 1,$$

$$\dot{p}(t) \in \mathrm{co}\{\eta : (\eta, p(t) + \int_{[S,t)} \gamma(s)\mu(ds), -q) \in$$

$$N_{\mathrm{Gr}\tilde{F}(t,\cdot)}(\bar{x}(t), \dot{\bar{x}}(t), L(t, \bar{x}(t), \dot{\bar{x}}(t)))\} \text{ a.e.,}$$

$$(p(S), -[p(T) + \int_{[S,T]} \gamma(s)\mu(ds)]) \in \lambda\partial l(\bar{x}(S), \bar{x}(T)) \text{ and } q = \lambda,$$

$$\gamma(t) \in \partial_x^> h(t, \bar{x}(t)) \quad \mu \text{ a.e.}$$

But

$$\begin{aligned}
\mathrm{Gr}\,\tilde{F}(t,.) &= \{(x, v, \alpha) : \alpha \geq L(t, x, v) + \Psi_{R(t)}(v)\} \\
&= \mathrm{epi}\{(x, v) \to L(t, x, v) + \Psi_{R(t)}(v)\} \text{ for all } t \in [S, T].
\end{aligned}$$

It follows that the condition

$$(w, p, -q) \in N_{\mathrm{Gr}\,\tilde{F}(t,.)}(\bar{x}(t), \dot{\bar{x}}(t), \dot{\bar{z}}(t))$$

implies $q \geq 0$ and

$$\begin{aligned}
(w, p) &\in q\partial(L(t, \bar{x}(t), \dot{\bar{x}}(t)) + \Psi_{R(t)}(\dot{\bar{x}}(t))) \\
&\subset q\partial L(t, \bar{x}(t), \dot{\bar{x}}(t)) + \{0\} \times N_{R(t)}(\dot{\bar{x}}(t)).
\end{aligned}$$

To justify the last inclusion we apply the Sum Rule (Theorem 5.4.1), taking note of the Lipschitz continuity of $L(t, ., .)$ on a neighborhood of $(\bar{x}(t), \dot{\bar{x}}(t))$. The differential inclusion for p may therefore be replaced by

$$\dot{p}(t) \in \mathrm{co}\{\eta : (\eta, p(t) + \int_{[S,t)} \gamma(s)\mu(ds)) \in \lambda\partial L(t, \bar{x}(t), \dot{\bar{x}}(t)))$$

$$+ \{0\} \times N_{R(t)}(\dot{\bar{x}}(t))\}.$$

Since $q = \lambda$ we have $\|p\|_{L^\infty} + \lambda + \|\mu\|_{T.V.} \neq 0$. By scaling the multipliers then we can arrange that $\|p\|_{L^\infty} + \lambda + \|\mu\|_{T.V.} = 1$. The proposition is proved. $\square$

Proof of Theorem 10.4.1. We deal first with the concluding assertions of the theorem. Suppose that a collection of multipliers (p, μ, γ) satisfies (ii). Then, for a.e. $t \in [S, T]$, there exists η such that

$$(\eta, q(t)) \in N_{\mathrm{Gr}\,F(t,.)}(\bar{x}(t), \dot{\bar{x}}(t)), \tag{10.10}$$

where $q(t) := p(t) + \int_{[S,t)} \gamma(s)\mu(ds)$. Since F is convex-valued, (10.10) implies

$$q(t) \cdot \dot{\bar{x}}(t) \geq q(t) \cdot v \quad \text{for all } v \in F(t, \bar{x}(t)) \quad \text{a.e.}$$

(See the beginning of the proof of Theorem 7.5.1.) This is Condition (II). Condition (I) follows from a pointwise application of the Dualization Theorem (Theorem 7.6.8), which tells us that, under the hypotheses currently in force,

$$\text{co}\{\xi : (\xi, q(t)) \in \partial \Psi_{\text{Gr } F(t,.)}(\bar{x}(t), \dot{\bar{x}}(t))\}$$
$$\subset \text{co}\{\eta : (-\eta, \dot{\bar{x}}(t)) \in H(t, \bar{x}(t), q(t))\}$$

and from the identity $\partial \Psi_{\text{Gr}F(t,.)} = N_{\text{Gr } F(t,.)}$.

Now for the remaining assertions of the theorem. We may assume without loss of generality that the function $k(.)$ of Hypothesis (G2)' ((a) or (b)) satisfies $k(t) \geq 1$ a.e. Define

$$R(t) := \dot{\bar{x}}(t) + (\epsilon/3)B.$$

Reduce the size of ϵ if necessary to ensure that $\bar{x}$ is minimizing with respect to arcs x satisfying $\|\bar{x} - x\|_{W^{1,1}} \leq \epsilon$ and that g is Lipschitz continuous on $(\bar{x}(\bar{S}), \bar{x}(\bar{T})) + \epsilon(B \times B)$ with Lipschitz constant k_g. Define $\mathcal{S} \subset W^{1,1}$ and $\tilde{g} : W^{1,1} \to R$:

$$\mathcal{S} := \{x \in W^{1,1} : \dot{x}(t) \in F(t, x(t)), \dot{x}(t) \in R(t) \text{ a.e.,}$$
$$h(t, x(t)) \leq 0 \text{ for all } t, (x(S), x(T)) \in C, \|x - \bar{x}\|_{W^{1,1}} \leq \epsilon\}$$

and

$$\tilde{g}(x) := g(x(S), x(T)).$$

Arguing as in the proof of Theorem 10.3.1 we can show that $\bar{x}$ is a minimizer for the problem

$$\begin{cases} \text{Minimize } \tilde{g}(x) + 2k_g \inf\{\|x - y\|_{W^{1,1}} : y \in \mathcal{S}\} \\ \text{over arcs } x \\ \dot{x}(t) \in R(t) \quad \text{a.e.} \\ h(t, x(t)) \leq 0 \quad \text{for all } t \in [S, T], \\ \|x - \bar{x}\|_{W^{1,1}} \leq \epsilon. \end{cases}$$

Again there are two possibilities to consider:

(a): there exist $\epsilon' \in (0, \epsilon)$ and $K' > 0$, such that if $x \in W^{1,1}$ satisfies $h(t, x(t)) \leq 0$ for all t, $\dot{x}(t) \in R(t)$ a.e. and $\|x - \bar{x}\|_{W^{1,1}} \leq \epsilon'$ then

$$\inf\{\|x - y\|_{W^{1,1}} : y \in \mathcal{S}\} \leq K'[d_C(x(S), x(T)) + \int_S^T \rho_F(t, x(t), \dot{x}(t))dt];$$

(b): there exists a sequence of arcs $\bar{x}_i$ with $\dot{\bar{x}}_i(t) \in R(t)$ a.e. and $h(t, \bar{x}_i(t)) \leq 0$ for all t such that $\bar{x}_i \to \bar{x}$ in $W^{1,1}$ and, for $i = 1, 2, \ldots,$

$$\int_S^T \rho_F(t, \bar{x}_i(t), \dot{\bar{x}}_i(t))dt + d_C(\bar{x}_i(S), \bar{x}_i(T))$$
$$< (2i)^{-1} \inf\{\|\bar{x}_i - y\|_{W^{1,1}} : y \in \mathcal{S}\}.$$

Consider first Case (a). Then $\bar{x}$ is a $W^{1,1}$ local minimizer of the problem

$$\left\{\begin{array}{l} \text{Minimize } g(x(S), x(T)) + 2k_g K'[d_C(x(S), x(T)) \\ \qquad\qquad\qquad\qquad + \int_S^T \rho_{F(t,.)}(x(t), \dot{x}(t))dt] \\ \text{over arcs } x \text{ satisfying} \\ \dot{x}(t) \in R(t) \quad \text{a.e.,} \\ h(t, x(t)) \leq 0 \quad \text{for all } t \in [S, T]. \end{array}\right.$$

Now apply Lemma 7.2.1 in which, for each $t \in [S, T]$, we identify $\Gamma(x)$ with $F(t, x)$. This tells us that $\rho_F(t, ., .)$ is Lipschitz continuous with Lipschitz constant $2k(t)$ on $(\bar{x}(t), \dot{\bar{x}}(t)) + (\epsilon/3B) \times (\epsilon/3B)$ if (G2)'(a) holds and on $(\bar{x}(t), \dot{\bar{x}}(t)) + (\epsilon(3\bar{K})^{-1}B) \times (\epsilon/3B)$ if (G2)'(b) holds. (Set the parameters of Lemma 7.2.1 $r = \epsilon k(t)$ and $K = k(t)$ if (G2)' holds, and $r = \epsilon$ and $K = \bar{K}$ if (G2)'' holds, and remember that $k(t) \geq 1$.)

We check that the hypotheses are met under which Proposition 10.4.2 applies to this problem. Appealing to Part (b) of Lemma 7.2.1, we deduce the existence of p, $\lambda \geq 0$, μ, and γ satisfying Conditions (i) to (iii) of the theorem statement. (Notice that $R(t)$ does not appear in these conditions because the constraint $\dot{x}(t) \in R(t)$ is inactive.) This deals with Case (a).

Consider next Case (b). Using arguments similar to those employed in the proof of Theorem 10.3.1, we show by means of Ekeland's theorem that there exists a $W^{1,1}$ strongly convergent sequence $x_i \to \bar{x}$ such that, for each i, x_i is a $W^{1,1}$ local minimizer for

$$\left\{\begin{array}{l} \text{Minimize } \int_S^T \rho_{F(t,.)}(x(t), \dot{x}(t))dt + d_C(x(S), x(T)) \\ \qquad\qquad + i^{-1}(|x(S) - x_i(S)| + \int_S^T |\dot{x}(t) - \dot{x}_i(t)|dt) \\ \text{over arcs } x \text{ satisfying} \\ \dot{x}(t) \in R(t) \quad \text{a.e.,} \\ h(t, x(t)) \leq 0 \quad \text{for all } t \in [S, T]. \end{array}\right.$$

Furthermore, $x_i \notin S$. This last property implies that

$$(x_i(S), x_i(T)) \notin C \quad \text{or} \quad \dot{x}_i(t) \notin F(t, x_i(t)) \text{ on a set of positive measure.} \tag{10.11}$$

Now apply Proposition 10.4.2 to the preceding problem, with reference to the $W^{1,1}$ local minimizer x_i, for $i = 1, 2, \ldots$. We conclude that there exist multipliers (p_i, λ_i, μ_i) and a Borel measurable function γ_i such that

(A) $\lambda_i + \|p_i\|_{L^\infty} + \|\mu_i\|_{T.V.} = 1$;

(B) $\dot{p}_i(t) \in \text{co}\{\eta : (\eta, p_i(t) + \int_{[S,t)} \gamma_i(s)\mu_i(ds)) \in \lambda_i \partial \rho_F(t, x_i(t), \dot{x}_i(t)) +$

$$\{0\} \times (\lambda_i i^{-1} B) + \{0\} \times N_{R(t)}(\dot{x}_i(t))\} \quad \text{a.e.;}$$

(C) $(p_i(S), -[p_i(T) + \int_{[S,T]} \gamma_i(s)\mu_i(ds)]) \in$

$$\lambda_i \partial d_C(x_i(S), x_i(T)) + \{\lambda_i i^{-1}\} \times \{0\};$$

(D) $\gamma_i(t) \in \partial_x^> h(t, x_i(t))$ μ_i a.e.

We deduce from Lemma 7.2.1 and (B) that $|\dot{p}_i(t)| \le k(t)$ a.e. for $i = 1, 2, \ldots$. It follows that the $\dot{p}_i$s are uniformly integrably bounded. Since the ps are uniformly bounded, we can therefore arrange by subsequence extraction that, for some $p \in W^{1,1}$,

$$p_i \to p \text{ uniformly} \quad \text{and} \quad \dot{p}_i \to \dot{p} \text{ weakly in } L^1.$$

We can ensure, by again extracting a suitable subsequence, that $\lambda_i \to \lambda'$ for some $\lambda' \ge 0$,

$$\mu_i \to \mu \text{ weakly}^* \quad \text{and} \quad \gamma_i d\mu_i \to \gamma d\mu \text{ weakly}^*$$

for some measure $\mu \in C^{\oplus}(S, T)$ and some μ-integrable function γ. The convergence analysis in the proof of Theorem 10.3.1 may now be reproduced to justify passing to the limit in the relationships (A) to (D). We thereby obtain

$$\|p\|_{L^\infty} + \lambda' + \|\mu\|_{T.V.} = 1;$$

$$\dot{p}(t) \in \text{co}\{w : (w, p(t) + \int_{[S,t)} \gamma(s)\mu(ds)) \in \lambda' \partial \rho_F(t, \bar{x}(t), \dot{\bar{x}}(t))\} \quad \text{a.e.};$$

$$(p(S), -[p(T) + \int_{[S,T]} \gamma(s)\mu(ds)]) \in \lambda' \partial d_C(\bar{x}(\bar{S}), \bar{x}(\bar{T})).$$

Suppose first that $0 \le \lambda' < 1$. In this case $\|p\|_{L^\infty} + \|\mu\|_{T.V.} > 0$. Scale the multipliers so that $\|p\|_{L^\infty} + \|\mu\|_{T.V.} = 1$. Recalling that

$$\partial \rho_{F(t,\cdot)}(\bar{x}(t), \dot{\bar{x}}(t)) \subset N_{\text{Gr } F(t,\cdot)}(\bar{x}(t), \dot{\bar{x}}(t))$$

and

$$\partial d_C(\bar{x}(S), \bar{x}(T)) \subset N_C(\bar{x}(S), \bar{x}(T)),$$

we see that p, μ, and γ satisfy Conditions (i) to (iv) of Theorem 10.4.1 with $\lambda = 0$.

The remaining case to consider is when $\lambda' = 1$. We complete the proof by showing that this cannot arise. Assume then $\lambda' = 1$. It follows that $p = 0$ and $\mu = 0$. In view of (10.11), we can arrange by subsequence extraction that either Condition (α) or (β) is satisfied:

(i): $(x_i(S), x_i(T)) \notin C$ for all i;

(ii): $\dot{x}_i(t) \notin F(t, x_i(t))$ on a set of positive measure for all i.

Suppose first that (i) is true. We deduce from (C) and the properties of the distance function that

$$|(p_i(S), -[p_i(T) + \int_{[S,T]} \gamma_i(s)\mu_i(ds)])| \ge \lambda_i(T - i^{-1}).$$

Passing to the limit we obtain

$$|(p(S), -[p(T) + \int_{[S,T]} \gamma(s)\mu(ds)])| \geq 1.$$

This contradicts $p = 0$ and $\mu = 0$.

Suppose finally (ii) is true. According to Lemma 7.2.3

$$|p_i(t) + \int_{[S,t)} \gamma_i(s)\mu_i(ds)| \geq \lambda_i \sqrt{5/16}$$

on a set of positive measure. Now, in view of the fact that $|\gamma_i(t)| \leq k_h$ μ a.e., we deduce that

$$\|p_i\|_{L^\infty} + k_h\|\mu_i\|_{T.V.} \geq \lambda_i \sqrt{5/16}.$$

In the limit we obtain

$$\|p\|_{L^\infty} + k_h\|\mu\|_{T.V.} \geq \sqrt{5/16}.$$

Once again we have arrived at a contradiction of the fact that $p = 0$ and $\mu = 0$.

It remains to observe that the additional information supplied by the theorem, concerning problems for which $F(t, x)$ is independent of t, is implied by Theorem 10.5.1. $\square$

10.5 Free Time Problems with State Constraints

In this section, we supply optimality conditions that are "state constrained" analogues of the free time necessary conditions of Chapter 8. As before, the extra conditions associated with the free end-times take the form of boundary conditions on the Hamiltonian. Attention focuses then on the following problem, in which the end-times are included among the choice variables.

$$(FT) \begin{cases} \text{Minimize } g(S, x(S), T, x(T)) \\ \text{over intervals } [S,T] \text{ and arcs } x \in W^{1,1}([S,T], R^n) \text{ satisfying} \\ \dot{x}(t) \in F(t, x(t)) \quad \text{a.e. } t \in [S,T], \\ (S, x(S), T, x(T)) \in C, \\ h(t, x(t)) \leq 0 \quad \text{for all } t \in [S,T]. \end{cases}$$

Here $g : R \times R^n \times R \times R^n \to R$ and $h : R \times R^n \to R$ are given functions, $F : R \times R^n \rightsquigarrow R^n$ is a given multifunction, and $C \subset R \times R^n \times R \times R^n$ is a given closed set.

The notion of a $W^{1,1}$ minimizer is the same as that used in the study of free-time problems in Chapter 8, adapted to require satisfaction of the state constraint.

The first set of necessary conditions addresses cases when the data are Lipschitz continuous with respect to time.

Theorem 10.5.1 *Let $([\bar{S}, \bar{T}], \bar{x})$ be a $W^{1,1}$ local minimizer for (FT) such that $\bar{T} > \bar{S}$. Assume that, for some $\delta > 0$, the following hypotheses are satisfied:*

(H1) g is Lipschitz continuous on a neighborhood of $(\bar{S}, \bar{x}(\bar{S}), \bar{T}, \bar{x}(\bar{T}))$ and C is a closed set.

(H2) $F(t, x)$ is nonempty for all $(t, x) \in R \times R^n$ and $\mathrm{Gr}\, F$ is closed.

(H3) There exist $k_F \in L^1$ and $\beta \geq 0$ such that, for a.e. $t \in [\bar{S}, \bar{T}]$,

$$F(t', x') \cap (\dot{\bar{x}}(t) + NB) \subset F(t'', x'') + (k_F(t) + \beta N)|(t', x') - (t'', x'')|B$$

for all $N \geq 0$, $(t', x'), (t'', x'') \in (t, \bar{x}(t)) + \delta B$.

(H4) h is upper semicontinuous and there exists a constant k_h such that, for a.e. $t \in [\bar{S}, \bar{T}]$,

$$|h(t', x') - h(t'', x'')| \leq k_h(|(t', x') - (t'', x'')|)$$

for all (t', x'), $t'', x'') \in (t, \bar{x}(t)) + \delta B$.

Then there exist absolutely continuous arcs

$$r \in W^{1,1}([\bar{S}, \bar{T}]; R) \quad \text{and} \quad p \in W^{1,1}([\bar{S}, \bar{T}]; R^n),$$

a nonnegative number λ, a measure $\mu \in C^{\oplus}(S, T)$, and a Borel measurable function $\gamma = (\gamma_0, \gamma_1) : [\bar{S}, \bar{T}] \to R^{1+n}$ such that

(i) $\lambda + \|p\|_{L^\infty} + \|\mu\|_{T.V.} = 1$;

(ii) $\gamma(s) \in \partial^> h(s, \bar{x}(s))$ μ a.e.;

(iii) $(-\dot{r}(t), \dot{p}(t)) \in \mathrm{co}\{(\alpha, \beta) : (\alpha, \beta, p(t) + \int_{[\bar{S}, t)} \gamma_1(s)\mu(ds))$

$$\in N_{\mathrm{Gr}\, F}(t, \bar{x}(t), \dot{\bar{x}}(t))\} \quad \text{a.e. } t \in [\bar{S}, \bar{T}];$$

(iv) $(p(t) + \int_{[\bar{S}, t)} \gamma_1(s)\mu(ds)) \cdot \dot{\bar{x}}(t)$

$$\geq (p(t) + \int_{[\bar{S}, t)} \gamma_1(s)\mu(ds)) \cdot v \text{ for all } v \in F(t, \bar{x}(t)) \text{ a.e. } t \in [\bar{S}, \bar{T}];$$

(v) $(-r(\bar{S}), p(\bar{S}), r(\bar{T}) - \int_{[\bar{S}, \bar{T}]} \gamma_0(s)\mu(ds), -p(\bar{T}) - \int_{[\bar{S}, \bar{T}]} \gamma_1(s)\mu(ds))$

$$\in \lambda \partial g(\bar{S}, \bar{x}(\bar{S}), \bar{T}, \bar{x}(\bar{T})) + N_C(\bar{S}, \bar{x}(\bar{S}), \bar{T}, \bar{x}(\bar{T}));$$

(vi) $r(t) = \int_{[\bar{S}, t)} \gamma_0(s)\mu(ds) + H(t, \bar{x}(t), p(t) + \int_{[\bar{S}, t)} \gamma_1(s)\mu(ds))$

a.e. $t \in [\bar{S}, \bar{T}]$.

Now suppose that

$$F(t, x) \text{ is convex for all } (t, x) \in R \times R^n.$$

and that Hypothesis (H3) has been replaced by the weaker hypothesis

(G3) there exist $k_F \in L^1 >, \epsilon > 0, \bar{K} \geq 0$ such that either

(a) $F(t', x') \cap (\dot{\bar{x}}(t) + \epsilon k_F(t)B) \subset F(t'', x'') + k_F(t)|(t', x') - (t'', x'')|B$
for all $(t', x'), (t'', x'') \in (t, \bar{x}(t)) + \epsilon B$, a.e. $t \in [\bar{S}, \bar{T}]$,

or

(b) $\left\{ \begin{array}{l} F(t', x') \cap (\dot{\bar{x}}(t) + \epsilon B) \subset F(t'', x'') + k_F(t)|(t', x') - (t'', x'')|B, \\ \inf\{|v - \dot{\bar{x}}(t)| : v \in F(t', x')\} \leq \bar{K}|(t', x') - (t, \bar{x}(t))| \end{array} \right.$
for all $(t', x'), (t'', x'') \in (t, \bar{x}(t)) + \epsilon B$, a.e. $t \in [\bar{S}, \bar{T}]$.

Then all the above assertions remain valid. Furthermore, Condition (iii) implies

(ii)' $(\dot{r}(t), -\dot{p}(t)) \in \mathrm{co}\{(\zeta, \eta) : ((\zeta, \eta), \dot{\bar{x}}(t)) \in$
$$\partial H(t, \bar{x}(t), p(t) + \int_{[\bar{S}, t)} \gamma_1(s)\mu(ds))\} \quad a.e.$$

Here,

$$\partial^> h(t, x) := \mathrm{co}\{\xi : \exists (t_i, x_i) \xrightarrow{h} (t, x) \text{ such that}$$
$$h(t_i, x_i) > 0 \text{ for all } i \text{ and } \nabla h(t_i, x_i) \to \xi\}$$

and

$$H(t, x, p) = \sup_{v \in F(t,x)} p \cdot v \quad .$$

The proof closely patterns that of Theorem 8.2.1 and is therefore omitted. The essential idea is to reformulate (FT) as a fixed time problem by means of a change of independent variable, apply known necessary conditions to the fixed time problem (necessary conditions for state constrained problems in this case), and to interpret these conditions in terms of the original problem.

Necessary conditions are also available for problems with measurably time-dependent data. For such problems, the boundary condition on the Hamiltonian is interpreted in the essential value sense, introduced in Chapter 8. Recall that, give a real-valued function f on an open interval I and a point $t \in I$, the essential value of f at t is the set

$$\mathrm{ess}_{\tau \to t} f(\tau) := [a^-, a^+],$$

where

$$a^- := \lim_{\delta \downarrow 0} \mathrm{ess\,inf}_{t - \delta \leq \tau \leq t + \delta} f(\tau)$$

and

$$a^+ := \lim_{\delta \downarrow 0} \mathrm{ess\,sup}_{t - \delta \leq \tau \leq t + \delta} f(\tau).$$

Theorem 10.5.2 *Let $([\bar{S}, \bar{T}], \bar{x})$ be a $W^{1,1}$ local minimizer for (FT). Assume that, for some $\delta > 0$, the data satisfy the following hypotheses.*

(H1) g is Lipschitz continuous on a neighborhood of $(\bar{S}, \bar{x}(\bar{S}), \bar{T}, \bar{x}(\bar{T}))$ and C is a closed set.

(H2) F is a nonempty closed set for all $(t, x) \in R \times R^n$, $\mathrm{Gr}\, F(t, .)$ is closed for each $t \in R$, and F is $\mathcal{L} \times \mathcal{B}^n$ measurable.

(H3) There exist $\beta \geq 0$ and $k_F \in L^1$ such that, for a.e. $t \in [\bar{S}, \bar{T}]$,

$$F(t, x') \cap (\dot{\bar{x}}(t) + NB) \subset F(t, x) + (k_F(t) + \beta N)|x' - x|B$$

for all $N \geq 0$, $x', x \in \bar{x}(t) + \delta B$.

(H4) h is upper semicontinuous and there exists a constant k_h such that, for a.e. $t \in [\bar{S}, \bar{T}]$,

$$|h(t, x) - h(t, x')| \leq k_h |x - x'|$$

for all $x, x' \in \bar{x}(t) + \delta B$.

(H5) There exist $c_0 \geq 0$, $k_0 \geq 0$, and $\delta_0 > 0$ such that, for a.e. $t \in [\bar{S} - \delta_0, \bar{S} + \delta_0]$,

$$\begin{cases} F(t, x) \in c_0 B & \text{for all } x \in \bar{x}(\bar{S}) + \delta B \\ F(t, x') \subset F(t, x'') + k_0 |x' - x''|B & \text{for all } x', x'' \in \bar{x}(\bar{S}) + \delta B, \end{cases}$$

and

$$h(\bar{S}, \bar{x}(\bar{S})) < 0.$$

(H6) There exist $c_1 \geq 0$, $k_1 \geq 0$, and $\delta_1 > 0$ such that, for a.e. $t \in [\bar{T} - \delta_1, \bar{T} + \delta_1]$,

$$\begin{cases} F(t, x) \in c_1 B & \text{for all } x \in \bar{x}(\bar{T}) + \delta B \\ F(t, x') \subset F(t, x'') + k_1 |x' - x''|B & \text{for all } x', x'' \in \bar{x}(\bar{T}) + \delta B, \end{cases}$$

and

$$h(\bar{T}, \bar{x}(\bar{T})) < 0.$$

Then there exist an absolutely continuous arc p, real numbers $\lambda \geq 0, \xi, \eta$, a measure $\mu \in C^{\oplus}(S, T)$, and a μ-integrable function γ such that

(i) $\lambda + \|p\|_{L^\infty} + \|\mu\|_{T.V.} = 1$;

(ii) $\dot{p}(t) \in \mathrm{co}\{\zeta : (\zeta, p(t) + \int_{[\bar{S}, t)} \gamma(s)\mu(ds)) \in N_{\mathrm{Gr} F(t, \cdot)}(\bar{x}(t), \dot{\bar{x}}(t))\}$ a.e.;

(iii) $(p(t) + \int_{[\bar{S}, t)} \gamma_1(s)\mu(ds)) \cdot \dot{\bar{x}}(t)$

$\geq (p(t) + \int_{[\bar{S}, t)} \gamma_1(s)\mu(ds)) \cdot v$ *for all $v \in F(t, \bar{x}(t))$ a.e. $t \in [\bar{S}, \bar{T}]$;*

(iv) $(-\xi, p(\bar{S}), \eta, -(p(\bar{T}) + \int_{[\bar{S},\bar{T}]} \gamma(s)\mu(ds)))$,

$$\in \lambda \partial g(\bar{S}, \bar{x}(\bar{S}), \bar{T}, \bar{x}(\bar{T})) + N_C(\bar{S}, \bar{x}(\bar{S}), \bar{T}, \bar{x}(\bar{T}));$$

(v) $\gamma(t) \in \partial_x^> h(t, \bar{x}(t))$ μ a.e.;

(vi) $\xi \in \operatorname{ess}_{t \to \bar{S}} H(t, \bar{x}(\bar{S}), p(\bar{S}))$;

(vii) $\eta \in \operatorname{ess}_{t \to \bar{T}} H(t, \bar{x}(\bar{T}), p(\bar{T}) + \int_{[\bar{S},\bar{T}]} \gamma(s)\mu(ds))$.

If the initial time is fixed (i.e., $C = \{(\bar{S}, x_0, T, x_1) : (x_0, T, x_1) \in \tilde{C}\}$ for some set $\tilde{C}$) then the above assertions (except (vi)) remain true when (H5) is dropped from the hypotheses. If the final time is fixed then the assertions (except (vii)) remain true when (H6) is dropped from the hypotheses.
   Now suppose that

$$F(t, x) \text{ is convex for all } (t, x) \in R \times R^n$$

and that (H3) has been replaced by the weaker hypothesis

(H3)' there exists $k_F \in L^1$, $\bar{K} > 0$, and $\epsilon > 0$ such that either

(a) $F(t, x') \cap (\dot{\bar{x}}(t) + \epsilon k_F(t)B) \subset F(t, x) + k(t)|x - x'|B$
   for all $x, x' \in \bar{x}(t) + \epsilon B$, a.e. $t \in [\bar{S}, \bar{T}]$,

or

(b) $\left\{ \begin{array}{l} F(t, x') \cap (\dot{\bar{x}}(t) + \epsilon B) \subset F(t, x) + k_F(t)|x - x'|B, \\ \inf\{|v - \dot{\bar{x}}(t)| : v \in F(t, x)\} \leq \bar{K}|x - \bar{x}(t)|, \end{array} \right.$
   for all $x, x' \in \bar{x}(t) + \epsilon B$, a.e. $t \in [\bar{S}, \bar{T}]$.

Then all the above assertions remain valid. Furthermore, Condition (ii) implies

(ii)' $-\dot{p}(t) \in \operatorname{co}\{\xi : (\xi, \dot{\bar{x}}(t)) \in \partial H(t, \bar{x}(t), p(t) + \int_{[S,t)} \gamma(s)\mu(ds))\}$ *a.e.*

Remark

The necessary conditions of Theorem 10.5.1 covering problems with Lipschitz time-dependent data are valid when the state constraint is active at the optimal end-times. It comes then as a surprise that, on passing to problems with measurably time-dependent data, we must invoke additional hypotheses (H5) and (H6), which include the requirement that the state constraints are *inactive* at the optimal end-times. This is not due to shortcomings of our analysis. Indeed a counter example in [127] demonstrates that the assertions of the theorem above are false in general if the state constraints are active at the optimal end-times. Modifications to Theorem 10.5.2, to allow for state constraints active at the optimal end-times, are covered in [127].

The proof of Theorem 10.5.2 is along similar lines to that of its state constraint-free counterparts Theorems 8.4.1 and 8.4.2. The underlying idea, we recall, was to use fixed time necessary conditions to derive necessary conditions for a simple free right end-time problem, which were used, in turn, to derive necessary conditions for (FT) in full generality. Applying this proof technique to state constrained problems, we make use of the fixed time necessary conditions of Section 10.4 in the first step. Details are supplied in [145].

10.6 Nondegenerate Necessary Conditions

We have seen in the preceding two sections how necessary conditions for state constraint-free problems can be adapted to a state constrained setting, by incorporating extra terms of the form $p + \int \gamma \mu(dt)$ ("multiplier terms" associated with the state constraint).

These necessary conditions provide useful information about minimizers in many cases. But there are cases of interest where they are unsatisfactory. In this section we elaborate on this point, and show how the necessary conditions can be supplemented to broaden their applicability.

Consider, once again, the free end-time optimal control problem of Section 10.5.

$$(FT) \begin{cases} \text{Minimize } g(S, x(S), T, x(T)) \\ \text{over intervals } [S,T] \text{ and arcs } x \in W^{1,1}([S,T], R^n) \text{ satisfying} \\ \dot{x}(t) \in F(t, x(t)) \quad \text{a.e.,} \\ (S, x(S), T, x(T)) \in C, \\ h(t, x(t)) \leq 0 \quad \text{for all } t \in [S,T], \end{cases}$$

in which $g : R \times R^n \times R \times R^n \to R$ and $h : R \times R^n \to R$ are given functions, $F : R \times R^n \leadsto R^n$ is a given multifunction, and $C \subset R \times R^n \times R \times R^n$ is a given closed set.

Subsequent analysis addresses problem (FT) in its full generality. But for clarity of exposition, we temporarily restrict attention to a special case in which $F(t, .)$ and $h(t, x)$ are independent of t. (Accordingly, we suppress t in the notation, writing $F(x)$ for $F(t, x)$ and $h(x)$ for $h(t, x)$ and $H(x, p)$ for the Hamiltonian.) We also assume that h is a continuously differentiable function, the underlying time interval and the left endpoint are fixed; i.e.,

$$C = \{\bar{S}\} \times \{x_0\} \times \{\bar{T}\} \times C_1,$$

for some $[\bar{S}, \bar{T}] \subset R$, $x_0 \in R^n$, and $C_1 \subset R^n$, and g depends only on the right endpoint of the state trajectory.

Take a minimizer $\bar{x}$. From Section 10.5 we already have at our disposal necessary conditions of optimality. It is convenient to focus on the following set of conditions, centered on the Hamiltonian Inclusion, which is a coarser version of the necessary conditions of Theorem 10.5.1. Under appropriate hypotheses, there exists an absolutely continuous arc p, $\lambda \geq 0$, a measure $\mu \in C^{\oplus}(\bar{S}, \bar{T})$, and a constant r such that

(i) $(p, \lambda, \eta) \neq 0$;

(ii) $(-\dot{p}(t), \dot{\bar{x}}(t)) \in \text{co } \partial H(\bar{x}(t), p(t) + \int_{[\bar{S},t]} \nabla h(\bar{x}(s))\mu(ds))$ a.e.;

(iii) $-p(\bar{T}) - \int_{[\bar{S},\bar{T}]} \nabla h(\bar{x}(s))\mu(ds) \in \lambda \partial g(\bar{x}(\bar{T})) + N_{C_1}(\bar{x}(\bar{T}))$;

(iv) $\text{supp}\{\mu\} \subset \{s \in [\bar{S}, \bar{T}] : h(\bar{x}(s)) = 0\}$;

(v) $H(\bar{x}(t), p(t) + \int_{[\bar{S},t)} \nabla h(\bar{x}(s))\mu(ds)) = r$ for all $t \in (\bar{S}, \bar{T})$.

Notice that (v) is an "everywhere" condition. It easily follows from the earlier a.e. condition, under the hypotheses we impose presently, namely, (H2) of Theorem 10.6.1. A case when the deficiencies of these necessary conditions are particularly in evidence is that when the initial state is fixed and lies in the boundary of the state constraint region. To explore this phenomenon, we suppose that

$$h(x_0) = 0. \tag{10.12}$$

Here, we find that conditions (i) to (v) are satisfied (for some p, λ, and μ) when $\bar{x}$ is *any* arc satisfying the constraints of (FT). A possible choice of multipliers is

$$(p \equiv -\nabla h(\bar{S}), \ \mu = \delta_{\{\bar{S}\}}, \ \lambda = 0) \tag{10.13}$$

($\delta_{\{\bar{S}\}}$ denotes the unit measure concentrated at $\{\bar{S}\}$). Provided $\nabla h(\bar{S}) \neq 0$, these multipliers are nonzero. Condition (iv) is satisfied, by (10.12). The remaining conditions (i) to (iii) and (v) are satisfied since

$$\int_{(\bar{S},t]} \nabla h(\bar{x}(s))\mu(ds) = 0 \quad \text{for } t \in (S, T].$$

Degeneracy of the "standard" conditions (i) to (v) in this case is associated with the fact that they hold not just for $\bar{x}$ a minimizer for (FT), but for $\bar{x}$ *any* feasible arc $\bar{x}$ with the following property: there exist no neighboring feasible arcs x such that the following conditions

(a) $g(x(T)) < g(\bar{x}(T))$,

(b) $\max_{t \in [S,T]} h(t, \bar{x}(t)) < 0$,

are simultaneously satisfied. This is easy to show.

The reason why the necessary conditions are degenerate in case (10.12) is that Condition (b) is satisfied by *every* feasible arc $\bar{x}$. Indeed, for any such arc,

$$\max_{t \in [S,T]} h(t, x(t)) \geq h(x_0) = 0.$$

The fact that the necessary conditions (i) to (iv) are automatically satisfied by feasible arcs renders them useless as necessary conditions of optimality in the case (10.12).

How should we deal with the degeneracy phenomenon? Extra necessary conditions are clearly required, which (in the case (10.12)) eliminate the uninteresting multipliers (10.13).

There are now a number of ways to do this. We focus on an extra condition to eliminate degeneracy, due to Arutyunov, Aseev, and Blagodat-Skikh. In the fixed endpoint, autonomous case now under consideration, this asserts that the constancy of the Hamiltonian condition (v) on the open interval (S, T) extends to the endpoints; i.e., condition (v) can be supplemented by

$$H(\bar{x}(\bar{S}), p(\bar{S})) = r \qquad (10.14)$$

and

$$H(\bar{x}(\bar{T}), p(\bar{T})) + \int_{[\bar{S}, \bar{T}]} \nabla h(\bar{x})(s) \mu(ds)) = r \,,$$

where r is the constant of Condition (v).

The significance of these conditions becomes evident when we impose the constraint qualification

$$\nabla h(x_0) \cdot v < 0 \quad \text{for some } v \in F(x_0). \qquad (10.15)$$

This is a condition on the data, requiring the existence of an admissible velocity driving the state into the interior of the state constraint set at x_0.

The constancy of the Hamiltonian condition, strengthened in this way, eliminates the degenerate multipliers (10.13). Indeed, for this choice of multipliers, $p(t) + \int_{[\bar{S}, t)} \nabla h(x(t)) \mu(dt) \equiv 0$ for $t \in (\bar{S}, \bar{T}]$, so $r = 0$ by (iv). But then, in view of (10.14) and (10.15), we arrive at the contradiction:

$$0 = H(\bar{x}(\bar{S}), p(\bar{S})) = - \min_{v \in F(x_0)} \nabla h(x_0) \cdot v > 0.$$

The following theorem is the end result of elaborating these ideas, to allow for time-dependent data and free end-times.

Theorem 10.6.1 Let $([\bar{S}, \bar{T}], \bar{x})$ be a strong local minimizer for (FT). Assume that, for some $\epsilon > 0$,

(H1) g is Lipschitz continuous on a neighborhood of $(\bar{S}, \bar{x}(\bar{S}), \bar{T}, \bar{x}(\bar{T}))$ and C is a closed set;

(H2) F takes as values nonempty, compact, convex sets, and there exist $K_f > 0$, $c_f > 0$ such that
$$\begin{cases} F(t', x') \subset F(t'', x'') + K_f(|t' - t''| + |x' - x''|)B, \\ F(t', x') \subset c_f B, \end{cases}$$
for all $(t', x'), (t'', x'') \in (t, \bar{x}(t)) + \epsilon B$, a.e. $t \in [\bar{S}, \bar{T}]$;

(H3) h is Lipschitz continuous on a neighborhood of $\operatorname{Gr}\bar{x}$.

Then there exist absolutely continuous arcs r and p, $\lambda \geq 0$, a measure $\mu \in C^{\oplus}(S,T)$, and a Borel measurable function $\gamma = (\gamma_0, \gamma_1) : [\bar{S}, \bar{T}] \to R^{1+n}$ such that

(i) $\lambda + \|p\|_{L^\infty} + \|\mu\|_{T.V.} = 1;$

(ii) $\gamma(s) \in \partial^> h(s, \bar{x}(s))$ μ a.e.;

(iii) $(\dot{r}(t), -\dot{p}(t), \dot{\bar{x}}(t)) \in \operatorname{co} \partial H(t, \bar{x}(t), p(t) + \int_{[\bar{S},t)} \gamma_1(s)\mu(ds))$ a.e.;

(iv) $(-r(\bar{S}), p(\bar{S}), r(\bar{T}) - \int_{[\bar{S},\bar{T}]} \gamma_0(s)\mu(ds), -p(\bar{T}) - \int_{[\bar{S},\bar{T}]} \gamma_1(s)\mu(ds))$
$\in \lambda \partial g(\bar{S}, \bar{x}(\bar{S}), \bar{T}, \bar{x}(\bar{T})) + N_{\tilde{C}}(\bar{S}, \bar{x}(\bar{S}), \bar{T}, \bar{x}(\bar{T}));$

(v) $r(\bar{S}) = H(\bar{S}, \bar{x}(\bar{S}), p(\bar{S}));$

(vi) $r(t) = \int_{[\bar{S},t)} \gamma_0(s)\mu(ds) + H(t, \bar{x}(t), p(t) + \int_{[\bar{S},t)} \gamma_1(s)\mu(ds))$
for all $t \in (\bar{S}, \bar{T});$

(vii) $r(\bar{T}) = \int_{[\bar{S},\bar{T}]} \gamma_0(s)\mu(ds) + H(t, \bar{x}(t), p(t) + \int_{[\bar{S},\bar{T}]} \gamma_1(s)\mu(ds)).$

In the above conditions, $\tilde{C}$ is the "effective" endpoint constraint set

$$\tilde{C} := C \cap \{(t_0, x_0, t_1, x_1) : h(t_0, x_0) \leq 0, h(t_1, x_1) \leq 0\}.$$

We emphasize that the new information supplied by this theorem is the existence of an absolutely continuous function $r : [\bar{S}, \bar{T}] \to R$ satisfying not only (vi), but Conditions (v) and (vii) also. For autonomous problems, it tells us that the constancy of the Hamiltonian on the open interval $(\bar{S}, \bar{T})$ extends to the endpoints.

The following corollary makes explicit the role of the extra necessary conditions in excluding trivial multipliers. Notice that the trivial multipliers (10.13),

$$(p \equiv -\nabla h(\bar{S}), \ \mu = \delta_{\{\bar{S}\}}, \ \lambda = 0)$$

of our earlier discussion violate the condition

$$\lambda + \int_{(\bar{S},\bar{T})} \mu(ds) + \|t \to p(t) + \int_{[\bar{S},t)} \nabla_x h(\bar{x}(s))\mu(ds)\|_{L^\infty} \neq 0$$

because μ is concentrated on the left end-time and

$$\|t \to (p(t) + \int_{[\bar{S},t)} \nabla_x h(\bar{x}(s))\mu(ds))\|_{L^\infty} = 0 \quad \text{a.e.}$$

The corollary, which asserts the existence of multipliers satisfying a time-varying nonsmooth version of this condition under a suitable constraint qualification, thereby guarantees the existence of multipliers distinct from the above trivial ones. (Notice that the multiplier r is zero for automous problems.)

Corollary 10.6.2 *Let* $([\bar{S}, \bar{T}], \bar{x})$ *be a strong local minimizer for (FT). Assume Hypotheses (H1) to (H3) of Theorem 10.6.1 are satisfied. Assume also that the following constraint qualification is satisfied.*

$$\gamma_0' + \min_{v \in F(\bar{S}, \bar{x}(\bar{S}))} \gamma_1' \cdot v < 0 \quad and \quad \gamma_0'' + \min_{v \in F(\bar{T}, \bar{x}(\bar{T}))} \gamma_1'' \cdot v < 0$$

for all $(\gamma_0', \gamma_1'), (\gamma_0'', \gamma_1'') \in \partial^> h(\bar{S}, \bar{x}(\bar{S}))$.

Then there exist absolutely continuous arcs r *and* p, $\lambda \geq 0$, *a measure* $\mu \in C^{\oplus}(S, T)$ *and a Borel measurable function* $\gamma = (\gamma_0, \gamma_1) : [\bar{S}, \bar{T}] \to R^{1+n}$ *such that*

(ii) $\gamma(s) \in \partial^> h(s, \bar{x}(s))$ μ *a.e.,*

(iii) $(\dot{r}(t), -\dot{p}(t)) \in \operatorname{co} \partial H(t, \bar{x}(t), p(t) + \int_{[\bar{S}, t)} \gamma_1(s)\mu(ds))$ *a.e.;*

(iv) $(-r(\bar{S}), p(\bar{S}), r(\bar{T}) - \int_{[\bar{S}, \bar{T}]} \gamma_0(s)\mu(ds), -p(\bar{T}) - \int_{[\bar{S}, \bar{T}]} \gamma_1(s)\mu(ds))$
$$\in \lambda \partial g(\bar{S}, \bar{x}(\bar{S}), \bar{T}, \bar{x}(\bar{T})) + N_{\tilde{C}}(\bar{S}, \bar{x}(\bar{S}), \bar{T}, \bar{x}(\bar{T})),$$

(v) $r(\bar{S}) = H(\bar{S}, \bar{x}(\bar{S}), p(\bar{S}))$;

(vi) $r(t) = \int_{[\bar{S}, t)} \gamma_0(s)\mu(ds) + H(t, \bar{x}(t), p(t) + \int_{[\bar{S}, t)} \gamma_1(s)\mu(ds))$,
    for all $t \in (\bar{S}, \bar{T})$;

(vii) $r(\bar{T}) = \int_{[\bar{S}, \bar{T}]} \gamma_0(s)\mu(ds) + H(t, \bar{x}(t), p(t) + \int_{[\bar{S}, \bar{T}]} \gamma_1(s)\mu(ds))$;

and also the strengthened nondegeneracy condition

(i)' $\lambda + \int_{(\bar{S}, \bar{T})} \mu(ds) + \|t \to p(t) + \int_{[\bar{S}, t)} \gamma_1(s)\mu(ds)\|_{L^\infty}$
$$+ \|t \to r(t) - \int_{[\bar{S}, t)} \gamma_0(s)\mu(ds)\|_{L^\infty} \neq 0.$$

Proof of Corollary 10.6.2. We deduce from Theorem 10.6.1 that there exist multipliers (λ, p, r, μ), not all zero, and a selector (γ_0, γ_1) satisfying conditions (ii) to (vii). We must show that (under the constraint qualification) condition (i)' is also satisfied. Suppose, to the contrary, (i)' fails to hold. Then $\lambda = 0$, $\mu = \alpha_0 \delta_{\bar{S}} + \alpha_1 \delta_{\bar{T}}$, and

$$p \equiv \alpha_0 \gamma_1', \qquad r \equiv \alpha_0 \gamma_0',$$

for some $\alpha_0, \alpha_1 \geq 0$ and some $(\gamma_0', \gamma_1') \in \partial^> h(\bar{S}, \bar{x}(\bar{S}))$. α_0 and α_1 cannot both be zero, for otherwise the multipliers are all zero. It follows then from Condition (v) that

$$\alpha_0 \gamma_0' = r(\bar{S}) = H(\bar{S}, \bar{x}(\bar{S}), p(\bar{S})) = -\alpha_0 \min_{v \in F(\bar{S}, \bar{x}(\bar{S}))} \gamma_1' \cdot v.$$

If $\alpha_0 > 0$, we conclude that

$$0 = \gamma_0' + \min_{v \in F(\bar{S}, \bar{x}(\bar{S}))} \gamma_1' \cdot v.$$

The right side of this equation is strictly negative, in view of the constraint qualification. We must therefore have $\alpha_0 = 0$. But then $\alpha_1 > 0$. Reasoning as above, we deduce from (vi) and (vii) that

$$0 = \gamma_0'' + \min_{v \in F(\bar{T}, \bar{x}(\bar{T}))} \gamma_1'' \cdot v,$$

for some $(\gamma_0'', \gamma_1'') \in \partial^> h(\bar{T}, \bar{x}(\bar{T}))$. But this contradicts the constraint qualification. Assertion (i)$'$ is confirmed. $\square$

Proof of Theorem 10.6.1. Let ϵ' be such that $([\bar{S}, \bar{T}], \bar{x})$ is a minimizer for (FT), with respect to arcs x satisfying

$$|S - \bar{S}| + |T - \bar{T}| + \|x^e - \bar{x}^e\|_{L^\infty} \leq \epsilon'.$$

($\bar{y}^e$, as usual, denotes the extension of an arc $\bar{y} : [S,T] \to R^n$ to all of $(-\infty, +\infty)$, by constant extrapolation.)

Take $K_i \uparrow \infty$ and for each i consider the optimal control problem:

$$(P_i) \begin{cases} \text{Minimize } J_i([S,T], x) \\ \text{over processes } ([S,T], x) \text{ satisfying} \\ \dot{x}(t) \in F(t, x(t)) \quad \text{a.e.,} \\ (S, x(S), T, x(T)) \in \tilde{C}, \\ |S - \bar{S}| + |T - \bar{T}| + \|x^e - \bar{x}^e\|_{L^\infty} \leq \epsilon'. \end{cases}$$

Here,

$$J_i([S,T], x) := g(S, x(S), T, x(T)) + K_i \int_S^T h^+(t, x(t))dt$$
$$+ \int_S^T |x(t) - \bar{x}^e(t)|^2 dt + |S - \bar{S}|^2 + |T - \bar{T}|^2,$$

in which

$$h^+(t, x) := \max\{h(t, x), 0\}.$$

Standard compactness arguments can be used to demonstrate that, for each i, (P_i) has a minimizer $([S_i, T_i], x_i)$ and that, along a subsequence, $S_i \to S$, $T_i \to T$, and $x_i^e(t) \to x^e(t)$ uniformly for some arc $x \in W^{1,1}$ satisfying the constraints of (FT). For each i,

$$g(\bar{S}, \bar{x}(\bar{S}), \bar{T}, \bar{x}(\bar{T})) \geq g(S_i, x_i(S_i), T_i, x_i(T_i)) + K_i \int_{S_i}^{T_i} h^+(t, x_i(t))dt$$

$$+ \int_{S_i}^{T_i} |x_i^e(t) - \bar{x}^e(t)|^2 dt + |S_i - \bar{S}|^2 + |T_i - \bar{T}|^2, \tag{10.16}$$

since $([S_i, T_i], x_i)$ is a minimizer for (P_i).

Dividing across this inequality by K_i and passing to the limit with the help of the Dominated Convergence Theorem, we deduce that

$$\int_S^T h^+(t, x(t))dt = \lim_{i \to \infty} \int_{S_i}^{T_i} h^+(t, x_i(t))dt = 0.$$

Since h is continuous, this implies $h(t, x(t)) \leq 0$ for all $t \in [S, T]$. It follows then that

$$g(S, x(S), T, x(T)) \geq g(\bar{S}, \bar{x}(\bar{S}), \bar{T}, \bar{x}(\bar{T})).$$

By (10.16), $\{K_i \int_{S_i}^{T_i} h^+(t, x_i(t))dt\}$ is a bounded sequence. So we may arrange, by subsequence extraction, that it has a limit. We conclude from (10.16) that

$$\lim_i K_i \int_{S_i}^{T_i} h^+(t, x_i(t))dt + \int_S^T |x_i^e(t) - \bar{x}^e(t)|^2 dt + |S_i - \bar{S}|^2 + |T_i - \bar{T}|^2$$
$$\leq g(\bar{S}, \bar{x}(\bar{S}), \bar{T}, \bar{x}(\bar{T})) - g(S, x(S), T, x(T)) \leq 0.$$

Since the terms on the left are positive and the x_i^es are bounded and uniformly Lipschitz continuous, we see that

$$x_i^e(t) \to \bar{x}^e(t) \text{ uniformly,} \quad S_i \to \bar{S}, \quad T_i \to \bar{T}$$

and

$$\int_{S_i}^{T_i} K_i h^+(t, x_i(t))dt \to 0.$$

It follows now from Egorov's Theorem that, after a further subsequence extraction,

$$K_i h^+(t, x_i^e(t)) \to 0 \quad \text{for a.e. } t \in [\bar{S}, \bar{T}]. \tag{10.17}$$

Fix i. Evidently the arc

$$([S_i, T_i], x_i, y_i(t) \equiv K_i \int_{S_i}^t h^+(s, x_i(s))ds + \int_{S_i}^t |x_i(s) - \bar{x}^e(s)|^2 ds)$$

is a strong local minimizer for

$$\begin{cases} \text{Minimize } g(S, x(S), T, x(T)) + y(T) + |S - \bar{S}|^2 + |T - \bar{T}|^2 \\ \text{over arcs } ([S, T], (x, y)) \text{ satisfying} \\ (\dot{x}(t), \dot{y}(t)) \in E(t, x(t)) \quad \text{a.e.,} \\ (S, x(S), y(S), T, x(T), y(T)) \in D, \end{cases}$$

where

$$E(t, x) := \{(v, w) : v \in F(t, x), w \geq K_i h^+(t, x) + |x - \bar{x}^e(t)|^2\}$$

and

$$D := \{(t_0, x_0, y_0, t_1, x_1, y_1) : (t_0, x_0, t_1, x_1) \in \tilde{C}, y_0 = 0\}.$$

For i sufficiently large, we may apply the necessary conditions of Theorem 8.2.1 to this problem. We thereby conclude that there exist arcs $r_i' \in W^{1,1}$ and $p_i' \in W^{1,1}$ and $\lambda_i \geq 0$ such that

$$\lambda_i + \|(r_i', p_i')\|_{L^\infty} = 1,$$
$$(\dot{r}_i'(t), -\dot{p}_i'(t), \dot{x}_i(t)) \in \text{co}\,\partial H(t, x_i(t), p_i'(t))$$
$$- \lambda_i K_i \partial h^+(t, x_i(t)) \times \{0\} + \delta_i' B \times \{0\}, \quad (10.18)$$
$$(-r_i'(S_i), p_i'(S_i), r_i'(T_i), -p_i'(T_i))$$
$$\in \lambda_i \partial g(S_i, x_i(S_i), T_i, x_i(T_i)) + N_{\tilde{C}}(S_i, x_i(S_i), T_i, x_i(T_i)) + \delta_i'' B,$$
$$r_i'(t) = H(t, x_i(t), p_i'(t))$$
$$- \lambda_i[K_i h^+(t, x_i(t)) + |x_i(t) - \bar{x}^e(t)|^2] \quad \text{a.e.} \quad (10.19)$$

for some sequences $\delta_i' \downarrow 0$ and $\delta_i'' \downarrow 0$. (δ_i' is a uniform bound on the Lipschitz constant for $(t', x') \to \lambda_i |x - \bar{x}^e(t)|^2$ near $(t, x^i(t))$ as t ranges over $[S_i, T_i]$ and $\delta_i'' = 2\lambda_i(|S_i - \bar{S}| + |T_i - \bar{T}|)$). But

$$\partial h^+(t, x) \subset \{(\alpha\tau, \alpha\xi) : (\tau, \xi) \in \partial^> h(t, x), \alpha \in [0, 1]\}$$

on a neighborhood of $\text{Gr}\,\bar{x}$. We deduce from (10.18), with the help of a measurable selection theorem, that

$$(\dot{r}_i'(t), -\dot{p}_i'(t), \dot{x}_i(t)) \in$$
$$\text{co}\partial H(t, x_i(t), p_i'(t)) - \lambda_i K_i \gamma_i(t) m_i(t) \times \{0\} + \delta_i' B \times \{0\} \text{ a.e.}$$

for some Borel measurable functions $\gamma_i = (\gamma_i^0, \gamma_i^1) : [S_i, T_i] \to R \times R^n$ and $m_i : [S_i, T_i] \to [0, 1]$ such that $\gamma_i(t) \in \partial^> h(t, x_i(t))$ for almost all $t \in \{s \in [S_i, T_i] : m_i(s) > 0\}$.

Now define

$$\mu_i(dt) := \lambda_i K_i m_i(t) dt,$$

$$r_i(t) = \begin{cases} r_i'(t) + \int_{[S_i, t)} \gamma_i^0(s) \mu_i(ds) & \text{for } t \in [S_i, T_i) \\ r_i'(T_i) + \int_{[S_i, T_i]} \gamma_i^0(s) \mu_i(ds) & \text{for } t = T_i \end{cases}$$

$$p_i(t) = \begin{cases} p_i'(t) - \int_{[S_i, t)} \gamma_i^1(s) \mu_i(ds) & \text{for } t \in [S_i, T_i) \\ p_i'(T_i) - \int_{[S_i, T_i]} \gamma_i^1(s) \mu_i(ds) & \text{for } t = T_i . \end{cases}$$

We have

$$(\dot{r}_i(t), -\dot{p}_i(t), \dot{x}_i(t)) \qquad\qquad\qquad\qquad (10.20)$$
$$\in \text{co}\,\partial H(t, x_i(t), p_i(t) + \int_{[S_i, t)} \gamma_i^1(s)\mu_i(ds)) + \delta_i' B \times \{0\} \quad \text{a.e.}$$

and

$$(-r_i(S_i), p_i(S_i), r_i(T_i) - \int_{[S_i,T_i]} \gamma_i^0 \mu_i(ds), -p_i(T_i) - \int_{[S_i,T_i]} \gamma_i^1(s)\mu_i(ds))$$

$$\in \lambda_i \partial g + N_{\tilde{C}} + \delta_i'' B. \tag{10.21}$$

(In the above relationships, ∂g and $N_{\tilde{C}}$ are evaluated at $(S_i, x_i(S_i), T_i, x_i(T_i))$.)
Recall that, by definition of $\tilde{C}$,

$$(t_0, x_0, t_1, x_1) \in \tilde{C} \quad \text{implies} \quad h^+(t_0, x_0) = 0 \text{ and } h^+(t_1, x_1) = 0,$$

for (t_0, x_0, t_1, x_1) close to $(\bar{S}, \bar{x}(\bar{S}), \bar{T}, \bar{x}(\bar{T}))$.
It follows that, for sufficiently large i, (10.19) implies

$$r_i(S_i) = H(S_i, x_i(S_i), p_i(S_i)) - \lambda_i |x_i(S_i) - \bar{x}^e(S_i)|^2, \tag{10.22}$$

$$r_i(t) = \int_{[S_i,t)} \gamma_i^0(s)\mu_i(ds) + H(t, x_i(t), p_i(t) + \int_{[S_i,t)} \gamma_i^1(s)\mu_i(ds)),$$

$$-\lambda_i[K_i h^+(t, x_i(t)) + |x_i(t) - \bar{x}^e(t)|^2] \text{ for all } t \in (S_i, T_i) \tag{10.23}$$

$$r_i(T_i) = \int_{[S_i,T_i]} \gamma_i^0(s)\mu_i(ds) + H(T_i, x_i(T_i), p_i(T_i) + \int_{[S_i,T_i]} \gamma_i^1(s)\mu_i(ds))$$

$$-\lambda_i |x_i(T_i) - \bar{x}^e(T_i)|^2. \tag{10.24}$$

Since

$$\lambda_i + \max_{t \in [S_i,T_i]} |(-r_i(t), p_i(t) + \int_{[S_i,t)} \gamma_i(s)\mu_i(ds))| \neq 0,$$

we can scale (p_i, λ_i, μ_i) so that

$$\lambda_i + \|(r_i, p_i)\|_{L^\infty} + \|\mu_i\|_{T.V.} = 1. \tag{10.25}$$

The p_is and r_is are sequences of functions that are uniformly bounded
and have a common Lipschitz constant. The functions γ_i are uniformly
bounded on $\text{supp}\{\mu_i\}$ and $\{\mu_i\}$ is uniformly bounded in total variation.
We may therefore arrange, by extracting subsequences, that

$$p_i^e \to p \text{ uniformly}, \ \lambda_i \to \lambda \quad \mu_i \to \mu \text{ weakly}^*, \text{ and } \gamma_i d\mu_i \to \gamma d\mu \text{ weakly}^*,$$

for some arc $p \in W^{1,1}$, $\lambda \geq 0$, some measure μ, and some Borel measurable
function $\gamma = (\gamma_0, \gamma_1)$. A by now familiar convergence analysis permits us
to pass to the limit in (10.20) through (10.25). All the assertions of the
theorem follow. Notice, in particular, that (10.22) and (10.24) supply, in
the limit, the extra "nondegenerate" optimality conditions (v) and (vii) of
the theorem statement. $\square$

10.7   Notes for Chapter 10

A nonsmooth version of the Hamiltonian Inclusion for optimal control problems with unilateral state constraints and dynamics modeled by a differential inclusion with convex-valued right side (convex velocity sets) was obtained by Clarke [38]. The validity of the Extended Euler–Lagrange Condition and the Extended Hamilton Condition, also for convex velocity sets, was established by Loewen and Rockafellar [96].

The Extended Euler–Lagrange Condition, for state constrained problems with unbounded, possibly nonconvex, velocity sets satisfying a one-sided Lipschitz continuity condition, was first derived by Vinter and Zheng [144]. The analysis in [144] is reproduced in this chapter. The key idea, to apply the smooth Maximum Principle to an approximating problem with state-free dynamics and pass to the limit, is that used in Chapter 7. But now, of course, the smooth Maximum Principle for state constrained problems is employed.

Free end-time necessary conditions for state constrained differential inclusion problems, with data measurable in time, were first derived by Clarke, Loewen, and Vinter [55]. Refinements, to allow for state constraints that are active at the optimal end-times, were investigated in [127].

The degeneracy phenomenon, namely, the fact that standard necessary conditions of optimality convey no useful information for certain state constrained optimal control problems of interest, was first noted in the Russian literature and has been the subject of subsequent extensive study. This topic is reviewed in [2]. The elegant, nondegenerate, necessary conditions of Section 10.6, expressed in terms of the boundary values of the Hamiltonian, were obtained by Arutyunev, Aseev, and Blagodat-Skikh [3], for problems with convex velocity sets and with data Lipschitz continuous with respect to the time variable. Refinements of these results, in particular necessary conditions, expressed in terms of the (partially convexified) Extended Hamiltonian Condition, are given in [4]. The nondegenerate necessary conditions of Ferreira and Vinter [64] allow measurable time dependence but place restrictions on the nature of the optimal state trajectories near the end-times, which are difficult to test in practice. Nondegenerate necessary conditions, in the form of the Extended Euler–Lagrange Condition, were recently derived by Rampazzo and Vinter [118], which do not presuppose any structural properties of minimizers and allow measurable time dependence, nonconvex velocity sets, and general endpoint constraints.

Chapter 11

Regularity of Minimizers

How often have I said to you that when you have eliminated the impossible, whatever remains, however improbable, must be the truth.

– Authur Conan Doyle, The Sign of Four

11.1 Introduction

In this chapter we seek information about regularity of minimizers. When do minimizing arcs have essentially bounded derivatives, higher-order derivatives, or other qualitative properties of interest in applications?

Minimizer regularity has a number of important implications. In the computation of minimizers, for example, the choice of efficient discretization schemes and numerical procedures depends on minimizer differentiability. Again, in Control Engineering applications, where an optimal control strategy is implemented digitally, the quality of the approximation of an ideal "continuous time" strategy depends on minimizer regularity. Minimizer regularity can also help us predict natural phenomena, governed by variational principles. In particular, studying the regularity of minimizers for variational problems arising in nonlinear elasticity, gives insights into mechanisms for material failure.

We focus here however on the fundamental role of regularity theory, discussed in Chapter 1, to provide a solid foundation for Tonelli's Direct Method. Recall that this involves the following steps.

1: Establish that the problem has a minimizer.

2: Derive necessary conditions for an arc to be a minimizer.

3: Search among "extremals" (arcs satisfying the necessary conditions) for an arc with minimum cost; this will be a minimizer.

The Direct Method fails when it is applied to any class of optimal control problems for which minimizers exist, but when necessary conditions of optimality are available only under hypotheses that the minimizers may possibly violate. An example of this phenomenon is given in Chapter 1. In this example, finding an arc that satisfies the necessary conditions yields a unique extremal. However this extremal is not a minimizer, because the minimizers are not located among the extremals.

Regularity analysis enters the picture because it helps us to identify classes of problems, for which all minimizers satisfy known necessary conditions of optimality. It thereby justifies searching for minimizers among extremals.

At the same time, we anticipate difficulties in investigating minimizer regularity, with a view to justifying the Direct Method. An obvious approach is to draw conclusions about minimizer regularity from the necessary conditions. Yet, to justify the Direct Method, we need to establish regularity of precisely those minimizers for which satisfaction of the necessary conditions is not guaranteed a priori.

Much of this chapter is devoted to deriving regularity properties of solutions to the Basic Problem in the Calculus of Variations, concerning the minimization of an integral functional (with finite Lagrangian) over absolutely continuous vector-valued arcs with fixed endpoints:

$$(BP) \begin{cases} \text{Minimize } \int_S^T L(t, x(t), \dot{x}(t)) dt \\ \text{over } x \in W^{1,1}([S,T]; R^n) \text{ satisfying} \\ (x(S), x(T)) = (x_0, x_1). \end{cases}$$

Here, $[S, T]$ is a fixed interval with $T > S$, $L : [S, T] \times R^n \times R^n \to R$ is a given function, and x_0, x_1 are given n-vectors.

No apology is made for limiting attention to fixed endpoint problems. If an arc $\bar{x}$ is a minimizer for any other kind of endpoint condition, then $\bar{x}$ is a minimizer also for the related fixed endpoint problem with $x_0 = \bar{x}(S)$, $x_1 = \bar{x}(T)$. So regularity properties for fixed endpoint problems imply the same regularity for all other kinds of boundary conditions.

On the other hand, the fact that the formulation (BP) covers only optimal control problems for which the velocity variable in unconstrained is a genuine shortcoming. We remedy it, to some extent, by studying some generalizations in the closing sections of the chapter, to allow for dynamic constraints.

Our starting point is the following well-known set of conditions (HE1) to (HE3) for the existence of minimizers to (BP). Because of their close affinity to conditions considered by Tonelli in his pioneering work on existence of minimizers, we refer to them as the *Tonelli Existence Hypotheses*.

Theorem 11.1.1 (Tonelli Existence Theorem) *Assume that the data for (BP) satisfy the following hypotheses.*

(E1) **(Convexity, etc.)** *L is bounded on bounded sets, $L(., x, v)$ is measurable for each (x, v), and $L(t, x, .)$ is convex for each (t, x).*

(E2) **(Uniform Local Lipschitz Continuity)** *For each $N > 0$, there exist $k_N > 0$ such that*

$$|L(t, x, u) - L(t, x', u')| \leq k_N |(x, v) - (x', v')|$$

for all $(x, v), (x', v') \in NB$ a.e. $t \in [S, T]$.

(E3) **(Coercivity)** *There exists $\alpha \geq 0$ and a convex function $\theta : [0, \infty) \to [0, \infty)$ satisfying $\lim_{r \to \infty} \theta(r)/r = +\infty$ such that*

$$L(t, x, v) \geq \theta(|v|) - \alpha|x| \quad \text{for all } (x, v) \in R^n \times R^n \text{ a.e. } t \in [S, T].$$

Then (BP) has a minimizer.

This is a special case of the existence theorem for the Generalized Bolza Problem proved in Chapter 2. (See Theorem 2.7.1.)

Chapter 7 provides the following necessary conditions of optimality for (BP). Here, a strong minimizer $\bar{x}$ has its customary meaning; namely, a minimizer over the the class of absolutely continuous arcs x satisfying the endpoint constraints and $||x - \bar{x}||_C \leq \epsilon$, for some $\epsilon > 0$.

Proposition 11.1.2 *Let $\bar{x}$ be a strong local minimizer for (BP). Assume that, in addition to (E1) and (E2), there exists $k(.) \in L^1$ and $\epsilon > 0$ such that*

$$|L(t, x, v) - L(t, x', u')| \leq k(t)|(x, v) - (x', v')| \tag{11.1}$$

for all $(x, v), (x', v') \in (\bar{x}(t), \dot{\bar{x}}(t)) + \epsilon B$, a.e.
Then, there exists an arc $p \in W^{1,1}([S, T]; R^n)$ such that

$$\dot{p}(t) \in co\{q : (q, p(t)) \in \partial L(t, \bar{x}(t), \dot{\bar{x}}(t))\} \quad a.e.$$

(where ∂L denotes the limiting subdifferential of $L(t, ., .)$) and

$$p(t) \cdot \bar{x}(t) - L(t, \bar{x}(t), \dot{\bar{x}}(t)) = \max_{v \in R^n}\{p(t) \cdot v - L(t, \bar{x}(t), v)\} \quad a.e. \tag{11.2}$$

Furthermore, if L is independent of t (write $L(x, v)$ in place of $L(t, x, v)$), then there exists a constant c such that

$$c = p(t) \cdot \dot{\bar{x}}(t) - L(\bar{x}(t), \dot{\bar{x}}(t)) \quad a.e.$$

Note that, since $L(t, x, .)$ is a convex function and the subdifferential for convex finite-valued functions on R^n coincides with the limiting subdifferential, (11.2) implies

$$p(t) \in \partial_v L(t, \bar{x}(t), \dot{\bar{x}}(t)) \quad a.e.$$

Proof. The arc $\left(\bar{x}, \bar{z}(t) = \int_T^t L(s, \bar{x}(s), \dot{\bar{x}}(s))ds\right)$ is a strong local minimizer for the optimal control problem

$$\begin{cases} \text{Minimize } z(T) \\ \text{over } (x, z) \in W^{1,1}([S, T]; R^{n+1}) \text{ satisfying} \\ (\dot{x}(t), \dot{z}(t)) \in F(t, x, z), \\ (x(S), z(S)) = (x_0, 0), \\ (x(T), z(T)) \in \{x_1\} \times R. \end{cases}$$

Here,
$$F(t, x, z) := \{(u, L(t, x, u)) : u \in \bar{u}(t) + B\}$$

The hypotheses hold under which the Extended Euler–Lagrange Condition (Theorem 7.5.1) is valid. By considering limits of proximal normal vectors we readily deduce that

$$(q, w, p, -r) \in N_{\mathrm{Gr}\, F(t,.)}(\bar{x}, \bar{z}, \dot{\bar{x}}, \dot{\bar{z}}) \quad \text{implies} \quad w = 0 \text{ and } (q, p) \in \partial(rL(t, \bar{x}, \dot{\bar{x}})).$$

It follows that there exist $p \in W^{1,1}([S, T]; R^n)$, $r \in R$, and $\lambda \geq 0$ such that

$$(p, \lambda) \neq 0,$$
$$\dot{p}(t) \in \mathrm{co}\,\{q : (q, p(t)) \in \partial_{x,v}\{rL(t, \bar{x}, \dot{\bar{x}})\}\} \quad \text{a.e.,}$$
$$r = \lambda,$$
$$p(t) = r\partial_v L(t, \bar{x}(t), \dot{\bar{x}}(t)) \quad \text{a.e.}$$

Theorem 7.5.1 also tells us that p can be chosen also to satisfy

$$c = p(t) \cdot \dot{\bar{x}}(t) - \lambda L(t, \bar{x}(t), \dot{\bar{x}}(t)) \quad \text{a.e.}$$

for some constant c, if L is independent of t.

Clearly, $\lambda = 0$ implies $p \equiv 0$. It follows that $\lambda \neq 0$. We can therefore arrange by scaling the multipliers that $\lambda = 1$. The assertions of the proposition follow. $\square$

Proposition 11.1.2 falls short of requirements, from the point of view of identifying a large class of problems, of type (BP), for which there exist minimizers and for which all minimizers satisfy known necessary conditions of optimality. This is because of the "extra" hypothesis (11.1) therein invoked, to justify the necessary conditions. (Note that Hypothesis (E2) does not imply (11.1) when $\dot{x}(.)$ fails to be essentially bounded.)

At the outset, we need to ask whether the presence of the troublesome Hypothesis (11.1) in Proposition 11.1.2 merely reflects a weakness in our analysis and can be dispensed with in situations in which the Tonelli Existence Hypotheses are satisfied, or whether it points to genuine restrictions on classes of problems for which the optimality conditions of Proposition 11.1.2 can be confirmed.

This question was resolved only recently by an example of Ball and Mizel, exposing the gap between the Tonelli Existence Hypotheses and hypotheses needed for the derivation of Euler–Lagrange type necessary conditions.

Example (Ball–Mizel)

$$\begin{cases} \text{Minimize } \int_0^1 \{r\dot{x}^2(t) + (x^3(t) - t^2)^2 \dot{x}^{14}(t)\}\, dt \text{ over} \\ x \in W^{1,1}([0, 1]; R) \text{ satisfying} \\ x(0) = 0, \quad x(1) = k. \end{cases}$$

Here, $r > 0$ and $k > 0$ are constants, linked by the relationship

$$r = (2k/3)^{12}(1 - k^3)(13k^3 - 7).$$

It can be shown that there exists $\epsilon > 0$ such that, for all $k \in (1 - \epsilon, 1)$, the arc

$$\bar{x}(t) := kt^{2/3} \tag{11.3}$$

is the unique minimizer for this problem. (This is by no means a straightforward undertaking!)

It is a simple matter to check by direct substitution that $\bar{x}$ satisfies, a.e., a pointwise version of the Euler Condition:

$$d/dt\, L_v(t, \bar{x}(t), \dot{\bar{x}}(t)) = L_x(t, \bar{x}(t), \dot{\bar{x}}(t)),$$

in which

$$L(t, x, v) = rv^2 + (x^3 - t^2)^2 v^{14}.$$

However this cannot be expressed as an Euler–Lagrange type condition in terms of an absolutely continuous adjoint arc p,

$$(\dot{p}(t), p(t)) = \nabla L((t, \bar{x}(t), \dot{\bar{x}}(t))),$$

because the only candidate for adjoint arc p, namely,

$$p(t) = -\alpha t^{-1/3}$$

(for some $\alpha \neq 0$ depending on k), is not an absolutely continuous function.

The train of thought behind this example, incidentally, is that the Lagrangian $(x^3 - t^2)^{2m}|\dot{x}|^{2k}$, for any integers $m > 0$ and $k > 0$, has a minimizer $x(t) = t^{2/3}$ with unbounded derivative. It fails however to exhibit "bad" behavior under the Tonelli Existence Hypotheses, because the coercivity condition (E3) is violated. To ensure satisfaction also of (E3) we now add a small quadratic coercive term $\epsilon|\dot{x}|^2$ to the Lagrangian. It then turns out that the values of m and k can be adjusted to ensure that $x(t) = kt^{3/2}$ is a pointwise solution to the Euler Equation, for some $k \in (0, 1)$.

The Ball–Mizel Example is of interest not only for confirming that the Tonelli Existence Hypotheses alone fail to ensure validity of standard necessary conditions such as the Euler–Lagrange Condition. It also helps us to predict the nature of minimizers under these hypotheses.

A salient feature of the minimizing arc (11.3) is that the minimizer is in some sense badly behaved only on some small subset of the underlying time interval $[0, 1]$, namely, $\{0\}$.

It is a remarkable fact that *all* minimizers to (BP) share this property. The key result of the chapter, the Generalized Tonelli Regularity Theorem, makes precise this assertion: it says that, merely under the Tonelli Existence Hypotheses, the bad behavior of each minimizer $\bar{x}$ is confined to a closed set of zero measure. In this context, $\bar{t}$ is defined to be a point of bad behavior for $\bar{x}$ if $\bar{x}$ fails to be Lipschitz continuous on any neighborhood of $\bar{t}$ in $[S, T]$.

We commented earlier on an inherent difficulty in establishing regularity properties of minimizers, namely, that we would like to exploit regularity implications of standard necessary conditions, but cannot do this in interesting cases when the hypotheses, under which these necessary conditions can be derived, are violated. This brings us to one of the most valuable aspects of the Generalized Tonelli Regularity Theorem. It permits us to derive a weakened "local" version of the Extended Euler–Lagrange Condition merely under the Tonelli Existence Hypotheses. We can combine information about minimizers inherent in this optimality condition with additional hypotheses on the data for (BP), to establish refined regularity properties for special classes of problems.

The role of the Generalized Tonelli Regularity Theorem to generate refined regularity theorems for special classes of problems is investigated at length in Section 12.4. A particularly striking result, first proved via the Generalized Tonelli Regularity Theorem, is Proposition 11.4.2 below.

> *Suppose that $L(t, x, v)$ is independent of t. Then, under the Tonelli Existence Hypotheses, all minimizers for (BP) are Lipschitz continuous.*

To paraphrase, existence of points of bad behavior is a phenomenon associated solely with nonautonomous problems.

A related topic to regularity analysis is investigation of the so-called *Lavrentiev Phenomenon*. This is the surprising fact that, for certain variational problems posed over spaces of arcs $x : [S, T] \rightarrow R^n$, the infimum cost over the space of Lipschitz continuous functions is strictly less than the infimum cost over the space of absolutely continuous functions; surprising because the Lipschitz functions comprise a dense subset of the space of absolutely continuous functions, with respect to the strong $W^{1,1}$ topology. Typical applications of the Generalized Tonelli Regularity Theorem identify classes of problems for which minimizers over the class of absolutely continuous functions are in fact Lipschitz continuous. The link is of course that all such applications inform us about special cases of (BP) for which the Lavrentiev Phenomenon cannot occur.

The concluding sections concern regularity of minimizers for optimal control problems with dynamic and pathwise constraints. We indicate how the analysis underlying the Generalized Tonelli Theorem can, in some cases, be extended to cover optimal control problems involving linear dynamic constraints. We also examine some situations in which the application of "standard" necessary conditions for optimal control problems with dynamic constraints and pathwise state constraints leads directly to regularity information.

The choice of topics in this chapter on minimizer regularity is motivated in part by a desire to illustrate the benefits of applying Nonsmooth Analysis, in smooth as well as nonsmooth settings. To the author's knowledge, the only available proof of the Generalized Tonelli Regularity Theorem, for vector-valued arcs, is via Nonsmooth Analysis, even when the data are smooth.

The proof techniques involve the construction of an auxiliary Lagrangian and the comparison of minimizers for the origin and auxiliary Lagrangians on suitable subintervals. Since we are unable to apply the standard necessary conditions to minimizers for the original problem, we establish regularity properties indirectly, by applying them to the problem with the auxiliary Lagrangian, which does satisfy the relevant hypotheses. So everything hinges on the construction of the auxiliary Lagrangian. This is most simply carried out by taking convex hulls of certain epigraph sets, operations that generate nonsmooth functions. Nonsmooth necessary conditions are then required, to analyze the solutions to the auxiliary problem, even if the original Lagrangian is smooth.

In the $n = 1$ case (scalar-valued arcs) and for strictly convex Lagrangians, Tonelli showed how to construct a smooth auxiliary Lagrangian with suitable properties. But the task of carrying out such an exercise for vector-valued arcs would appear to be a formidable one and, since nonsmooth necessary conditions are now available, unnecessary.

11.2   Tonelli Regularity

In a landmark paper published in 1915, Tonelli [134] developed a technique for establishing regularity properties of solutions to (BP), under hypotheses similar to (E1) through (E3), based on the construction of auxiliary Lagrangians. The regularity properties of minimizers proved in this chapter, which go beyond Tonelli's results in a number of significant respects, are the fruits of combining Tonelli's technique and modern nonsmooth analytical methods.

A cornerstone of Tonelli's regularity analysis is the concept of a regular point.

Definition 11.2.1 *Take a function $y : [S, T] \to R^n$ and a point $\tau \in [S, T]$. We say τ is a* regular point *of y if*

$$\liminf_{\substack{S \leq s_i \leq \tau \leq t_i \leq T \\ s_i \to \tau, t_i \to \tau, s_i \neq t_i}} \frac{|y(t_i) - y(s_i)|}{|t_i - s_i|} < +\infty. \tag{11.4}$$

Otherwise expressed, τ is a regular point of y if there exist $s_i \to \tau$, $t_i \to \tau$, and $K > 0$ such that, for all i

$$s_i \neq t_i \quad and \quad S \leq s_i < t_i \leq T$$

and

$$|y(t_i) - y(s_i)| \leq K|t_i - s_i|.$$

Because the definition involves limits of difference quotients, it would at first sight appear that assuming y is regular at t comes close to assuming that it is Lipschitz continuous near t. In fact the definition of "regular point" is much less restrictive than this. For example, $t = 0$ is a regular point of $y(s) = |s|^{1/2}$, $-1 \le s \le +1$, because

$$\frac{|y(t_i) - y(s_i)|}{|t_i - s_i|} = 0 \quad \text{for all } i,$$

when we choose $s_i = -i^{-1}$ and $t_i = +i^{-1}$ for $i = 1, 2, \ldots$. Yet this function has unbounded slope near $t = 0$.

Theorem 11.2.2 (Generalized Tonelli Regularity Theorem) *Assume that the Tonelli Existence Hypotheses (E1) to (E3) are satisfied. A solution to (BP) exists. Let $\bar{x}$ be any strong local minimizer for (BP). Take any regular point τ for $\bar{x}$. Then there exists a relatively open subinterval $I \subset [S, T]$ with the properties: $\tau \in I$ and the restriction of $\bar{x}$ to I is Lipschitz continuous.*

We defer proof of the theorem until the next section. Here we examine some of its implications.

Take an arbitrary strong local minimizer $\bar{x}$ for (BP). Let D be the subset of $[S, T]$ comprising points at which $\bar{x}$ is differentiable. Since $\bar{x}$ is an absolutely continuous function, it is differentiable almost everywhere. Consequently D has full measure.

Take any $t \in D$. Then, since $\bar{x}$ is differentiable at t,

$$\lim_i \frac{|\bar{x}(t_i) - \bar{x}(s_i)|}{|t_i - s_i|} < +\infty$$

for some sequences $t_i \to t$ and $s_i \to s$ such that $S \le s_i \le t \le t_i \le T$ and $t_i > s_i$ for each i. We see that t is a regular point of $\bar{x}$. According to the preceding theorem, we can choose a relatively open interval $I_t \subset [S, T]$, containing t, such that x is Lipschitz continuous on I_t.

Now set

$$\Omega = \cup_{t \in D} I_t.$$

Ω is a relatively open set because it is a union of relatively open sets. It has full measure because the subset D has full measure. We see also that $\bar{x}$ is locally Lipschitz continuous on Ω, in the sense that for any $t \in \Omega$ we can find a neighborhood of t (I_t serves the purpose) on which $\bar{x}$ is Lipschitz continuous. We have drawn the following important conclusions from Theorem 11.2.2.

Corollary 11.2.3 *Assume the Tonelli Existence Hypotheses (E1)–(E3) are satisfied. Take any strong local minimizer $\bar{x}$ for (BP). Then there exists a relatively open set of full measure, Ω, such that $\bar{x}$ is locally Lipschitz continuous on Ω.*

For a given absolutely continuous function $\bar{x}$, let $\Omega_{\max}(\bar{x})$ be the union of all relatively open sets Ω with the property that $\bar{x}$ is locally Lipschitz continuous on Ω. Then $\Omega_{\max}$ is a relatively open subset of $[S,T]$, of full measure, and $\bar{x}$ is locally Lipschitz continuous on $\Omega_{\max}(\bar{x})$. We define

$$S := [S,T]\backslash\Omega_{\max}(\bar{x})$$

to be the *Tonelli Set* for $\bar{x}$. The Tonelli set, which we think of as the set of times at which $\bar{x}$ exhibits bad behavior can be alternatively described as the set of points t in $[S,T]$ such that $\bar{x}$ has unbounded slope in the vicinity of t.

Summarizing and extending these results, we arrive at

Theorem 11.2.4 *Assume that the data for (BP) satisfy the Tonelli Existence Hypotheses (E1) to (E3). Take any strong local minimizer $\bar{x}$. Then the Tonelli set S for $\bar{x}$ is a (possibly empty) closed set of zero measure. We have:*

(i) *given any closed subinterval $I \subset [S,T]\backslash S$, $\bar{x}$ is Lipschitz continuous on I;*

(ii) *given any closed interval $I \subset [S,T]\backslash S$, there exists $p \in W^{1,1}(I;R^n)$ such that, for a.e. $t \in I$,*

$$\dot{p}(t) \in \mathrm{co}\,\{q : (q,p(t)) \in \partial L(t,\bar{x}(t),\dot{\bar{x}}(t))\}$$

and

$$p(t) \cdot \dot{\bar{x}}(t) - L(t,\bar{x}(t),\dot{\bar{x}}(t)) = \max_{v \in R^n}\{p(t) \cdot v - L(t,\bar{x}(t),v)\};$$

(iii) *suppose in addition to (E1) through (E3) that, for each $t \in [S,T]$ and $w \in R^n$, $L(.,\bar{x}(t),w)$ is continuous at t and $L(t,\bar{x}(t),.)$ is strictly convex. Then $\bar{x}$ is continuously differentiable on $[S,T]\backslash S$;*

(iv) *suppose in addition to the hypotheses of (iii) that, for each $t \in [S,T]$, the function L is C^r on a neighborhood of $(t,\bar{x}(t),\dot{\bar{x}}(t))$ (for some integer $r \geq 2$) and $L_{vv}(t,\bar{x}(t),\dot{\bar{x}}(t)) > 0$. Then $\bar{x}$ is r-times continuously differentiable on $[S,T]\backslash S$.*

In (iii) and (iv), the assertion $\bar{x}$ is C^r on $[S,b)$ is taken to mean that $\bar{x}$ is of class C^r on the interior of the open set (S,b) and all derivatives of order r or less have limits as $t \downarrow S$. We interpret $\bar{x}$ is C^r on $(b,T]$, etc., likewise.

Proof. We know already that S is a closed set of zero measure, off which $\bar{x}$ is locally Lipschitz continuous.

(i): Take a closed interval $I \subset [S,T]\backslash S$. Since $\bar{x}$ is locally Lipschitz continuous on the compact interval I, $\bar{x}$ is (globally) Lipschitz continuous on I.

(ii): Take any closed interval $[a,b] \subset [S,T]\backslash S$. Then $\bar{x}$ restricted to $[a,b]$ is a strong local minimizer for (BP) with endpoint constraints $x(a) = \bar{x}(a)$ and $x(b) = \bar{x}(b)$. Because $\bar{x}$ is Lipschitz continuous on $[a,b]$ and the relevant Lipschitz continuity hypothesis (11.1) is satisfied, the asserted necessary conditions follow from Proposition 11.1.2.

(iii): Take any point t in the relatively open set $[S,T]\backslash S$. Then there exists some relatively open interval containing t, with endpoints a and b and such that $[a,b] \subset [S,T]\backslash S$. We know L is Lipschitz continuous on $[a,b]$ and that there exists $p \in W^{1,1}([a,b];R^n)$ such that, for a.e. $t \in [a,b]$,

$$p(t) \cdot \dot{\bar{x}}(t) - L(t,\bar{x}(t),\dot{\bar{x}}(t)) \geq p(t) \cdot v - L(t,\bar{x}(t),v) \quad \text{for all } v \in R^n. \quad (11.5)$$

If the bounded function $\dot{\bar{x}}$ is not almost everywhere equal to a continuous function on $[a,b]$ there exists $r \in [a,b]$ and sequences $\{c_i\}$ and $\{d_i\}$ converging to r in $[a,b]$, such that (11.5) is satisfied for $t = r$ and $t = c_i$, $t = d_i$, $i = 1, 2, \ldots$, the limits

$$\alpha := \lim_{i \to \infty} \dot{x}(c_i), \qquad \beta = \lim_{i \to \infty} \dot{x}(d_i)$$

exist and $\alpha \neq \beta$.

Fix $v \in R^n$. Under the additional hypotheses we have

$$L(r,\bar{x}(r),v) = \lim_{i \to \infty} L(c_i,\bar{x}(c_i),v),$$
$$L(r,\bar{x}(r),\alpha) = \lim_{i \to \infty} L(c_i,\bar{x}(c_i),\dot{\bar{x}}(c_i)).$$

Setting $t = c_i$ in (11.5) and passing to the limit as $i \to \infty$ gives

$$p(r) \cdot v - L(r,\bar{x}(r),v) \leq p(r) \cdot \alpha - L(r,\bar{x}(r),\alpha).$$

Using the same arguments in relation to the sequence $\{d_i\}$, we arrive at

$$p(r) \cdot v - L(r,\bar{x}(r),v) \leq p(r) \cdot \beta - L(r,\bar{x}(r),\beta).$$

These inequalities are satisfied for arbitrary points $v \in R^n$. We have shown that the function

$$v \to p(r) \cdot v - L(r,\bar{x}(r),v)$$

achieves its maximum at the points $v = \alpha$ and $v = \beta$. Since it is strictly concave, we deduce that $\alpha = \beta$. From this contradiction we conclude that an arbitrary point $t \in [S,T]\backslash\mathcal{S}$ is contained in a relatively open interval on which $\dot{\bar{x}}$ is almost everywhere equal to a continuous function. It follows that $\dot{\bar{x}}$ is continous on $[S,T]\backslash\mathcal{S}$.

(iv): Take any point $s \in [S,T]\backslash\mathcal{S}$. Then s is contained in some relatively open subinterval of $[S,T]$ with endpoints a and b, such that $[a,b] \subset [S,T]\backslash\mathcal{S}$. $\bar{x}$, restricted to $[a,b]$, is a strong local minimizer to (BP) for "endpoint data" $(a, \bar{x}(a), b, \bar{x}(b))$. Since, by (iii), this subarc has an essentially bounded derivative, we deduce from Proposition 11.1.2 the existence of a number d such that

$$L_v(t, \bar{x}(t), \dot{\bar{x}}(t)) = d + \int_a^t L_x(\sigma, \bar{x}(\sigma), \dot{\bar{x}}(\sigma)) d\sigma \qquad (11.6)$$

for all $t \in [a,b]$. The right side is C^1 and so is $(t,v) \to L_v(t, \bar{x}(t), v)$. Since $L_{vv} > 0$, it follows from the Implicit Function Theorem that $\dot{\bar{x}}$ is C^1, from which it follows that $\bar{x}$ is C^2. (11.6) therefore implies

$$d^2\bar{x}(t)/dt^2 = [L_{vv}]^{-1}\{L_x - L_{vt} - L_{vx} \cdot \dot{\bar{x}}(t)\}, \qquad (11.7)$$

on (a,b). (The derivatives are evaluated at $(t, \bar{x}(t), \dot{\bar{x}}(t))$.) It follows that $d^2\bar{x}/dt^2$ is continuous (and has limits at a and b). Now suppose we know that $\bar{x}$ is C^{r-1} on $[a,b]$ and that L is C^r (for some integer $r \geq 2$). The right side of (11.7) is C^{r-2}, and therefore $d^2\bar{x}/dt^2$ is C^{r-2}. It follows that $\bar{x}$ restricted to I is C^r (and if $d^{r-1}\bar{x}/dt^{r-1}$ has limits at the endpoints of I, then so does $d^r\bar{x}/dt^r$.) The fact that $\bar{x}$ is C^r on a relatively open subinterval containing s now follows by induction. $\square$

Thus far, we have placed no restrictions on the dimension n of the state variable. If $n = 1$, then minimizing arcs have the surprising property that they are differentiable at *every* point in $[S,T]$, even on their Tonelli sets. To justify such assertions however, we must allow derivatives that take values $+\infty$ or $-\infty$.

Theorem 11.2.5 *Assume that, in addition to the Tonelli Existence Hypotheses (E1) through (E3), for each $t \in [S,T]$ and $w \in R^n$, $L(., \bar{x}(t), w)$ is continuous at t and $L(t, \bar{x}(t), .)$ is strictly convex. Suppose further that*

$$n = 1.$$

Let $\bar{x}$ be a minimizer. Then $\bar{x}$ is everywhere differentiable on $[S,T]$, in the sense that the following limit exists (finite or infinite) for each $\tau \in [S,T]$.

$$\lim_{\substack{t \to \tau \\ a \leq t \leq b}} \frac{\bar{x}(t) - \bar{x}(\tau)}{t - \tau}. \qquad (11.8)$$

Proof. If τ is such that the left side of (11.4) is finite then, as we have shown, $\bar{x}$ is C^1 near τ. In this case, the limit (11.8) exists and is finite. So we may assume that the left side of (11.4) is infinite.

Suppose that $\tau = S$. The limit (11.8) can fail to exist only if

$$\lim\sup_{t\downarrow S} \frac{\bar{x}(t) - \bar{x}(S)}{t - S} = +\infty \quad \text{and} \quad \lim\inf_{t\downarrow S} \frac{\bar{x}(t) - \bar{x}(S)}{t - S} = -\infty.$$
(11.9)

But since $\bar{x}$ is continuous, we deduce from these two relationships that there exist points t, arbitrarily close to S, with $\bar{x}(t) = \bar{x}(S)$. It follows that the left side of (11.4) is finite, a contradiction. We show similarly that limit (11.8) exists also when $\tau = T$.

It remains to consider the case when τ lies in (S, T) and the left side of (11.4) is infinite. Reasoning as above, we justify restricting attention to the case when

$$\lim_{t\downarrow\tau} \frac{\bar{x}(t) - \bar{x}(\tau)}{t - \tau} = +\infty \quad \text{and} \quad \lim_{s\uparrow\tau} \frac{\bar{x}(s) - \bar{x}(\tau)}{s - \tau} = -\infty. \quad (11.10)$$

(The related case, in which the limits from right and left are $-\infty$ and $+\infty$, respectively, is treated analogously.)

Fix $\epsilon > 0$ such that $[\tau - \epsilon, \tau + \epsilon] \subset [S, T]$. We claim that there exists $\delta > 0$ with the following property: corresponding to any $r \in (0, \delta)$, a point $s \in (\tau - \epsilon, \tau)$ can be found such that $\bar{x}(s) = \bar{x}(\tau) + r$. Indeed, if this were not the case, we could deduce from the continuity of $\bar{x}$ that $\bar{x}(s) \leq \bar{x}(\tau)$ for all $s \in (\tau - \epsilon, \tau)$. This contradicts the second condition in (11.10).

Arguing in similar fashion, we can show that (possibly after reducing the size of δ) there exists $t \in (\tau, \tau + \epsilon)$ and $r \in (0, \delta)$ for which $\bar{x}(t) = \bar{x}(\tau) + r$, for otherwise we obtain a contradiction of the first condition in (11.10). We know that $\tau \in (s, t)$. Also, $\bar{x}(s) = \bar{x}(t)$ and $|t-s| \leq 2\epsilon$. Since ϵ is an arbitrary positive number, the left side of (11.4) is zero, again a contradiction. $\square$

11.3   Proof of The Generalized Tonelli Regularity Theorem

We prove Theorem 11.2.2. Existence of minimizers is assured by Theorem 11.1.1. Take any strong local minimizer $\bar{x}$. Note at the outset that, without loss of generality, we can assume:

(E4): $\theta : [0, \infty) \to R$ is a nonnegative-valued, nondecreasing function.

Indeed, if θ fails to satisfy this condition we can replace it by

$$\tilde{\theta}(r) := \inf\{\theta(r') : r' \geq r\} - \inf\{\theta(r') : r' \geq 0\}.$$

It is a straightforward task to deduce from the convexity and superlinear growth of $L(t, x, .)$ that this new function meets the requirements of Hypothesis (E3) and also satisfies (E4), for the Lagrangian

$$L(t, x, v) - \inf\{\theta(r') : r' \geq 0\}.$$

Since minimizers are unaffected by the addition of a constant to L, we have confirmed that we can add (E4) to the hypotheses.

We can also assume

(E5): In condition (E3), the inequality is strict and $\alpha = 0$.

To see this, take $k > 0$ such that $\|\bar{x}\|_{L^\infty} < k$. Consider a new problem, in which L in problem (BP) is replaced by

$$L'(t, x, v) := \max\{L(t, x, v), -\alpha k + \theta(|v|)\} + \alpha k + 1.$$

Hypotheses (E1) through (E2) continue to be satisfied (with $\theta = \tilde{\theta}$ and $\alpha = 0$). However the inequality in (E3) is now strict. We have $L' \geq L + \alpha k + 1$ everywhere and $L'(t, x, v) = L(t, x, v) + \alpha k + 1$ for all $v \in R^n$ and points (t, x) in some tube about $\bar{x}$. So $\bar{x}$ remains a strong local minimizer and the assertions of the theorem for L' imply those for the original L. This confirms that we can add (E5) to the hypotheses.

Take a regular point $\tau \in [S, T]$ of $\bar{x}$. Then there exist sequences $s_i \to \tau$, $t_i \to \tau$ such that $S \leq s_i \leq \tau \leq t_i \leq T$ for all i, and

$$\lim_{i \to \infty} \frac{|\bar{x}(t_i) - \bar{x}(s_i)|}{|t_i - s_i|} < \infty. \tag{11.11}$$

Since $\bar{x}$ is continuous, we can arrange (by decreasing each s_i and increasing each t_i if necessary) that, for each i, $t_i > \tau$ if $\tau < T$ and $s_i < \tau$ if $\tau > S$. This means that, for each i, τ is contained in a relatively open subinterval of $[S, T]$ with endpoints s_i and t_i.

For each i let $y : [s_i, t_i] \to R^n$ be the linear interpolation of the endpoints of $\bar{x}$ restricted to $[s_i, t_i]$:

$$y_i(t) := \bar{x}(s_i) + \frac{t - s_i}{|t_i - s_i|}(\bar{x}(t_i) - \bar{x}(s_i)).$$

Lemma 11.3.1 *There exist constants R_0 and M such that*

$$\|y_i\|_{L^\infty} \leq M, \qquad \|\bar{x}\|_{L^\infty} \leq M \quad \text{for all } i$$

and

$$\|\dot{y}_i\|_{L^\infty} < R_0 \quad \text{for all } i.$$

Furthermore, if $\{z_i \in W^{1,1}([s_i, t_i]; R^n)\}$ is any sequence of arcs such that, for each i,

$$z_i(s_i) = \bar{x}(s_i) \quad \text{and} \quad z_i(t_i) = \bar{x}(t_i)$$

and

$$\frac{1}{2}\int_{s_i}^{t_i}\theta(\dot{z}_i(t))dt \ \leq \ \int_{s_i}^{t_i}L(t,y_i(t),\dot{y}_i(t))dt,$$

then

$$\|z_i\|_{L^\infty} \leq M \quad \text{for all } i.$$

Proof. The fact that M and R_0 can be chosen to satisfy all the stated conditions is obvious, with the exception of $\|z_i\|_{L^\infty} \leq M$. To show that this condition can also be satisfied, we choose $\alpha > 0$ such that $\theta(\alpha') > \alpha'$ whenever $\alpha' \geq \alpha$. Then, for any i and $t \in [s_i, t_i]$,

$$
\begin{aligned}
|z_i(t)| \ &\leq \ |z_i(s_i)| + \int_{s_i}^{t_i}|\dot{z}_i(s)|ds \\
&\leq \ |z_i(s_i)| + \alpha|t_i - s_i| + \int_{s_i}^{t_i}\theta(\dot{z}_i(s))ds \\
&\leq \ |z_i(s_i)| + \alpha|t_i - s_i| + \int_{s_i}^{t_i}L(t,y_i(s),\dot{y}_i(s))ds.
\end{aligned}
$$

We have now merely to note that all terms on the right side of this inequality are bounded by a constant which does not depend on i or $t \in [s_i, t_i]$. □

Define

$$c_0 \ := \ \max\{|L(t,x,v)| : t \in [S,T], |x| \leq M \text{ and } |v| \leq R_0\},$$

and the function $d : R \to R$,

$$d(\sigma) \ := \ \inf\{|v| : p \in \partial_v L(t,x,v),\ t \in [S,T],\ |x| \leq M,\ |p| \geq \sigma\}.$$

We deduce from (E2) that

$$\lim_{\sigma\to\infty}d(\sigma) \ = \ +\infty. \tag{11.12}$$

Fix $\epsilon > 0$. Choose $R_1 \geq 0$ to satisfy

$$\theta \circ d\left(\frac{\theta(r')}{2r'} - \frac{c_0}{r'} - 2\epsilon\right) \ > \ 2c_0 \tag{11.13}$$

whenever $r' \geq R_1$, $t \in [S,T]$, and $|x| \leq M$. This is possible by (11.12) and since θ has superlinear growth. Define

$$c_1 \ := \ \max\{|L(t,x,v)| : t \in [S,T], |x| \leq M, |v| \leq R_1\} \tag{11.14}$$

and

$$\sigma_1 \ := \ \max\{|\xi| : \xi \in \partial_v L(t,x,v)|,\ t \in [S,T],\ |x| \leq M,\ |v| \leq R_1\}. \tag{11.15}$$

Choose $R_2 > R_1$ such that

$$\frac{1}{2}\theta(r) \geq \sigma_1(R_1 + r) \quad \text{if } r \geq R_2.\qquad(11.16)$$

Set

$$\phi(v) := \frac{1}{2}\max\{\theta(|v|), \theta(R_2)\}$$

and, for each $(t, x) \in [S, T] \times R^n$ and $v \in R^n$, define

$$\tilde{L}(t, x, v) := \inf\{\alpha : (v, \alpha) \in \text{co} \left[\text{epi}\,\phi \cup \text{epi}\{L(t, x, .) + \Psi_{R_2 B}\}\right]\}.$$

The expression on the right summarizes the following construction.

> $\tilde{L}(t, x, .)$ is the function with epigraph set E, where E is the convex hull of the unions of the epigraph sets of ϕ and of the function $v \to L(t, x, v)$, restricted to the ball $R_2 B$.

An alternative representation is as follows.

$$\tilde{L}(t, x, v) = \inf\{\lambda L(t, x, u) + (1 - \lambda)\phi(w) :$$
$$0 \leq \lambda \leq 1, |u| \leq R_2 \text{ and } \lambda u + (1 - \lambda)w = v\}.$$

This "auxiliary Lagrangian" $\tilde{L}$ has the following properties.

Lemma 11.3.2

(a) $\tilde{L}(t, x, v)$ is locally bounded, measurable in t, and convex in v.

(b) $\tilde{L}(t, x, v)$ is locally Lipschitz continuous in (x, v) uniformly in $t \in [S, T]$.

(c) $\tilde{L}(t, x, v) \geq \theta(|v|)/2$ for all (t, x, v).

(d) For all $t \in [S, T]$ and $x \in MB$ we have

$$\begin{aligned}
\tilde{L}(t, x, v) &= L(t, x, v) \quad \text{if } |v| \leq R_1, \\
\tilde{L}(t, x, v) &\leq L(t, x, v) \quad \text{if } |v| \leq R_2, \\
\tilde{L}(t, x, v) &< L(t, x, v) \quad \text{if } |v| > R_2.
\end{aligned}$$

(e) For $(t, x) \in [S, T] \times R^n$,

$$\tilde{L}(t, x, v) = \theta(|v|)/2 \quad \text{if } |v| \geq R_2.$$

Proof.

(a): $\tilde{L}$ is convex in v, by construction. It is locally bounded since $0 \leq \tilde{L} \leq \phi$ and ϕ is locally bounded. To see that $\tilde{L}$ is measurable in t we use the fact that, for fixed x and v, $\tilde{L}$ can be expressed as a pointwise infimum of a countable family of measurable functions

$$t \to \lambda L(t, x, v) + (1 - \lambda)\phi(w)$$

obtained by allowing (λ, v, w) to range over a countable dense subset of $[0, 1] \times R^n \times R^n$. (a) has been proved.

(c): This property follows from the facts that both L and ϕ satisfy the desired inequality and that $v \to \frac{1}{2}\theta(|v|)$ is a convex function.

(e): Take arbitrary points $w' \in R^n$, $|w'| > R_2$, and $(t, x) \in [S, T] \times R^n$. Choose $\zeta \in \partial\phi(w')$. The subgradient inequality for convex functions gives

$$\phi(w) - \phi(w') - \zeta \cdot (w - w') \geq 0 \quad \text{for all } w \in R^n. \tag{11.17}$$

Since θ is continuous and strictly increasing on $[R_2, \infty)$, $\phi(w)$ and $\frac{1}{2}\theta(|w|)$ coincide on a neighborhood of $w = w'$. It follows that ζ is a subgradient also of $w \to \frac{1}{2}\theta(|w|)$ at $w = w'$ and, for all $u \in R^n$ and $(t, x) \in [S, T] \times R^n$ we have

$$L(t, x, u) - \phi(w') - \zeta \cdot (u - w')$$
$$\geq \frac{1}{2}\theta(|u|) - \frac{1}{2}\theta(|w'|) - \zeta \cdot (u - w') \geq 0. \tag{11.18}$$

From (11.17) and (11.18) we deduce that

$$\tilde{L}(t, x, v) = \inf\{\lambda L(t, x, u) + (1 - \lambda)\phi(w) :$$
$$\lambda \in [0, 1], |u| \leq R_2, v = \lambda u + (1 - \lambda)w\}$$
$$\geq \phi(w') - \zeta \cdot (v - w').$$

Setting $v = w'$ we see that $\tilde{L}(t, x, w') \geq \phi(w')$. But then $\tilde{L}(t, x, v) = \frac{1}{2}\theta(|v|)$ in the region $\{v : |v| > R_2\}$, since ϕ and $v \to \frac{1}{2}\theta(|v|)$ coincide here and ϕ majorizes $\tilde{L}$. This remains true in the region $\{v : |v \geq R_2\}$, by the continuity properties of convex functions.

(b): Take any $k_1 > 0$. Let K be a Lipschitz constant for $x \to L(t, x, v)$ uniformly valid for $t \in [S, T]$, $|v| \leq R_2$, and $|x| \leq k_1$. Take $x_1, x_2 \in R^n$ such that $|x_1|$ and $|x_2| \leq k_1$. Then for any $\delta > 0$, $t \in [S, T]$, and w we can choose u, w, and λ in the definition of $\tilde{L}$ such that

$$\tilde{L}(t, x_1, v) \leq \lambda L(t, x_1, u) + (1 - \lambda)\phi(w)$$
$$\leq \lambda L(t, x_2, u) + K|x_1 - x_2| + (1 - \lambda)\phi(w)$$
$$\leq \tilde{L}(t, x_2, v) + K|x_1 - x_2| + \delta.$$

Since x_1 and x_2 are interchangeable and $\delta > 0$ is arbitrary, it follows that $x \to \tilde{L}(t, x, v)$ has Lipschitz constant at most K on $k_1 B$, for all $t \in [S, T]$ and $w \in R^n$.

Now take $k_2 \geq R_2$. We show that $v \to \tilde{L}(t, x, v)$ is Lipschitz continuous in the region $k_2 B$, uniformly over $(t, x) \in [S, T] \times R^n$. It follows that $\tilde{L}$ is locally Lipschitz continuous jointly in the variables x, w, uniformly in $t \in [S, T]$, since it has this property with respect to these variables individually.

Choose $(t, x) \in [S, T] \times R^n$, and let $v \to p \cdot v + q$ be an arbitrary, nonconstant, affine function that is majorized by $v \to \tilde{L}(t, x, v)$. By (e), we must have

$$p \cdot v + q \leq \frac{1}{2}\theta(|v|)$$

for all v such that $|v| \geq k_2 B$. Setting $v = (k_2 + 1)p/|p|$ we obtain

$$(k_2 + 1)|p| + q \leq \theta(k_2 + 1). \tag{11.19}$$

However since $\tilde{L} \geq 0$ we also have

$$\tilde{L}(t, x, v) - p \cdot v - q \geq -|p|k_2 - q \tag{11.20}$$

for $v \in k_2 B$. (11.19) and (11.20) yield

$$\tilde{L}(t, x, v) - p \cdot v - q \geq -|p| - \frac{1}{2}\theta(k_2 + 1)$$

for $v \in k_2 B$. Set $K_1 = \frac{1}{2}\theta(k_2 + 1) + 1$. It follows that

$$\tilde{L}(t, x, v) - p \cdot v - q \geq 1 \tag{11.21}$$

for $v \in k_2 B$ and $|p| \geq K_1$.

Now the function $v \to \tilde{L}(t, x, v)$ is expressible as the pointwise supremum of affine functionals majorized by $\tilde{L}$. However inequality (11.21) tells us that, to evaluate the pointwise supremum in the region $|v| \leq k_2$, we can restrict attention to affine functions with Lipschitz constant at most K_1. It follows then that $v \to \tilde{L}(t, x, v)$ has Lipschitz constant at most K_1 in this region, uniformly with respect to $(t, x) \in [S, T] \times R^n$.

(d): The cases $|v| \geq R_2$ and $R_2 > |v| > R_1$ follow from (e) since L majorizes $\tilde{L}$ and L strictly majorizes $u \to \frac{1}{2}\theta(|u|)$. It remains to show that $L(t, x, v) = \tilde{L}(t, x, v)$ for all $(t, x) \in [S, T] \times R^n$ and $v \in R_1 B$.

Take (t, x) and v as above and choose $\zeta \in \partial_v L(t, x, v)$. By (11.16) and the definition of the constants c_1 and σ_1 (see (11.14) and (11.15)) we have that

$$\phi(w) \geq \frac{1}{2}\theta(|w|) \geq \sigma_1(R_1 + |w|) + c_1 \geq L(t, x, v) + \zeta \cdot (w - v)$$

for all points w which satisfy $|w| \geq R_2$. On the other hand, we also know that

$$\phi(w) \geq \frac{1}{2}\theta(|R_2|) \geq \sigma_1(R_1 + R_2) + c_1 \geq L(t, x, v) + \zeta \cdot (w - v)$$

for all points w which satisfy $|w| \leq R_2$. By the subgradient inequality however

$$L(t, z, u) - L(t, z, v) - \zeta \cdot (u - v) \geq 0$$

for all $u \in R^n$. Scaling and adding these inequalities, we arrive at

$$
\begin{aligned}
\tilde{L}(t, x, v') &= \inf\{\lambda L(t, x, u) + (1 - \lambda)\phi(w) : \\
&\qquad 0 \leq \lambda \leq 1,\ |u| \leq R_2,\ v' = \lambda u + (1 - \lambda)w\} \\
&\leq L(t, x, v) + \zeta \cdot (v' - v)
\end{aligned}
$$

for all points $v' \in R^n$. Setting $v' = v$ yields $\tilde{L}(t, x, v) \geq L(t, x, v)$. Since however L majorizes $\tilde{L}$ we can replace inequality here by equality. This is what we set out to prove. $\square$

Consider the optimization problems (P_i), $i = 1, 2\ldots$,

$$
\left\{
\begin{aligned}
&\text{Minimize } \int_{s_i}^{t_i} \tilde{L}(t, x(t), \dot{x}(t))dt \\
&\qquad \text{over } x \in W^{1,1}([s_i, t_i]; R^n) \text{ satisfying} \\
&x(s_i) = \bar{x}(s_i),\ x(t_i) = \bar{x}(t_i).
\end{aligned}
\right.
$$

In view of Properties (a) to (c) of the auxiliary Lagrangian, we deduce from Theorem 11.1.1 that (P_i) has a minimizer, which we denote by x_i, for each i. Notice that, for each i,

$$
\begin{aligned}
\frac{1}{2} \int_{s_i}^{t_i} \theta(|\dot{x}_i(s)|)ds &\leq \int_{s_i}^{t_i} \tilde{L}(t, x_i(t), \dot{x}_i(t))dt \\
&\leq \int_{s_i}^{t_i} \tilde{L}(t, y_i(t), \dot{y}_i(t))dt \\
&\leq \int_{s_i}^{t_i} L(t, y_i(t), \dot{y}_i(t))dt,
\end{aligned}
$$

by Properties (c) and (d) of $\tilde{L}$ and since $||\dot{y}_i||_{L^\infty} \leq R_1$.

We conclude from Lemma 11.3.1 that

$$||x_i||_{L^\infty} \leq M \quad \text{for all } i.$$

It is straightforward to confirm the hypotheses under which the necessary conditions of Theorem 11.1.2 are valid, with reference to the minimizer x_i. Notice the crucial role of Property (e) of $\tilde{L}$ which, together with (b), ensures that $\tilde{L}$ satisfies the Lipschitz continuity hypothesis: there exists $\beta > 0$ such that

$$|\tilde{L}(t, x, v) - \tilde{L}(t, x', v)| \leq \beta|x - x'| \tag{11.22}$$

for all $v \in R^n$ and all (t, x), (t, x') in some tube about x_i.

We deduce existence of an arc $p_i \in W^{1,1}$ such that

$$\dot{p}_i(t) \in \text{co } \partial_x \tilde{L}(t, x_i(t), \dot{x}_i(t)) \tag{11.23}$$
$$p_i(t) \in \partial_v \tilde{L}(t, x_i(t), \dot{x}_i(t)). \tag{11.24}$$

We deduce from (11.22) and (11.23) that

$$|p_i(t) - p_i(s_i)| \le \beta |t_i - s_i|$$

for all i. Let i_0 be the smallest integer such that

$$\beta |t_j - s_j| \le \epsilon \quad \text{for all } j \ge i_0.$$

Choose any $i \ge i_0$. Then, by (11.24),

$$p_i(s_i) \in \partial_v \tilde{L}(t, x_i(t), \dot{x}_i(t)) + \epsilon B \quad \text{a.e. } t \in [s_i, t_i]. \tag{11.25}$$

We claim that

$$\|\dot{x}_i\|_{L^\infty} \le R_1. \tag{11.26}$$

Indeed, assume to the contrary. Then for all points $\bar{t}$ in some subset of $[s_i, t_i]$ of positive measure we have

$$|\dot{x}_i(\bar{t})| > R_1.$$

According to (11.25) then, we can arrange that

$$\tilde{L}(\bar{t}, x_i(\bar{t}), 0) - \tilde{L}(\bar{t}, x_i(\bar{t}), \dot{x}_i(\bar{t})) \ge -p_i(s_i) \cdot \dot{x}_i(\bar{t}) - \epsilon |\dot{x}_i(\bar{t})|.$$

So,

$$|p_i(s_i)| \cdot |\dot{x}_i(\bar{t})| \ge \tilde{L}(\bar{t}, x_i(\bar{t}), \dot{x}_i(\bar{t})) - \tilde{L}(\bar{t}, x_i(\bar{t}), 0) - \epsilon |\dot{x}_i(\bar{t})|.$$

It follows that

$$|p_i(s_i)| \ge \frac{\theta(r)}{2r} - \frac{c_0}{r} - \epsilon,$$

for some $r > R_1$. By (11.25), for a.e. $t \in [s_i, t_i]$,

$$|\dot{x}_i(t)| \ge d(|p_i(s_i)| - \epsilon) \ge d\left(\frac{\theta(r)}{2r} - \frac{c_0}{r} - 2\epsilon\right).$$

Since θ is a monotone function,

$$\frac{1}{2}\theta(|\dot{x}_i(t)|) \ge \frac{1}{2}\theta \circ d\left(\frac{\theta(r)}{2r} - \frac{c_0}{r} - 2\epsilon\right) > c_0 \quad \text{a.e.}$$

It follows now from Properties (c) and (d) of $\tilde{L}$ (see Lemma 11.3.2) that

$$
\begin{aligned}
\int_{s_i}^{t_i} \tilde{L}(t, x_i(t), \dot{x}_i(t))dt \ &\geq \ \frac{1}{2}\int_{s_i}^{t_i} \theta(|\dot{x}_i(t)|)dt \\
&> \ |t_i - s_i|c_0 \\
&\geq \ \int_{s_i}^{t_i} L(t, y_i(t), \dot{y}_i(t))dt \\
&= \ \int_{s_i}^{t_i} \tilde{L}(t, y_i(t), \dot{y}_i(t))dt.
\end{aligned}
$$

But this contradicts the optimality of x_i. Condition (11.26) is verified.

We claim finally that

$$
\|\dot{\bar{x}}(t)\|_{L^\infty([s_i,t_i];R^n)} \ \leq \ R_2. \tag{11.27}
$$

This will imply that $\dot{\bar{x}}$ is locally essentially bounded on $[s_i, t_i]$.

Suppose, to the contrary, that there exists a subset $\mathcal{D} \subset [s_i, t_i]$ of positive measure such that

$$
|\dot{\bar{x}}(t)| \ > \ R_2 \quad \text{for all } t \in \mathcal{D}.
$$

Since $\|\bar{x}\|_{L^\infty} \leq M$, we have from Property (d) of L that

$$
\tilde{L}(t, \bar{x}(t), \dot{\bar{x}}(t)) \ < \ L(t, \bar{x}(t), \dot{\bar{x}}(t)) \quad \text{for } t \in \mathcal{D}. \tag{11.28}
$$

Then

$$
\int_{s_i}^{t_i} L(t, \bar{x}(t), \dot{\bar{x}}(t))dt \ \leq \ \int_{s_i}^{t_i} L(t, x_i(t), \dot{x}_i(t))dt
$$

(by optimality of $\bar{x}$)

$$
= \ \int_{s_i}^{t_i} \tilde{L}(t, x_i(t), \dot{x}_i(t))dt
$$

(since $\|\dot{x}_i\|_{L^\infty([s_i,t_i];R^n)} \leq R_1$ and by Property (d) of $\tilde{L}$)

$$
\leq \ \int_{s_i}^{t_i} \tilde{L}(t, \bar{x}(t), \dot{\bar{x}}(t))dt
$$

(by optimality of x_i)

$$
< \ \int_{s_i}^{t_i} L(t, \bar{x}(t), \dot{\bar{x}}(t))dt
$$

(by (11.28) and since $L \geq \tilde{L}$).

From this contradiction, we deduce that (11.27) is true. We have confirmed that $\bar{x}$ is essentially bounded on the relatively open subinterval $[s_i, t_i]$ containing τ. $\square$

11.4   Lipschitz Continuous Minimizers

Our aim in this section is to explore the implications of the Generalized Tonelli Regularity Theorem for particular classes of problems. The idea is to provide a more detailed, qualitative description of minimizers than that of Section 11.2, when the Tonelli Existence Hypotheses are supplemented by additional hypotheses.

The Generalized Tonelli Regularity Theorem is a very fruitful source of refined regularity theorems, supplying information about the Tonelli set in special cases. It can be used, for example, to show that for a large class of problems with polynomial Lagrangians, the Tonelli set is a countable set with a finite number of accumulation points [47].

We concentrate here, however, on just one application area: identifying hypotheses that, when added to the Tonelli Existence Hypotheses, ensure that all minimizers of (BP) over the class of absolutely continuous functions are, in fact, Lipschitz continuous. An equivalent property is that the Tonelli set is empty.

The significance of establishing that a minimizer for (BP) is Lipschitz continuous is that the hypotheses are then met under which the standard necessary conditions, such as those summarized as Proposition 11.1.2, can be used to investigate minimizers in detail. On the other hand, it rules out pathological behavior, which might otherwise give rise to difficulties in the computation of minimizers, associated with the Lavrentiev phenomenon.

We make repeated use of the following lemma, which gives sufficient conditions under which an interval on which a minimizer is Lipschitz continuous can be extended to include an endpoint of $[S, T]$.

Lemma 11.4.1 *Assume the Tonelli Existence Hypotheses (E1) to (E3). Let $\bar{x}$ be a minimizer for (BP) and let $\mathcal{S}$ be the Tonelli set of $\bar{x}$. Take $\bar{t} \in [S, T)\backslash\mathcal{S}$. Suppose that there exists $k > 0$ with the following property: for any $t' \in (\bar{t}, T)$ such that $[\bar{t}, t'] \subset [S, T]\backslash\mathcal{S}$, there exists $p \in W^{1,1}([\bar{t}, t']; R^n)$ satisfying*

$$\dot{p}(t) \in \mathrm{co}\{q : (q, p(t)) \in \partial L(t, \bar{x}(t), \dot{\bar{x}}(t))\} \quad a.e.,$$
$$|p(t)| \le k \quad for\ all\ t \in [t, t'].$$

Then $\dot{\bar{x}}$ is essentially bounded on $[\bar{t}, T]$.

Proof. Define

$$\tau_{\max} := \sup\{\tau \in (\bar{t}, T] : \dot{\bar{x}} \text{ is essentially bounded on } [\bar{t}, \tau]\}.$$

(The set over which the supremum is taken is nonempty, by Theorem 11.2.2.)

Choose a sequence $\{t_i\}$ in $(\bar{t}, t_{\max})$ such that $t_i \uparrow t_{\max}$. According to the hypotheses, for each i there exists $p_i \in W^{1,1}([\bar{t}, t_i]; R^n)$ and $k \ge 0$ such that

$$\dot{p}_i(t) \in \mathrm{co}\{q : (q, p_i(t)) \in \partial L(t, \bar{x}(t), \dot{\bar{x}}(t))\} \quad \text{a.e. } t \in [\bar{t}, t_i] \qquad (11.29)$$

and
$$|p_i(t)| \leq k \quad \text{for all } t \in [\bar{t}, t_i].\tag{11.30}$$

Since $L(t, x, .)$ is convex, (11.29) implies that

$$p_i(t) \in \partial_v L(t, \bar{x}(t), \dot{\bar{x}}(t)) \quad \text{a.e. } t \in [\bar{t}, t_i].\tag{11.31}$$

We claim that $\dot{\bar{x}}$ is essentially bounded on $[\bar{t}, \tau_{\max}]$. If this were not the case, there would exist a point σ_i in $(\bar{t}, t_i)$ for $i = 1, 2, \ldots$, such that $\sigma_i \uparrow \tau_{\max}$, (11.31) is satisfied at $t = \sigma_i$, and

$$|\dot{\bar{x}}_i(\sigma_i)| \to +\infty.\tag{11.32}$$

Since $L(\sigma_i, \bar{x}(\sigma_i), .)$ is convex, (11.31) implies

$$\begin{aligned}
p_i(\sigma_i) \cdot \dot{\bar{x}}(\sigma_i) - L(\sigma_i, \bar{x}(\sigma_i), \dot{\bar{x}}(\sigma_i)) &\geq \min_{|v| \leq 1}\{p_i(\sigma_i) \cdot v - L(\sigma_i, \bar{x}(\sigma_i), v)\} \\
&\geq -|p_i(\sigma_i)| - \alpha
\end{aligned}$$

for $i = 1, 2, \ldots$, where

$$\alpha := \sup\{L(\sigma, \bar{x}(\sigma), v) : \sigma \in [S, T], |v| \leq 1\}.$$

It follows that, for $i = 1, 2, \ldots$,

$$L(\sigma_i, \bar{x}(\sigma_i), \dot{\bar{x}}(\sigma_i))/|\dot{\bar{x}}(\sigma_i)| \leq |p_i(\sigma_i)|(1 + |\dot{\bar{x}}(\sigma_i)|) + \alpha.$$

As $i \to \infty$, the left side of this inequality has limit $+\infty$ by (11.32). But the right side is bounded above, in view of (11.30). From this contradiction we deduce that $\dot{\bar{x}}$ is essentially bounded on $[\bar{t}, \tau_{\max}]$.

If $\tau_{\max} < T$, then $\tau_{\max}$ is a regular point of $\bar{x}$ and $\dot{\bar{x}}$ is essentially bounded on a relatively open neighborhood of $\tau_{\max}$ in $[S, T]$. This contradicts the defining property of $\tau_{\max}$. It follows that $\dot{\bar{x}}$ is essentially bounded on all of $[\bar{t}, T]$. $\square$

Our first application of the lemma is to show that, if L does not depend on time (the "autonomous" case), then all minimizers are Lipschitz continuous.

Proposition 11.4.2 *Assume that, in addition to the Tonelli Existence Hypotheses (E1) to (E3), $L(t, x, v)$ is independent of t. Then all strong local minimizers for (P) are Lipschitz continuous.*

Proof. Write $L(x, v)$ in place of $L(t, x, v)$. Take a strong local minimizer $\bar{x}$. Choose a regular point $\bar{t} \in (S, T)$ of $\bar{x}$. (This is possible since the regular points have full measure.) We show that $\dot{\bar{x}}$ is essentially bounded on $[\bar{t}, T]$. A similar argument can be used to confirm that $\dot{\bar{x}}$ is essentially bounded on $[S, \bar{t}]$. It follows that $\bar{x}$ is Lipschitz continuous on $[S, T]$.

Take any $t' \in (\bar{t}, T]$ such that $\dot{\bar{x}}$ is essentially bounded on $[\bar{t}, t']$. Because L is independent of t, we can apply the necessary conditions of Proposition 11.1.2, including the constancy of the Hamiltonian condition, to the optimal subarc $\bar{x}$ restricted to $[\bar{t}, t']$. We know then that there exist an arc $p \in W^{1,1}([\bar{t}, t']; R^n)$ and a number c such that, for a.e. $t \in [\bar{t}, t']$,

$$c = p(t) \cdot \dot{\bar{x}}(t) - L(\bar{x}(t), \dot{\bar{x}}(t)) \qquad (11.33)$$
$$p(t) \in \partial_v L(\bar{x}(t), \dot{\bar{x}}(t)). \qquad (11.34)$$

Fix $\epsilon > 0$ such that $\dot{\bar{x}}$ is essentially bounded on $[\bar{t}, \bar{t} + \epsilon]$. It follows from the local Lipschitz continuity of L and (11.34) that there exists a constant $k_1 > 0$, independent of t', such that

$$|p(t)| \leq k_1 \quad \text{a.e. } t \in [\bar{t}, t'] \cap [\bar{t}, \bar{t} + \epsilon].$$

But then (11.33) implies that there exists k_2 (independent of t') such that

$$|c| \leq k_2.$$

We deduce from (11.33) and (11.34) that, for a.e. $t \in [\bar{t}, t']$,

$$\begin{aligned}
c &= p(t) \cdot \dot{\bar{x}} - L(\bar{x}(t), \dot{\bar{x}}(t)) \\
&\geq \max_{|v| \leq 1} \{p(t) \cdot v - L(\bar{x}(t), v)\} \\
&\geq |p(t)| - \alpha,
\end{aligned}$$

where α, defined by

$$\alpha := \inf\{L(\bar{x}(t), v) : S \leq t \leq T, |v| \leq 1\},$$

does not depend on t'.

Since p is continuous, we have

$$|p(t)| \leq \alpha + c \quad \text{for all } t \in [\bar{t}, t'].$$

We see that p is bounded on $[\bar{t}, t']$ by a constant which does not depend on t'. It follows from Lemma 11.4.1 that $\dot{\bar{x}}$ is essentially bounded on $[\bar{t}, T]$. $\square$

Another case when Lemma 11.4.1 can be used to establish Lipschitz continuity of minimizers is when $L(t, x, v)$ is convex, jointly in (x, v).

Proposition 11.4.3 *Assume, in addition to the Tonelli Existence Hypotheses (E1) to (E3), that $L(t, x, v)$ is convex in (x, v) for each $t \in [S, T]$. Then all strong local minimizers for (BP) are Lipschitz continuous.*

Proof. Take a strong local minimizer $\bar{x}$. Choose a regular point $\tau \in (S, T)$ of $\bar{x}$. As in the proof of the previous proposition, we content ourselves with showing that $\dot{\bar{x}}$ is essentially bounded on $[\tau, T]$. The demonstration that the same is true on $[S, \tau]$ is along precisely similar lines.

Take any $t' \in (\bar{t}, T]$ such that $\dot{\bar{x}}$ is essentially bounded on $[\bar{t}, t']$. We know from Theorem 7.3.1 that there exists $p \in W^{1,1}([\bar{t}, t']; R^n)$ such that

$$(\dot{p}(t), p(t)) \in \partial L(t, \bar{x}(t), \dot{\bar{x}}(t)) \quad \text{a.e. } t \in [\bar{t}, t'], \tag{11.35}$$

where $\partial L(t, ., .)$ is the subdifferential in the usual sense of convex analysis.

Since $\dot{\bar{x}}$ is essentially bounded on a neighborhood of $\bar{t}$, $L(t, ., .)$ is locally Lipschitz continuous uniformly in t, and p is continuous, (11.35) implies that there exists $k_1 > 0$ (independent of t') such that

$$|p(\bar{t})| \le k_1. \tag{11.36}$$

Since the "convex" subdifferential is employed, (11.35) implies that, for a.e. $t \in [\bar{t}, t']$,

$$L(t, y, v) - L(t, \bar{x}(t), \dot{\bar{x}}(t))$$
$$\ge (y - \bar{x}(t)) \cdot \dot{p}(t) + (v - \dot{\bar{x}}(t)) \cdot p(t) \quad \text{for all } y \in R^n, \, v \in R^n.$$

By examining the implications of this inequality when $y = \bar{x} + u$ and $v = 0$, for an arbitrary unit vector u, we deduce that, for a.e. $t \in [\bar{t}, t']$,

$$|\dot{p}(t)| = \max_{|u| \le 1} \dot{p} \cdot u$$
$$\le \max_{|u| \le 1} L(t, \bar{x}(t) + u, 0) - L(t, \bar{x}(t), \dot{\bar{x}}(t)) + p(t) \cdot \dot{\bar{x}}(t).$$

It follows that, for some integrable functions, γ_1 and γ_2 that do not depend on t',

$$|\dot{p}(t)| \le \gamma_1(t)|p(t)| + \gamma_2(t) \quad \text{a.e. } t \in [\bar{t}, t'].$$

We deduce from this inequality and (11.36), with the help of Gronwall's Inequality, that there exists $k_2 > 0$, independent of t', such that

$$|p(t)| \le k_2 \quad \text{for all } t \in [\bar{t}, t'].$$

It follows now from Lemma 11.4.1 that $\dot{\bar{x}}$ is essentially bounded on $[\bar{t}, T]$.
□

Finally, we illustrate how the Tonelli Regularity Theorem can be used to reduce the hypotheses under which Euler–Lagrange type necessary conditions of optimality have traditionally been derived. As in the previous applications, this is done via the intermediary of Lemma 11.4.1.

Proposition 11.4.4 *Take a strong local minimizer $\bar{x}$ for (BP). Assume that, in addition to the Tonelli Existence Hypotheses (E1) to (E3), there exist integrable functions $c(.)$ and $\gamma(.)$ such that, for a.e. $t \in [S, T]$,*

$$\sup_{\xi \in P_x[\text{co}\partial L](t)} |\xi| \le c(t) \inf_{\eta \in \partial_v L} |\eta| + \gamma(t), \tag{11.37}$$

where $P_x[\text{co}\partial L](t)$ denotes the projection of $\text{co}\partial L$ onto the first coordinate

$$P_x[\text{co}\partial L](t) := \{\xi : (\xi, \eta) \in \text{co}\,\partial L(t, \bar{x}(t), \dot{\bar{x}}(t)) \text{ for some } \eta \in R^n\}$$

and $\partial_v L$ is evaluated at $(t, \bar{x}(t), \dot{\bar{x}}(t))$.
Then $\bar{x}$ is Lipschitz continuous.

Notice that the supplementary hypothesis (11.37) reduces to

$$|L_x(t, \bar{x}(t), \dot{\bar{x}}(t))| \leq c(t)|L_v(t, \bar{x}(t), \dot{\bar{x}}(t))| + \gamma(t) \quad \text{a.e. } t \in [S, T]$$

when L is smooth.

This hypothesis is less restrictive than that invoked in standard necessary conditions, such as those of Proposition 11.1.2, in certain respects. First, the presence of the nonnegative term $c(t)|L_v|$ on the right side reduces the severity of the inequality. Second, the inequality is required to hold precisely along the strong local minimizer $\bar{x}(.)$, not over a tube about $\bar{x}(.)$.

Proof. Choose a regular point $\tau \in (S, T)$ of $\bar{x}$. As usual, we show merely that $\dot{\bar{x}}$ is essentially bounded on $[\tau, T]$, since a similar analysis can be used to demonstrate the essential boundedness of $\dot{\bar{x}}$ also on $[S, \tau]$.

Take any $t' \in (\bar{t}, T]$ such that $\dot{\bar{x}}$ is essentially bounded on $[\bar{t}, t']$. The necessary conditions of Proposition 11.1.2 supply $p \in W^{1,1}([\bar{t}, t']; R^n)$ such that

$$\dot{p}(t) \in \text{co}\{q : (q, p(t)) \in \partial L(t, \bar{x}(t), \dot{\bar{x}}(t))\} \quad \text{a.e. } t \in [\bar{t}, t'].$$

This implies that, for a.e. $t \in [\bar{t}, t']$,

$$\dot{p}(t) \in P_x[\text{co}\partial L](t)$$

and

$$p(t) \in \partial_v L(t, \bar{x}(t), \dot{\bar{x}}(t)). \tag{11.38}$$

Since $L(t, ., .)$ is locally Lipschitz continuous uniformly in t, $\bar{x}$ is essentially bounded on a neighborhood of $\bar{t}$, and p is continuous, we deduce from (11.38) that there exists some k_1 (independent of t'), such that

$$|p(\bar{t})| \leq k_1.$$

It follows from supplementary hypothesis (11.37) that

$$|\dot{p}(t)| \leq c(t)|p(t)| + \gamma(t) \quad \text{a.e. } t \in [\bar{t}, t'].$$

We now deduce from Gronwall's Inequality (Lemma 2.4.4) that there exists $k_2 > 0$ (independent of t') such that

$$|p(t)| \leq k_2 \quad \text{for all } t \in [\bar{t}, t'].$$

It follows now from Lemma 11.4.1 that $\dot{\bar{x}}$ is essentially bounded on $[\bar{t}, T]$.
□

11.5   Autonomous Variational Problems with State Constraints

In the preceding section, we illustrated the role of the Generalized Tonelli Regularity Theorem in establishing that minimizers of the Basic Problem (BP) over the space of absolutely continuous functions are Lipschitz continuous in certain cases of interest. One such case is when $L(t, x, v)$ is time-independent (the "autonomous" case).

An alternative proof that minimizers are Lipschitz continuous in the autonomous case, due to Butazzo et al., is based on an application of the Maximum Principle and time reparameterization. This approach has the merit of simplicity. Besides, the hypotheses which it is necessary to impose are weaker and it extends to problems with pathwise state constraints.

Consider the optimization problem

$$(CV) \begin{cases} \text{Minimize } \int_S^T L(x(t), \dot{x}(t))dt \\ \text{over arcs } x \in W^{1,1}([S, T]; R^n) \text{ satisfying} \\ x(S) = x_0, \quad x(T) = x_1, \\ x(t) \in A \quad \text{for all } t \in [S, T], \end{cases}$$

the data for which comprise an interval $[S, T]$, a function $L : R^n \times R^n \to R$, points x_0 x_1, and a set $A \subset R^n$.

We lay stress on the fact that the integrand L does not depend on t and on the presence of the pathwise state constraint

$$x(t) \in A \quad \text{for all } t \in [S, T].$$

The formulation does however preclude a dynamic constraint.

Theorem 11.5.1 *Let $\bar{x}$ be a minimizer for (CV). Assume that*

(a) L is Borel measurable and is bounded on bounded sets;

(b) $L(x, .)$ is convex for all $x \in R^n$;

(c) there exists an increasing function $\theta : [0, \infty) \to [0, \infty)$ such that

$$\lim_{\alpha \to \infty} \theta(\alpha)/\alpha = +\infty$$

and a constant β such that

$$L(x, v) > \theta(|v|) - \beta|v|.$$

Then $\bar{x}$ is Lipschitz continuous.

Proof. Consider the optimal control problem

$$
\begin{cases}
\text{Minimize } \int_S^T l(s, v(s))ds \text{ over measurable functions} \\
\qquad\qquad v : [S, T] \to R \text{ and } y \in W^{1,1}([S, T]; R) \text{ satisfying} \\
\dot{y}(s) = v \quad \text{a.e. } t \in [S, T], \\
v(s) \in [0.5, 1.5], \\
y(S) = S, \quad y(T) = T,
\end{cases}
$$

in which

$$
l(s, v) := L(\bar{x}(s), \dot{\bar{x}}(s)/v)v.
$$

Here we regard y as a state variable and v as a control variable. Notice that the cost function is well defined under the hypotheses since $l(.,.)$ is a Borel measurable function such that $s \to l(s, v)$ is minorized by a common integrable function, as v ranges over $[0.5, 1.5]$.

We claim that

$$
(\ \bar{y}(s) \equiv s, \ \bar{v}(s) \equiv 1)
$$

is a minimizer. To see this note that, given any process (y, v) satisfying the constraints of the optimal control problem, the value of the cost function can be expressed in terms of a new independent variable $t = \tau(s)$, where $\tau : [S, T] \to [S, T]$ is the strictly increasing Lipschitz continuous function with Lipschitz continuous inverse:

$$
\tau(s) = S + \int_S^s v(\sigma)d\sigma.
$$

This gives

$$
\int_S^T l(s, v(s))ds = \int_S^T L\left(\bar{x}(s), \frac{\dot{\bar{x}}(s)}{v(s)}\right)v(s)ds = \int_S^T L(z(t), \dot{z}(t))dt
$$

in which $z(t) = (\bar{x} \circ \tau^{-1})(t)$. Of course we have

$$
\int_S^T l(s, \bar{v}(s))ds = \int_S^T L(\bar{x}(t), \dot{\bar{x}}(t))dt.
$$

But z is a $W^{1,1}$ function satisfying $z(t) \in A$ for all $t \in [S, T]$, $z(S) = x_0$, and $z(T) = x_1$. It follows now from the optimality of $\bar{x}$ that

$$
\int_S^T l(s, v(s))ds \geq \int_S^T l(s, \bar{v}(s))ds.
$$

The claim is confirmed.

We also require the fact that, for each s, $l(s, .)$ is convex on $(0, \infty)$. Take any $v_1, v_2 \in (0, \infty)$ and $\epsilon \in (0, 1)$. To establish this property, it suffices to show that

$$
\epsilon l(s, v_1) + (1 - \epsilon)l(s, v_2) \geq l(s, \epsilon v_1 + (1 - \epsilon)v_2). \tag{11.39}
$$

But, by the convexity of $L(\bar{x}(s), .)$,

$$\frac{\epsilon v_1}{\epsilon v_1 + (1-\epsilon)v_2} L(\frac{\dot{\bar{x}}}{v_1}) + \frac{(1-\epsilon)v_2}{\epsilon v_1 + (1-\epsilon)v_2} L(\frac{\dot{\bar{x}}}{v_2}) \geq L(\frac{\dot{\bar{x}}}{\epsilon v_1 + (1-\epsilon)v_2}).$$

(To simplify notation, we have suppressed the "x" argument in L.) Multiplying across by $\epsilon v_1 + (1-\epsilon)v_2$ and noting the definition of $l(s,v)$, we arrive at (11.39). Convexity is proved.

The final step is to apply the Maximum Principle (Theorem 6.2.1) to the above optimal control problem, after reducing it to a terminal cost problem by state augmentation. The rather modest regularity hypotheses imposed on L suffice to justify this, since the optimal control problem is state free. A simple contradiction argument establishes that the cost multiplier must be nonzero. We can therefore arrange by scaling that it takes value 1. We deduce that there exists a number p such that

$$v \longmapsto pv - l(s,v)$$

is maximized at $v = 1$, over the neighborhood $(0.5, 1.5)$, for all points s in a subset $D \subset [S,T]$ having full measure. By the convexity of $l(s,.)$ however, this local maximization property in fact implies global maximization; i.e.,

$$p - l(s,1) \geq pv - l(s,v), \quad \text{for all } v \in (0,\infty) \quad \text{a.e.}$$

Now take $v = 1 + |\dot{\bar{x}}(s)|$. This gives

$$L(\bar{x}(s), \dot{\bar{x}}(s)) \leq$$
$$p - p(1 + |\dot{\bar{x}}(s)|) + L\left(\bar{x}(s), \dot{\bar{x}}(s)/(1 + |\dot{\bar{x}}(s)|)\right)(1 + |\dot{\bar{x}}(s)|) \quad \text{a.e.}$$

By the coercivity hypothesis on L however

$$\theta(|\dot{\bar{x}}(s)|) - \beta|\dot{\bar{x}}(s)| \leq L(\bar{x}(s), \dot{\bar{x}}(s)) \quad \text{a.e.}$$

Setting $k = \| L\left(\bar{x}, \dot{\bar{x}}/(1 + |\dot{\bar{x}}|)\right)\|_{L^\infty}$ and $k_1 = k - p + \beta$, we deduce from the last two inequalities that

$$\theta(|\dot{\bar{x}}(s)|) \leq k_1|\dot{\bar{x}}(s)| + k \quad \text{a.e.} \tag{11.40}$$

Now according to the properties of θ, we may choose $K, \bar{\alpha} > 0$ such that $\theta(\alpha) \geq K\alpha$ if $\alpha \geq \bar{\alpha}$ and $(K - k_1)\bar{\alpha} > k$. We show that

$$|\dot{\bar{x}}(s)| \geq \bar{\alpha} \quad \text{a.e.}$$

Suppose, to the contrary, there exists a point $s \in [S,T]$ such that (11.40) is satisfied and $|\dot{\bar{x}}(s)| \geq \bar{\alpha}$. Then certainly

$$(K - k_1)|\dot{\bar{x}}(s)| \geq (K - k_1)\bar{\alpha} > k.$$

So

$$K|\dot{\bar{x}}(s)| > k_1|\dot{\bar{x}}(s)| + k.$$

On the other hand (11.40) yields

$$K|\dot{\bar{x}}(s)| \leq \theta(|\dot{\bar{x}}(s)|) \leq k_1|\dot{\bar{x}}(s)| + k.$$

From this contradiction we deduce that such a point cannot exist. $\square$

11.6 Bounded Controls

Up to now, we have restricted attention to regularity properties of min-imizers for variational problems with no dynamic constraints. Our inves-tigations have centered on conditions under which derivatives are locally essentially bounded on some relatively open subset of full measure and on the implications of such conditions. What regularity properties of minimiz-ers can be established for problems with nonlinear dynamic constraints, under standard hypotheses guaranteeing existence of a minimizer? General results are lacking. Our earlier analysis adapts however to yield conditions for optimal controls to be locally essentially bounded (and hence for op-timal state trajectories to have locally essentially bounded derivatives) in the case of a time-invariant linear dynamic constraint:

$$\dot{x}(t) = Ax(t) + Bu(t) + d(t).$$

Consider the problem:

$$(L) \begin{cases} \text{Minimize } \int_S^T L(t, x(t), u(t))dt \\ \text{over } x \in W^{1,1}([S,T];R^n) \text{ and measurable } u : [S,T] \to R^m \text{ satisfying} \\ \dot{x}(t) = Ax(t) + Bu(t) + d(t) \quad \text{a.e.,} \\ x(S) = x_0, \ x(T) = x_1, \end{cases}$$

the data for which comprise an interval $[S,T]$, functions $L : [S,T] \times R^n \times R^m \to R$ and $d : [S,T] \to R^n$, matrices $A \in R^{n \times n}$ and $B \in R^{n \times m}$, and points $x_0, x_1 \in R^n$.

Theorem 11.6.1 *Suppose that the data for (L) satisfy the following hy-potheses.*

(H1): $L(t, x, u)$ is bounded on bounded sets, measurable in t, and convex in u. d is an integrable function.

(H2): For each bounded set $M \subset R^n \times R^m$, there exists a constant K such that for all $t \in [S,T]$ and (x_1, u_1), $(x_2, u_2) \in M$,

$$|L(t, x_1, u_1) - L(t, x_2, u_2)| \leq K|(x_1 - x_2, u_1 - u_2)|.$$

*(H3): There exist a number $c \geq 0$ and a convex function $\theta : [0, \infty) \to [0, \infty)$
such that $\theta(r)/r \to \infty$ as $r \to \infty$ and*

$$L(t, x, v) \geq -c|x| + \theta(|v|)$$

for all $(t, x, v) \in [S, T] \times R^n \times R^m$.

*Suppose that there exists a process that satisfies the endpoint constraints
for (L).*

*Take any minimizer $(\bar{x}, \bar{u})$. (Under the hypotheses a minimizer exists.)
Then there exists a closed subset $\Omega \subset [S, T]$ of zero measure with the following property: for any $t' \in [S, T] \backslash \Omega$, $\bar{u}$ is essentially bounded on a relative neighborhood of t'. Furthermore, there exists a measurable function
$p : [S, T] \to R^n$ that is locally Lipschitz continuous on $[S, T] \backslash \Omega$, such that*

$$-\dot{p}(t) \in p(t)A - \mathrm{co}\, \partial_x L(t, \bar{x}(t), \bar{u}(t)) \quad a.e.$$

and

$$\mathcal{H}(t, \bar{x}(t), \bar{u}(t), p(t)) = \max_{u \in R^m} \mathcal{H}(t, \bar{x}(t), u, p(t)) \quad a.e.,$$

where $\mathcal{H}$ denotes

$$\mathcal{H}(t, x, u, p) := p \cdot (Ax + Bu) - L(t, x, u)$$

and $\partial_x L$ denotes the limiting subgradient with respect to x.

We deduce from Theorem 2.7.1 that (L) has a minimizer. However the earlier derived Maximum Principle, Theorem 6.2.1, cannot be applied to this problem, because the hypotheses of Theorem 11.6.1, which are tailored to the requirements of existence theory, do not imply those of Theorem 6.2.1. Theorem 11.6.1 asserts validity of a weaker form of Maximum Principle, in which the adjoint arc is not required to be absolutely continuous, but is required instead to be merely locally Lipschitz continuous on a relatively open subset of $[S, T]$, of full measure.

Theorem 11.6.1 is proved by reducing problem (L) to a variational problem without a dynamic constraint, but in which the the cost integrand depends on x and its higher derivatives $Dx, \ldots, D^{\tilde{n}-1}x$, for some $\tilde{n} \leq n$, and then carrying out a similiar, but more intricate, analysis to that of Section 11.3. Details are given in [52].

Not surprisingly, Theorem 11.6.1 serves as a stepping stone to proving essential boundedness of optimal controls in special cases. One such case we now consider. In applications of Optimal Control to control system design, it is usually necessary to take account of magnitude constraints on control variables, which reflect actuator limitations, safety considerations, or permissible regions of the control variable space for validity of the dynamic model. The presence of constraints can, however, greatly complicate the computation of optimal controls. The standard technique for bypassing these difficulties is to drop the constraint and to add, instead, a term $\epsilon|u|^r$ to the cost integrand that penalizes excessive control action. In quadratic cost control it is known that inclusion of this penalty term ensures that optimal controls are bounded, for any $\epsilon > 0$ and $r = 2$. Furthermore, the larger ϵ, the smaller is the uniform bound on optimal controls. This is shown by direct calculation, an approach that is not possible for (L) in general. We can however use Theorem 11.6.1 to assess how large the exponent r in the penalty term must be, at least to ensure boundedness of controls.

Proposition 11.6.2 Let $(\bar{x}, \bar{u})$ be a minimizer for (L). Assume that Hypotheses (H1) to (H3) of Theorem 11.6.1 are satisfied and that L has the form

$$L(t, x, u) = L_1(t, x, u) + \epsilon|u|^r$$

in which $L_1 : [S, T] \times R^n \times R^m \to R$ is a given nonnegative-valued function and $r \geq 1$ and $\epsilon > 0$ are given numbers. Suppose further that:

given any compact set $D \subset [S, T] \times R^n$, there exists a number c such that

$$\max\{|a| + |b| : (a, b) \in co\partial L(t, x, u)\} \leq c(1 + |u|^r)$$

for all $(t, x) \in D$ and $u \in R^m$.

Then $\bar{u}$ is essentially bounded on $[S, T]$.

A similar analysis to that of Section 11.4 can be used to prove the proposition. See [52] for details.

Recall the Ball–Mizel example, which can be reformulated as an optimal control problem:

$$\begin{cases} \text{Minimize } \int_0^1 \left\{ (x^3(t) - t^2)^2 |u(t)|^{14} + \epsilon|u(t)|^2 \right\} dt \\ \text{over } x \in W^{1,1}([0, 1]; R) \text{ and measurable } u : [0, 1] \to R \text{ satisfying} \\ \dot{x}(t) = u(t) \quad \text{a.e.}, \\ x(0) = 0, \quad x(1) = k. \end{cases}$$

It follows from the discussion of Section 11.1, that there is a unique optimal control for this problem, namely, the unbounded function $u(t) = kt^{-1/3}$, for suitable choices of the constants $k > 0$ and $\epsilon > 0$. Theorem 11.6.2 provides an engineering perspective on this problem. It tells us that $r = 2$ is too small a value for the exponent in the penalty term $\epsilon|u(t)|^r$, to ensure boundedness of optimal controls. On the other hand, it is a straightforward matter to check by applying Proposition 11.6.2 that optimal controls are bounded if the penalty term is taken to be

$$\epsilon \int_S^T |u(t)|^r dt$$

for any $\epsilon > 0$ and $r \geq 14$.

11.7   Lipschitz Continuous Controls

In certain cases, necessary conditions of optimality can be used directly to establish regularity properties of minimizers. Indeed, examining implications of known necessary conditions is the traditional approach to regularity analysis. A simple example is Hilbert's proof that arcs satisfying the Euler–Lagrange Condition for the Basic Problem in the Calculus of Variations are automatically of class C^r if the Lagrangian $L(t, x, v)$ is of class C^r $(r \geq 2)$ and strictly convex in v. (The main steps are reproduced in the above proof of Assertion (iv) of Theorem 11.2.4.) We illustrate the method by giving conditions under which a control function satisfying the state constrained Maximum Principle of Chapter 8 is Lipschitz continuous.

Consider the optimal control problem

$$(R) \begin{cases} \text{Minimize } l(x(S), x(T)) + \int_S^T [L(t, x(t)) + \tfrac{1}{2} u^T(t) R u(t)] dt \\ \text{over arcs } x \in W^{1,1}([S, T]; R^n) \text{ and measurable functions} \\ u : [S, T] \to R^n \\ \text{satisfying} \\ \dot{x}(t) = f(t, x(t)) + G(t, x(t)) u(t) \quad \text{a.e.,} \\ h(x(t)) \leq 0 \quad \text{for all } t \in [S, T], \\ (x(S), x(T)) \in C, \end{cases}$$

with data an interval $[S, T]$, functions $L : [S, T] \times R^n \to R$, $l : R^n \times R^n \to R$, $f : [S, T] \times R^n \to R^n$, $G : [S, T] \times R^n \to R^{n \times m}$, $h : R^n \to R$, a closed set $C \subset R^n \times R^n$, and a symmetric $m \times m$ matrix R.

It is to be expected that control functions for (R) satisfying the Maximum Principle (with nonzero cost multiplier) are Lipschitz continuous, when the corresponding state trajectories are interior. In this case, the regularity property follows directly from the Generalized Weierstrass Condition and strict convexity of the Unmaximized Hamiltonian with respect to the control variable. The fact that optimal controls are Lipschitz continuous also for problems with active state constraints comes, on the other hand, as something of a surprise. It means that, for the class of problems here considered, optimal state trajectories do not instantly change direction when they strike the boundary of the state constraint set.

Write

$$\mathcal{H}(t, x, p, u) := p \cdot [f(t, x) + G(t, x)u] - [L(t, x) + (1/2)u^T R u]$$

and, for $x(.) \in W^{1,1}$,

$$I(x(.)) := \{t \in [S, T] : h(x(t)) = 0\}.$$

We say that a process $(\bar{x}, \bar{u})$, satisfying the constraints of (R) is a *normal extremal* if there exist $p \in W^{1,1}([S, T]; R^n)$ and $\mu \in C^*(S, T)$ such that

$$-\dot{p}(t) \in \partial_x \mathcal{H}(t, \bar{x}(t), p(t) + \int_{[S,t)} \nabla h(\bar{x}(s)) \mu(ds), \bar{u}(t)) \quad \text{a.e.,}$$

$$(p(S), -[p(T) + \int_{[S,T]} \nabla h(\bar{x}(t))\mu(dt)]) \in \partial l(\bar{x}(S), \bar{x}(T)) + N_C(\bar{x}(S), \bar{x}(T)),$$

$$\text{supp}\,\{\mu\} \subset I(\bar{x}(.)),$$

$$\mathcal{H}(t, \bar{x}(t), p(t) + \int_{[S,t)} \nabla h(\bar{x}(s))\mu(ds), \bar{u}(t)) \qquad (11.41)$$

$$= \max_{u \in R^n} \mathcal{H}(t, \bar{x}(t), p(t) + \int_{[S,t)} \nabla h(\bar{x}(s))\mu(ds), u) \quad \text{a.e.}$$

In other words, a normal extremal is a feasible process, for which the Maximum Principle is satisfied with cost multiplier $\lambda = 1$.

Theorem 11.7.1 (Lipschitz Continuity of Optimal Controls) *Take a normal extremal $(\bar{x}, \bar{u})$ for (R). Assume that*

(H1): L, f, G, and l are locally Lipschitz continuous;

(H2) h is of class $C^{1,1}$; i.e., h is everywhere differentiable with a derivative that is locally Lipschitz continuous;

(H3) $\nabla h^T(\bar{x}(t))G(t, \bar{x}(t)) \neq 0$ for all $t \in I(\bar{x}(.))$;

(H4) R is positive definite.

Then $\bar{u}$ is Lipschitz continuous.

Remarks

(i): The proof to follow can be adapted to allow for multiple state constraints. Extensions are also possible to cover problems in which the cost integrand is strictly convex in the control variable, but possibly nonquadratic.

(ii): The normality hypothesis in Theorem 11.7.1 is of an intrinsic nature. Sufficient conditions, open to direct verification, can be given for normality. These typically require that local approximations to the dynamics are controllable in some sense, by means of controls which maintain strict feasibility of the pathwise state constraint.

(iii): Hager [78] and Malanowski [99] have shown that optimal controls are Lipschitz continuous (and have estimated the Lipschitz constants of optimal controls and Lagrange multipliers) for certain classes of problems involving control constraints as well as state constraints. Theorem 11.7.1 does not cover problems with control constraints but, on the other hand, departs from [78] and [99] by allowing nonsmooth data and also initial states that lie in the state constraint boundary.

Proof. In view of the special structure of the Hamiltonian, we deduce from (11.41) that

$$\bar{u}(t) = R^{-1}G^T(t,\bar{x}(t)) \left[p(t) + \int_{[S,t)} \nabla h(\bar{x}(s))\mu(ds) \right] \quad \text{a.e.} \quad (11.42)$$

It is immediately evident from this expression that $\bar{u}$ is essentially bounded. It follows also that $\bar{u}$ can be chosen to have left and right limits on (S,T) and to be continuous at $t = S$ and $t = T$. We deduce from the state and adjoint equation that p and x are both Lipschitz continuous.

Step 1: We show that μ has no atoms in (S,T) and, for any $t \in (S,T) \cap I(\bar{x}(.))$,

$$\nabla h^T(\bar{x}(t)) \left[f + GR^{-1}G^T \left(p(t) + \int_{[S,t)} \nabla h(\bar{x}(s))\mu(ds) \right) \right] = 0. \quad (11.43)$$

(In this relationship, f and G are evaluated at $(t,\bar{x}(t))$.)

Take any point $t \in (S,T) \cap I(\bar{x}(.))$. Then the fact that $\bar{x}(t)$ satisfies the state constraint permits us to conclude that, for all $\delta > 0$ sufficiently small,

$$\delta^{-1}(h(\bar{x}(t+\delta)) - h(\bar{x}(t))) \leq 0,$$

$$\delta^{-1}(h(\bar{x}(t)) - h(\bar{x}(t-\delta))) \geq 0.$$

Passing to the limit as $\delta \downarrow 0$, we deduce that

$$\nabla h^T(\bar{x}(t)) \left[f + GR^{-1}G^T \left(p(t) + \int_{[S,t]} \nabla h(\bar{x}(s))\mu(ds) \right) \right] \leq 0,$$

$$\nabla h^T(\bar{x}(t)) \left[f + GR^{-1}G^T \left(p(t) + \int_{[S,t)} \nabla h(\bar{x}(s))\mu(ds) \right) \right] \geq 0.$$

Subtracting these inequalities gives

$$\nabla h^T(\bar{x}(t))G(t,\bar{x}(t))R^{-1}G^T(t,\bar{x}(t))\nabla h(\bar{x}(t))\mu(\{t\}) \leq 0.$$

Since $\nabla h^T(\bar{x}(t))G(t,\bar{x}(t))R^{-1}G^T(t,\bar{x}(t))\nabla h(\bar{x}(t)) > 0$, it follows that

$$\mu(\{t\}) = 0.$$

These relationships also imply (11.43). Since the support of μ is contained in $I(\bar{x}(.))$, we conclude that μ has no atoms in (S,T).

Step 2: We show that $t \to \int_{[S,t)} \mu(ds)$ is Lipschitz continuous on (S,T). As $\bar{u}(.)$ is continuous at $t = S$ and $t = T$, the Lipschitz continuity of $\bar{u}(.)$ on $[S,T]$ then follows directly from (11.42).

Assume to the contrary that the function is not Lipschitz continuous on (S, T). Then there exists $K_i \uparrow \infty$ and a sequence of intervals $\{[s_i, t_i]\}$ in (S, T) such that, for each i,

$$s_i \neq t_i \quad \text{and} \quad \int_{s_i}^{t_i} \mu(ds) = K_i|t_i - s_i|. \tag{11.44}$$

Since supp $\{\mu\} \subset I(\bar{x}(.))$, it follows that $[s_i, t_i] \cap I(\bar{x}(.)) \neq \emptyset$. Furthermore, we can arrange by increasing s_i and decreasing t_i if necessary that

$$s_i, t_i \in (S, T) \cap I(\bar{x}(.)).$$

In view of (11.44), we can ensure by subsequence extraction that either

(A): $\int_{s_i}^{(s_i + t_i)/2} \mu(ds) \geq \frac{1}{2} \int_{s_i}^{t_i} \mu(ds), \quad$ for all i

or

(B): $\int_{(s_i + t_i)/2}^{t_i} \mu(ds) \geq \frac{1}{2} \int_{s_i}^{t_i} \mu(ds), \quad$ for all i.

Assume first Case (A). Under the hypotheses and since ∇h is continuous, there exists $\beta > 0$ such that, for each i sufficiently large,

$$\nabla h^T(\bar{x}(t))GR^{-1}G^T\nabla h(\bar{x}(s)) > \beta \quad \text{for all } s, t \in [s_i, t_i].$$

Since $h(\bar{x}(s_i)) = 0$,

$$
\begin{aligned}
h(\bar{x}(t_i)) &= 0 + \int_{s_i}^{t_i} \tfrac{d}{dt} h(\bar{x}(s))ds \\
&= \int_{s_i}^{t_i} \nabla h^T(\bar{x}(t)) \left[f + GR^{-1}G^T\left(p(t) + \int_{[S,t)} \nabla h(\bar{x}(s))\mu(ds)\right)\right] dt \\
&= \int_{s_i}^{t_i} [D_i(t) + E_i(t)]dt,
\end{aligned}
$$

where

$$D_i(t) := \nabla h^T(\bar{x}(t))\left[f + GR^{-1}G^T\left(p(t) + \int_{[S,s_i)} \nabla h(\bar{x}(s))\mu(ds)\right)\right]$$

and

$$E_i(t) := \nabla h^T(\bar{x}(t))GR^{-1}G^T\int_{[s_i,t)} \nabla h(\bar{x}(s))\mu(ds).$$

Under the hypotheses, the functions $D_i : [s_i, t_i] \to R$, $i = 1, \ldots$, are Lipschitz continuous with a common local Lipschitz constant (write it K). Also, by (11.43),

$$D_i(s_i) = 0 \quad \text{for all } i.$$

It follows that

$$\int_{s_i}^{t_i} D_i(t)dt = \int_{s_i}^{t_i} (t_i - t)\frac{d}{dt} D_i(t)\, dt \geq -K\frac{(t_i - s_i)^2}{2}.$$

Also,

$$
\begin{aligned}
\int_{s_i}^{t_i} E_i(t)dt &= \int_{s_i}^{t_i} \nabla h^T(\bar{x}(t)) GR^{-1}G^T \int_{[s_i,t)} \nabla h(\bar{x}(s))\mu(ds)dt \\
&\geq \beta \int_{s_i}^{t_i} \int_{s_i}^{t} \mu(ds)dt \\
&= \beta[-\int_{s_i}^{t_i}(t-s_i)\mu(dt) + |t_i - s_i| \int_{s_i}^{t_i} \mu(dt)] \\
&= \beta \int_{s_i}^{t_i}(t_i - t)\mu(dt).
\end{aligned}
$$

But by (A),

$$
\begin{aligned}
\int_{s_i}^{t_i}(t_i - t)\mu(dt) &= \int_{s_i}^{(s_i+t_i)/2}(t_i - t)\mu(dt) + \int_{(s_i+t_i)/2}^{t_i}(t_i - t)\mu(dt) \\
&\geq \left(t_i - \frac{s_i+t_i}{2}\right)\int_{s_i}^{(s_i+t_i)/2}\mu(dt) + 0 \\
&\geq \frac{t_i - s_i}{2}\frac{K_i}{2}(t_i - s_i).
\end{aligned}
$$

Therefore,

$$
h(\bar{x}(t_i)) \geq -K\frac{(t_i - s_i)^2}{2} + \beta\frac{K_i}{4}(t_i - s_i)^2 \quad \text{for all } i.
$$

Since $K_i \uparrow \infty$ it follows that $h(\bar{x}(t_i)) > 0$, for i sufficiently large. This contradicts the fact that $\bar{x}$ satisfies the state constraint.

Similar reasoning leads to a contradiction in Case (B) also. Specifically, we examine the properties of the functions

$$
\tilde{D}_i(t) := \nabla h^T(\bar{x}(t))\left[f(t,\bar{x}(t)) + GR^{-1}G^T\left(p(t) + \int_{[S,t_i)} \nabla h(\bar{x}(s))\mu(ds)\right)\right]
$$

and

$$
\tilde{E}_i(t) := -\nabla h^T(\bar{x}(t))GR^{-1}G^T\int_{[t,t_i)} \nabla h(\bar{x}(s))\mu(ds)
$$

in place of D_i and E_i and show that, for i sufficiently large,

$$
\int_{s_i}^{t_i}[\tilde{D}_i + \tilde{E}_i]dt < 0.
$$

Since $h_j(\bar{x}(s_i)) = 0$ we have

$$
h_i(\bar{x}(t_i)) = 0 - \int_{s_i}^{t_i}[\tilde{D}_i + \tilde{E}_i]dt > 0,
$$

which is not possible. $\square$

11.8   Notes for Chapter 11

Regularity properties of minimizers for variational problems in one independent variable were studied extensively by Tonelli in the early years of the century, for inherent interest, no doubt, but also motivated by a desire to fill the gap between hypotheses for existence of minimizers and hypotheses needed to derive first-order necessary conditions of optimality and thereby to validate the Direct Method. The regularity issue has remained a central one in multidimensional Calculus of Variations (see [73]). However this aspect of Tonelli's legacy, notably the discovery that under the hypotheses of Existence Theory bad behavior can be confined to a closed set of zero measure (the Tonelli set), was unaccountably overlooked when one-dimensional Calculus of Variations evolved into Optimal Control. Interest in Tonelli Regularity was reactivated by Ball and Mizel, who saw potential applications to the field of Nonlinear Elasticity in which material failure can be associated with the existence of nonempty Tonelli sets. Ball and Mizel studied the structure of Tonelli sets and gave the first examples of problems satisfying the hypotheses of Existence Theory and yet having nonempty Tonelli sets [13], including the Ball–Mizel Example of Section 11.1. A brief proof by Vinter and Clarke that the Ball–Mizel Example exhibits the pathological behavior of interest, based on the construction of a nonsmooth verification function, appears in [13].

In a series of papers Clarke and Vinter brought together Tonelli's proof techniques, based on the application of necessary conditions to suitably regular auxiliary Lagrangians, and methods of nonsmooth analysis, to explore further the properties of minimizers under the hypotheses of Existence Theory. The scope for constructing nonsmooth Lagrangians adds greatly to the flexibility of the approach. [45] generalizes Tonelli's earlier results [134] on the structure of Tonelli sets, to allow for vector-valued arcs and nonsmooth Lagrangians, and to eliminate the need for strict convexity. [45] supplied the first proof that minimizers for autonomous problems satisfying the hypotheses of Existence Theory are Lipschitz continuous. Other sufficient conditions for Lipschitz continuity are proved, a sample of which are reproduced in this chapter. [46] concerns properties of the Tonelli set for noncoercive problems. In [47] it is established that Tonelli sets for polynomial Lagrangians are countable, with a finite number of accumulation points. Generalizations to variational problems involving higher derivatives and to optimal control problems with affine dynamic constraints are provided in [51] and [52], respectively.

The use of time reparameterization to supply an independent proof that minimizers for autonomous problems satisfying the hypotheses of Tonelli Existence Theory are Lipschitz continuous originates with Ambrosio et al. [1]. We provide a streamlined proof based on this approach, devised by Clarke [39].

The final section of the chapter concerns the direct application of necessary conditions to establish regularity of optimal controls. A significant early advance was Hager's proof of Lipschitz continuity of optimal controls, for optimal control problems with affine dynamics, a smooth, coercive cost integrand jointly convex with respect to state and control variables, and with unilateral state and control constraints satisfying an independence condition [78]. Extensions to allow for nonlinear dynamics were carried out by Malanowski [99]. A simple independent proof of Hager's regularity theorem, in the case of linear quadratic problems with affine state constraints, based on discrete approximations, was provided by Dontchev and Hager [58]. A more refined regularity analysis of this class of problems was undertaken by Dontchev and Kolmanovsky [59], who have given conditions for optimal controls to be piecewise analytic. The assertions of Theorem 11.7.1 are those of Malanowski's regularity theorem in the case of a single state constraint, no control constraints, and a quadratic control term in the cost. The proof, which allows nonsmooth data and a milder constraint qualification, strong normality, is new.

Chapter 12

Dynamic Programming

> ... *In place of determining the optimal sequence of decisions from the* fixed *state of the system, we wish to determine the optimal decision to be made at* any *state of the system. Only if we know the latter, do we understand the intrinsic structure of the solution.*
>
> — Richard Bellman, *Dynamic Programming*

12.1 Introduction

Consider the optimal control problem:

$$(P) \begin{cases} \text{Minimize } g(x(T)) \\ \text{over arcs } x \in W^{1,1}([S,T]; R^n) \text{ satisfying} \\ \dot{x}(t) \in F(t, x(t)) \quad \text{a.e.,} \\ x(S) = x_0, \end{cases}$$

the data for which comprise an interval $[S, T] \subset R$, a function $g : R^n \to R \cup \{+\infty\}$, a multifunction $F : [S, T] \times R^n \rightsquigarrow R^n$, and a point $x_0 \in R^n$.

Notice that the cost function g is allowed to take value $+\infty$. Implicit in this formulation then is the endpoint constraint:

$$x(T) \in C,$$

where

$$C := \{x \in R^n : g(x) < \infty\}.$$

Dynamic Programming, as it relates to the above problem, concerns the relationship between minimizers and the infimum cost of (P) and, on the other hand, solutions to the Hamilton–Jacobi Equation:

$$\phi_t(t, x) + min_{v \in F(t,x)} \phi_x(t, x) \cdot v = 0 \qquad \text{for all } (t, x) \in D \text{ (12.1)}$$
$$\phi(T, x) = g(x) \quad \text{for all } x \in D_1. \quad \text{(12.2)}$$

Here, D and D_1 are given subsets of $[S, T] \times R^n$ and R^n, respectively. (12.1) can alternatively be expressed in terms of the Hamiltonian

$$H(t, x, p) := \sup_{v \in F(t,x)} p \cdot v,$$

thus

$$\phi_t(t, x) - H(t, x, -\phi_x(t, x)) = 0 \qquad \text{for all } (t, x) \in D.$$

The link between the optimal control problem (P) and the Hamilton–Jacobi Equation is the value function $V : [S, T] \to R \cup \{+\infty\}$: for each $(t, x) \in [S, T] \times R^n$, $V(t, x)$ is defined to be the infimum cost for the problem

$$(P_{t,x}) \quad \begin{cases} \text{Minimize } g(y(T)) \\ \text{over arcs } y \in W^{1,1}([t, T]; R^n) \text{ satisfying} \\ \dot{y}(s) \in F(s, y(s)) \quad \text{a.e.,} \\ y(t) = x. \end{cases}$$

We write this relationship

$$V(t, x) = \inf(P_{t,x}).$$

The elementary theory (see Chapter 1) tells us that, if V is a C^1 function then, under appropriate hypotheses on the data for (P), V is a solution to (12.1) and (12.2) when $D = (S, T) \times R^n$ and $D_1 = R^n$.

Of course knowledge of the value function V provides the minimum cost for (P): it is simply V evaluated at (S, x_0). But in favorable circumstances, it also supplies information about minimizers. Suppose that V is a C^1 function and also that, for each (t, x),

$$\chi(t, x) := \{v \in F(t, x) : V_x(t, x) \cdot v = \min_{v' \in F(t,x)} V_x(t, x) \cdot v'\} \qquad (12.3)$$

is nonempty and single-valued. Finally suppose that

$$\dot{y}(s) = \chi(s, y(s)), \quad y(t) = x \qquad (12.4)$$

has a $W^{1,1}$ solution y on $[t, T]$, for any $(t, x) \in [S, T] \times R^n$. Then y is a minimizer for $(P_{t,x})$. Indeed, the calculation

$$\begin{aligned} V(t, x) &= V(T, y(T)) - \int_t^T \frac{d}{ds} V(s, y(s)) ds \\ &= V(T, y(T)) - \int_t^T [V_t(s, y(s)) + V_x(s, y(s)) \cdot \chi(s, y(s))] ds \\ &= V(T, y(T)) - \int_t^T [V_t(s, y(s)) + \min_{v \in F(t,x)} V_x(s, y(s)) \cdot v] ds \\ &= V(T, y(T)) - 0 = g(y(T)) \end{aligned}$$

shows that y has cost $V(t, x)$; i.e., y is a minimizer for $(P_{t,x})$. In particular, selecting $(t, x) = (S, x_0)$ and solving the differential equation (12.4) yields a minimizer for (P). An advantage to this approach is that it supplies the minimizer in "feedback" form, favored in control engineering applications.

In a sense then, Dynamic Programming reduces the optimal control problem to one of solving a partial differential equation.

In current Dynamic Programming research, the role of the Hamilton–Jacobi Equation in characterizing the value function is emphasized. But the value function as an object of interest in its own right is a relative newcomer to Variational Analysis. Traditionally, the Hamilton–Jacobi Equation has had a different role: that of providing sufficient conditions of optimality, to test whether a putative minimizer (arrived at by finding an arc that satisfies some set of necessary conditions, say) is truly a minimizer. Using the Hamilton–Jacobi Equation in this spirit is called the Carathéodory method. The essential character of the approach is captured by the following sufficient condition of optimality.

Let $\bar{x}$ be an arc that satisfies the constraints of (P). Then $\bar{x}$ is a strong local minimizer if, for some $\epsilon > 0$, a C^1 function[1] $\phi : T(\bar{x}, \epsilon) \to R$ can be found satisfying the Hamilton–Jacobi Equation (12.1) and (12.2) with $D = \operatorname{int} T(\bar{x}, \epsilon)$ and $D_1 = (\bar{x}(T) + \epsilon B) \cap \operatorname{dom} g$ and if

$$\phi(S, x_0) = g(\bar{x}(T)). \tag{12.5}$$

In the above, $T(\bar{x}, \epsilon)$ denotes the ϵ-tube about $\bar{x}$:

$$T(\bar{x}, \epsilon) := \{(t, y) \in [S, T] \times R^n : y \in \bar{x}(t) + \epsilon B\}.$$

To justify these assertions, we have merely to note that, for any arc x satisfying the constraints of (P) and also the condition $\|x - \bar{x}\|_{L^\infty} < \epsilon$, we have

$$
\begin{aligned}
\phi(S, x_0) &= \phi(T, x(T)) - \int_t^T \frac{d}{ds} \phi(s, x(s)) ds \\
&= \phi(T, y(T)) - \int_t^T [\phi_t(s, x(s)) + \phi_x(s, x(s)) \cdot \dot{x}(s)] ds \\
&\leq g(x(T)) + 0.
\end{aligned}
$$

This inequality combines with (12.5) to confirm that strong local optimality of $\bar{x}$. Appropriately, functions ϕ used in this way are called *Verification Functions*. The application of the Carathéodory method is illustrated, for example, in L. C. Young's book [152], where it is used to solve a number of classical problems in the Calculus of Variations. For many of these problems, it is comparatively straightforward to determine a candidate for minimizer by solving Euler's Equation (a necessary condition for an arc to

---

[1] Given a closed set $A \subset R^k$ and a function $\psi : A \to R^m$, we say that ψ is a C^1 function if it is continuous, if it is of class C^1 on the interior of A, and if $\nabla \psi$ extends, as a continuous function, to all of A.

be a minimizer). Confirming this arc is truly a minimizer can be accomplished in favorable circumstances by constructing a verification function.

An important feature of verification functions is that they do not have to be the value function for (P). In many cases, *finite* verification functions serve to confirm the local minimality of a particular arc, even when the value function is infinite at some points in its domain. This flexibility can simplify the task of finding verification functions in specific applications.

How have these ideas evolved in recent years? Consider the relationship between the Hamilton–Jacobi Equation and the value function. The first issue to be settled, if we are to regard solving the Hamilton–Jacobi Equation as a means to generating the value function, is whether the Hamilton–Jacobi Equation has a unique solution that coincides with the value function. This question is a challenging one because, for many optimal control problems of interest, the value function is not continuously differentiable. A simple example of this phenomenon was discussed in the introduction. At the outset then, we must come up with a suitable concept of generalized solution to the Hamilton–Jacobi Equation. Minimal requirements are:

(i) the (possibly nondifferentiable) value function is a generalized solution,

and

(ii) there exists a unique generalized solution.

Viscosity solutions, as introduced in the early 1980s, met these requirements in situations where the value function is continuous: a continous function $\phi : [S, T] \times R^n \to R$ is said to be a *continuous viscosity solution* of

$$\phi_t(t, x) - H(t, x, -\phi_x(t, x)) = 0 \quad \text{for all } (t, x) \in (S, T) \times R^n$$

if the following two conditions are satisfied.

(a) For any point $(t, x) \in (S, T) \times R^n$ and any C^1 function $w : R \times R^n \to R$ such that $(t', x') \to \phi(t', x') - w(t', x')$ has a local minimum at (t, x) we have

$$\eta^0 - H(t, x, -\eta^1) \leq 0,$$

where $(\eta^0, \eta^1) = \nabla w(t, x)$

and

(b) For any point $(t, x) \in (S, T) \times R^n$ and any C^1 function $w : R \times R^n \to R$ such that $(t', x') \to \phi(t', x') - w(t', x')$ has a local maximum at (t, x) we have

$$\eta^0 - H(t, x, -\eta^1) \geq 0,$$

where $(\eta^0, \eta^1) = \nabla w(t, x)$.

Notice that two one-sided conditions have replaced the original "equality" condition. This is a hallmark of viscosity solutions.

Innovative methods in the analysis of nonlinear partial differential equations (Viscosity Techniques) established that, under hypotheses that include uniform continuity of the value function, the value function is the unique uniformly continuous, generalized solution to the Hamilton–Jacobi Equation (in the viscosity sense). In fact the scope of Viscosity Techniques extends beyond problem (P); they have been used to characterize value functions as "solutions" to Hamilton–Jacobi type equations associated with a wide variety of dynamic optimization and differential games problems.

These developments, based on *continuous* generalized solutions of the Hamilton–Jacobi Equation, left unanswered the question of how to characterize the value function for optimal control problems with a right endpoint constraint. The cost functions for such problems are extended-valued, discontinuous functions and give rise to extended-valued, discontinuous value functions.

In the context of problems with right endpoint constraints it is not tenable then to invoke continuity of the value function. The value function is lower semicontinuous, however, under unrestrictive verifiable hypotheses on the data. This suggests that we should aim to characterize value functions in terms of lower semicontinuous solutions to the Hamilton–Jacobi Equation, appropriately defined.

The centerpiece of this chapter is a framework for studying lower semicontinuous "solutions" to the Hamilton–Jacobi Equation which relates them to the corresponding value function. It is based on the concept of lower semicontinuous proximal solutions to the Hamilton–Jacobi Equation.

A lower semicontinuous function $\phi : [S, T] \times R^n \to R \cup \{+\infty\}$ is said to be a proximal solution to the Hamilton–Jacobi Equation if at each point $(t, x) \in (S, T) \times R^n$ such that $\partial^P \phi(t, x)$ is nonempty,

$$\eta^0 - H(t, x, -\eta^1) \ = \ 0 \quad \text{for all } (\eta^0, \eta^1) \in \partial^P \phi(t, x).$$

Notice that Condition (a) in the earlier definition of the "continuous" viscosity solution implies

$$\eta^0 + H(t, x, \eta^1) \le 0 \quad \text{for all } (\eta^0, \eta^1) \in \partial^P \phi(t, x)$$

at every point $(t, x) \in (S, T) \times R^n$ such that $\partial^P \phi(t, x)$ is nonempty, because $(\eta^0, \eta^1) \in \partial^P \phi(t, x)$ means that (t, x) is a local minimizer of $\phi - w$, where w is the quadratic function

$$w(t', x') \ = \ (t', x') \cdot (\eta^0, \eta^1) - M(|(t', x') - (t, x)|^2),$$

for some $M > 0$.

We see that this definition of "lower semicontinuous" generalized solution is arrived at by discarding condition (b) in the definition of viscosity solutions, relaxing condition (a) somewhat, and replacing inequality by equality.

The link between lower semicontinuous solutions to the Hamilton–Jacobi Equation and value functions is achieved, not by studying the partial differential equation directly, but as a byproduct of theorems on the invariance properties of solutions to differential inclusions. Our use of control theoretic ideas is consistent with analytical techniques employed elsewhere in this book, and illustrates further areas of application of generalized subdifferentials.

The invariance theorems, on which the analysis is based, are of great independent interest and have been applied in many areas of dynamic systems theory and in nonlinear analysis. (The systematic use of invariance theorems, such as those proved in this chapter, is called Viability Theory).

Of course showing that the value function is a generalized solution of the Hamilton–Jacobi Equation falls somewhat short of obtaining detailed information about minimizers for (P). What is required is an analysis of the feedback map χ of (12.3), or some nonsmooth analogue of it, and also means of interpreting solutions to the "closed loop equation" $\dot{x} \in \chi(t, x)$ (we can expect the right side to be multivalued) which generates minimizers for (P). This important area of study, *Optimal Synthesis* as it is called, requires rather different analytical techniques to those used elsewhere in this book and is not entered into here.

Recent research in Carathéodory's method has been largely concerned with the inverse problem: what concept of verification function should we adopt in order that, corresponding to every local minimizer, there exists a verification function? This area of research aims at simplifying the task of finding a verification function to confirm the optimality of a specified putative minimizer: the smaller the class of verification functions, one of which can be guaranteed to confirm the optimality of a given minimizer, the narrower will be the search and therefore the "better" the inverse theorem. The main result in this area, proved in this chapter, is that, under a mild nondegeneracy hypothesis, there always exists a Lipschitz continuous verification function, even in situations when there are right endpoint constraints and when the value function (which is after all the obvious choice for verification function) is not even continuous!

Another topic covered in this chapter is the interpretation of adjoint arcs in terms of generalized gradients of value functions. Connections are thereby made between the necessary conditions of earlier chapters and Dynamic Programming.

The relevant relationships were sketched in Chapter 1, where the adjoint arcs were associated with the Maximum Principle. We review them, in the context of the optimal control problem (P). Now, gradients of the value function are related to the adjoint arc appearing in the Extended Euler–Lagrange Condition.

Let $\bar{x}$ be a minimizer. To simplify the analysis, let us suppose that V is a C^2 function, that g is a C^1 function, and that $F(t,.)$ has a closed graph. We deduce from the fact that $g(.)$ coincides with $V(T,.)$ that

$$V_x(T, \bar{x}(T)) = g_x(\bar{x}(T)).$$

Since $\bar{x}$ is a minimizer, we have $V(S, x_0) = g(\bar{x}(T)) = V(T, \bar{x}(T))$. It then follows from (12.1) and the identity

$$V(S, x_0) = V(T, \bar{x}(T)) - \int_S^T \{V_t(t, \bar{x}(t)) + V_x(t, \bar{x}(t)) \cdot \dot{\bar{x}}(t)\}dt$$

that

$$\begin{aligned} V_t(t, \bar{x}(t)) + V_x(t, \bar{x}(t)) \cdot \dot{\bar{x}}(t) = \\ \min[V_t(t, x) + V_x(t, x) \cdot v \; : \; x \in R^n, \; v \in F(t, x)] \quad \text{a.e.} \end{aligned}$$

Clearly,

$$-V_x(t, \bar{x}(t)) \cdot \dot{\bar{x}}(t) = \max_{v \in F(t, \bar{x}(t))} (-V_x(t, \bar{x}(t))) \cdot v \quad \text{a.e.}$$

Furthermore, from the Multiplier Rule, we have that

$$-(V_{tx}(t, \bar{x}(t)) + V_{xx}(t, \bar{x}(t)) \cdot \dot{\bar{x}}(t), V_x(t, \bar{x}(t))) \in N_{\mathrm{Gr}\, F(t,.)}(\bar{x}(t), \dot{\bar{x}})(t) \quad \text{a.e.,}$$

Now define

$$\begin{aligned} p(t) &:= -V_x(t, \bar{x}(t)), \\ h(t) &:= \max_{v \in F(t, \bar{x}(t))} p(t) \cdot v \; (= p(t) \cdot \dot{\bar{x}}(t)). \end{aligned}$$

Since $\dot{p}(t) = -V_{tx}(t, \bar{x}(t)) - V_{xx}(t, \bar{x}(t)) \cdot \dot{\bar{x}}(t)$ we deduce from these relationships that

$$\begin{aligned} (\dot{p}(t), p(t)) &\in N_{\mathrm{Gr}\, F(t,.)}(\bar{x}(t), \dot{\bar{x}}(t)) \quad \text{a.e.,} \\ p(t) \cdot \dot{\bar{x}}(t) &= \max_{v \in F(t, \bar{x}(t))} p(t) \cdot v \quad \text{a.e.,} \\ -p(T) &= g_x(\bar{x}(T)), \\ (h(t), -p(t)) &= \nabla V(t, \bar{x}(t)) \quad \text{a.e.} \end{aligned} \tag{12.6}$$

We have arrived at a set of necessary conditions including the Euler–Lagrange Condition, in which the Hamiltonian, evaluated along the optimal trajectory, and adjoint arc are interpreted in terms of gradients of the value function. In general, V is not a C^2 function and the above arguments cannot be justified. We do however derive, by means of a more sophisticated analysis, necessary conditions of optimality involving an adjoint arc p that satisfies a nonsmooth version of (12.6), namely,

$$(h(t), -p(t)) \in \mathrm{co}\, \partial V(t, \bar{x}(t)) \quad \text{a.e.} \tag{12.7}$$

We conclude this introduction with two elementary relationships satisfied by the value function for (P), ones to which frequent reference is made. Take points s_1, $s_2 \in [S, T]$ with $s_1 \leq s_2$, a minimizer $\bar{x} : [s_1, T] \to R^n$ (for problem $(P_{s_1, \bar{x}(s_1)})$) and also an F-trajectory $x : [s_1, T] \to R^n$. Then

$$V(s_1, x(s_1)) \quad \leq \quad V(s_2, x(s_2)) \tag{12.8}$$
$$V(s_1, \bar{x}(s_1)) \quad = \quad V(s_2, \bar{x}(s_2)). \tag{12.9}$$

These relationships are collectively referred to as the *Principle of Optimality*, which can be paraphrased as

the value function for (P) is nondecreasing along an arbitrary F-trajectory, and is constant along a minimizing F-trajectory.

We validate this principle. Suppose that (12.8) is false. Then there exists $\epsilon > 0$ such that
$$V(s_1, x(s_1)) > V(s_2, x(s_2)) + \epsilon.$$
By definition of V, there exists an F-trajectory $z : [s_2, T] \to R^n$ such that $z(s_2) = x(s_2)$ and
$$V(s_2, x(s_2)) > g(z(T)) - \epsilon/2.$$
But then the F-trajectory $y : [S, T] \to R^n$, defined as

$$y(t) \; = \; \begin{cases} x(t) & s_1 \leq t < s_2 \\ z(t) & s_2 \leq t \leq T \end{cases}$$

satisfies

$$\begin{aligned} V(s_1, x(s_1)) \leq g(y(T)) = g(z(T)) \quad &< \quad V(s_2, x(s_2)) + \epsilon/2 \\ &< \quad V(s_1, x(s_1)) - \epsilon/2. \end{aligned}$$

From this contradiction we deduce that (12.8) is true.

As for (12.9), we note that, again by definition of V,

$$V(s_2, \bar{x}(s_2)) \leq g(\bar{x}(T)) = V(s_1, \bar{x}(s_1)).$$

This combines with Condition (12.8) to give Condition (12.9).

12.2   Invariance Theorems

Invariance theorems concern solutions to a differential inclusion, that satisfy a specified constraint. Theorems giving conditions for existence of at least one solution satisfying the constraint are called *weak invariance theorems* or *viability theorems*. Those asserting that *all* solutions satisfy the constraint are called *strong invariance theorems*.

These theorems have far-reaching implications in stability theory, Dynamic Programming, Robust Controller Design, and Differential Games (to name but a few applications areas). Here, however, we concentrate on versions of the theorems suitable for characterizing value functions in Optimal Control.

The starting point for all the important results in this section is a theorem providing conditions on an "autonomous" multifunction $F : R^n \rightsquigarrow R^n$ and a closed set $D \subset R^n$ such that, for a given initial state x_0 in D, the differential inclusion and accompanying constraint:

$$\begin{cases} \dot{x} \in F(x) & \text{a.e. } t \in [S, \infty) \\ x(S) = x_0 \\ x(t) \in D & \text{for all } t \in [S, \infty) \end{cases}$$

have a locally absolutely continuous solution. Typically, the hypotheses invoked in theorems of this nature include the requirement that, corresponding to any point $x \in D$, $F(x)$ contains a tangent vector to D; i.e., there is an admissible velocity pointing into the set D. In the following theorem, this "inward pointing" condition takes the form: for every point $x \in D$ such that $N_D^P(x)$ is nonempty,

$$\min_{v \in F(t,x)} \xi \cdot v \leq 0 \quad \text{for all } \xi \in N_D^P(x).$$

The condition imposes restrictions on (F, D) only at points x in $\mathcal{D}$:

$$\mathcal{D} := \{x \in D : N_D^P(x) \neq \emptyset\}.$$

$\mathcal{D}$ can be a rather sparse subset of D; it is surprising that such an economical condition suffices to guarantee existence of an F-trajectory satisfying the constraint.

Theorem 12.2.1 (Weak Invariance Theorem for Autonomous Systems) *Take a multifunction $F : R^n \rightsquigarrow R^n$, a number S, and a closed set $D \subset R^n$. Assume:*

(i) Gr F is closed and $F(x)$ is a nonempty convex set for each $x \in R^n$;

(ii) there exists $c > 0$ such that

$$F(x) \subset c(1 + |x|)B \quad \text{for all } x \in R^n.$$

Assume further that, for every $x \in D$ such that $N_D^P(x)$ is nonempty,

$$\min_{v \in F(x)} \zeta \cdot v \leq 0 \quad \text{for all } \zeta \in N_D^P(x).$$

Then, given any $x_0 \in D$ and $S \in R$, there exists a locally absolutely continuous function x satisfying

$$\begin{cases} \dot{x}(t) \in F(t, x(t)) & \text{a.e. } t \in [S, +\infty), \\ x(S) = x_0, \\ x(t) \in D & \text{for all } t \in [S, +\infty). \end{cases}$$

Proof. It suffices to prove existence of an arc x satisfying the stated conditions on an arbitrary finite interval $[S,T]$, since we can then generate an arc on $[0,+\infty)$ by concatenating a countable number of such arcs.

Observe that the hypothesis (ii) in the theorem statement can be replaced by the stronger hypothesis:

(H') there exists $K > 0$ such that $F(x) \subset KB$ for all $x \in R^n$,

without loss of generality. Indeed, choose any r satisfying

$$r > \left[e^{c|T-S|} (|x_0| + c|T-S|) \right].$$

Define

$$\tilde{F}(x) := \begin{cases} F(x) & \text{if } |x| \leq r \\ \text{co} \left((F(rx/|x|) \cup \{0\}) \right) & \text{if } |x| \geq r. \end{cases}$$

$\tilde{F}$ and D satisfy the hypotheses of the theorem statement, and also (H'), in which $K = c(1+r)$. If the assertions of the theorem are valid under the stronger hypotheses then, there exists an $\tilde{F}$-trajectory z such that $z(S) = x_0$ and $z(t) \in D$ for all $t \in [S,T]$. But

$$|\dot{z}| \leq c(1 + |z|),$$

so by Gronwall's Inequality (Lemma 2.3.4)

$$|z(t)| \leq e^{c|T-S|} (|x_0| + c|T-S|) < r.$$

Since F and $\tilde{F}$ coincide on rB, z is also an F-trajectory. It follows that the assertions of the "finite interval" version of the theorem are true with $x = z$. So we can assume (H').

Fix an integer $m > 0$. Let $\{t_0 = S, t_1, \ldots, t_m = T\}$ be a uniform partition of $[S,T]$. Set $h_m := |T-S|/m$.

Define sequences $\{x_0^m, \ldots, x_m^m\}$, $\{v_0^m, \ldots, v_{m-1}^m\}$, and $\{y_0^m, \ldots, y_{m-1}^m\}$ recursively as follows: take $x_0^m = x_0$. If, for some $i \in \{0, \ldots, m-1\}$, x_i^m is given then choose $y_i^m \in D$ to satisfy

$$|x_i^m - y_i^m| = \inf\{|x_i^m - y| : y \in D\}.$$

We now make use of the fact that, since y_i^m is a closest point to x_i^m in D, $x_i^m - y_i^m \in N_D^P(y_i^m)$. Under the hypotheses then, $v_i^m \in F(y_i)$ can be chosen to satisfy

$$v_i^m \cdot (x_i^m - y_i^m) \leq 0. \tag{12.10}$$

Finally, set

$$x_{i+1}^m = x_i^m + h_m v_i^m.$$

Now we define $z^m : [S,T] \to R^n$ to be the polygonal arc whose graph is the linear interpolant between the node points $\{(t_0^m, x_0^m), \ldots, (t_m^m, x_m^m)\}$. We have, for $i = 0, 1, \ldots, m-1$,

$$z^m(t) = x_i^m + (t - t_i^m)v_i^m \quad \text{for all } t \in [t_i^m, t_{i+1}^m].$$

Since $|v_0^m| \leq K$ and $x_0^m \in D$,

$$d_D^2(x_1^m) \leq |x_1^m - x_0^m|^2 = h_m^2 |v_0|^2 \leq K^2(h_m)^2. \tag{12.11}$$

Fix i. Since $y_i^m \in D$, we have

$$\begin{aligned} d_D^2(x_i^m) &\leq |x_i^m - y_{i-1}^m|^2 \\ &= |x_{i-1}^m - y_{i-1}^m|^2 + |x_i^m - x_{i-1}^m|^2 + 2\left(x_{i-1}^m - y_{i-1}^m\right) \cdot \left(x_i^m - x_{i-1}^m\right) \\ &\leq d_D^2(x_{i-1}^m) + K^2(h_m)^2 + 0 \end{aligned} \tag{12.12}$$

(by (12.10)).

Relationships (12.11) and (12.12), which are valid for arbitrary $i \in \{1, \ldots, m\}$, combine to tell us that

$$d_D^2(x_i^m) \leq mK^2(h_m)^2 = |T - S|^2 K^2 m^{-1}, \tag{12.13}$$

for $i = 0, \ldots, m$.

The functions $\{z^m\}_{m=1}^\infty$ are uniformly bounded and have common Lipschitz constant K. By the Compactness of Trajectories Theorem (Theorem 2.4.3), there exists a Lipschitz continuous arc z such that, following the extraction of a suitable subsequence, we have

$$z^m(.) \to z(.) \quad \text{uniformly as } m \to \infty \tag{12.14}$$

and

$$\dot{z}^m(.) \to \dot{z}(.) \quad \text{weakly in } L^1.$$

In view of (12.13) and the continuity of the distance function, we have

$$d_D(z(t)) = 0 \quad \text{for all } t \in [S, T].$$

Since D is closed, $z(t) \in D$ for all $t \in [S, T]$.

It remains to check that z is an F-trajectory. Suppose to the contrary that $\dot{z}(t) \notin F(z(t))$ on a set of positive measure. We deduce from the upper semicontinuity of F that there exists $\delta > 0$ such that

$$\dot{z}(t) \notin F_\delta(t)$$

on a set of positive measure. Here

$$F_\delta(t) := \bar{\text{co}} \{v : v \in F(\xi) \text{ for some } \xi \in z(t) + \delta B\} + \delta B.$$

Arguing as in the proof of the Compactness of Trajectories Theorem (Theorem 2.4.3), we can find some $p \in R^n$ and a subset $A \subset [S, T]$ of positive measure such that

$$p \cdot \dot{z}(t) > h(t) \quad \text{for all } t \in A. \tag{12.15}$$

Here, h is the bounded measurable function

$$h(t) = \max_{v \in F_\delta(t)} p \cdot v.$$

By (12.14) however,

$$\dot{z}^m(t) \in F_\delta(t) \quad \text{a.e.}$$

for all m sufficiently large. This implies that

$$p \cdot \dot{z}^m(t) \le h(t) \quad \text{a.e.}$$

for all m sufficiently large. By weak convergence of $\{\dot{z}^m\}$ however,

$$\int_A h(t)dt \ge \lim_{m \to \infty} \int_A p \cdot \dot{z}^m(t)dt = \int_A p \cdot \dot{z}(t)dt.$$

This contradicts (12.15). It follows that z is an F-trajectory. $\square$

Theorem 12.2.2 (Weak Invariance Theorem for Time-Varying Systems) *Take multifunctions $F : [S,T] \times R^n \rightsquigarrow R^n$, $P : [S,T] \rightsquigarrow R^n$, a number $\bar{S} \in [S,T]$, and a point $x_0 \in P(\bar{S})$. Assume that*

(i) Gr F *is closed and F takes as values nonempty convex sets;*

(ii) there exists $c > 0$ such that

$$F(t,x) \subset c\,(1 + |x|)\,B \quad \text{for all } (t,x) \in [S,T] \times R^n;$$

(iii) Gr P *is closed and, in the case $\bar{S} = S$,*

$$x_0 \in \limsup_{t \downarrow S} P(t). \tag{12.16}$$

Assume further that, for every $(t,x) \in$ Gr $P \cap ((S,T) \times R^n)$ such that $N^P_{\text{Gr}\,P}(t,x)$ is nonempty,

$$\zeta^0 + \min_{e \in F(t,x)} \zeta^1 \cdot e \le 0 \quad \text{for all } (\zeta^0, \zeta^1) \in N^P_{\text{Gr}\,P}(t,x).$$

Then there exists a locally absolutely continuous function x satisfying

$$\begin{cases} \dot{x}(t) \in F(t,x(t)) & \text{a.e. } t \in [\bar{S},T], \\ x(\bar{S}) = x_0, \\ x(t) \in P(t) & \text{for all } t \in [\bar{S},T]. \end{cases}$$

Proof. Define $\tilde{F} : R \times R^n \rightsquigarrow R^{1+n}$, $\tilde{P} : R \rightsquigarrow R^n$, and $\tilde{D} \subset R^{1+n}$ to be

$$\tilde{F}(t,x) := \begin{cases} \{1\} \times F(t,x) & \text{for } \bar{S} < t < T \\ \text{co}\,(\{(0,0)\} \cup (\{1\} \times F(S,x))) & \text{for } t \le \bar{S} \\ \text{co}\,(\{(0,0)\} \cup (\{1\} \times F(T,x))) & \text{for } t \ge T, \end{cases}$$

$$\tilde{P}(t) := \begin{cases} P(t) & \text{for } \bar{S} < t < T \\ P(S) & \text{for } t \leq \bar{S} \\ P(T) & \text{for } t \geq T \end{cases}$$

and

$$\tilde{D} := \text{Gr } \tilde{P}.$$

According to (12.16), we can choose sequences $\{t_0^i\} \subset (S,T)$ and $\{x_0^i\} \subset R^n$ such that $t_0^i \to \bar{S}$, $x_0^i \to x_0$, and $x_0^i \in P(t_0^i)$ for all i. (In the case $\bar{S} \in (S,T]$ we choose $t_0^i = \bar{S}$ and $x_0^i = x_0$ for all i.)

Fix i and apply the Autonomous Weak Invariance Theorem (Theorem 12.2.1) to $(\tilde{F}, \tilde{D})$, with initial time t_0^i and initial state (t_0^i, x_0). (It is a straightforward matter to check that the relevant hypotheses are satisfied.) This supplies an arc $(\tau^i(.), x^i(.)) : [t_0^i, T] \to R \times R^n$ such that

$$\begin{cases} (\dot{\tau}^i(t), \dot{x}^i(t)) & \in & \tilde{F}(\tau^i(t), x^i(t)) \quad \text{a.e. } t \in [t_0^i, +\infty) \\ (\tau^i, x^i)(t_0^i) & = & (t_0^i, x_0^i) \\ (\tau^i(t), x^i(t)) & \in & \tilde{D} \quad \text{for all } t \in [t_0^i, +\infty). \end{cases}$$

Clearly,

$$0 \leq \dot{\tau}^i(t) \leq 1.$$

It follows that

$$S < t_0^i \leq \tau(t) < T \quad \text{a.e. } t \in [t_0^i, T).$$

By the definition of $\tilde{F}$ then,

$$\dot{\tau}^i(t) = 1 \quad \text{a.e. } t \in [t_0^i, T].$$

We conclude that $\tau^i(t) = t$ for all $t \in [t_0^i, T]$, whence x^i satisfies

$$\begin{aligned} \dot{x}^i(t) & \in & F(t, x^i(t)) \quad \text{a.e. } t \in [t_0^i, T] \\ x^i(t) & \in & P(t) \quad \text{for all } t \in [t_0^i, T]. \end{aligned}$$

Now extend the domain of x^i to $[\bar{S}, +\infty)$ by constant extrapolation from the right and restrict the resulting function to $[\bar{S}, T]$. This last function we henceforth denote x_i.

Consider the sequence of arcs $x_i(.) : [\bar{S}, T] \to R^n$. The x_is are uniformly bounded and the $\dot{x}$s are uniformly integrably bounded. According to the Compactness of Trajectories Theorem (Theorem 2.4.3), there exists an absolutely continuous arc x, which is an F-trajectory on $[S, T]$ and

$$x^i(t) \to x(t) \quad \text{for all } t \in [\bar{S}, T].$$

Because the x^i's are equicontinuous, $x^i(t_0^i) = x_0^i$ for each i and P has a closed graph, it follows that the F-trajectory x satisfies the conditions $x(\bar{S}) = x_0$ and $x(t) \in P(t)$ for all $t \in [\bar{S}, T]$. This is what we set out to prove. $\square$

We now consider strong invariance properties of differential inclusions. In contrast to weak invariance theorems, in which the existence of some F-trajectory is asserted under regularity hypotheses on F which merely require F to have a closed graph, strong invariance theorems typically invoke stronger, "Lipschitz" regularity hypotheses concerning F.

The first strong invariance theorem we state covers autonomous differential inclusions and constraints. The proof, due to Clarke and Ledyaev, is remarkable for its simplicity and for the fact that it covers cases in which F is possibly nonconvex-valued and fails even to have closed values.

Theorem 12.2.3 (An Autonomous Strong Invariance Theorem)
Take a multifunction $F : R^n \rightsquigarrow R^n$ and a closed set $D \subset R^n$. Assume that

(i) F takes as values nonempty sets;

(ii) there exists $K > 0$ such that

$$F(x') \subset F(x'') + K|x' - x''|B \quad \text{for all } x', x'' \in R^n.$$

Suppose furthermore that for each $x \in D$ such that $N_D^P(x)$ is nonempty

$$\sup_{v \in F(x)} \zeta \cdot v \leq 0 \quad \text{for all } \zeta \in N_D^P(x). \tag{12.17}$$

Then for any Lipschitz continuous function $x(.) : [S, T] \to R^n$ satisfying

$$\begin{cases} \dot{x}(t) \in F(x(t)) & a.e. \ t \in [S, T] \\ x(S) = x_0 \end{cases}$$

we have

$$x(t) \in D \quad \text{for all } t \in [S, T].$$

Proof. Since $x(.)$ is Lipschitz continuous and $d_D^2(.)$ is locally Lipschitz continuous, the function $s \to d_D^2(x(s))$ is Lipschitz continuous. Choose a point $t \in (S, T)$ such that $x(.)$ and $s \to d_D^2(x(s))$ are differentiable at t and also $\dot{x}(t) \in F(x(t))$. The set of such points has full measure.

Let z be a closest point to $x(t)$ in D. Then $z \in D$ and $x(t) - z \in N_D^P(z)$. Under the hypotheses, there exists $v \in F(z)$ such that

$$|v - \dot{x}(t)| \leq K|x(t) - z|.$$

In view of (12.17),

$$(x(t) - z) \cdot v \leq 0.$$

Since

$$\begin{aligned}
(x(t) - z) \cdot \dot{x}(t) &= (x(t) - z) \cdot v + (x(t) - z) \cdot (\dot{x}(t) - v) \\
&\leq 0 + |x(t) - z| \cdot |\dot{x}(t) - v| \leq K|x(t) - z|^2,
\end{aligned}$$

it follows that

$$(x(t) - z) \cdot \dot{x}(t) \leq K d_D^2(x(t)). \tag{12.18}$$

Now

$$d_D^2(x(t)) = |x(t) - z|^2$$

and, for any $\delta \in (0, T - t)$,

$$d_D^2(x(t + \delta)) = \inf_{y \in D} |x(t + \delta) - y|^2 \leq |x(t + \delta) - z|^2.$$

We conclude that

$$
\begin{aligned}
(d/dt) d_D^2(x(t)) &= \lim_{\delta \downarrow 0} \delta^{-1} \left[d_D^2(x(t + \delta)) - d_D^2(x(t)) \right] \\
&\leq \lim_{\delta \downarrow 0} \delta^{-1} \left[\psi(x(t + \delta)) - \psi(x(t)) \right] \\
&= \nabla \psi \cdot \dot{x}(t) = 2(x(t) - z(t)) \cdot \dot{x}(t)
\end{aligned}
$$

in which $\psi : R^n \to R$ is the analytic function

$$\psi(x) := |x - z|^2.$$

We deduce from (12.18) that

$$(d/dt) d_D^2(x(t)) \leq 2K d_D^2(x(t)).$$

This inequality holds for almost every $t \in [S, T]$. Since $d_D^2(x(S)) = 0$, we conclude from Gronwall's Inequality (Lemma 2.3.4) that

$$d_D^2(x(t)) = 0 \quad \text{for all } t \in [S, T].$$

Since D is closed, this implies that $x(t) \in D$ for all $t \in [S, T]$. $\square$

The Autonomous Strong Invariance Theorem can be extended to cover time-dependent multifunctions F, by means of state augmentation. This is possible however only if F is Lipschitz continuous with respect to the time variable (since state augmentation accords the time variable the status of a state variable component, and therefore requires that F satisfy the same regularity hypotheses with respect to the time variable as it does with respect to the original state variable). The following strong invariance theorem does not require F to be Lipschitz continuous with respect to the time variable (although it invokes additional hypotheses on F in other respects).

Theorem 12.2.4 (Strong Invariance Theorem for Time-Varying Systems) *Take multifunctions $F : [S, T] \times R^n \rightsquigarrow R^n$ and $P : [S, T] \rightsquigarrow R^n$. Take also a number $\bar{S} \in [S, T)$ and an F-trajectory $x(.) : [\bar{S}, T] \to R^n$ such that*

$$x(\bar{S}) \in P(\bar{S}).$$

Assume that, for some constant $\delta > 0$,

(i) F is continuous and takes as values nonempty, closed convex sets.

(ii) There exists $c > 0$ such that

$$F(t,x) \subset cB \quad \text{for all } x \in x(t) + \delta B \quad a.e.$$

(iii) There exists $k > 0$ such that

$$F(t,x') \subset F(t,x'') + k|x' - x''|B \quad \text{for all } x',x'' \in x(t) + \delta B \quad a.e.$$

(iv) $\operatorname{Gr} P$ is closed and, if $\bar{S} = S$,

$$x(\bar{S}) \in \limsup_{t\downarrow\bar{S}} P(t).$$

Assume further that, for every $(t,x) \in \operatorname{Gr} P \cap T_\delta(x(.)) \cap ((S,T) \times R^n)$ at which $N^P_{\operatorname{Gr} P}(t,x)$ is nonempty,

$$\zeta^0 + \max_{v \in F(t,x)} \zeta^i \cdot v \le 0 \quad \text{for all } (\zeta^0, \zeta^1) \in N^P_{\operatorname{Gr} P}(t,x).$$

Then

$$x(t) \in P(t) \quad \text{for all } t \in [\bar{S}, T].$$

Proof. Assume to begin with that $x(.)$ is a C^1 function. Since F has a closed graph, it follows that

$$\dot{x}(t) \in F(t, x(t)),$$

for *all* $t \in [\bar{S}, T]$. Under the hypotheses, we can choose sequences $\{t_0^i\} \subset (S,T)$ and $\{x_0^i\} \subset R^n$ such that $t_0^i \to \bar{S}$, $x_0^i \to x(\bar{S})$, and $x_0^i \in P(t_0^i)$ for all i. (In the case $\bar{S} \in (S,T]$ we set $t_0^i = \bar{S}$ and $x_0^i = x(\bar{S})$ for all i.) By discarding initial terms in the sequences, we can arrange that

$$|x_0^i - x(t_0^i)|e^{|k(T-S)|} < \delta. \tag{12.19}$$

Define

$$\tilde{F}(t,x) := \{v \in F(t,x) : |v-\dot{x}(t)| \le k|x-x(t)|\} \quad \text{for all } (t,x) \in [\bar{S},T]\times R^n.$$

Notice that, in view of the Lipschitz continuity hypothesis on F,

$$\tilde{F}(t,x) \ne \emptyset \quad \text{when } t \in [\bar{S},T] \text{ and } |x - x(t)| \le \delta.$$

Now define

$$\Gamma(t,x) := \begin{cases} \{1\} \times \tilde{F}(t,x) & \text{if } S < t < T \text{ and } |x - x(t)| < \delta \\ \operatorname{co}\left\{\left(\{1\} \times \tilde{F}(t,x(t) + \operatorname{tr}_\delta(x - x(t)))\right) \cup \{(0,0)\}\right\} & \text{otherwise,} \end{cases}$$

where

$$\mathrm{tr}_\delta(x) := \begin{cases} x & \text{if } |x| \le \delta \\ \delta x/|x| & \text{if } |x| > \delta. \end{cases}$$

Define also $D \subset [S,T] \times R^n$ to be

$$D := \mathrm{Gr}\, P.$$

Fix i. It is a straightforward matter to check that the hypotheses of the Time-Invariant Weak Invariance Theorem (Theorem 12.2.1) are satisfied, with reference to Γ, D, x_0^i, and the initial time t_0^i. (Notice that, since $\dot{y}$ is assumed to be continuous, the multifunction Γ has a closed graph.)

We deduce existence of a Lipschitz continuous function $(\tau_i(t), x_i(t))$, $t_0^i \le t \le T$, satisfying

$$\begin{aligned}
(\dot{\tau}_i(t), \dot{x}_i(t)) &\in \Gamma(\tau_i(t), x_i(t)) \quad \text{a.e. } t \in [t_0^i, T], \\
(\tau_i(t_0^i), x_i(t_0^i)) &= (t_0^i, x_0^i), \\
(\tau_i(t), x_i(t)) &\in D \quad \text{for all } t \in [t_0^i, T].
\end{aligned}$$

Define

$$E^i = \{t \in [t_0^i, T] : |x_i(t) - x(t)| \ge \delta\}$$

and

$$t_{\max}^i := \begin{cases} \inf\{t : t \in E_i\} & \text{if } E_i \ne \emptyset \\ T & \text{if } E_i = \emptyset. \end{cases}$$

Since $|x_i(t_0^i) - x(t_0^i)| < \delta$ and x_i and x are continuous, we have $t_0^i < t_{\max} \le T$. Noting that $0 \le \dot{\tau}_i(t) \le 1$, we deduce that

$$S < \tau_i(t) < T \quad \text{and} \quad |x_i(t) - x(t)| < \delta$$

for all $t \in [t_0^i, t_{\max}^i)$. But then $\tau_i(t) = t$ on $[t_0^i, t_{\max}^i]$ and

$$\dot{x}_i(t) \in \{v \in F(t, x_i(t)) : |v - \dot{x}(t)| \le K|x_i(t) - x(t)|\}.$$

We conclude that

$$\begin{aligned}
\dot{x}_i(t) &\in F(t, x_i(t)) \quad \text{a.e. } t \in [t_0^i, t_{\max}] \\
x_i(t) &\in P(t) \quad \text{for all } t \in [t_0^i, t_{\max}] \\
|\dot{x}_i(t) - \dot{x}(t)| &\le k|x_i(t) - x(t)| \quad \text{a.e. } t \in [t_0^i, t_{\max}].
\end{aligned}$$

It follows now from Gronwall's Inequality (Lemma 2.3.4) that

$$|x_i(t) - x(t)| \le e^{k|T - \bar{S}|}|x_0^i - x(t_0^i)| < \delta \quad \text{for all } t \in [t_0^i, t_{\max}]. \quad (12.20)$$

Since $|x_i(t_{\max}) - x(t_{\max})| = \delta$ if $t_{\max} < T$, we conclude that $t_{\max} = T$. But then x_i is an F-trajectory on $[t_0^i, T]$ satisfying

$$|x_i(t) - x(t)| \le e^{k|T - S|}|x_0^i - x(t_0^i)| \quad \text{for all } t \in [t_0^i, T].$$

Now extend x_i to $[\bar{S}, T]$ by constant extrapolation from the right. The extended arc we continue to write as x_i. Consider the sequence of arcs $\{x_i\}$ so constructed. The x_is are uniformly bounded and have a common Lipschitz constant. By extracting a subsequence, we can arrange that $x_i \to x'$ uniformly, for some Lipschitz continuous function x'. Since P has closed graph,

$$x'(t) \in P(t) \quad \text{for all } t \in [\bar{S}, T].$$

With the help of the Compactness of Trajectories Theorem (Theorem 2.4.3), we can show that x' is an F-trajectory. By (12.20)

$$|x'(t) - x(t)| \le e^{k|T-S|}|x'(\bar{S}) - x(\bar{S})| = 0 \quad \text{for all } t \in [\bar{S}, T].$$

It follows that $x'(t) = x(t)$ for $t \in [\bar{S}, T]$ and therefore

$$x(t) \in P(t) \quad \text{for all } t \in [\bar{S}, T].$$

Thus far we have assumed that $x(.)$ is a continuously differentiable F-trajectory on $[\bar{S}, T]$. Now suppose that x is merely absolutely continuous. Let $\{v_j\}$ be a sequence of continuous functions such that $\|v_j - \dot{y}\|_{L^1} \to 0$ as $j \to \infty$. Set $z_j(t) := x(\bar{S}) + \int_{\bar{S}}^{t} v_j(s)ds$. Then $z_j(.) \to x(.)$ uniformly, and each $z_j(.)$ is continuously differentiable. According to the Generalized Filippov Existence Theorem (Theorem 2.3.3) there exists a C^1 F-trajectory $x_j : [\bar{S}, T] \to R$, for each j, and

$$x_j \to x \quad \text{uniformly.}$$

But, for j sufficiently large, the preceding analysis can be applied with x_j replacing $x(.)$ (and with the constant δ reduced in size). It follows that

$$x_j(t) \in P(t) \quad \text{for all } t \in [S, T].$$

Since P has as values closed sets, we obtain in the limit as $j \to \infty$,

$$x(t) \in P(t) \quad \text{for all } t \in [S, T].$$

$\square$

12.3   The Value Function and Generalized Solutions of the Hamilton–Jacobi Equation

With the apparatus of invariance now at hand, we are ready to return to the optimal control problem:

$$(P) \begin{cases} \text{Minimize } g(x(T)) \\ \text{over arcs } x \in W^{1,1}([S,T]; R^n) \text{ satisfying} \\ \dot{x}(t) \in F(t, x(t)) \quad \text{a.e.,} \\ x(S) = x_0, \end{cases}$$

in which $g : R^n \to R \cup \{+\infty\}$ is a given function, $[S, T]$ is a given interval, $F : [S, T] \times R^n \rightsquigarrow R^n$ is a given multifunction, and $x_0 \in R^n$ a given point.

Recall that the value function for (P), $V : [S, T] \times R^n \to R \cup \{+\infty\}$, is the function

$$V(t, x) := \inf (P_{t,x}) \quad \text{for all } (t, x) \in [S, T] \times R^n,$$

in which $\inf (P_{t,x})$ denotes the infimum cost of the optimization problem:

$$(P_{t,x}) \begin{cases} \text{Minimize } g(y(T)) \\ \text{over arcs } y \in W^{1,1}([t, T]; R^n) \text{ satisfying} \\ \dot{y}(s) \in F(t, y(s)) \quad \text{a.e.,} \\ x(t) = x; \end{cases}$$

thus

$$V(t, x) := \inf(P_{t,x}) \quad \text{for all } (t, x) \in [S, T] \times R^n.$$

Our goal is to characterize V as a generalized solution, appropriately defined, to the Hamilton–Jacobi Equation:

$$\phi_t(t, x) + \min_{v \in F(t,x)} \phi_x(t, x) \cdot v = 0$$

$$\text{for all } (t, x) \in (S, T) \times R^n \quad (12.21)$$

$$\phi(T, x) = g(x) \quad \text{for all } x \in R^n, \quad (12.22)$$

in various senses.

To achieve this, we use the properties of another construct of nonsmooth analysis.

Given a set $A \subset R^k$, a function $\psi : A \to R \cup \{+\infty\}$, a point $x \in \operatorname{dom} \psi$, and a vector w, the *Lower Dini Directional Derivative of ψ at x in the direction w* is defined to be:

$$D^\dagger \psi(x; w) := \liminf_{h_i \downarrow 0, w_i \to w} \{h_i^{-1}(\psi(x + h_i w_i) - \psi(x))\}.$$

(For purposes of evaluating the limit on the right side, $\psi(x + h_i w_i)$ is assigned the value $+\infty$ when $x + h_i w_i \notin A$.

The following lemma, which provides alternative characterizations of the Lower Dini Derivative in the case $A = [S, T] \times R^n$ and also the case when ϕ is locally Lipschitz continuous, is a straightforward consequence of the definitions and is stated without proof. (The differences are in the choice of sequences used to evaluate the "lim inf".)

Lemma 12.3.1 *Take a function* $\phi : [S, T] \times R^n \to R \cup \{+\infty\}$, *a point* $(t, x) \in ([S, T] \times R^n) \cap \operatorname{dom} \phi$, *and a vector* $v \in R^n$. *We have*

(i) $D^\dagger \phi((t, x); (1, v))$

$\quad = \liminf\{h_i^{-1}(\phi(t + h_i, x + h_i v_i) - \phi(t, x)) : h_i \downarrow 0, v_i \to v\};$

(ii) if ϕ is Lipschitz continuous on a neighborhood of (t,x), then

$$D^\uparrow\phi((t,x);(1,v)) = \liminf\{h_i^{-1}(\phi(t+h_i, x+h_iv) - \phi(t,x)) \ : \ h_i \downarrow 0\}.$$

Two notions of "generalized solution" to the Hamilton–Jacobi Equation (12.21) and (12.22) are now made precise.

Definition 12.3.2 *A function $\phi : [S,T] \times R^n \to R \cup \{+\infty\}$ is a* lower Dini *solution to (12.21) and (12.22) if*

(i) $\inf_{v \in F(t,x)} D^\uparrow\phi((t,x);(1,v)) \le 0$

$$\text{for all } (t,x) \in ([S,T) \times R^n) \cap \operatorname{dom}\phi;$$

(ii) $\sup_{v \in F(t,x)} D^\uparrow\phi((t,x);(-1,-v)) \le 0$

$$\text{for all } (t,x) \in ((S,T] \times R^n) \cap \operatorname{dom}\phi;$$

(iii) $\phi(T,x) = g(x)$ *for all $x \in R^n$.*

Definition 12.3.3 *A function $\phi : [S,T] \times R^n \to R \cup \{+\infty\}$ is a* proximal *solution to (12.21) and (12.22) if*

(i) for every $(t,x) \in ((S,T) \times R^n) \cap \operatorname{dom}\phi$ such that $\partial^P\phi(t,x)$ is nonempty,

$$\eta^0 + \inf_{v \in F(t,x)} \eta^1 \cdot v = 0 \quad \text{for all } (\eta^0,\eta^1) \in \partial^P\phi(t,x);$$

(ii) $\phi(S,x) \ge \liminf_{t' \downarrow S, x' \to x} \phi(t',x')$ and

$$\phi(T,x) \ge \liminf_{t' \uparrow T, x' \to x} \phi(t',x') \text{ for all } x \in R^n;$$

(iii) $\phi(T,x) = g(x)$ for all $x \in R^n$.

These definitions are consistent with the classical notion of "solution" to (12.21) since any C^1 function $\phi : [S,T] \times R^n$ satisfying

$$\phi_t(t,x) + \min_{v \in F(t,x)} \phi_x(t,x) \cdot v = 0 \quad \text{for all } (t,x) \in (S,T) \times R^n,$$

is automatically a lower Dini solution and also a proximal solution in the sense just defined. These relationships follow directly from the definitions.

The following proposition relates these two solution concepts.

Proposition 12.3.4 *Suppose that ϕ is a lower Dini solution to (12.21) and (12.22). Then ϕ is also a proximal solution to (12.21) and (12.22).*

Proof. Let ϕ be a lower Dini solution. Take any $x \in R^n$. Then the condition

$$\phi(S,x) \ge \liminf_{t' \downarrow S, x' \to x} \phi(t',x') \tag{12.23}$$

is certainly satisfied if $(S, x) \notin \operatorname{dom} \phi$. Suppose that $(S, x) \in \operatorname{dom} \phi$. It follows from the definition of the lower Dini solution that there exist a vector $v \in F(S, x)$ and sequences $h_i \downarrow 0$ and $v_i \to v$ such that

$$\lim_i h_i^{-1}(\phi(S + h_i, x + h_i v_i) - \phi(S, x)) \leq 0.$$

But then

$$\phi(S, x) \geq \lim_i \phi(t_i, x_i),$$

where, for each i, $t_i = S + h_i$ and $x_i = x + h_i v_i$. (12.23) has been confirmed in this case also. In similar fashion, we deduce from the second defining relationship of lower Dini solutions that

$$\phi(T, x) \geq \lim_{t' \uparrow T, x' \to x} \phi(t', x').$$

Now suppose that $(t, x) \in ((S, T) \times R^n) \cap \operatorname{dom} \phi$ and $(\eta^0, \eta^1) \in \partial^P \phi(t, x)$. This means that there exist $M > 0$ and $\epsilon > 0$ such that

$$\eta^0(t' - t) + \eta^1 \cdot (x' - x) \leq \phi(t', x') - \phi(t, x) + M(|t' - t|^2 + |x' - x|^2)$$
$$\text{for all } (t', x') \in (t, x) + \epsilon B.$$

Since ϕ is a lower Dini solution of (12.21) and (12.22), there exist sequences $h_i \downarrow 0$ and $\{v_i\}$ in $F(t, x)$ such that

$$\lim_i h_i^{-1}(\phi(t + h_i, x + h_i v_i) - \phi(t, x)) \leq 0.$$

Careful study of the definition of the lower Dini solution reveals that we can also arrange that

$$h_i |v_i|^2 \to 0.$$

Setting $(t', x') = (t + h_i, x + h_i v_i)$, we see that, for i sufficiently large,

$$\eta^0 + \eta^1 \cdot v_i \leq h_i^{-1}(\phi(t + h_i, x + h_i v_i) - \phi(t, x)) + M h_i(1 + |v_i|^2).$$

It follows that

$$\eta^0 + \inf_{v \in F(t,x)} \eta^1 \cdot v \leq \eta^0 + \limsup_i \eta^1 \cdot v_i \leq 0. \tag{12.24}$$

Now choose $\{v_i\}$ be a minimizing sequence for $v \to \eta^1 \cdot v$ over $F(t, x)$. According to the definition of the lower Dini solution, we can choose a sequence $h_i \downarrow 0$ such that $h_i |v_i|^2 \to 0$ and

$$\limsup_{i \to \infty} h_i^{-1}(\phi(t - h_i, x - h_i v_i) - \phi(t, x)) \leq 0.$$

Setting $(t', x') = (t - h_i, x - h_i v_i)$, we obtain

$$-(\eta^0 + \eta^1 \cdot v_i) \leq h_i^{-1}(\phi(t - h_i, x - h_i v_i) - \phi(t, x)) + M h_i(1 + |v_i|^2),$$

for i sufficiently large. From these relationships we deduce that

$$\eta^0 + \inf_{v \in F(t,x)} \eta^1 \cdot v = \eta^0 + \lim_i \eta^1 \cdot v_i \geq 0.$$

This inequality combines with (12.24) to give

$$\eta^0 + \inf_{v \in F(t,x)} \eta^1 \cdot v = 0.$$

The proof is concluded. $\square$

The following proposition assembles some useful facts about the value function.

Proposition 12.3.5 *Assume that the data for (P) satisfy the following hypotheses.*

(i) *F is a continuous multifunction and has as values nonempty, closed, and convex sets.*

(ii) *There exists $c > 0$ such that*

$$F(t, x) \subset c(1 + |x|)B \quad \text{for all } (t, x) \in [S, T] \times R^n.$$

(iii) *There exists $k > 0$ such that*

$$F(t, x) \subset F(t, x') + k|x - x'|B \quad \text{for all } (t, x) \in [S, T] \times R^n.$$

(iv) *g is lower semicontinuous.*

Then

(a) *V is lower semicontinuous and $V(t, x) > -\infty$ for all $(t, x) \in [S, T] \times R^n$.*

(b) *V is locally Lipschitz continuous if g is locally Lipschitz continuous.*

Proof.

(a) Fix $(t, x) \in [S, T] \times R^n$. Then, by Proposition 2.6.2, we know that $(P_{t,x})$ has a minimizer y. (We allow the possibly that $(P_{t,x})$ has infinite cost.) It follows that

$$V(t, x) \ (= \ g(y(T)) \ ) \ > \ -\infty.$$

Now take any sequence $\{(t_i, x_i)\}$ in $[S, T] \times R^n$, such that $(t_i, x_i) \to (t, x)$ and $\lim_i V(t_i, x_i)$ exists. To prove that V is lower semicontinuous we must show that

$$V(t, x) \ \leq \ \lim_i V(t_i, x_i).$$

For each i, (P_{t_i,x_i}) has a minimizer, which we write $y_i : [t_i, T] \to R^n$. Of course $y_i(t_i) = x_i$. Extend y_i to all $[S, T]$ by constant extrapolation from the right on $[S, t_i]$.

We deduce from Gronwall's Inequality (Lemma 2.3.4) that both the y_is and their derivatives are uniformly bounded. The Compactness of Trajectories Theorem (Theorem 2.4.3) ensures that, following extraction of a subsequence, we have $y_i \to y$, uniformly, for some Lipschitz continuous arc whose restriction to $[t, T]$ is an F-trajectory. Since the $\dot{y}_i$s are uniformly bounded,

$$y(t) = \lim_i y_i(t_i) = \lim_i x_i = x.$$

But then,

$$V(t, x) \leq g(y(T)) = \lim_i g(y_i(T)) = \lim_i V(t_i, x_i).$$

It follows that V is lower semicontinuous.

(b) Assume that g is locally Lipschitz continuous. It follows from the Generalized Filippov Existence Theorem (Theorem 2.3.3) that V is everywhere finite. Fix $(\bar{t}, \bar{x}) \in [S, T] \times R^n$. Choose $\epsilon > 0$ and write

$$\mathcal{N} := \{(t, x) \in [S, T] \times R^n : |t - \bar{t}| < \epsilon, |x - \bar{x}| < \epsilon\}.$$

It follows from Gronwall's Inequality (Lemma. 2.3.4) that constants $c_1, k_1 > 0$ can be found, with the following properties: for any sub interval $[t_1, t_2] \subset [S, T]$ and F-trajectory $x : [t_1, t_2] \to R^n$ such that $Gr\{x(.)\} \cap \mathcal{N} \neq \emptyset$, we have

$$|x(t)| \leq c_1 \quad \text{and} \quad |x(t) - x(s)| \leq k_1|t - s|$$

for all subintervals $[s, t] \subset [t_1, t_2]$.

Let k_g be a Lipschitz constant for g on $c_1 B$. Choose any $(t, x), (t', x') \in \mathcal{N}$ such that $t \leq t'$. Let y be a minimizer for $(P_{t,x})$. By the Generalized Filippov Existence Theorem (Theorem 2.3.3), there exists an F-trajectory $\tilde{y} : [t, T] \to R^n$ such that $\tilde{y}(t) = x'$ and

$$\|\tilde{y} - y(t)\|_{L^\infty} \leq r|x - x'|$$

in which $r = e^{k|T-S|}$. But then

$$\begin{aligned} V(t, x') \leq g(\tilde{y}(T)) &\leq g(y(T)) + k_g\|\tilde{y} - y\|_{L^\infty} \\ &\leq V(t, x) + k_g r|x - x'|. \end{aligned}$$

Combining this inequality with that obtained by interchanging the roles of x and x', we arrive at

$$|V(t, x) - V(t, x')| \leq k_g r|x - x'|.$$

Since y is a minimizer for $(P_{t,x})$, we deduce from the Principle of Optimality that

$$V(t,x) = V(t',y(t')) = g(x(T)).$$

But then

$$
\begin{aligned}
|V(t,x) - V(t',x')| &\leq |V(t,x) - V(t',y(t'))| + |V(t',y(t')) - V(t',x')| \\
&= 0 + k_g r|y(t') - x'| \leq k_g r(|x - x'| + |x - y(t')|) \\
&\leq k_g r(|x - x'| + c_1|t - t'|) \\
&\leq \sqrt{2} k_g r(1 + c_1)|(x,t) - (x',t')|.
\end{aligned}
$$

Since (t,x) and (t',x') are arbitrary points in the relative neighborhood $\mathcal{N}$ of $(\bar{t},\bar{x})$, it follows that V is locally Lipschitz continuous. $\square$

These properties are now used to show that the value function is a generalized solution of the Hamilton–Jacobi Equation, in either the lower Dini or the proximal normal sense.

Proposition 12.3.6 *Let V be the value function for (P). Assume that Hypotheses (i) to (iv) of Proposition 12.3.5 are satisfied. Then V is a lower Dini solution of (12.21) and (12.22). It follows that V is also a proximal solution of (12.21) and (12.22).*

Proof. Take any point $(t,x) \in \operatorname{dom} V \cap ([S,T] \times R^n)$. Let y be a minimizer for $(P_{t,x})$. Then, by the Principle of Optimality,

$$V(t + \delta, y(t + \delta)) - V(t,x) = 0 \qquad (12.25)$$

for all $\delta \in (0, T - t]$. Since y is an F-trajectory

$$\delta^{-1}(y(t + \delta) - y(t)) = \delta^{-1} \int_t^{t+\delta} \dot{y}(s) ds$$

for all $\delta \in (0, T - t]$ and

$$\dot{y}(s) \in F(s,y(s)) \qquad \text{a.e. } s \in [t,T].$$

It is a straightforward matter to deduce from the continuity properties of F that there exist measurable functions $\xi : [t,T] \to R^n$ and $\eta : [t,T] \to R^n$ such that $\lim_{s \downarrow t} |\eta(s)| = 0$,

$$
\begin{aligned}
\xi(s) &\in F(t,x) && \text{a.e. } s \in [t,T] \\
\dot{y}(s) &= \xi(s) + \eta(s) && \text{a.e. } s \in [t,T].
\end{aligned}
$$

Now take an arbitrary sequence $\delta_i \downarrow 0$. For each i (sufficiently large) we define

$$v_i := \delta_i^{-1} \int_t^{t+\delta_i} \dot{y}(s) ds.$$

Because y is a Lipschitz continuous function, it follows that $\{v_i\}$ is a bounded sequence. By restricting attention to a subsequence then, we can arrange that

$$v_i \to v \quad \text{as } i \to \infty$$

for some $v \in R^n$. Noting that $\eta(s) \to 0$ as $s \downarrow 0$, we conclude that

$$v = \lim_i \tilde{v}_i,$$

where

$$\tilde{v}_i = \delta_i^{-1} \int_t^{t+\delta_i} \xi(s)\,ds.$$

Since $\xi(s) \in F(t,x)$ a.e. and $F(t,x)$ is a closed convex set, it can be shown, with the help of the Separation Theorem, that

$$\tilde{v}_i \in F(t,x) \quad \text{for each } i.$$

But then $v\ (= \lim_i \tilde{v}_i) \in F(t,x)$.

For each i, $y(t+\delta_i) = x + \delta_i v_i$. From (12.25),

$$\delta_i^{-1}[V(t+\delta_i, x+\delta_i v_i) - V(t,x)] = 0.$$

Since $v_i \to v$ and $v \in F(t,x)$, it follows that

$$\inf_{v' \in F(t,x)} D^\uparrow V((t,x),(1,v')) \le \lim_i \delta_i^{-1}[V(t+\delta_i, x+\delta_i v_i) - V(t,x)] = 0.$$

Now take any $(t,x) \in (S,T] \times R^n$. Choose any $v \in F(t,x)$. Take a sequence $\delta_i \downarrow 0$ such that $\delta_i < |t-S|$ for all i. By using as "comparison function" $z(s) = x + (s-t)v$, $S \le s \le t$, we deduce from the Generalized Filippov Existence Theorem (Theorem 2.3.3) that there exists an F-trajectory y on $[S,t]$ such that $y(t) = x$ and

$$|y(t-\delta_i) - [x - \delta_i v]| \le K \int_{t-\delta_i}^t \eta(s)\,ds \quad \text{for all } i,$$

where $K = \exp(k|T-s|)$ and

$$\eta(\sigma) := d_{F(\sigma, x+(\sigma-t)v)}(v) \quad \text{for } \sigma \in [S,t].$$

We have

$$|v - v_i| \le K\delta_i^{-1} \int_{t-\delta_i}^t \eta(s)\,ds,$$

in which

$$v_i := \delta_i^{-1}(x - y(t-\delta_i)) \quad \text{for each } i.$$

It follows from the continuity properties of F, however, that $\eta(s) \to 0$ as $s \uparrow t$. We deduce that $v_i \to v$. By the Principle of Optimality,

$$V(t-\delta_i, x-\delta_i v_i) \le V(t, y(t)) \quad \text{for all } i.$$

This condition implies

$$\delta_i^{-1}(V(t-\delta_i, x-\delta_i v_i) - V(t,x)) \leq 0 \quad \text{for all } i.$$

But then,

$$D^\uparrow V((t,x), (-1,-v)) \left(\leq \limsup_{i\to\infty} \delta_i^{-1}(V(t-\delta_i, x-\delta_i v_i) - V(t,x))\right) \leq 0$$

for all $v \in F(t,x)$. It follows that

$$\sup_{v \in F(t,x)} D^\uparrow V((t,x), (-1,-v)) \leq 0.$$

This concludes the proof of the proposition. $\square$

We have shown that V is a generalized solution to the Hamilton–Jacobi Equation ((12.21) and (12.22)). We now show that V is the *only* generalized solution. This important property of the value function is an immediate consequence of the following theorem.

Theorem 12.3.7 (Characterization of Lower Semicontinuous Value Functions) *Let $V : [S,T] \times R^n \to R \cup \{+\infty\}$ be a given lower semicontinuous function. Assume that the data for (P) satisfy the following hypotheses.*

(H1) F is a continuous multifunction, that takes as values nonempty, closed convex sets.

(H2) There exists $c > 0$ such that

$$F(t,x) \subset c(1+|x|)B \quad \text{for all } (t,x) \in [S,T], \times R^n.$$

(H3) There exists $k > 0$ such that

$$F(t,x) \subset F(t,x') + k|x - x'|B \quad \text{for all } (t,x) \in [S,T] \times R^n.$$

(H4) g is lower semicontinuous.

Then assertions (a) to (c) below are equivalent:

(a) V is the value function for (P).

(b) (i) $\inf_{v \in F(t,x)} D^\uparrow V((t,x); (1,v)) \leq 0$

   for all $(t,x) \in ([S,T) \times R^n) \cap \operatorname{dom} V$;

(ii) $\sup_{v \in F(t,x)} D^\uparrow V((t,x); (-1,-v)) \leq 0$

   for all $(t,x) \in ((S,T] \times R^n) \cap \operatorname{dom} V$;

(iii) $V(T,x) = g(x) \quad$ for all $x \in R^n$.

(c) (i)' For every $(t, x) \in ((S, T) \times R^n) \cap \mathrm{dom}\, V$ such that $\partial^P V(t, x)$ is nonempty

$$\eta^0 + \inf_{v \in F(t,x)} \eta^1 \cdot v = 0 \quad \text{for all } (\eta^0, \eta^1) \in \partial^P V(t, x);$$

(ii)' $V(S, x) \geq \liminf_{t \downarrow S, x' \to x} V(t', x')$

and

$V(T, x) \geq \liminf_{t \uparrow T, x' \to x} V(t', x')$

for all $x \in R^n$;

(iii)' $V(T, x) = g(x) \quad$ for all $x \in R^n$.

Proof. It has already been shown that

$$\text{(a)} \Rightarrow \text{(b)} \quad \text{and} \quad \text{(b)} \Rightarrow \text{(c)}.$$

It remains to supply the missing link: (c) $\Rightarrow$ (a). Assume then that (c) is true. Take an arbitrary point $(\bar{t}, \bar{x}) \in [S, T) \times R^n$. We must show that

$$V(\bar{t}, \bar{x}) = \inf(\mathrm{P}_{\bar{t}, \bar{x}}).$$

Step 1: We show that

$$V(\bar{t}, \bar{x}) \leq \inf(\mathrm{P}_{\bar{t}, \bar{x}}).$$

Let y be an arbitrary F-trajectory on $[\bar{t}, T]$. Clearly, it suffices to establish

$$V(\bar{t}, \bar{x}) \leq g(y(T)).$$

We can assume, without loss of generality, that $g(y(T)) < +\infty$ for, otherwise, the relationship holds automatically. Define

$$\tilde{y}(s) := y(T - s), \quad \tilde{F}(s, x) := -F(T - s, x) \text{ and } \tilde{V}(s, x) := V(T - s, x)$$

for $(s, x) \in [0, T - \bar{t}] \times R^n$. It is a straightforward matter to check that $\tilde{y}$ satisfies

$$\begin{cases} \dot{\tilde{y}}(s) \in \tilde{F}(s, \tilde{y}(s)) & \text{a.e. } t \in [0, T - \bar{t}] \\ \tilde{y}(0) = y(T). \end{cases}$$

We readily deduce from the fact that V satisfies Condition (c)(i)' that

$$\xi^0 + \max_{v \in \tilde{F}(s,x)} \xi^1 \cdot v \leq 0 \tag{12.26}$$

for all $(s, x) \in ((0, T - \bar{t}) \times R^n) \cap \mathrm{dom}\, \tilde{V}$ and all $(\xi^0, \xi^1) \in \partial^P \tilde{V}(s, x)$. Condition (c)(ii)' implies that

$$\tilde{V}(0, \tilde{y}(0)) = \liminf_{s' \downarrow 0, x' \to \tilde{y}(0)} \tilde{V}(s', x'). \tag{12.27}$$

Now define the multifunctions $\tilde{\Gamma} : [0, T - \bar{t}] \times R^{n+1} \rightsquigarrow R^{n+1}$ and $\tilde{P} : [0, T - \bar{t}] \rightsquigarrow R^{n+1}$ to be

$$\tilde{\Gamma}(s, (z, a)) := \tilde{F}(s, z) \times \{0\},$$

$$\tilde{P}(s) := \{(x, \alpha) : \tilde{V}(s, x) \leq \alpha\}$$

respectively. We apply the Strong Invariance Theorem (Theorem 12.2.4) to the differential inclusion and accompanying constraint

$$\begin{cases} (\dot{z}, \dot{a}) \in \tilde{\Gamma}(s, (z, a)) & \text{a.e. } s \in [0, T - \bar{t}], \\ z(0) = y(T), \quad a(0) = \tilde{V}(0, \tilde{y}(0)), \\ (z(s), a(s)) \in \tilde{P}(s) & \text{for all } s \in [0, T - \bar{t}]. \end{cases}$$

(Note that the boundary condition on $a(.)$ makes sense, since

$$\tilde{V}(0, \tilde{y}(0)) \ (= g(y(T))) < +\infty.)$$

To do so, we need first to check that the relevant hypotheses are satisfied. According to (12.27), we can find $s_i \downarrow 0$, $x_i \to y(T)$ such that $\tilde{V}(s_i, x_i) \to \tilde{V}(0, y(T))$. Writing $a_i := \tilde{V}(s_i, x_i)$, we see that $(x_i, a_i) \to (y(T), \tilde{V}(0, y(T)))$ and $(s_i, (x_i, a_i)) \in \tilde{P}(t_i)$ for all i. This confirms that

$$(z(0), a(0)) \in \limsup_{s' \downarrow 0} \tilde{P}(s').$$

$\tilde{\Gamma}(.)$ and $\tilde{P}(.)$ both clearly possess the required closure and regularity properties for application of the Strong Invariance Theorem.

It remains to check the "inward-pointing" hypothesis. With this objective in mind, we consider any point $(s, z, \alpha) \in \operatorname{Gr} \tilde{P}$, satisfying $0 < s < T - S$, and any vector

$$(\zeta^0, \zeta^1, -\lambda) \in N_{\operatorname{Gr} \tilde{P}}^P(s, (z, \alpha)). \tag{12.28}$$

We must show that

$$\zeta^0 + (\zeta^1, -\lambda) \cdot \tilde{v} \leq 0 \quad \text{for all } \tilde{v} \in \tilde{\Gamma}(s, (z, a)).$$

This last condition can be expressed as

$$\zeta^0 + \max_{v \in \tilde{F}(s, x)} \zeta^1 \cdot v \leq 0. \tag{12.29}$$

Since $\operatorname{Gr} \tilde{P}$ is an epigraph set, $\lambda \geq 0$. We deduce from (12.28) and the proximal normal inequality that

$$(\zeta^0, \zeta^1, -\lambda) \in N_{\operatorname{epi} \tilde{V}}^P(s, z, \tilde{V}(s, z)).$$

If $\lambda > 0$, then

$$(\zeta^0, \zeta^1) \in \lambda \partial^P \tilde{V}(s, z).$$

In this case, inequality (12.29) follows from (i)', since

$$(\zeta^0, \zeta^1) \in \lambda \partial^P \tilde{V}(s, z) \quad \text{implies} \quad (-\zeta^0, \zeta^1) \in \lambda \partial^P V(T - s, z)$$

and

$$v \in \tilde{F}(s, x) \quad \text{implies} \quad -v \in \tilde{F}(T - s, x).$$

Consider now the case $\lambda = 0$. Take any $u \in \tilde{F}(s, z)$. We can find sequences $(s_i, z_i) \xrightarrow{\tilde{V}} (s, z)$ and $(\zeta_i^0, \zeta_i^1, -\lambda_i) \to (\zeta^0, \zeta^1, 0)$, such that $\lambda_i > 0$ and

$$(\zeta_i^0, \zeta_i^1, -\lambda_i) \in N_{\text{epi}\,\tilde{V}}^P(s_i, z_i, \tilde{V}(s_i, z_i))$$

for each i. In view of the continuity properties of $\tilde{F}$, there exists $u_i \to u$ such that $u_i \in \tilde{F}(s_i, z_i)$ for all i. For each i,

$$(\zeta_i^0, \zeta_i^1) \in \lambda_i \partial \tilde{V}(s_i, z_i).$$

We deduce from (c)(i)' that

$$\zeta_i^0 + \zeta_i^1 \cdot u_i \leq \zeta_i^0 + \max_{v \in \tilde{F}(s_i, z_i)} \zeta_i^1 \cdot v \leq 0.$$

Passing to the limit as $i \to \infty$, we deduce that

$$\zeta^0 + \zeta^1 \cdot u \leq 0.$$

Since u is an arbitrary point on $\tilde{F}(s, x)$, we conclude that

$$\zeta^0 + \max_{u \in \tilde{F}(s, x)} \zeta^1 \cdot u \leq 0.$$

We have confirmed (12.29) in this case also.

The Strong Invariance Theorem tells us that the $\tilde{\Gamma}$-trajectory $(\tilde{y}, a \equiv \tilde{V}(0, \tilde{y}(0)))$ satisfies

$$(\tilde{y}(s), a(s)) \in \tilde{P}(s) \quad \text{for all } s \in [0, T - \bar{t}].$$

Setting $s = T - \bar{t}$, we deduce that

$$\tilde{V}(0, \tilde{0}) = a(0) = a(T - \bar{t}) \geq \tilde{V}(T - \bar{t}, \tilde{y}(T - \bar{t})).$$

Since $V(T, y(T)) = \tilde{V}(0, \tilde{y}(0))$ and $V(\bar{t}, y(\bar{t})) = \tilde{V}(T - \bar{t}, \tilde{y}(T - \bar{t}))$, it follows that

$$g(y(T)) = V(T, y(T)) \geq V(\bar{t}, y(\bar{t})) = V(\bar{t}, \bar{x}).$$

This is the desired inequality.

Step 2: We show that

$$V(\bar{t}, \bar{x}) \geq g(x(T)),$$

for some F-trajectory x on $[\bar{t}, T]$ which satisfies $x(\bar{t}) = \bar{x}$. The proof will then be complete, because this condition combines with that derived in Step 1 to give

$$V(\bar{t}, \bar{x}) = \inf(\mathrm{P}_{\bar{t}, \bar{x}}).$$

We can assume that $V(\bar{t}, \bar{x}) < +\infty$, since otherwise the inequality is automatically satisfied. Now define $\Gamma : [\bar{t}, T] \times R^{n+1} \leadsto R^{n+1}$ and $P : [\bar{t}, T] \leadsto R^{n+1}$:

$$\Gamma(t, x) := F(t, x) \times \{0\}, \qquad P(t) := \{(x, \alpha) \in R^{n+1} : V(t, x) \le \alpha\}.$$

We apply the Weak Invariance Theorem to

$$\begin{cases} (\dot{x}, \dot{a}) & \in & \Gamma(t, x) \quad \text{a.e. } t \in [\bar{t}, T] \\ (x(t), a(t)) & \in & P(t) \quad \text{for all } t \in [\bar{t}, T] \\ (x(\bar{t}), a(\bar{t})) & = & (\bar{x}, V(\bar{t}, \bar{x})). \end{cases}$$

To begin, we check that the relevant hypotheses are satisfied.

Condition (c)(ii)$'$ of the theorem statement ensures that, if $\bar{t} = S$, then

$$(x(\bar{t}), a(\bar{t})) \in \limsup_{t \downarrow \bar{t}} P(t).$$

Suppose that $(t, x, \alpha) \in \mathrm{Gr}\, P$, $\bar{t} < t < T$, and $(\zeta^0, \zeta^1, -\lambda) \in N^P_{\mathrm{Gr}\, P}(t, x, \alpha)$. We must show that

$$\zeta^0 + \min_{u \in \Gamma(t, x)} (\zeta^1, -\lambda) \cdot u \le 0.$$

This inequality can be expressed

$$\zeta^0 + \min_{u \in F(t, x)} \zeta^1 \cdot u \le 0. \tag{12.30}$$

Because P takes as values epigraph sets, we have $\lambda \ge 0$ and

$$(\zeta^0, \zeta^1, -\lambda) \in N^P_{\mathrm{epi}\, V}((t, x), V(t, x)).$$

If $\lambda > 0$, then

$$(\zeta^0, \zeta^1) \in \lambda \partial^P V(t, x).$$

In this case (12.30) is valid, because V satisfies Condition (c)(i)$'$.

Finally, suppose that $\lambda = 0$. In this case, there exist sequences $(t_i, x_i) \overset{V}{\to} (t, x)$ and $(\zeta^0_i, \zeta^1_i, -\lambda_i) \to (\zeta^0, \zeta^1, 0)$, such that $\lambda_i > 0$ and

$$(\zeta^0_i, \zeta^1_i, -\lambda_i) \in N^P_{\mathrm{epi}\, V}((t_i, x_i), V(t_i, x_i)) \quad \text{for } i = 1, 2, \ldots.$$

It can be deduced from Condition (c)(i)$'$ that

$$\zeta^0_i + \min_{v \in F(t_i, x_i)} \zeta^1_i \cdot v \le 0 \quad \text{for } i = 1, 2, \ldots.$$

Let u_i minimize $\zeta_i^1 \cdot v$ over $v \in F(t_i, x_i)$, for each i. Then, under the hypotheses on F, $\{u_i\}$ is a bounded sequence. By restricting attention to an appropriate subsequence, we can arrange that $u_i \to u$ as $i \to \infty$. By continuity of F, $u \in F(t, x)$. Passing to the limit as $i \to \infty$, we deduce that

$$\zeta^0 + \min_{v \in F(t,x)} \zeta^1 \cdot v \ = \ \zeta^0 + \zeta^1 \cdot u \ = \ \lim_{i \to \infty} \zeta_i^0 + \zeta_i^1 \cdot v_i \ \leq \ 0.$$

Condition (12.30) has been confirmed in this case also.

The multifunctions $\Gamma(.)$ and $P(.)$ have the requisite closure and regularity properties for application of the Time-Dependent Weak Invariance Theorem (Theorem 12.2.2). This supplies an F-trajectory $x : [\bar{t}, T] \to R^n$ and a constant function $a(.)$ such that $x(\bar{t}) = \bar{x}$, $a(\bar{t}) = V(\bar{t}, \bar{x})$, and

$$(x(t), a(t)) \in \operatorname{epi} V \quad \text{for all } t \in [\bar{t}, T].$$

Then

$$g(x(T)) \ = \ V(T, x(T)) \ \leq \ a(T) \ = \ a(\bar{t}) \ = \ V(\bar{t}, \bar{x}) \quad \text{for all } t \in [S, T].$$

Step 2 is complete. $\square$

12.4 Local Verification Theorems

A traditional role of the Hamilton–Jacobi Equation in Variational Analysis has been to provide sufficient conditions of optimality. Consider the Optimal Control Problem

$$(Q) \left\{ \begin{array}{l} \text{Minimize } g(x(T)) \\ \text{over arcs } x \in W^{1,1}([S, T]; R^n) \text{ satisfying} \\ \dot{x}(t) \in F(t, x(t)) \quad \text{a.e.,} \\ x(S) = x_0, \\ x(T) \in C, \end{array} \right.$$

the data for which comprise: a function $g : R^n \to R$, a subinterval $[S, T] \subset R$, a multifunction $F : [S, T] \times R^n \rightsquigarrow R^n$, a point $x_0 \in R^n$, and a closed set $C \subset R^n$.

It is convenient at this stage to adopt a formulation in which the terminal cost function g is assumed to be finite-valued and constraints on terminal values of state trajectories are expressed as a set inclusion $x(T) \in C$.

Take an arc $x \in W^{1,1}$ satisfying the constraints of problem (Q). A smooth local verification function for $\bar{x}$ is a C^1 function $\phi : [S, T] \times R^n \to R$ with the properties: there exists $\delta > 0$ such that

$$\phi_t(t, x) + \min_{v \in F(t,x)} \phi_x(t, x) \cdot v \ \geq \ 0 \quad \text{for all } (t, x) \in \operatorname{int} T(\bar{x}, \delta)$$

$$\phi(T, x) \ \leq \ g(x) \quad \text{for all } x \in C \cap (\bar{x}(T) + \delta B)$$
$$\phi(0, x_0) \ = \ g(\bar{x}(T)).$$

Here $T(\bar{x}, \delta)$ is the δ tube about $\bar{x}$:

$$T(\bar{x}, \delta) := \{(t, x) \in [S, T] \times R^n : |x - \bar{x}(t)| \leq \delta\}.$$

Existence of a smooth verification function for $\bar{x}$ is a sufficient condition for $\bar{x}$ to be a strong local minimizer for (Q). It is of interest to know how broad the class of optimal control problems is, for which the local optimality of local minimizers can be confirmed by some verification function. Inverse verification theorems, which provide conditions for the existence of verification functions associated with a minimizer address this issue.

Unfortunately, continuously differentiable verification theorems provide a rather restrictive framework for inverse verification theorems. For optimal control problems whose solutions exhibit bang-bang behavior indeed, nonexistence of continuously differentiable verification functions is to be expected.

Clearly, we need a broader concept of smooth local verification functions, which can still be used in sufficient conditions of optimality, but whose existence can be guaranteed under unrestrictive hypotheses.

These considerations lead to the following definition, which makes reference to the data for problem (Q) above and an arc $\bar{x} \in W^{1,1}([S, T] \times R^n)$.

Definition 12.4.1 *A function* $\phi : T(\bar{x}, \delta) \to R \cup \{+\infty\}$ *is called a lower semicontinuous local verification function for* $\bar{x}$ *(with parameter* $\delta > 0$*) if* ϕ *is lower semicontinuous and the following conditions are satisfied.*

(i) *For every* $(t, x) \in \operatorname{int} T(\bar{x}, \delta)$ *such that* $\partial^P \phi(t, x)$ *is nonempty,*

$$\eta^0 + \min_{v \in F(t,x)} \eta^1 \cdot v \geq 0 \quad \text{for all } (\eta^0, \eta^1) \in \partial^P \phi(t, x).$$

(ii) $\phi(T, x) \leq g(x)$ *for all* $x \in C$.

(iii) $\liminf_{t' \uparrow T, x' \to x} \phi(t', x') = \phi(T, x)$ *for all* $x \in C \cap (\bar{x}(T) + \delta B)$.

(iv) $\phi(S, x_0) = g(\bar{x}(T))$.

The next proposition tells us that sufficient conditions for local minimality can be formulated in terms of lower semicontinuous local verification functions. Furthermore, the existence of a lower semicontinuous verification function confirming optimality of a given local minimizer is guaranteed, under unrestrictive hypotheses on the data.

Proposition 12.4.2 *Let the arc* $\bar{x} \in W^{1,1}([S, T]; R^n)$ *satisfy the constraints of problem* (Q). *Assume that, for some* $\epsilon > 0$, *we have*

(i) F *is a continuous multifunction that takes as values nonempty, closed, convex sets;*

(ii) there exists $c > 0$ such that

$$F(t, x) \subset cB \qquad \text{for all } (t, x) \in T(\bar{x}, \epsilon);$$

(iii) there exists $k > 0$ such that

$$F(t, x') \subset F(t, x'') + k|x' - x''|B \qquad \text{for all } (t, x'), (t, x'') \in T(\bar{x}, \epsilon);$$

(iv) g is lower semicontinuous on $\bar{x}(T) + \epsilon B$.

We have

(a) *suppose that there exists a lower semicontinous local verification function for $\bar{x}$. Then $\bar{x}$ is a strong local minimizer for (Q).*

Conversely,

(b) *suppose that $\bar{x}$ is a local minimizer for (Q) and that $|g(.)|$ is bounded on $\bar{x}(T) + \epsilon B$. Then there exists a lower semicontinuous local verification function for $\bar{x}$.*

Proof.

(a): Suppose that there exists a lower semicontinuous local verification function for $\bar{x}$ (with parameter δ). Reduce the size of δ if necessary, to ensure that $\delta < \epsilon$.

Let x be any F-trajectory satisfying the constraints for (Q) and such that $\|x - \bar{x}\|_{L^\infty} \le \delta$. It is required to show that

$$g(x(T)) \ge g(\bar{x}(T)).$$

Apply the Strong Invariance Theorem to the differential inclusion

$$\begin{cases} (\dot{z}, \dot{a}) & \in \quad -F(T - s, z) \times \{0\} \quad \text{a.e. } s \in [0, T - S] \\ (z(0), a(0)) & = \quad (x(T), \phi(T, x(T))) \end{cases}$$

and the accompanying constraint

$$(z(s), a(s)) \in \{(z, \alpha) : \phi(T - s, z) \le \alpha\}.$$

This gives

$$\phi(S, x_0) \le \phi(T, x(T)).$$

(The arguments involved are almost identical to those employed in the proof of Theorem 12.3.7.) But then

$$\phi(S, x_0) \, (\le \phi(T, x(T)) \,) \le g(x(T)).$$

Since $\phi(S, x(0)) = g(\bar{x}(T))$, $\bar{x}$ is a strong local minimizer as claimed.

(b): Now suppose that $\bar{x}$ is a strong local minimizer. Choose $\epsilon_1 \in (0, \epsilon)$ such that $\bar{x}$ is a minimizer with respect to arcs x satisfying the constraints of (Q) together with $||x - \bar{x}||_{L^\infty} \leq \epsilon_1$. Choose $\epsilon' \in (0, \epsilon_1)$. Now consider the optimal control problem $(\tilde{P})$ (this involves a constant k, whose value is set presently):

$$(\tilde{P}) \begin{cases} \text{Minimize } \tilde{g}(x(T)) + ky(T) \\ \text{over arcs } (x, y) \in W^{1,1}([S, T]; R^{n+1}) \text{ satisfying} \\ (\dot{x}(t), \dot{y}(t)) \in \tilde{F}(t, x(t)) \times \{\tilde{L}(t, x(t))\} \quad \text{a.e.,} \\ (x(S), y(S)) = (x_0, 0), \\ (x(T), y(T)) \in C \times R. \end{cases}$$

Here $\tilde{g}$ and $\tilde{F}$ are "localized" versions of g and F, namely,

$$\tilde{g}(x) := g(\bar{x}(T) + tr_{\epsilon_1}(x - \bar{x}(T))), \qquad \tilde{F}(t, x) := F(t, \bar{x}(t) + tr_{\epsilon_1}(x - \bar{x}(t)))$$

and

$$\tilde{L}(t, x) := \max\{|x - \bar{x}(t)| - \epsilon', 0\}$$

in which

$$tr_\alpha(z) := \begin{cases} z & \text{if } |z| \leq \alpha \\ \alpha z/|z| & \text{if } |z| > \alpha. \end{cases}$$

Choose constants $\kappa > 0$ and $k > 0$ that satisfy

$$\kappa > \max\{|g(x)| \; : \; x \in \bar{x}(T) + \epsilon_1 B\},$$

$$k > |\epsilon_1 - \epsilon'|^{-1}|\epsilon_1 + \epsilon'|^{-1}16c\kappa.$$

Claim $\bar{x}$ is a global minimizer for $(\tilde{P})$.

To verify this, take any other arc satisfying the constraints of $(\tilde{P})$. It must be shown that:

$$\tilde{g}(x(T)) + k \int_S^T \tilde{L}(t, x(t))dt \geq \tilde{g}(\bar{x}(T)) + k \int_S^T \tilde{L}(t, \bar{x}(t))dt \; (= g(\bar{x}(T))).$$

There are two cases to consider.

(A): $||x - \bar{x}||_{L^\infty} < \epsilon_1$. In this case, $\tilde{F}(t, x(t)) = F(t, x(t))$ for all $t \in [S, T]$ and $\tilde{g}(x(T)) = g(x(T))$. So x is actually an F-trajectory. But then, by local optimality of $\bar{x}$, we have

$$\tilde{g}(x(T)) + k \int_S^T L(t, x(t))dt$$

$$\geq g(x(T)) \geq g(\bar{x}(T)) = \tilde{g}(\bar{x}(T)) + k \int_S^T L(t, \bar{x}(t))dt,$$

as required.

(B): There exists $\bar{t} \in [S, T]$ such that

$$|x(\bar{t}) - \bar{x}(\bar{t})| = \epsilon_1.$$

In this case, since $t \to |x(t) - \bar{x}(t)|$ is Lipschitz continuous with Lipschitz constant $2c$ and $|x(S) - \bar{x}(S)| = 0$, we have

$$\text{meas}\,\{t \,:\, |x(t) - \bar{x}(t)| \geq (\epsilon' + \epsilon_1)/2\} \geq (\epsilon_1 - \epsilon')/(4c).$$

But then

$$\tilde{g}(x(T)) + k \int_S^T \tilde{L}(t, x(t))dt$$

$$= \tilde{g}(x(T)) + k \int_S^T \max\{|x(t) - \bar{x}(t)| - \epsilon', 0\}dt$$

$$\geq -\kappa + (k/8c)(\epsilon_1 + \epsilon')(\epsilon_1 - \epsilon') > \kappa$$

$$> g(\bar{x}(T)).$$

The inequality holds in this case too then; the claim is confirmed.

Now let $\tilde{V}(t, x, y)$ be the value function for $(\tilde{P})$. Since $(\tilde{P})$ is, in effect, a reformulation of an optimal control problem with endpoint and integral cost terms as an optimal control problem with endpoint cost term alone, $\tilde{V}(t, x, y)$ can be expressed

$$\tilde{V}(t, x, y) = \phi(t, x) + y$$

for some function $\phi : [S, T] \times R^n \to R \cup \{+\infty\}$.

Notice that $F \equiv \tilde{F}$ and $\tilde{L} \equiv 0$ on $T(\bar{x}, \epsilon')$ and $\tilde{g} \equiv g$ on $\bar{x}(T) + \epsilon'B$. Now we use information about value functions supplied by Theorem 12.3.4, in which the terminal constraint functional is taken to be the extended-valued function

$$\tilde{g}(x(T)) + ky(T) + \Psi_C(x(T)).$$

We deduce that ϕ is lower semicontinuous and, for every $(t, x) \in (\text{int}\, T(\bar{x}, \epsilon')) \cap \text{dom}\,\phi$ such that $\partial^P \phi(t, x)$ is nonempty,

$$\eta^0 + \min_{v \in F(t, x)} \eta^1 \cdot v = 0 \quad \text{for all } (\eta^0, \eta^1) \in \partial^P \phi(t, x),$$

$$\phi(T, x) = g(x) \quad \text{for all } x \in (\bar{x}(T) + \epsilon'B) \cap C,$$

$$\liminf_{t' \uparrow T, x' \to x} \phi(t', x') = \phi(T, x) \quad \text{for all } x \in C \cap (\bar{x}(T) + \epsilon'B).$$

Finally, since $\bar{x}$ is a minimizer for $(\tilde{P})$,

$$\phi(S, x_0) = g(\bar{x}(T)).$$

We have confirmed that there exists some lower semicontinuous, local verification function for $\bar{x}$, namely, ϕ. $\square$

The preceding theorem tells us that any local minimizer can, in principle, be confirmed as such by some lower semicontinuous, local verification function. There follows an inverse verification theorem that aims to simplify the task of finding a local verification function. It gives conditions under which the search can be confined to the class of functions that are *Lipschitz continuous* on a tube about the putative minimizer $\bar{x}$. An important point is that we can establish existence of Lipschitz continuous local verification functions, even in the presence of an endpoint constraint

$$x(T) \in C.$$

In such cases, Lipschitz continuous functions are typically *distinct* from the value function V for the original problem. (If, for example, a strong local minimizer $\bar{x}$ is such that $\bar{x}(T) \in \text{bdy } C$, then the value function will take value $+\infty$ at some points in any tube about $\bar{x}$ and so certainly cannot serve as a Lipschitz continuous local verification function.)

The conditions under which there exists a Lipschitz continuous local verification function include a constraint qualification, requiring that necessary conditions of optimality, in the form of the Hamiltonian inclusion, are "normal" at the locally minimizing arc $\bar{x}$ under examination.

(CQ) For any $p \in W^{1,1}([S,T]; R^n)$ and $\lambda \geq 0$ such that

$$\begin{cases} \|p\|_{L^\infty} + \lambda \neq 0, \\ (-\dot{p}(t), \dot{\bar{x}}(t)) \in \text{co}\, \partial H(t, \bar{x}(t), p(t)) \quad \text{a.e.,} \\ -p(T) \in \lambda \partial g(\bar{x}(T)) + N_C(\bar{x}(T)), \end{cases}$$

we have $\lambda \neq 0$.

Here, as usual, $H(t, x, p)$ is the Hamiltonian

$$H(t, x, p) = \max_{v \in F(t,x)} p \cdot v.$$

Theorem 12.4.3 (Existence of Lipschitz continuous verification functions) *Let the arc $\bar{x} \in W^{1,1}([S,T]; R^n)$ satisfy the constraints of problem (Q). Assume that, for some $\epsilon > 0$,*

(i) the multifunction F is continuous and takes as values nonempty, closed, convex sets;

(ii) there exists $k > 0$ such that

$$F(t, x') \subset F(t, x'') + k|x' - x''|B \quad \text{for all } (t, x'), (t, x'') \in T(\bar{x}, \bar{\epsilon});$$

(iii) there exists $c > 0$ such that

$$F(t,x) \quad \subset \quad cB \qquad for\ all\ (t,x) \in T(\bar{x}, \epsilon);$$

(iii) g is Lipschitz continuous on $\bar{x}(T) + \epsilon B$.

We have

(a) *if there exists a Lipschitz continuous local verification function for $\bar{x}$, then $\bar{x}$ is a strong local minimizer for (Q);*

(b) *if $\bar{x}$ is a strong local minimizer and (CQ) is satisfied, then there exists a Lipschitz continuous local verification function for $\bar{x}$.*

Proof.

(a): This is, of course, just a special case of Theorem 12.4.2, already proved.

(b): Let $\bar{x}$ be a strong local minimizer for (Q). We construct a Lipschitz continuous local verification function for $\bar{x}$.

Choose a constant $\bar{\epsilon} \in (0, \epsilon)$, such that $\bar{x}$ is a minimizer with respect to all arcs satisfying the constraints of (Q) and also $||x - \bar{x}||_{L^\infty} \leq \bar{\epsilon}$. Let κ be a constant such that $|g(x)| \leq \kappa$ for $x \in \bar{x}(T) + \bar{\epsilon}B$. Take any sequence $\epsilon_i \downarrow 0$ such that $\epsilon_i \in (0, \bar{\epsilon})$ for all i. Take also sequences $k_i \uparrow +\infty$, $K_i \uparrow +\infty$ such that

$$k_i > 64c\kappa\epsilon_i^{-2} \text{ for each } i \quad \text{and} \quad K_i/k_i \to \infty \text{ as } i \to \infty.$$

For each i, define

$$L_i(t,x) := \max\{|x - \bar{x}(t)| - \epsilon_i/2, 0\}.$$

Define also

$$\tilde{g}(x) := g(\bar{x}(T) + \mathrm{tr}_{\bar{\epsilon}}(x - \bar{x}(T)), \qquad \tilde{F}(t,x) := F(t, \bar{x}(t) + \mathrm{tr}_{\bar{\epsilon}}(x - \bar{x}(t)))$$

For each i, consider the optimal control problem $(\tilde{P}_i)$:

$$(\tilde{P}_i) \begin{cases} \text{Minimize } \tilde{g}(x(T)) + k_i z(T) + K_i d_C(x(T)) \\ \text{over arcs } (x,z) \in W^{1,1}([S,T]; R^{n+1}) \text{ satisfying} \\ (\dot{x}(t), \dot{z}(t)) \in \tilde{F}(t, x(t)) \times L_i(t, x(t))\} \quad \text{a.e.,} \\ (x(S), z(S)) = (x_0, 0). \end{cases}$$

We show presently that,

$$(\bar{x}, z(t) \equiv 0) \text{ is a minimizer for } (P_{i_0}) \tag{12.31}$$

for some index value i_0.

Completion of the proof is then a straightforward matter. Indeed, let $V(t, x, z)$ be the value function for $(\tilde{P}_{i_0})$. Clearly

$$V(t, x, z) = \phi(t, x) + z$$

for some function $\phi : [S, T] \times R^n \to R \cup \{+\infty\}$. We also note that

$$\tilde{F}(t, x) = F(t, x) \text{ and } L(t, x) = 0 \text{ for } (t, x) \in T(\bar{x}, \epsilon_{i_0}/2)$$

and

$$\tilde{g}(x) + K_{i_0} d_C(x) = g(x) \quad \text{for } x \in C \cap (\bar{x}(T) + \epsilon_{i_0}/2\, B).$$

The information about value functions supplied by Proposition 12.3.4, applied to $V(t, x, z)$, tells us that ϕ is a local verification function for (P_{i_0}). (P_{i_0}) has a Lipschitz continuous terminal cost function, however, and there is no right endpoint constraint. It follows from Proposition 12.3.5 that V (and therefore ϕ also) is locally Lipschitz continuous. This is what we set out to prove.

It remains then to confirm (12.31). Here we make use of a contradiction argument. Suppose that, for each i, $\bar{x}$ is not a minimizer for (P_i).

Under the hypotheses it is known that, for each i, (P_i) has a minimizer: write it

$$\left(x_i(.), t \to k_i \int_S^t L_i(s, x_i(s)) ds\right).$$

Because it is assumed that $\bar{x}$ is not a minimizer for (P_i), we have

$$\tilde{g}(x_i(T)) + k_i \int_S^T L_i(t, x_i(t)) dt + K_i d_C(x_i(T)) < \tilde{g}(\bar{x}(T)) + 0 + 0 = g(\bar{x}(T)).$$
$$(12.32)$$

We now note two important properties of x_i:

$$\|x_i - \bar{x}\|_{L^\infty} \leq \epsilon_i \qquad (12.33)$$
$$x_i(T) \notin C. \qquad (12.34)$$

Indeed, if (12.33) is not true, then there exists $\bar{t} \in [S, T]$ such that $|x_i(\bar{t}) - \bar{x}(\bar{t})| = \epsilon_i$. Since $t \to |x_i(t) - \bar{x}(t)|$ is Lipschitz continuous with Lipschitz constant $2c$, it follows that

$$\text{meas}\{t \in [S, T] : |x_i(t) - \bar{x}(t)| > (3/4)\epsilon_i\} \geq \epsilon_i/(8c).$$

But then

$$\tilde{g}(x_i(T)) + k_i \int_S^T L_i(t, x_i(t)) dt + K_i d_C(x_i(T))$$
$$> -\kappa + k_i \epsilon_i^2/(32c) > \kappa > g(\bar{x}(T)),$$

in contradiction of (12.32). So (12.33) is true.

Suppose (12.34) is not true. Then x_i satisfies the constraints of (Q) and also $||x_i - \bar{x}|| \le \bar{\epsilon}$. So, by local optimality of $\bar{x}$,

$$g(x_i(T)) \ge g(\bar{x}(T)).$$

But then, in view of (12.32),

$$\tilde{g}(x_i(T)) + k_i \int_S^T L_i(t, x_i(t))dt + K_i d_C(x_i(T))$$
$$> \tilde{g}(x_i(t)) + 0 + 0 = g(x_i(T)) \ge g(\bar{x}(T)).$$

This contradicts (12.32). So (12.34) is true.

Now apply the necessary conditions of Theorem 7.5.1 to problem (Q), with reference to the strong local minimizer x_i. These provide an adjoint arc $p_i \in W^{1,1}([S,T]; R^n)$ and $\lambda \ge 0$ satisfying:

$$||p_i||_{L^\infty} + \lambda_i = 1,$$
$$(-\dot{p}_i(t), x_i(t)) \in \text{co}\, \partial H(t, x_i(t), p_i(t)) + \lambda k_i/K_i \text{co}\, \partial L(t, x_i(t)),$$
$$-p_i(T) \in \lambda_i(1/K_i)\partial \tilde{g}(x_i(T)) + \lambda_i \partial d_C(x_i(T)).$$

Here, $H(t, x, p)$ is the Hamiltonian for the original problem. Notice that, before applying the necessary conditions, we have scaled the cost by $(1/K_i)$.

Since $x_i(T) \notin C$,

$$\partial d_C(x_i(t)) \in \{\xi \in R^n : |\xi| = 1\}.$$

We deduce that
$$|p_i(T)| \ge \lambda_i(1 - k_g/K_i).$$

(k_g is a Lipschitz constant for g on $\bar{x}(T) + \bar{\epsilon}B$.)

A by now familiar convergence analysis can be used to justify passing to the limit in the above relationships: along a subsequence, $p_i \to p$ uniformly and $\lambda_i \to \alpha$, for some $p \in W^{1,1}([S,T]; R^n)$ and $\alpha \ge 0$, satisfying

$$||p||_{L^\infty} + \alpha = 1, \qquad (12.35)$$
$$(-\dot{p}(t), \dot{\bar{x}}(t)) \in \text{co}\, \partial H(t, \bar{x}(t), p(t)) \quad \text{a.e.,}$$
$$-p(T) \in \alpha d_C(\bar{x}(T)) \subset N_C(\bar{x}(T)),$$
$$||p||_{L^\infty} \ge \alpha.$$

Notice that $p \ne 0$ since, if $p = 0$, then $\alpha = 0$ which contradicts (12.35). We have exhibited an adjoint arc violating the constraint qualification (CQ). It follows that the original claim, namely, that $\bar{x}$ is minimizer for (P_i) for some i, is true. The proof is complete. $\square$

12.5   Adjoint Arcs and Gradients of the Value Function

In this section we again consider the optimal control problem of Section 12.1:

$$(P) \begin{cases} \text{Minimize } g(x(T)) \\ \text{over arcs } x \in W^{1,1}([S,T];R^n) \text{ satisfying} \\ \dot{x}(t) \in F(t,x(t)) \quad \text{a.e.,} \\ x(0) = x_0, \end{cases}$$

the data for which comprise a function $g : R^n \to R$, an interval $[S,T] \subset R$, a multifunction $F : [S,T] \times R^n \rightsquigarrow R^n$, and a vector $x_0 \in R^n$.

Note, however, that g is taken to be finite-valued. Endpoint constraints are therefore excluded.

Let $\bar{x}$ be a minimizer. Then, under appropriate hypotheses on the data, the necessary conditions of Theorem 7.4.1 assert the existence of an adjoint arc $p \in W^{1,1}([S,T];R^n)$ such that

$$\begin{aligned}
\dot{p}(t) &\in \text{co}\,\{q : (q,p(t)) \in N_{\text{Gr }F(t,.)}(\bar{x}(t),\dot{\bar{x}}(t))\} \quad \text{a.e.,} \\
p(t) \cdot \dot{\bar{x}}(t) &= h(t) \quad \text{a.e.,} \\
-p(T) &\in \partial g(\bar{x}(T)),
\end{aligned}$$

where

$$h(t) := \max_{v \in F(t,\bar{x}(t))} p(t) \cdot v .$$

(Since there are no right endpoint constraints, we are justified in assuming that the cost multiplier is $\lambda = 1$.)

Let V be the value function for (P). We invoke hypotheses under which V is locally Lipschitz continuous. Earlier discussion leads us to expect that the adjoint arc $p(.)$ (and $h(.)$) will be related to V according to

$$(h(t), -p(t)) \in \text{co}\,\partial V(t,\bar{x}) \quad \text{a.e. } t \in (S,T). \tag{12.36}$$

(To be more precise, we would expect this inclusion to hold for some kind of subdifferential of V; with hindsight we adopt the convexified limiting subdifferential.)

The aim of this section is to confirm the "sensitivity relation" (12.36). Before entering into the details of the arguments involved, we briefly examine two different approaches that hold out hope of simpler proof techniques, if only to dismiss them. The first is to take an arbitrary selector of the multifunction $t \to \text{co}\,\partial V(t,\bar{x}(t))$, partitioned as $(h(t), -p(t))$, such that $p(.)$ is absolutely continuous, and attempt to show that it is an adjoint arc. The other is to take an arbitrary adjoint arc and attempt to show that it satisfies the sensitivity relation (12.36). Their appeal is that they involve simply checking the properties of a plausible candidate for adjoint arc. The inadequacies of these approaches, at least for problems with nonsmooth data, are made evident by the following example, which reveals that absolutely continuous functions $(h(.), -p(.))$, chosen from the set of selectors for $t \to \operatorname{co} \partial V(t, \bar{x}(t))$, may fail to generate adjoint arcs and, on the other hand, adjoint arcs may fail to satisfy the sensitivity relation.

Example 12.1

$$\left\{ \begin{array}{l} \text{Minimize } g(x(1)) \text{ over } x \in W^{1,1}([0,1]; R) \text{ satisfying} \\ \dot{x}(t) \in \{x(t)u \: : \: u \in [0,1]\}, \\ x(0) = 0, \end{array} \right.$$

where

$$g(\xi) := \left\{ \begin{array}{ll} -\xi & \text{if } \xi > 0 \\ -e^{0.5}\xi & \text{if } \xi \leq 0. \end{array} \right.$$

The only admissible arc for this problem is $\bar{x}(t) \equiv 0$. This then is the minimizer. The value function $V(.,.)$ is easily calculated:

$$V(t,\xi) = \left\{ \begin{array}{ll} -e^{(1-t)}\xi & \text{if } \xi > 0 \\ -e^{0.5}\xi & \text{if } \xi \leq 0. \end{array} \right.$$

Evidently

$$\operatorname{co} \partial V(t,0) = \{0\} \times \operatorname{co}\{-e^{0.5}, -e^{1-t}\}.$$

The Hamiltonian is

$$H(t,x,p) = \max\{px, 0\}.$$

For this problem an adjoint arc is any absolutely continuous function $p(.)$ that satisfies

$$-\dot{p}(t) = \alpha(t)p(t) \quad \text{a.e. } t \in [0,1] \tag{12.37}$$

and

$$p(1) \in [1, e^{0.5}], \tag{12.38}$$

for some measurable function $\alpha(.) \: : \: [0,1] \to [0,1]$.

Notice that $p(.) \equiv e^{0.5}$ is an adjoint arc (it corresponds to the choice $\alpha(.) \equiv 0$) that satisfies

$$(H(t,\bar{x}(t),p(t)), -p(t)) \in \operatorname{co} \partial V(t, \bar{x}(t)) \quad \text{for all } t \in [0,1].$$

So, for this example there is an adjoint arc satisfying the anticipated relationship.

However, $p_1(.) \equiv 1$ is also an adjoint arc (it too corresponds to the choice $\alpha(.) \equiv 0$) with the property that

$$(H(t,\bar{x}(t),p_1(t)), -p_1(t)) \notin \operatorname{co} \partial V(t, \bar{x}(t)) \quad \text{for all } t \in [0,1).$$

This means that an adjoint arc exists that, on a set of full measure, fails to satisfy the sensitivity relation.

On the other hand, for any number $\omega > 0$, $p_2(.)$ given by

$$p_2(t) = e^{0.5} + (e^{1-t} - e^{0.5})\sin(\omega t)$$

is a continuously differentiable function that satisfies

$$(H(t, \bar{x}(t), p_2(t)), -p_2(t)) \in \text{co}\, \partial V(t, \bar{x}(t)) \quad \text{for all } t \in [0, 1].$$

We note however that

$$dp_2/dt(0) = \omega e^{0.5}(e^{0.5} - 1).$$

By Gronwall's Inequality (Lemma 2.3.4), if $p_2(.)$ is also a solution to (12.37) and (12.38) then $p_2(0) \le e^{1.5}$. Hence $\dot{p}_2(0) \le e^{1.5}$. It follows that $p_2(.)$ cannot be an adjoint arc if $\omega > e(e^{0.5} - 1)^{-1}$.

We have shown that there also exists an absolutely continuous function which satisfies the sensitivity relation but which is not an adjoint arc.

In order to pick out a special adjoint arc satisfying the sensitivity relation, even in situations such as that described in the above example, we make use of quite different techniques, the flavor of which we now attempt to convey. The underlying ideas become more transparent when we switch to problems having a traditional Pontryagin formulation and when we assume that the value function V is a C^1 solution to the Hamilton–Jacobi Equation. Consider for the time being then:

$$\begin{cases} \text{Minimize } g(x(T)) \text{ over } x \in W^{1,1}([S, T]; R^n) \\ \qquad\qquad \text{and measurable functions } u : [S, T] \to R^n \text{ satisfying} \\ \dot{x} = f(t, x(t), u(t)) \qquad \text{a.e. } [S, T], \\ u(t) \in U \qquad \text{a.e. } [S, T], \\ x(S) = x_0, \end{cases}$$

where $[S, T]$ is a given time interval, $g : R^n \to R$ and $f : [S, T] \times R^n \times R^m \to R^m$ are given functions, $U \subset R^m$ is a given set, and $x_0 \in R^n$ is a given point.

Let $(\bar{x}, \bar{u})$ be a minimizer. Under appropriate hypotheses, the value function satisfies

$$V_t(t, \xi) + V_x(t, \xi) \cdot f(t, \xi, u) \ge 0 \qquad\qquad\qquad (12.39)$$
$$\text{for all } (t, \xi) \in (S, T) \times R^n \text{ and } u \in U.$$

Also,

$$V_t(t, \bar{x}(t)) + V_x(t, \bar{x}(t)) \cdot f(t, \bar{x}(t), \bar{u}(t)) = 0 \qquad \text{a.e. } t \in [S, T]. \quad (12.40)$$

Let $u : [S, T] \to R^m$, $e : [S, T] \to R^n$, $v : [S, T] \to R^n$, and $w : [S, T] \to R$ be arbitrary measurable functions and $x(.)$ be an absolutely continuous arc such that

$$\dot{x}(t) \;=\; (1 + w(t))(f(t, x(t), u(t)) + v(t)) \quad \text{a.e.}$$
$$(u(t), w(t), v(t)) \;\in\; U \times [-1, +1] \times B \quad \text{a.e.}$$

Inserting $(\xi, u) = (x(t), u(t))$ into (12.39), multiplying across the inequality by the nonnegative number $(1 + w(t))$ and adding and subtracting terms, we obtain the inequality

$$V_t(t, x(t)) + V_x(t, x(t)) \cdot (f(t, x(t), u(t)) + v)(1 + w(t))$$
$$+\, \eta(t, x(t), w(t), v(t)) \;\geq\; 0 \quad \text{a.e.}$$

Here, η is the function

$$\eta(t, x, w, v) \;:=\; -(1 + w)V_x(t, x) \cdot v + wV_t.$$

Noting that the first two terms on the left of this inequality can be written $dV(t, x(t))/dt$ and also the boundary condition $V(T, .) = g(.)$, we deduce that

$$J(x(.), u(.), v(.), w(.)) \;\geq\; 0,$$

in which J is defined to be the functional

$$J(x(.), u(.), v(.), w(.)) \;:=$$
$$g(x(T)) - V(S, x(S)) + \int_S^T \eta(t, x(t), w(t), v(t))dt.$$

Similar reasoning applied to (12.40) gives

$$J(\bar{x}(.), \bar{u}(.), \bar{v}(.) \equiv 0, \bar{w}(.) \equiv 0) \;=\; 0.$$

We deduce that $(\bar{x}(.), \bar{u}(.), \bar{v}(.) \equiv 0, \bar{w}(.) \equiv 0)$ is a strong local minimizer for the optimization problem:

$$\left\{ \begin{array}{l} \text{Minimize } g(x(T)) - V(S, x(S)) + \int_S^T \eta(t, x(t), w(t), v(t))dt \\ \text{over } x \in W^{1,1} \text{ and measurable functions } (u, v, w) : [S, T] \to R^{m+n+1} \\ \text{satisfying} \\ \dot{x}(t) \;=\; (1 + w(t))(f(t, x(t), u(t)) + v(t)) \quad \text{a.e.}, \\ (u(t), w(t), v(t)) \;\in\; U \times [-1, +1] \times B \quad \text{a.e.} \end{array} \right.$$

This is an optimal control problem to which the Maximum Principle is applicable, with reference to the strong local minimizer $(\bar{x}, (\bar{u}, \bar{v} \equiv 0, \bar{w} \equiv 0))$. It turns out that the adjoint arc p for this derived problem associated with the cost multiplier $\lambda = 1$ (we can make such a choice of cost multiplier since the optimal control problem has free right endpoint) is also an adjoint arc for the original optimal control problem. But the Weierstrass Condition for the derived problem gives us the extra information that, for almost every t,

$$d(t, \bar{w}(t) = 0, \bar{v}(t) = 0) \;=\; \max\{d(t, w, v) \;:\; |w| \leq 1, |v| \leq 1\},$$

where

$$\begin{aligned} d(t, w, v) &:= (1 + w)[p(t) \cdot f(t, \bar{x}(t), \bar{u}(t)) - V_t t(t, \bar{x}(t))] \\ &\quad + (1 + w)[p(t) + V_x(t, \bar{x}(t))] \cdot v. \end{aligned}$$

In particular, $d(t, 0, 0) \leq d(t, w, 0)$ for every $w \in B$ and $d(t, 0, 0) \leq d(t, 0, w)$ for every $v \in B$, from which we conclude that

$$(p(t) \cdot f(t, \bar{x}(t), \bar{u}(t)), -p(t)) \;=\; (V_t(t, \bar{x}(t)), V_x(t, \bar{x}(t))) \quad \text{a.e.}$$

This is precisely the sensitivity relation.

What we have done is to modify the problem so that $(\bar{x}, \bar{u})$ becomes a minimizer with respect to a richer class of controls. It is reasonable to expect that necessary conditions of optimality for the derived problem will convey more information about $(\bar{x}, \bar{u})$ than those for the original problem; the extra information is the sensitivity relation.

The reader will be justifiably sceptical at this stage as to whether these elementary arguments can be modified to handle situations where V is possibly nonsmooth, since then the cost integrand in the derived problem, which involves derivatives of V, may be discontinuous with respect to the state variable and altogether unsuitable for application of the standard first-order necessary conditions. We get round this difficulty by replacing the offending function $\eta(t, x, v, w)$ by an approximation that does not depend on the state variable at all.

We are ready to prove

Theorem 12.5.1 (Sensitivity Relationships) *Let $\bar{x}$ be a minimizer for (P). Assume that the following hypotheses are satisfied.*

(i) *F is a continuous multifunction that takes as values nonempty, closed, convex sets.*

(ii) *There exists $c > 0$ such that*

$$F(t, x) \subset c(1 + |x|)B \quad \text{for all } (t, x) \in [S, T] \times R^n.$$

(iii) *There exists $k > 0$ such that*

$$F(t, x) \subset F(t, x') + k|x - x'|B \quad \text{for all } (t, x), (t, x') \in [S, T] \times R^n.$$

(iv) *g is locally Lipschitz continuous.*

*Let V be the value function for problem (P). Then there exists an arc
$p \in W^{1,1}([S,T]; R^n)$ such that*

$$\dot{p}(t) \in \text{co}\{q : (q, p(t)) \in N_{\text{Gr}\, F(t,.)}(\bar{x}(t), \dot{\bar{x}}(t))\},$$

$$p(t) \cdot \dot{\bar{x}}(t) = \max_{v \in F(t, \bar{x}(t))} p(t) \cdot v \quad a.e. \ t \in [S,T],$$

$$-p(T) \in \partial g(\bar{x}(T)),$$

$$\left(\max_{v \in F(t, \bar{x}(t))} p(t) \cdot v, -p(t)\right) \in \text{co}\, \partial V(t, \bar{x}(t)) \quad a.e. \ t \in [S,T],$$

$$p(S) \in \partial_x(-V)(S, \bar{x}(S)).$$

Proof. According to Proposition 12.3.5, V is a locally Lipschitz continuous
function. For $\epsilon \in (0,1)$ we define $G_\epsilon : [S,T] \rightsquigarrow R^{1+n}$ and $\sigma_\epsilon : R \times R^n \times R \to R$ to be

$$G_\epsilon(t) = \{(\alpha, \beta) : R^{1+n} : (\alpha, \beta) \in \text{co}\, \partial V(s, y)$$
$$\text{for some } (s, y) \in ((t, \bar{x}(t)) + \epsilon B) \cap ((S, T) \times R^n)\}$$

and

$$\sigma_\epsilon(t, v, w) = \sup_{(\alpha, \beta) \in G_\epsilon(t)} (\alpha, \beta) \cdot (w, -(1+w)v).$$

It is a straightforward matter to check that, for fixed $\epsilon \in (0,1)$, $\sigma_\epsilon(.,.,.)$
is upper semicontinuous and bounded on bounded sets and $\sigma_\epsilon(t,.,.)$ is
continuous for each $t \in [S,T]$.

Now consider the following optimal control problem.

$$(P_\epsilon) \begin{cases} \text{Minimize } J(x(S), x(T), y(T)) \\ \text{over arcs } (x, y) \in W^{1,1}([S,T]; R^{n+1}) \text{ satisfying} \\ (\dot{x}(t), \dot{y}(t)) \in \tilde{F}_\epsilon(t, x(t)) \quad \text{a.e.,} \\ (x(S), y(S)) \in R^n \times \{0\} \end{cases}$$

in which

$$\tilde{F}_\epsilon(t, x) := \{((e+v)(1+w), \sigma_\epsilon(t, v, w)) : e \in F(t, x), v \in \epsilon B, w \in \epsilon B\}$$

and

$$J(x_0, x_1, y_1) := g(x_1) - V(S, x_0) + y_1.$$

We verify:

Claim   For any $\epsilon \in (0,1)$, $(\bar{x}, \bar{y} \equiv 0)$ is a minimizer for (P_ϵ).

To confirm the claim, take any arc (x, y) satisfying the constraints of (P_ϵ).
Notice to begin with that at every point $t \in (S, T)$ which is a Lebesgue
point of $s \to \dot{x}(s)$ and which is a differentiability point of the Lipschitz

continuous function $s \rightarrow V(s, x(s))$ (such points comprise a set of full measure) we have

$$
\begin{aligned}
d/dt V(t, x(t)) &= \lim_{h \downarrow 0} h^{-1}[V(t + h, x + \int_t^{t+h} \dot{x}(s)ds) - V(t, x)], \\
&= \lim_{h \downarrow 0} h^{-1}[V(t + h, x + \dot{x}(t)h) - V(t, x)] \\
&= D^\uparrow V((t, x(t)); (1, \dot{x}(t))).
\end{aligned}
\tag{12.41}
$$

Since (y, x) satisfies the constraints of (P_ϵ), it can be deduced from the Measurable Selection Theorem that there exist measurable functions $e(.)$, $v(.)$, and $w(.)$ such that, for all points $t \in (S, T)$ in a set of full measure, we have

$$
\begin{aligned}
\dot{y}(t) &= \sigma_\epsilon(t, v(t), w(t)), \\
\dot{x}(t) &= (e(t) + v(t))(1 + w(t)), \\
e(t) &\in F(t, x(t)), \\
|v(t)|, |w(t)| &\leq \epsilon.
\end{aligned}
$$

Since V is a locally Lipschitz continuous lower Dini generalized solution to the Hamilton–Jacobi Equation for (P), for all points $t \in (S, T)$ in a set of full measure the following relationships, in which we write $x(t)$, $e(t)$, and $w(t)$ briefly as x, e, and w, are valid.

$$
\begin{aligned}
0 \leq\ & \liminf_{h \downarrow 0}[V(t + h, x + he) - V(t, x)]h^{-1}(1 + w) \\
& \text{(we have used the fact that } (1 + w) \geq 0) \\
\leq\ & \liminf_{h \downarrow 0}[V(t + h(1 + w), x + h(1 + w)e) - V(t, x)]h^{-1} \\
& \text{(by positive homogeneity)} \\
\leq\ & \liminf_{h \downarrow 0}[V(t + h, x + h(e + v)(1 + w)) - V(t, x)]h^{-1} \\
& + \limsup_{h \downarrow 0}[V(t_h + hw, x_h - hv(1 + w)) - V(t_h, x_h)]h^{-1} \\
& \text{(in which } t_h := t + h \text{ and } x_h := x + h(e + v)(1 + w)) \\
\leq\ & D^\uparrow V((t, x); (1, (e + v)(1 + w))) + D^0 V((t, x); (w, -v(1 + w))).
\end{aligned}
$$

In the final expression $D^0 V$ denotes the generalized directional derivative.

We deduce from the fact that the generalized directional derivative $D^0 V$ is the support function of $\operatorname{co} \partial V$ that

$$
\begin{aligned}
& D^0 V(t, x); (w, -v(1 + w)) \\
& \leq \sup_{(\alpha, \beta) \in G_\epsilon(t)} (\alpha, \beta) \cdot (w, -v(1 + w)) = \sigma_\epsilon(t, v, w).
\end{aligned}
$$

It follows

$$
D^\uparrow V((t, x(t)); (1, (e(t) + v(t))(1 + w(t)))) + \sigma_\epsilon(t, v(t), w(t)) \geq 0 \text{ a.e.}
$$

But then, since $V(T, .) = g(.)$ and in view of (12.41),

$$
\begin{aligned}
J(x(S), x(T), y(T)) &= g(x(T)) - V(S, x(S)) + \int_S^T \sigma_\epsilon(t, v(t), w(t)) dt \\
&= V(T, x(T)) - V(S, x(S)) + \int_S^T \sigma_\epsilon(t, v(t), w(t)) dt \\
&= \int_S^T [D^\uparrow V((t, x(t)); (1, \dot{x}(t))) + \sigma_\epsilon(t, v(t), w(t))] dt \\
&\geq 0 = J(\bar{x}(S), \bar{x}(T), \bar{y}(T)).
\end{aligned}
$$

The claim is verified.

We now apply the necessary conditions of Theorem 7.4.1 to (P_ϵ), with reference to the minimizer $(\bar{x}, \bar{y} \equiv 0)$. Since the problem has a free right endpoint, we may set the cost multiplier $\lambda = 1$. It follows that there exists an adjoint arc p such that

$$
\dot{p}(t) \in
$$
$$
\text{co} \{ q : (q, p(t), -1) \in N_{\mathrm{Gr}\, \tilde{F}_\epsilon(t, .)}(\bar{x}(t), \dot{\bar{x}}(t), 0) \} \text{ a.e.,} \quad (12.42)
$$
$$
p(t) \cdot \dot{\bar{x}}(t) = \max \{ p(t) \cdot (e + v)(1 + w) - \sigma_\epsilon(t, v, w)
$$
$$
: e \in F(t, \bar{x}(t)), |v| \leq \epsilon, |w| \leq \epsilon \} \text{ a.e.,} \quad (12.43)
$$
$$
-p(T) \in \partial g(\bar{x}(T)),
$$
$$
p(S) \in \partial_x(-V)(S, \bar{x}(S)).
$$

An analysis of proximal normals approximating vectors in the limiting normal cone $N_{\mathrm{Gr}\, \tilde{F}_\epsilon}$ permits us to conclude from (12.42) that

$$
\dot{p}(t) \in \text{co} \{ q : (q, p(t)) \in N_{\mathrm{Gr}\, F(t, .)}(\bar{x}(t), \dot{\bar{x}}(t)) \}.
$$

Recalling the definition of σ_ϵ, we deduce from (12.43) that for every $v \in \epsilon B$ and $w \in \epsilon B$ we have

$$
\inf_{(\alpha, \beta) \in G_\epsilon(t)} \{ wH(t, \bar{x}(t), p(t)) + p(t) \cdot v(1 + w) - \alpha w + \beta \cdot v(1 + w) \} \leq 0 \text{ a.e.}
$$

Here H is the Hamiltonian for the original problem, namely,

$$
H(t, x, p) := \max_{v \in F(t, x)} p \cdot v .
$$

Since $(1 + w) > 0$ for each $w \in \epsilon B$, the preceding inequality is preserved if, inside the brackets, we divide by $(1 + w)$. Now make the substitutions $w' = w/(1 + w)$ and $v' = -v$. A constant $\bar{\epsilon} \in (0, 1)$ can be found with the following property: for all $(w', v') \in \bar{\epsilon} B$,

$$
(H(t, \bar{x}(t), p(t)), -p(t)) \cdot (w', v') \leq \sup \{ (\alpha, \beta) \cdot (w', v') : (\alpha, \beta) \in G_\epsilon \} \quad \text{a.e.}
$$

It follows that

$$(H(t, \bar{x}(t), p(t)), -p(t)) \in \text{co}\, G_\epsilon(t). \tag{12.44}$$

Thus far, $\epsilon \in (0, 1)$ has been treated as a constant. Now choose a sequence $\epsilon_i \downarrow 0$. For each i, write p_i in place of p and ϵ_i in place of ϵ in the above relationships. The p_is are uniformly bounded and are Lipschitz continuous with a common Lipschitz constant. So, by extracting a subsequence if necessary, we can arrange that $p_i \to p$ uniformly, for some Lipschitz continuous function p.

A by now familiar convergence analysis can be used to show that the following conditions are satisfied.

$$\dot{p}(t) \in \text{co}\, \{q : (q, p(t)) \in N_{\text{Gr}\, F(t,.)}(\bar{x}(t), \dot{\bar{x}}(t))\}, \quad \text{a.e.} \tag{12.45}$$
$$-p(T) \in \partial g(\bar{x}(T)),$$
$$-p(S) \in \partial_x(-V)(S, \bar{x}(S)).$$

Since F is convex-valued, (12.45) implies that

$$p(t) \cdot \dot{\bar{x}}(t) = \max_{v \in F(t, \bar{x})} p(t) \cdot v \quad \text{a.e.}$$

Write $H(t, \bar{x}(t), p(t))$ briefly as $h(t)$. From (12.44) we obtain in the limit the information that, for some subinterval $\mathcal{S} \subset [S, T]$ of full measure,

$$(h(t), -p(t)) \in \cap_{\delta > 0} W_\delta(t) \quad \text{for all } t \in \mathcal{S}. \tag{12.46}$$

Here,

$$W_\delta(t) := \bar{\text{co}}\, \{e : e \in \text{co}\partial V(s, y)$$
$$\text{for some } (s, y) \in ((t, \bar{x}(t)) + \delta B) \cap ((S, T) \times R^n)\}.$$

Finally, we use these relationships to show that

$$(h(t), -p(t)) \in \text{co}\, \partial V(t, \bar{x}(t)) \quad \text{for all } t \in \mathcal{S}. \tag{12.47}$$

This completes the proof.

Suppose to the contrary that (12.47) is false at some $t \in \mathcal{S}$. Then we can strictly separate the point $(h(t), -p(t))$ and the closed convex set $\text{co}\, \partial V(t, \bar{x}(t))$. In other words, there exist $\alpha \in R$, $\beta \in R^n$, and $\gamma > 0$ such that

$$\alpha h(t) - p(t) \cdot \beta - \gamma > \max\{\alpha \tau + \xi \cdot \beta : (\tau, \xi) \in \text{co}\, \partial V(t, \bar{x}(t))\}$$
$$= D^0 V((t, \bar{x}(t)); (\alpha, \beta)).$$

Again we have used here the fact that $D^0 V$ is the support function of $\text{co}\, \partial V$.

However D^0V is upper semicontinuous with respect to all its arguments. It follows that, for some $\delta_1 > 0$,

$$\alpha h(t) - p(t) \cdot \beta - \gamma/2 > D^0V((s,y);(\alpha,\beta))$$

for all points $(s,y) \in ((t,\bar{x}(t)) + \delta_1 B) \cap ([S,T] \times R^n)$. We conclude that

$$\alpha h(t) - p(t) \cdot \beta - \gamma/2 > \sup_{(\tau,\xi) \in W_{\delta_1}(t)} \alpha\tau + \xi \cdot \beta.$$

But then

$$(h(t), -p(t)) \notin \bar{co}\, W_{\delta_1}(t),$$

contradicting (12.46). It follows that (12.47) is true. $\square$

12.6   State Constrained Problems

We investigate generalizations of the earlier characterization of the value function in terms of solutions to the Hamilton–Jacobi Equation, to allow for pathwise state constraints. Consider then the optimal control problem

$$(SC) \begin{cases} \text{Minimize } g(x(T)) \\ \text{over arcs } x \in W^{1,1}([S,T];R^n) \text{ satisfying} \\ \dot{x}(t) \in F(t,x(t)) \quad \text{a.e.,} \\ x(t) \in A \quad \text{for all } t \in [S,T], \\ x(S) = x_0, \end{cases}$$

with data an interval $[S,T] \subset R^n$, a function $g : R^n \to R \cup \{+\infty\}$, a multifunction $F : [S,T] \times R^n \leadsto R^n$, a closed set $A \subset R^n$, and a point $x_0 \in R^n$.

Write $(SC_{t,x})$ for problem (SC) when the initial data (t,x) replace $(0,x_0)$. Denote by $V : [S,T] \times A \to R \cup \{+\infty\}$ the value function for (SC). That is, for each $(t,x) \in [S,T] \times A$, $V(t,x)$ is the infimum cost of $(SC_{t,x})$.

It is to be expected that the value function V for (S) will be the unique solution (appropriately defined) to the Hamilton–Jacobi Equation

$$V_t + \min_{v \in F(t,x)} V_x \cdot v = 0 \quad \text{for } (t,x) \in (S,T) \times A,$$

accompanied by a boundary condition on $\{T\} \times A$ and also a boundary condition on $(S,T) \times \text{bdy } A$.

Under the hypotheses here imposed, it turns out that the appropriate boundary condition on $(S,T) \times \text{bdy } A$ is a boundary inequality

$$V_t + \min_{v \in F(t,x)} V_x \cdot v \leq 0 \quad \text{for } (t,x) \in (S,T) \times \text{bdy } A,$$

suitably interpreted.

The following theorem provides a characterization of the value function for state constrained problems along the above lines is possible, when a number of conditions are satisfied, concerning the state constraint set A and the nature of its interaction with F.

The first condition requires that the state constraint set is expressible as the intersection of a finite number of smooth functional inequality constraint sets; i.e.,

$$A \; = \; \cap_{j=1}^{r} \{x : h_j(x) \leq 0\} \tag{12.48}$$

for a finite family of $C^{1,1}$ functions $\{h_j : R^n \to R\}_{j=1}^{r}$. ($C^{1,1}$ denotes the class of C^1 functions with locally Lipschitz continuous gradients.)

The second requires that there exist admissible outward-pointing velocities at points in bdy A. (This is the constraint qualification (CQ) below.)

The "active set" of index values $I(x)$, at a point $x \in A$, is

$$I(x) \; := \; \{j \in (1, \ldots, r) : h_j(x) \geq 0\}.$$

Theorem 12.6.1 (Characterization of Lower Semicontinuous Value Functions: State Constrained Problems) *Take a function $V : [S,T] \times A \to R \cup \{+\infty\}$. Assume that the following hypotheses are satisfied.*

(H1) F is a continuous multifunction that takes as values nonempty, closed, convex sets.

(H2) There exists $c > 0$ such that

$$F(t,x) \subset c(1 + |x|)B \quad for \ all \ (t,x) \in [S,T] \times R^n.$$

(H3) There exists $k_F > 0$ such that

$$F(t,x) \subset F(t,x') + k_F|x - x'|B \quad for \ all \ t \in [S,T], \ x, x' \in R^n \times R^n.$$

(H4) g is lower semicontinuous.

Assume furthermore that

(CQ) For each point $(t,x) \in [S,T] \times$ bdy A there exists $v \in F(t,x)$ such that

$$\nabla h_j(x) \cdot v > 0 \quad for \ all \ j \in I(x).$$

Then assertions (a) through (c) below are equivalent.

(a) V is the value function for (SC).

(b) V is lower semicontinuous and

(i) for all $(t, x) \in ([S, T) \times A) \cap \operatorname{dom} V$

$$\inf_{v \in F(t,x)} D^\uparrow V((t, x); (1, v)) \leq 0;$$

(ii) for all $(t, x) \in ((S, T] \times \operatorname{int} A) \cap \operatorname{dom} V$

$$\sup_{v \in F(t,x)} D^\uparrow V((t, x); (-1, -v)) \leq 0;$$

(iii) for all $x \in A$

$$\liminf_{\{(t', x') \to (T, x): t' < T, x' \in \operatorname{int} A\}} V(t', x') = V(T, x) = g(x).$$

(c) V is lower semicontinuous and

(i) for all $(t, x) \in ((S, T) \times \operatorname{int} A) \cap \operatorname{dom} V$, $(\eta^0, \eta^1) \in \partial^P V(t, x)$

$$\eta^0 + \inf_{v \in F(t,x)} \eta^1 \cdot v = 0;$$

(ii) for all $(t, x) \in ((S, T) \times \operatorname{bdy} A) \cap \operatorname{dom} V$, $(\eta^0, \eta^1) \in \partial^P V(t, x)$

$$\eta^0 + \inf_{v \in F(t,x)} \eta^1 \cdot v \leq 0;$$

(iii) for all $x \in A$,

$$\liminf_{\{(t', x') \to (S, x): t' > S\}} V(t', x') = V(T, x)$$

and

$$\liminf_{\{(t', x') \to (T, x): t' < T, x' \in \operatorname{int} A\}} V(t', x') = V(T, x) = g(x).$$

Proof. Establishing the relationships (a) $\Rightarrow$ (b) and (b) $\Rightarrow$ (c) is straightforward. We sketch the proof of (c) $\Rightarrow$ (a).

Take an arbitrary function $V : [S, T] \times A \to R \cup \{+\infty\}$ satisfying Condition (c) and a point $(\bar{t}, \bar{x}) \in [S, T] \times A$. It suffices to show that

(i): There exists a feasible F-trajectory $\bar{x}(.)$ for $(S_{\bar{t}, \bar{x}})$ such that

$$V(\bar{t}, \bar{x}) \geq g(\bar{x}(T))$$

and

(ii): for every feasible F-trajectory $x(.)$ for $(SC_{\bar{t}, \bar{x}})$,

$$V(\bar{t}, \bar{x}) \leq g(x(T)),$$

since these inequalities imply

$$V(\bar{t}, \bar{x}) \; = \; \inf (SC_{\bar{t}, \bar{x}}).$$

We prove (i). Define $\tilde{V} : [S, T] \times R^n \to R \cup \{+\infty\}$

$$\tilde{V}(t, x) \; := \; \begin{cases} V(t, x) & \text{if } (t, x) \in [S, T] \times A \\ +\infty & \text{otherwise}. \end{cases}$$

It is easy to check that $\tilde{V}$ is a lower semicontinuous function that satisfies Conditions (c)(i)$'$ and (c)(ii)$'$ of Theorem 12.3.7. But these conditions imply the existence of a feasible F-trajectory $\bar{x}(.)$ for $(SC_{\bar{t}, \bar{x}})$ satisfying

$$V(\bar{t}, \bar{x}) \; (\; = \; \tilde{V}(\bar{t}, \bar{x}) \;) \; \geq \; g(\bar{x}(T)),$$

as required. (The arguments are those used in the proof of Theorem 12.3.7.)

We prove (ii). Take any feasible F-trajectory x for $(SC_{\bar{t}, \bar{x}})$. We must show that

$$V(\bar{t}, \bar{x}) \; \leq \; g(x(T)).$$

We can assume that $g(x(T)) < +\infty$ since, otherwise, the inequality is automatically satisfied.

It is shown in [70], with the help of (CQ) and the second relationship in (c)(iii), that there exist $\delta_i \downarrow 0$, $t_i \uparrow T$, and a sequence of feasible F-trajectories $\{x_i(.) : [\bar{t}, t_i] \to R^n\}$ such that

$$x_i(t) + \delta_i B \; \subset \; \text{int } A \quad \text{for all } t \in [\bar{t}, t_i],$$

$$x_i^e(.) \; \to \; x^e(.) \quad \text{uniformly,} \tag{12.49}$$

and

$$V(t_i, x_i(t_i)) \; \to \; g(x(T)). \tag{12.50}$$

($x^e(.)$ denotes the extension of $x(.)$ to all of $(-\infty, +\infty)$, by constant extrapolation. x_i^e is likewise defined.)

By Condition c(i), for each i

$$\eta^0 + \inf_{v \in F(t, x)} \eta^1 \cdot v = 0 \quad \text{for all } (\eta^0, \eta^1) \in \partial^P V(t, x),$$

at all points (t, x) in a tube about x_i, such that $t > \bar{t}$. We now apply the Strong Invariance Theorem (Theorem 12.2.4) as in the proof of Theorem 12.3.7, to show that

$$V(\bar{t}, x_i(\bar{t})) \; \leq \; V(t_i, x_i(t_i)).$$

This is possible because the hypotheses of the Strong Invariance Theorem involve only properties of the multifunction $P(.)$ on a tube about the F-trajectory whose invariance (with respect to $P(.)$) is under consideration. By (12.49), (12.50), and the lower semicontinuity of V, we have

$$V(\bar{t}, \bar{x}(\bar{t})) \leq \liminf_{i \to \infty} V(\bar{t}, x_i(\bar{t})) \leq \lim_{i \to \infty} V(t_i, x_i(t_i)) = g(x(T)).$$

This is the required inequality. The proof is complete. □

12.7   Notes for Chapter 12

Dynamic Programming, initiated by Richard Bellman [17] in the 1950s, is a large field and we focus here on just a few significant topics. These include the relationship between the value function and generalized solutions of the Hamilton–Jacobi Equation for deterministic optimal control problems on a fixed, finite time interval, existence of verification functions, and the interpretation of adjoint arcs as gradients of value functions.

In his 1942 paper, Nagumo studied conditions under which a differential equation has a solution that evolves in a given closed set [109]. Extensions to differential inclusions, supplying conditions for the existence of a solution to a differential equation evolving in a given closed set and, on the other hand, conditions for all solutions to evolve in the set, have been of continuing interest since the 1970s. Following Clarke et al. [53], we refer to them as weak and strong invariance theorems, respectively. These are the tools used in this chapter to characterize value functions as generalized solutions to the Hamilton–Jacobi Equation.

Weak invariance theorems can be regarded as existence theorems for solutions to differential inclusions on general closed domains — a viewpoint stressed by Deimling [57]. Weak invariance theorems are widely referred to as viability theorems (Aubin's terminology). They are the cornerstone of Viability Theory [8], in which broad issues are addressed, relating to existence of invariant trajectories (nature of "viability domains," etc.).

Nagumo showed that, under an inward-pointing hypothesis expressed in terms of the Bouligand tangent cone, there exists an invariant trajectory for a differential equation that can be constructed as a limit of polygonal arcs satisfying the given state constraint at mesh points. An abstract compactness argument assures the existence of node points for the polygonal arcs, with appropriate limiting properties.

Generalizations to differential inclusions were considered by a number of authors. Under an inward-pointing hypothesis involving the Clarke tangent cone, existence of invariant trajectories (possessing also a monotonicity property) was established by Clarke [30] for Lipschitz multifunctions and for continuous multifunctions by Aubin and Clarke [10]. Weak invariance theorems for upper semicontinuous, convex-valued multifunctions (and, more generally, for multifunctions "with memory") were proved by Haddad [77], under an inward-pointing hypothesis involving the Bouligand tangent cone.

The Bouligand contingent cone condition is given a central role in Aubin's book [8]. This is because it is in some sense necessary for weak invariance and because it can be shown to be equivalent to various, apparently less restrictive, conditions that arise in applications.

We follow a different approach to building up sets of conditions for weak invariance, mapped out by Clarke et al. in [53]. It is to hypothesize an inward-pointing condition, expressed in terms of the proximal normal cone of the state constraint set (wherever it is nonempty), to construct a polygonal arc on a uniform mesh by "proximal aiming" and obtain an invariant arc in the limit as the mesh size tends to zero. The advantage of the proximal aiming approach is simplicity, regarding both the proof of weak invariance theorems and also the investigation of alternative sets of conditions for weak invariance. In earlier proofs of weak invariance, taking inspiration from Nagumo's original ideas, abstract compactness arguments are required to select a special nonuniform mesh for construction of polygonal arcs, to ensure convergence. The proximal aiming approach, by contrast, is "robust" regarding mesh selection — we can, for convenience, choose a uniform mesh. The other point is that proximal aiming yields weak invariance theorems under an inward-pointing hypothesis formulated in terms of the proximal normal cone, which yield as direct corollaries weak invariance under other commonly encountered formulations of the hypothesis, because the proximal normal formulation is the weakest among them.

The proof given here of the the strong invariance theorem for data continuous in time, based on application of the weak invariance theorem to a modified differential inclusion which in some sense penalizes deviations from the arc under consideration, draws on ideas from [53]. A simple alternative proof, but one which also requires application of a theorem on the existence of Lipschitz parameterizations of multifunctions, is implicit in [68].

Gonzalez' paper [74] establishing that locally Lipschitz value functions are almost everywhere solutions of the Hamilton–Jacobi Equation is typical of the early literature linking nonsmooth value functions and the Hamilton–Jacobi Equation. See also [18]. Clarke and Offin showed that locally Lipschitz continuous value functions are generalized solutions of the Hamilton–Jacobi Equation, not in an almost everywhere sense but according to a new definition of generalized solution (based on generalized gradients) that looked ahead to the "pointwise" concepts of generalized solution that dominate the contemporary field [35],[112]. Information about uniqueness of solutions was lacking at this stage, although a characterization of locally Lipschitz continuous value functions in terms of the Hamilton–Jacobi Equation was achieved, namely, as the upper envelope of almost everywhere solutions of the Hamilton–Jacobi Inequality [74]. In a key advance [56], Crandall and Lions introduced the concept of viscosity solutions and associated techniques for establishing that the value function is the unique viscosity solution to the Hamilton–Jacobi Equation in the class of uniformly continuous functions. Bardi and Capuzzo-Dolcetta's book [14] is a recent expository text on viscosity methods in Optimal Control, Games Theory, robust nonlinear controller design, and discrete approximation. See also [66].

An important subsequent development [16] was Barron and Jensen's characterization of lower semicontinuous, finite-valued value functions as unique viscosity solutions to the Hamilton–Jacobi Equation for certain classes of problems. These authors adopted a new, simple "single differential" definition of viscosity solution to replace former definitions involving two partial differential inequalities.

The viscosity approach to characterizing value functions is to show that the relevant Hamilton–Jacobi Equation has a unique solution (in a generalized "viscosity" sense) and then to show that the value function is also a generalized solution.

An alternative approach is to use invariance theorems to show directly that an arbitrary generalized solution simultaneously majorizes and minorizes the value function and, therefore, coincides with it. This approach is system theoretic, to the extent that it involves the construction of state trajectories and the analysis of the monotonicity properties of state trajectories. It has its roots in interpretations of generalized solutions to Hamilton–Jacobi equations and Lyapunov inequalities by Clarke (later distilled in [43]) and Aubin [7], predating viscosity solutions, used to establish basic properties of verification functions and Lyapunov functions in a nonsmooth setting, and also in early work (referenced in Subbotin's book [131]) in the Games Theory literature.

In [68], Frankowska gave an independent system theoretic proof of Barron and Jensen's characterization of lower semicontinuous value functions (for the class of problems considered in this chapter), based on applications of a weak invariance theorem forward in time and a strong invariance theorem backward in time. This is the proof technique used in Section 12.3. (The results of Section 12.3, expressed in terms of proximal solutions to the Hamilton–Jacobi Equation, differ slightly from those in [68], in which a more restrictive "strict subdifferential" notion of solution is employed.)

Some extensions of the "invariance approach," to allow for free end-times and also unbounded Fs, are provided in [151] and [72], respectively.

A notable feature of the system theoretic approach is that it provides a characterization of *extended-valued* lower semicontinuous value functions. (In Barron and Jensen's framework, value functions are required to be finite-valued). This is convenient from the point of view of treating finite horizon optimal control problems with endpoint constraints, where value functions are inevitably extended-valued, to take account of initial conditions from which the terminal constraint set cannot be reached. (However "hyperbolic tangent" transformation techniques are described in [14] for treating endpoint constraint problems via nonextended-valued generalized solutions to the Hamilton–Jacobi Equation.)

The results of Section 12.3, relating value functions and generalized solutions to the Hamilton–Jacobi Equation, generalize, to some extent, to allow for measurably time-dependent data [142], [71] and [119]. In this context, it is known that a lower Dini solution to the Hamilton–Jacobi Equation V, in the class of lower semicontinuous functions, is the value function, only under the regularity hypothesis that $t \to V(t,.)$ is epicontinuous. (It should be mentioned that this regularity hypothesis automatically satisfied by the value function.) Whether a related characterization involving proximal normal solutions is valid, or whether the "extra" regularity hypothesis can be dispensed with for problems with measurably time-dependent data, remain open questions.

Inverse verification theorems give conditions under which the existence of a verification function is guaranteed, confirming the optimality of a putative minimizer. In contrast to investigations of the relationship of the value function and solutions to the Hamilton–Jacobi Equation, where we seek as unrestrictive conditions on value functions as possible for validity of the stated characterizations, improvements to inverse verification theorems are achieved by *restricting* the class of verification functions considered: they narrow the search for a verification function to suit a particular application. The link between normality hypotheses and the existence of Lipschitz continuous verification functions for optimal control problems with endpoint constraints was established in [35] and elaborated in [43].

No mention is made in this chapter of a powerful approach to the derivation of inverse verification theorems, based on the application of convex analysis. It allows free end-times and general endpoint and pathwise state constraints. The approach, which has roots in L. C. Young's Theory of Flows for parametric problems in the Calculus of Variations, has been employed by Fleming, Ioffe, Klötzler, Vinter, and others. (See [139].) Here, in effect, a dual problem to the optimal control problem at hand is set up, in which a Hamilton–Jacobi Inequality features as a constraint. In this context, a verification theorem involves a sequence of smooth functions (a maximizing sequence for the dual problem) satisfying a Hamilton–Jacobi Inequality.

The Sensitivity Relationship, providing an interpretaton of the adjoint arc, the Hamiltonian and gradients of the value function, is implicit in the early heuristic "Dynamic Programming" proof of the Maximum Principle outlined in Chapter 1. A rigorous analysis, confirming the sensitivity relationship for some adjoint arc, selected from the set of all possible adjoint arcs satisfying nonsmooth necessary conditions, and the Hamiltonian is given in [138]. An earlier proof of the Sensitivity Relationship for the adjoint variable alone appears in [48].

In [130] Soner characterized value functions for optimal control problems with state constraints as unique viscosity-type solutions to the Hamilton–Jacobi Equation, under hypotheses ensuring continuity of the value function. These hypotheses included an inward-pointing condition. The material of Section 12.6 on this topic which, unusually, invokes an *outward*-pointing condition and allows lower semicontinous value functions, draws on [70]. (See also [69].)

References

[1] L. AMBROSIO, O. ASCENTI, and G. BUTAZZO, Lipschitz regularity for minimizers of integral functionals with highly discontinuous integrands, *J. Math. An. Appl.*, **142**, 1989, pp. 301–316.

[2] A. V. ARUTYUNOV and S. M. ASEEV, Investigation of the degeneracy phenomenon of the maximum principle for optimal control problems with state constraints, *SIAM J. Control Optim.*, **35**, 1977, pp. 930–952.

[3] A. V. ARUTYUNOV, S. M. ASEEV, and V. I. BLAGODATSKIKH, First order necessary conditions in the problem of optimal control of a differential inclusion with phase constraints, *Russian Acad. Sci. Sb. Math.*, **79**, 1994, pp. 117–139.

[4] S.M. ASEEV, Method of smooth approximations in the theory of necessary conditions for differential inclusions, preprint.

[5] H. ATTOUCH, *Variational Convergence for Functions and Operators*, Applicable Mathematics Series, Pitman, London, 1984.

[6] J.-P. AUBIN, *Applied Functional Analysis*, Wiley Interscience, New York, 1978.

[7] J.-P. AUBIN, Contingent derivatives of set-valued maps and existence of solutions to nonlinear inclusions and differential inclusions, in *Advances in Mathematics, Supplementary Studies*, Ed. L. Nachbin, 1981, pp. 160–232.

[8] J.-P. AUBIN, *Viability Theory*, Birkhäuser, Boston, 1991.

[9] J.-P. AUBIN and A. CELLINA, *Differential Inclusions*, Springer Verlag, Berlin, 1984.

[10] J.-P. AUBIN and F. H. CLARKE, Monotone invariant solutions to differential inclusions, *J. London Math. Soc.*, **16**, 1977, pp. 357–366.

[11] J.-P. AUBIN and I. EKELAND, *Applied Nonlinear Analysis*, Wiley-Interscience, New York, 1984.

[12] J.-P. AUBIN and H. FRANKOWSKA, *Set-Valued Analysis*, Birkhäuser, Boston, 1990.

494   References

[13] J. BALL and V. MIZEL, One-dimensional variational problems whose minimizers do not satisfy the Euler-Lagrange equation, *Arch. Rat. Mech. Anal.*, **90**, 1985, pp. 325–388.

[14] M. BARDI and I. CAPUZZO-DOLCETTA, *Optimal Control and Viscosity Solutions of Hamilton–Jacobi Equations*, Birkhäuser, Boston, 1997.

[15] G. BARLES, *Solutions de Viscosité des Equations de Hamilton–Jacobi*, vol. 17 of Mathématiques et Applications, Springer, Paris, 1994.

[16] E. N. BARRON and R. JENSEN, Semicontinuous viscosity solutions for Hamilton–Jacobi equations with convex Hamiltonians, *Comm. Partial Differential Eq.*, **15**, 1990, pp. 1713–1742.

[17] R. BELLMAN, *Dynamic Programming*, Princeton University Press, Princeton, NJ, 1957.

[18] S. H. BENTON, *The Hamilton Jacobi Equation: A Global Approach*, Academic Press, New York, 1977.

[19] L. D. BERKOVITZ, *Optimal Control Theory*, Springer Verlag, New York, 1974.

[20] D. BESSIS, Y. S. LEDYAEV and R. B. VINTER, Dualization of the Euler and Hamiltonian Inclusions, *Nonlinear Anal.*, to appear.

[21] BILLINGSLEY, *Convergence of Probability Measures*, John Wiley and Sons, New York, 1968.

[22] V. G. BOLTYANSKII, The maximum principle in the theory of optimal processes (in Russian), *Dokl. Akad. Nauk SSSR*, **119**, 1958, pp. 1070–1073.

[23] J. M. BORWEIN and D. PREISS, A smooth variational principle with applications to subdifferentiability and to differentiability of convex functions, *Trans. Amer. Math. Soc.*, **303**, 1987, pp. 517–527.

[24] J. M. BORWEIN and Q. J. ZHU, A survey of subdifferential calculus with applications, SIAM Rev., to appear.

[25] A. E. BRYSON and Y.-C. HO, *Applied Optimal Control*, Blaisdell, New York, 1969, and (in revised addition) Halstead Press, New York, 1975.

[26] R. BULIRSCH, F. MONTRONE, and H. J. PESCH, Abort landing in the presence of windshear as a minimax optimal problem, Part 1: Necessary conditions and Part 2: Multiple shooting and homotopy, *J. Opt. Theory Appl.*, **70**, 1991, pp. 1–23, 223–254.

[27] C. CASTAING and M. VALADIER, *Convex Analysis and Measurable Multifunctions*, Springer Lecture Notes in Mathematics, vol. 580, Springer Verlag, New York, 1977.

[28] L. CESARI, *Optimization – Theory and Applications: Problems with Ordinary Differential Equations*, Springer Verlag, New York, 1983.

[29] F. H. CLARKE, Necessary conditions for nonsmooth problems in optimal control and the calculus of variations, Ph.D. dissertation, University of Seattle, W, 1973.

[30] F. H. CLARKE, Generalized gradients and applications, *Trans. Amer. Math. Soc.*, **205**, 1975, pp. 247–262 .

[31] F. H CLARKE, Necessary conditions for a general control problem in *Calculus of Variations and Control Theory*, Ed. D. L. Russell, Academic Press, New York, 1976, pp. 259–278.

[32] F. H. CLARKE, The maximum principle under minimal hypotheses, *SIAM J. Control Optim.*, **14**, 1976, pp. 1078–1091.

[33] F. H. CLARKE, A new approach to Lagrange multipliers, *Math. Oper. Res.*, **1**, 1976, pp. 165–174.

[34] F. H. CLARKE, Optimal solutions to differential inclusions, *J. Optim. Theory Appl.*, **19**, 1976, pp. 469–478.

[35] F. H. CLARKE, The applicability of the Hamilton–Jacobi verification technique, *Proceedings of the Tenth IFIP Conference*, New York, 1981, Eds. R.F. Drenick and F. Kozin, System Modeling and Optimization Ser., **38**, Springer Verlag, New York, 1982, pp. 88–94.

[36] F. H. CLARKE, Perturbed optimal control problems, *IEEE Trans. Automat. Control*, **31**, 1986, pp. 535–542.

[37] F. H. CLARKE, *Methods of Dynamic and Nonsmooth Optimization*, CBMS/NSF Regional Conf. Ser. in Appl. Math. vol. 57, SIAM, Philadelphia, 1989.

[38] F. H. CLARKE, *Optimization and Nonsmooth Analysis*, Wiley–Interscience, New York, 1983; reprinted as vol. 5 of Classics in Applied Mathematics, SIAM, Philadelphia, 1990.

[39] F. H. CLARKE, A decoupling principle in the calculus of variations, *J. Math. Anal. Appl.*, **172**, 1993, pp. 92–105.

[40] F. H. CLARKE and Y. S. LEDYAEV, Mean value inequalities, *Proc. Amer. Math. Soc.*, **122**, 1994, pp. 1075–1083.

[41] F. H. CLARKE and Y. S. LEDYAEV, Mean value inequalities in Hilbert space, *Trans. Amer. Math. Soc.*, **344**, 1994, pp. 307–324.

[42] F. H CLARKE and P. D. LOEWEN, The value function in optimal control: Sensitivity, controllability and time-optimality, *SIAM J. Control Optim.*, **24**, 1986, pp. 243–263.

[43] F. H CLARKE and R. B. VINTER, Local optimality conditions and Lipschitzian solutions to the Hamilton–Jacobi equation, *SIAM J. Control Optim.*, **21**, 1983, pp. 856–870.

[44] F. H CLARKE and R. B. VINTER, On the conditions under which the Euler equation or the maximum principle hold, *Appl. Math. Optim.*, **12**, 1984, pp. 73–79.

[45] F. H. CLARKE and R. B. VINTER, Regularity properties of solutions to the basic problem in the calculus of variations, *Trans. Amer. Math. Soc.*, **289**, 1985, pp. 73–98.

[46] F. H CLARKE and R. B. VINTER, Existence and regularity in the small in the calculus of variations, *J. Differential Eq.*, **59**, 1985, pp. 336–354.

[47] F. H CLARKE and R. B. VINTER, Regularity of solutions to variational problems with polynomial Lagrangians, *Bull. Polish Acad. Sci.*, **34**, 1986, pp. 73–81.

[48] F. H CLARKE and R. B. VINTER, The relationship between the maximum principle and dynamic programming, *SIAM J. Control Optim.*, **25**, 1987, pp. 1291–1311.

[49] F. H CLARKE and R. B. VINTER, Optimal multiprocesses, *SIAM J. Control Optim.*, **27**, 1989, pp. 1072–1091.

[50] F. H CLARKE and R. B. VINTER, Applications of optimal multiprocesses, *SIAM J. Control Optim.*, **27**, 1989, pp. 1048–1071.

[51] F. H CLARKE and R. B. VINTER, A regularity theory for problems in the calculus of variations with higher order derivatives, *Trans. Amer. Math. Soc.*, **320**, 1990, pp. 227–251.

[52] F. H CLARKE and R. B. VINTER, Regularity properties of optimal controls, *SIAM J. Control Optim.*, **28**, 1990, pp. 980–997.

[53] F. H. CLARKE, Y. S. LEDYAEV, R. J. STERN and P. R. WOLENSKI, Qualitative properties of trajectories of control systems: A survey, *J. Dynam. Control Systems*, **1**, 1995, pp. 1–47.

[54] F. H. CLARKE, Y. S. LEDYAEV, R. J. STERN and P. R. WOLENSKI, *Nonsmooth Analysis and Control Theory*, Graduate Texts in Mathematics vol. 178, Springer Verlag, New York, 1998.

[55] F. H CLARKE, P. D. LOEWEN and R. B. VINTER, Differential inclusions with free time, *Ann. de l'Inst. Henri Poincaré (An. Nonlin.)*, **5**, 1989, pp. 573–593.

[56] M. G. CRANDALL and P. L. LIONS, Viscosity solutions of Hamilton–Jacobi equations, *Trans. Amer. Math. Soc.*, **277**, 1983, pp. 1–42.

[57] K. DEIMLING, *Multivalued Differential Equations*, de Gruyter, Berlin, 1992.

[58] A. L. DONTCHEV and W. W. HAGER, A new approach to Lipschitz continuity in state constrained optimal control, *Syst. and Control Letters*, to appear.

[59] A. L. DONTCHEV and I. KOLMANOVSKY, On regularity of optimal control, in *Recent Developments in Optimization*, Proceedings of the French–German Conference on Optimization, Eds. R. Durier, C. Michelot, Lecture Notes in Economics and Mathematical Systems, **429**, Springer Verlag, Berlin, 1995, pp. 125–135.

[60] A. J. DUBOVITSKII and A. A. MILYUTIN, Extremal problems in the presence of restrictions, *U.S.S.R. Comput. Math. and Math. Phys.*, **5**, 1965, pp. 1–80.

[61] N. DUNFORD and J. T. SCHWARTZ, *Linear Operators. Part I: General Theory*, Interscience, London, 1958, reissued by Wiley–Interscience (Wiley Classics Library), 1988.

[62] I. EKELAND, On the variational principle, *J. Math. Anal. Appl.*, **47**, 1974, pp. 324–353.

[63] I. EKELAND, Nonconvex minimization problems, *Bull. Amer. Math. Soc. (N.S.)*, **1**, 1979, pp. 443–474.

[64] M. M. A. FERREIRA and R. B. VINTER, When is the maximum principle for state-constrainted problems degenerate?, *J. Math. Anal. Appl.*, **187**, 1994, pp. 432–467.

[65] W. H. FLEMING and R. W. RISHEL, *Deterministic and Stochastic Optimal Control*, Springer Verlag, New York, 1975.

[66] W. H. FLEMING and H. M. SONER, *Controlled Markov Processes and Viscosity Solutions*, Springer Verlag, New York, 1993.

[67] H. FRANKOWSKA, The maximum principle for an optimal solution to a differential inclusion with end point constraints, *SIAM J. Control Optim.*, **25**, 1987, pp. 145–157.

[68] H. FRANKOWSKA, Lower semicontinuous solutions of the Hamilton–Jacobi equation, *SIAM J. Control Optim.*, **31**, 1993, pp. 257–272.

[69] H. FRANKOWSKA and M. PLASKACZ, Semicontinuous solutions of Hamilton–Jacobi equations with state constraints, in *Differential Inclusions and Optimal Control, Lecture Notes in Nonlinear Analysis*, J. Schauder Center for Nonlinear Studies, vol. 2, Eds. J. Andres, L. Gorniewicz, and P. Nistri, 1998, pp. 145–161.

[70] H. FRANKOWSKA and R. B. VINTER, Existence of neighbouring feasible trajectories: Applications to dynamic programming for state constrained optimal control problems, *J. Optim. Theory Appl.*, to appear.

[71] H. FRANKOWSKA, M. PLASKACZ, and T. RZEZUCHOWSKI, Measurable viability theorems and the Hamilton–Jacobi–Bellman equation, *J. Diff. Eq.*, to appear.

[72] G. N. GALBRAITH, Applications of variational analysis to optimal trajectories and nonsmooth Hamilton–Jacobi theory, Ph.D. dissertation, University of Seattle, W, 1999.

[73] M. GIAQUINTA, *Multiple Integrals in the Calculus of Variations and Nonlinear Elliptic Systems*, Princeton University Press, Princeton, NJ, 1983.

[74] M. R. GONZALEZ, Sur l'existence d'une solution maximale de l'equation de Hamilton Jacobi, *C. R. Acad. Sci.*, **282**, 1976, pp. 1287–1280.

[75] R. V. GAMKRELIDZE, Optimal control processes with restricted phase coordinates (in Russian), *Izv. Akad. Nauk SSSR, Ser. Math.*, **24**, 1960, pp. 315–356.

[76] R. V. GAMKRELIDZE, On sliding optimal regimes, *Soviet Math. Dokl.*, **3**, 1962, pp. 559–561.

[77] G. HADDAD, Monotone trajectories of differential inclusions with memory, *Israel J. Math.*, **39**, 1981, pp. 83–100.

[78] W. W. HAGER, Lipschitz continuity for constrained processes, *SIAM J. Control and Optim.*, **17**, 1979, pp. 321–338.

[79] H. HALKIN, On the necessary condition for the optimal control of nonlinear systems, *J. Analyse Math.*, **12**, 1964, pp. 1–82.

[80] H. HALKIN, Implicit functions and optimization problems without continuous differentiability of the data, *SIAM J. Control*, **12**, 1974, pp. 239–236.

[81] H HERMES and J. P. LASALLE, *Functional Analysis and Time Optimal Control*, Academic Press, New York, 1969.

[82] A. D. IOFFE, Approximate subdifferentials and applications I: The finite dimensional theory, *Trans. Amer. Math. Soc.*, **281**, 1984, pp. 389–416.

[83] A. D. IOFFE, Necessary conditions in nonsmooth optimization, *Math. Oper. Res.*, **9**, 1984, pp. 159–189.

[84] A. D. IOFFE, Proximal analysis and approximate subdifferentials, *J. London Math. Soc.*, **41**, 1990, pp. 175–192.

[85] A. D. IOFFE, A Lagrange multiplier rule with small convex-valued subdifferentials for non-smooth problems of mathematical programming involving equality and non-functional constraints, *Math. Prog.*, **72**, 1993, pp. 137–145.

[86] A. D. IOFFE, Euler-Lagrange and Hamiltonian formalisms in dynamic optimization, *Trans. Amer. Math. Soc.*, **349**, 1997, pp. 2871–2900.

[87] A. D. IOFFE and R. T. ROCKAFELLAR, The Euler and Weierstrass conditions for nonsmooth variational problems, *Calc. Var. Partial Differential Eq.*, **4**, 1996, pp. 59–87.

[88] A. D. IOFFE and V. M TIHOMIROV, *Theory of Extremal Problems*, North-Holland, Amsterdam, 1979.

[89] B. KASKOSZ and S. LOJASIEWICZ, A maximum principle for generalized control systems, *Nonlinear Anal. Theory Meth. Appl.*, **19**, 1992, pp. 109–130.

[90] P. KOKOTOVIC and M. ARCAK, Constructive nonlinear control: Progress in the 90s, *Automatica*, to appear.

[91] A. Y. KRUGER, Properties of generalized differentials, *Siberian Math. J.*, **26**, 1985, pp. 822–832.

[92] G. LEBOURG, Valeur moyenne pour gradient généralisé, *Comptes Rondus de l'Académie des Sciences de Paris*, **281**, 1975, pp. 795–797.

[93] G. LEITMAN, *The Calculus of Variations and Optimal Control, Mathematical Concepts and Methods in Science and Engineering*, vol. 24, Plenum Press, New York, 1981.

[94] P. D. LOEWEN, *Optimal Control via Nonsmooth Analysis*, CRM Proc. Lecture Notes vol. 2, American Mathematical Society, Providence, RI, 1993.

[95] P. D. LOEWEN, A mean value theorem for Fréchet subgradients, *Nonlinear Anal. Theory Meth. Appl.*, **23**, 1995, pp. 1365–1381.

[96] P. D. Loewen and R. T. Rockafellar, Optimal control of unbounded differential inclusions, *SIAM J. Control Optim.*, **32**, 1994, pp. 442–470.

[97] P. D. Loewen and R. T. Rockafellar, New necessary conditions for the generalized problem of Bolza, *SIAM J. Control Optim.*, **34**, 1996, pp. 1496–1511.

[98] P. D. Loewen and R. B. Vinter, Pontryagin-type necessary conditions for differential inclusion problems, *Syst. Control Letters*, **9**, 1987, pp. 263–265.

[99] K. Malanowski, On the regularity of solutions to optimal control problems for systems linear with respect to control variable, *Arch. Auto. i Telemech.*, **23**, 1978, pp. 227–241.

[100] K. Malanowski and H. Maurer, Sensitivity analysis for state constrained optimal control problems, *Discrete Contin. Dynam. Syst.*, **4**, 1998, pp. 241–272.

[101] D. Q Mayne and E. Polak, An exact penalty function algorithm for control problems with control and terminal equality constraints, Parts I and II, *J. Optim. Theory Appl.*, **32**, 1980, pp. 211–246, 345–363.

[102] A. A. Milyutin and N. P. Osmolovskii, *Calculus of Variations and Optimal Control*, American Mathematical Society (Translations of Mathematical Monographs), Providence, RI, 1998.

[103] M. Morari and E. Zafiriou, *Robust Process Control*, Prentice Hall, Englewood Cliffs, 1989.

[104] B. S. Mordukhovich, Maximum principle in the optimal time control problem with non-smooth constraints, *Prikl. Math. Mech.*, **40**, 1976, pp. 1004–1023.

[105] B. S. Mordukhovich, Metric approximations and necessary optimality conditions for general classes of nonsmooth extremal problems, *Soviet Math. Doklady*, **22**, 1980, pp. 526–530.

[106] B. S. Mordukhovich, Complete characterizaton of openness, metric regularity, and Lipschitz properties of multifunctions, *Trans. Amer. Math. Soc.*, **340**, 1993, pp. 1–36.

[107] B. S. Mordukhovich, Generalized differential calculus for nonsmooth and set-valued mappings, *J. Math. Anal. Appl.*, **183**, 1994, pp. 250–282.

[108] B. S. Mordukhovich, Discrete approximations and refined Euler–Lagrange conditions for non-convex differential inclusions, *SIAM J. Control Optim.*, **33**, 1995, pp. 882–915.

[109] M. NAGUMO, Uber die lage der integralkurven gewöhnlicher differentialgleichungen, *Proc. Phys. Math. Soc. Japan*, **24**, 1942, pp. 551–559.

[110] L. W. NEUSTADT, A general theory of extremals, *J. Comp. and Sci.*, **3**, 1969, pp. 57–92.

[111] L. W. NEUSTADT, *Optimization*, Princeton University Press, Princeton, NJ, 1976.

[112] D. OFFIN, A Hamilton–Jacobi approach to the differential inclusion problem, M.Sc. Thesis, University of British Columbia, Canada, 1979.

[113] E. POLAK, *Optimization: Algorithms and Consistent Approximations*, Springer Verlag, New York, 1997.

[114] L. S. PONTRYAGIN, V. G. BOLTYANSKII, R. V. GAMKRELIDZE and E. F. MISCHENKO, *The Mathematical Theory of Optimal Processes*, K. N. Tririgoff, Transl., Ed. L. W. Neustadt, Wiley, New York, 1962.

[115] R. PYTLAK, Runge–Kutta based procedure for optimal control of differential-algebraic equations, *J. Optim. Theory Appl.*, **97**, 1998, pp. 675–705.

[116] R. PYTLAK and R. B. VINTER, A feasible directions algorithm for optimal control problems with state and control constraints: Convergence analysis, *SIAM J. Control Optim.*, **36**, 1998, pp. 1999–2019.

[117] R. J. QIN and T. A. BADGWELL, An overview of industrial model predictive control applications, *Proceedings of the Workshop on Nonlinear Model Predictive Control*, Ascona, Switzerland, 1998.

[118] F. RAMPAZZO and R. B. VINTER, Degenerate optimal control problems with state constraints, *SIAM J. Control Optim.*, to appear.

[119] A. E. RAPAPORT and R. B. VINTER, Invariance properties of time measurable differential inclusions and dynamic programming, *J. Dynam. Control Syst.*, **2**, 1996, pp. 423–448.

[120] R. T. ROCKAFELLAR, Generalized Hamiltonian equations for convex problems of Lagrange, *Pacific. J. Math.*, **33**, 1970, pp. 411–428.

[121] R. T. ROCKAFELLAR, Existence and duality theorems for convex problems of Bolza, *Trans. Amer. Math. Soc.*, **159**, 1971, pp. 1–40.

[122] R. T. ROCKAFELLAR, Existence theorems for general control problems of Bolza and Lagrange, *Adv. in Math.*, **15**, 1975, pp. 312–333.

[123] R. T. ROCKAFELLAR, Proximal subgradients, marginal values and augmented Lagrangians in nonconvex optimzation, *Math. Oper. Res.*, **6**, 1982, pp. 424–436.

[124] R. T. ROCKAFELLAR, Equivalent subgradient versions of Hamiltonian and Euler–Lagrange equations in variational analysis, *SIAM J. Control Optim.*, **34**, 1996, pp. 1300–1314.

[125] R. T. ROCKAFELLAR and R. J.-B. WETS, *Variational Analysis*, Grundlehren der Mathematischen Wissenschaften vol. 317, Springer Verlag, New York, 1998.

[126] J. ROSENBLUETH and R. B. VINTER, Relaxation procedures for time delay systems, *J. Math. Anal. and Appl.*, **162**, 1991, pp. 542–563.

[127] J. D. L. ROWLAND and R. B. VINTER, Dynamic optimization problems with free time and active state constraints, *SIAM J. Control Optim.*, **31**, 1993, pp. 677–697.

[128] G. N. SILVA and R. B. VINTER, Necessary conditions for optimal impulsive control problems, *SIAM J. Control Optim.*, **35**, 1998, 1829–1846.

[129] G. V. SMIRNOV, Discrete approximations and optimal solutions to differential inclusion, *Cybernetics*, **27**, 1991, pp. 101–107.

[130] H. M SONER, Optimal control with state-space constraints, *SIAM J. Control Optim.*, **24**, 1986, pp. 552–561.

[131] A. I. SUBBOTIN, *Generalized Solutions of First-Order PDEs*, Birkhäuser, Boston, 1995.

[132] H. J. SUSSMANN, Geometry and optimal control, in *Mathematical Control Theory*, Ed. J. Baillieul and J. C. Willems, Springer Verlag, New York, 1999, pp. 140–194.

[133] H. J. SUSSMANN and J. C. WILLEMS, 300 years of optimal control: from the brachystochrone to the maximum principle, *IEEE Control Syst. Mag.*, June 1997.

[134] L. TONELLI, Sur une méthode direct du calcul des variations, *Rend. Circ. Math. Palermo*, **39**, 1915, pp. 233–264.

[135] L. TONELLI, *Fondamenti di Calcolo delle Variazioni vol. 1 and 2*, Zanichelli, Bologna, 1921, 1923.

[136] J. L. TROUTMAN, *Variational Calculus with Elementary Convexity*, Springer Verlag, New York, 1983.

[137] H. D. TUAN, On controllability and extremality in nonconvex differential inclusions, J. Optim. Theory Appl., **85**, 1995, pp. 435–472.

[138] R. B. VINTER, New results on the relationship between dynamic programming and the maximum principle, *Math. Control Signals and Syst.*, **1**, 1988, pp. 97–105.

[139] R. B. VINTER, Convex duality and nonlinear optimal control, *SIAM J. Control Optim.*, **31**, 1993, pp. 518–538.

[140] R. B. VINTER and G. PAPPAS, A maximum principle for non-smooth optimal control problems with state constraints, *J. Math. Anal. Appl.*, **89**, 1982, pp. 212–232.

[141] R. B. VINTER and F. L. PEREIRA, A maximum principle for optimal processes with discontinuous trajectories, *SIAM J. Control Optim.*, **26**, 1988, pp. 205–229.

[142] R. B. VINTER and P. WOLENSKI, Hamilton–Jacobi theory for optimal control problems with data measurable in time, *SIAM J. Control Optim.*, **28**, 1990, pp. 1404–1419.

[143] R. B. VINTER and H. ZHENG, The extended Euler–Lagrange condition for nonconvex variational problems, *SIAM J. Control Optim.*, **35**, 1997, pp. 56–77.

[144] R. B. VINTER and H. ZHENG, The extended Euler Lagrange condition for nonconvex variational problems with state constraints, *Trans. Amer. Math. Soc.*, **350**, 1998, pp. 1181–1204.

[145] R. B. VINTER and H. ZHENG, Necessary conditions for free end-time, measurably time dependent optimal control problems with state constraints, *J. Set-Valued Anal.*, to appear.

[146] D. H. WAGNER, Survey of measurable selection theorems, *SIAM. J. Control and Optim.*, **15**, 1977, pp. 859–903.

[147] J. WARGA, *Optimal Control of Differential and Functional Equations*, Academic Press, New York, 1972.

[148] J. WARGA, Derivate containers, inverse functions and controllability, in *Calculus of Variations and Control Theory*, Ed. D. L. Russell, Academic Press, New York, 1976.

[149] J. WARGA, Fat homeomorphisms and unbounded derivate containers, *J. Math. Anal. Appl.*, **81**, 1981, pp. 545–560 .

[150] J. WARGA, Optimization and controllability without differentiability assumptions, *SIAM J. Control Optim.*, **21**, 1983, pp. 239–260.

[151] P. R. WOLENSKI and Y. Z HUANG, Proximal analysis and the minimal time function, *SIAM J. Control Optim.*, to appear.

[152] L. C. YOUNG, *Lectures on the Calculus of Variations and Optimal Control Theory*, Saunders, Philadelphia, 1991.

[153] J. ZABZYK, *Mathematical Control Theory: An Introduction*, Birkhäuser, Boston, 1992.

[154] D. ZAGRODNY, Approximate mean value theorem for upper subderivatives, *Nonlinear Anal. Theory Meth. and Appl.*, **12**, 1988, pp. 1413–1428.

[155] V. ZEIDAN, Second order admissible directions and generalized coupled points for optimal control problems, *Nonlinear Anal. Theory Meth. and Appl.*, **25**, 1996, pp. 479–507.

[156] Q. J. ZHU, Necessary optimality conditions for nonconvex differential inclusion with endpoint constraints, *J. Differential Eq.*, **124**, 1996, pp. 186–204.

Index